*Volumes already published*

Volume 1: Trees I (1986)
Volume 2: Crops I (1986)
Volume 3: Potato (1987)
Volume 4: Medicinal and Aromatic Plants I (1988)
Volume 5: Trees II (1989)
Volume 6: Crops II (1988)
Volume 7: Medicinal and Aromatic Plants II (1989)
Volume 8: Plant Protoplasts and Genetic Engineering I (1989)
Volume 9: Plant Protoplasts and Genetic Engineering II (1989)
Volume 10: Legumes and Oilseed Crops I (1990)
Volume 11: Somaclonal Variation in Crop Improvement I (1990)
Volume 12: Haploids in Crop Improvement I (1990)
Volume 13: Wheat (1990)
Volume 14: Rice (1991)
Volume 15: Medicinal and Aromatic Plants III (1991)
Volume 16: Trees III (1991)
Volume 17: High-Tech and Micropropagation I (1991)
Volume 18: High-Tech and Micropropagation II (1992)
Volume 19: High-Tech and Micropropagation III (1992)
Volume 20: High-Tech and Micropropagation IV (1992)
Volume 21: Medicinal and Aromatic Plants IV (1993)
Volume 22: Plant Protoplasts and Genetic Engineering III (1993)
Volume 23: Plant Protoplasts and Genetic Engineering IV (1993)
Volume 24: Medicinal and Aromatic Plants V (1993)
Volume 25: Maize (1994)
Volume 26: Medicinal and Aromatic Plants VI (1994)
Volume 27: Somatic Hybridization in Crop Improvement I (1994)
Volume 28: Medicinal and Aromatic Plants VII (1994)

*Volumes in preparation*

Volume 29: Plant Protoplasts and Genetic Engineering V (1994)
Volume 30: Somatic Embryogenesis and Synthetic Seed I (1995)
Volume 31: Somatic Embryogenesis and Synthetic Seed II (1995)
Volume 32: Cryopreservation of Plant Germplasm I (1995)
Volume 33: Medicinal and Aromatic Plants VIII
Volume 34: Plant Protoplasts and Genetic Engineering VI
Volume 35: Trees IV
Volume 36: Somaclonal Variation in Crop Improvement II

# Biotechnology in Agriculture and Forestry 28

## *Medicinal and Aromatic Plants VII*

Edited by Y. P. S. Bajaj

With 217 Figures

Springer-Verlag
Berlin Heidelberg GmbH

Professor Dr. Y. P. S. Bajaj
A-137
New Friends Colony
New Delhi 110065, India

ISBN 978-3-662-30371-9 ISBN 978-3-662-30369-6 (eBook)
DOI 10.1007/978-3-662-30369-6

Library of Congress Cataloging-in-Publication Data (Revised for Vol. 7) Medicinal and aromatic plants. (Biotechnology in agriculture and forestry; 4, 7, 15, 21, 24, 26, 28) Includes bibliographies and indexes. 1. Medicinal plants - Biotechnology. 2. Aromatic plants - Biotechnology. 3. Plant cell culture. 4. Material medica, Vegetable. I. Bajaj, Y. P. S., 1936- . II. Series: Biotechnology in agriculture and forestry; 4, etc. TP248.27.P55M43 1988 660'.6 88-3059 ISBN O-387-18414-7 (U.S.: v. 1)


Originally published by Springer-Verlag Berlin Heidelberg New York in 1994
Softcover reprint of the hardcover 1st edition 1994



Typesetting: Macmillan India Limited, Bangalore-25
SPIN: 10126426 31/3130/SPS - 5 4 3 2 1 0 - Printed on acid-free paper

*Dedicated to*
*Inderjit, Sumon, and Gitanjali Bajaj*

# Preface

The series of books on the biotechnology of *Medicinal and Aromatic Plants* provides a survey of the literature, focusing on recent information and the state of the art in tissue culture and the in vitro production of secondary metabolites. This book, *Medicinal and Aromatic Plants VII*, like the previous six volumes published in 1988, 1989, 1991, 1993 and 1994, is unique in its approach. It comprises 28 chapters dealing with the distribution, importance, conventional propagation, micropropagation, tissue culture studies, and the in vitro production of important medicinal and pharmaceutical compounds in various species of *Aesculus*, *Althaea*, *Baptisia*, *Berberis*, *Beta*, *Bowiea*, *Camptotheca*, *Chrysanthellum*, *Citrus*, *Claviceps*, *Coleonema*, *Dianthus*, *Dunaliella*, *Epimedium*, *Euphorbia*, *Forsythia*, *Gomphrena*, *Larix*, *Lobelia*, *Medicago*, *Papaver*, *Phytolacca*, *Pueraria*, *Santalum*, *Santolina*, *Sapium*, *Tabebuia*, and *Tripterygium*. This book is tailored to the needs of advanced students, teachers, and research scientists in the field of pharmacy, plant tissue culture, phytochemistry, biochemical engineering, and plant biotechnology in general.

New Delhi, July 1994

Professor Y. P. S. Bajaj
Series Editor

# Contents

# List of Contributors

AKKAN, Z., Institut für Lebensmitteltechnologie und Analytische Chemie der Technischen Universität München, 85350 Freising, Germany

ALFERMANN, A. W., Institut für Entwicklungs- und Molekularbiologie der Pflanzen, Heinrich-Heine-Universität Düsseldorf, Universitätsstr. 1, 40225 Düsseldorf, Germany

ANDREU, F., Plant Biotechnology Laboratory, EA-1370, Faculty of Pharmacy, University of Tours, 2bis Bd Tonnellé, 37000 Tours, France

ARAKAWA, H., Department of Applied Biological Sciences, Faculty of Agriculture, Saga University, 1 Honjo, Saga 840, Japan

ASAKAWA, Y., Faculty of Pharmaceutical Sciences, Tokushima Bunri University, 180 Yamashiro-cho, Tokushima City 770, Japan

BANTHORPE, D. V., Department of Chemistry, University College London, 20 Gordon Street, London WC1H 0AJ, United Kingdom

BERGER, R. G., Institut für Lebensmittelchemie der Universität Hannover, Wunstorfer Str. 14, 30453 Hannover, Germany

BUITELAAR, R. M., Agrotechnological Research Institute ATO-DLO, P.O. Box 17, 6700 AA Wageningen, The Netherlands

CAVIGLIA, A. M., Istituto Botanico Hanbury, Università di Genova, 16136 Genova, Italy

CHÉNIEUX, J. C., Plant Biotechnology Laboratory, EA-1370, Faculty of Pharmacy, University of Tours, 2bis Bd Tonnellé, 37000 Tours, France

CHU, E. P., Laboratory of Plant Physiology and Biochemistry, Institute of Botany, C.P. 4005, 01061-970 São Paulo, SP, Brazil

DEL RIO, J. A., Department of Plant Biology (Plant Physiology), University of Murcia, Campus de Espinardo, 30071 Murcia, Spain

DELIU, C., Institute of Biological Research, 48 Republicii Street, 3400 Cluj-Napoca, Romania

DEWICK, P. M., Department of Pharmaceutical Sciences, University of Nottingham, Nottingham NG7 2RD, United Kingdom

DREWES, F. E., NU Research Unit for Plant Growth and Development, Department of Botany, University of Natal, P.O. Box 375, Pietermaritzburg 3200, Republic of South Africa

EBIZUKA, Y., Faculty of Pharmaceutical Sciences, University of Tokyo, 7-3-1 Hongo, Bunkyo-ku, Tokyo 113, Japan

EDWARDS, R., Department of Biological Sciences, University of Durham, South Road, Durham DH1 3LE, United Kingdom

FERNANDES-FERREIRA, M., Departamento de Biologia, Universidade do Minho, 4719 Braga Codex, Portugal

FIGUEIREDO-RIBEIRO, R. C. L., Laboratory of Plant Physiology and Biochemistry, Institute of Botany, C.P. 4005, 01061-970 São Paulo, SP, Brazil

FINNIE, J.F., NU Research Unit for Plant Growth and Development, Department of Botany, University of Natal, P.O. Box 375, Pietermaritzburg 3200, Republic of South Africa

GASPAR, TH., Hormonologie Fondamentale at Appliquée, Université de Liège, Institut de Botanique B22, Sart Tilman, 4000 Liège, Belgium

GASTALDO, P., Istituto Botanico Hanbury, Università di Genova, 16136 Genova, Italy

GHIRAN, D., Medicinal and Pharmacy University, Faculty of Pharmacy, Pharmaceutical Chemistry Department, 41 V. Babes Street, 3400 Cluj-Napoca, Romania

HAKAMATSUKA, T., Faculty of Pharmaceutical Sciences, University of Tokyo, 7-3-1 Hongo, Bunkyo-ku, Tokyo 113, Japan

HARAZIM, P., GALENA – Pharmaceutical Works, 74700 Opava, Czech Republic

HARNISCHFEGER, G., Fa. Schaper and Brümmer, Bahnhofstr. 35, 38259 Salzgitter, Germany

HOOK, I. L. I., Department of Pharmacognosy, School of Pharmacy, Trinity College Dublin, 18 Shrewsbury Road, Dublin 4, Ireland

IINUMA, M., Department of Pharmacognosy, Gifu Pharmaceutical University, 6-1 Mitahora-higashi 5 chome, Gifu 502, Japan

IONKOVA, I., Department of Pharmacognosy, Faculty of Pharmacy, Medical Academy, Dunav Str. 2, 1000 Sofia, Bulgaria

ISHIMARU, K., Department of Applied Biological Sciences, Faculty of Agriculture, Saga University, 1 Honjo, Saga 840, Japan

KONNO, H., Research Institute for Bioresources, Okayama University, Kurashiki 710, Japan

KŘEN, V., Institute of Microbiology, Laboratory of Metabolites of Eukaryotic Microorganisms, Academy of Sciences of the Czech Republic, Vídeňská 1083, 14220 Prague 4, Czech Republic

LECKY, R., Department of Pharmacognosy, College of Pharmacy, Trinity College, Dublin, Ireland

LIEBEREI, R., Institut für Angewandte Botanik, Universität Hamburg, Marseiller Str. 7, 20355 Hamburg, Germany

MALINKA, Z., GALENA – Pharmaceutical Works, 74700 Opava, Czech Republic

MATERN, U., Plant Biochemistry, Biological Institute II, University of Freiburg, Schaenzlestr. 1, 79104 Freiburg, Germany

MERCIER, H., Laboratory of Physiology and Tissue Culture of Plants, Department of Botany, University of São Paulo, C.P. 11461, 05422 - 970 São Paulo, SP, Brazil

MEVENKAMP, G., Gemüsebaubetrieb Rudolf Behr, Ohlendorfer Str. 83, 21220 Ohlendorf, Germany

NABETA, K., Department of Bioresource Chemistry, Obihiro University of Agriculture and Veterinary Medicine, Inada-cho, Obihiro 080, Japan

NEERA, S., Department of Applied Biological Sciences, Faculty of Agriculture, Saga University, 1 Honjo, Saga 840, Japan

NOMA, Y., Faculty of Domestic Sciences, Tokushima Bunri University, 180 Yamashiro-cho, Tokushima City 770, Japan

NOVAIS, J. M., Laboratório de Engenharia Bioquimica, Instituto Superior Técnico, Av. Rovisco Pais, 1000 Lisboa, Portugal

ORTUÑO, A., Department of Plant Biology (Plant Physiology), University of Murcia, Campus de Espinardo, 30071 Murcia, Spain

PAIS, M. S. S., Departamento de Biologia Vegetal, Faculdade de Ciências de Lisboa, 1294 Lisboa Codex, Portugal

PARRY, A.D., Department of Biological Sciences, University of Durham, South Road, Durham DH1 3LE, United Kingdom

PETIT-PALY, G., Plant Biotechnology Laboratory, EA-1370, Faculty of Pharmacy, University of Tours, 2bis Bd Tonnellé, 37000 Tours, France

PROFUMO, P., Istituto Botanico Hanbury, Università di Genova, 16136 Genova, Italy

REIL, G., Institut für Lebensmitteltechnologie und Analytische Chemie der Technischen Universität München, 85350 Freising, Germany

RIDEAU, M., Plant Biotechnology Laboratory, EA-1370, Faculty of Pharmacy, University of Tours, 2bis Bd Tonnellé, 37000 Tours, France

SANKAWA, U., Faculty of Pharmaceutical Sciences, University of Tokyo, 7-3-1 Hongo, Bunkyo-ku, Tokyo 113, Japan

SHERIDAN, H., Department of Pharmacognosy, School of Pharmacy, Trinity College Dublin, 18 Shrewsbury Road, Dublin 4, Ireland

SZÖKE, E., Institute of Pharmacognosy, Semmelweis University of Medicine, Üllöi Street 26, 1085 Budapest, Hungary

TAKAYAMA, S., Department of Biological Science and Technology, Tokai University, 317 Nishino, Numazu, Shizuoka 410-03, Japan

TAMAS, M., Medicinal and Pharmacy University, Faculty of Pharmacy, Pharmaceutical Botany Department, 8 V. Babes Street, 3400 Cluj-Napoca, Romania

UEDA, S., Faculty of Pharmaceutical Sciences, Kyoto University, Sakyo-ku, Kyoto 606-01, Japan

VALLURI, J. V., Department of Biological Sciences, Marshall University, Huntington, West Virginia 25701, USA

VAN HENGEL, A. J., Agricultural University Wageningen, Department of Molecular Biology, Dreijenlaan 3, 6703 HA Wageningen, The Netherlands

VAN STADEN, J., NU Research Unit for Plant Growth and Development, Department of Botany, University of Natal, P.O. Box 375, Pietermaritzburg 3200, Republic of South Africa

VIEIRA, C. C. J., Laboratory of Plant Physiology and Biochemistry, Institute of Botany, C.P. 4005, 01061-970 São Paulo, SP, Brazil

WICHERS, H. J., Agrotechnological Research Institute ATO-DLO, P.O. Box 17, 6700 AA Wageningen, The Netherlands

YAMAMOTO, H., Department of Pharmacognosy, Gifu Pharmaceutical University, 6-1 Mitahora-higashi 5 chome, Gifu 502, Japan

YAMASAKI, Y., Research Institute for Bioresources, Okayama University, Kurashiki 710, Japan

# I *Aesculus hippocastanum* L. (Horse Chestnut): In Vitro Culture and Production of Aescin

P. GASTALDO, A. M. CAVIGLIA, and P. PROFUMO[1]

## 1 General Account

### 1.1 Distribution and Importance

*Aesculus hippocastanum* L. family Hippocastanaceae is a large deciduous tree up to 25–30 m high with a broad crown (Fig. 1). It is distributed in parts of the Balkan peninsula (Yugoslavia, Greece, Albania) and in one area in eastern Bulgaria (Tutin et al. 1968). It was introduced into central Europe in the 16th century and was extensively planted for ornament or shade in European and non-European countries. It is locally naturalized in western and central Europe. The fruit is a large, fleshy, spiny capsule, dehiscing into two to three valves on reaching full ripeness. It contains one to four shiny brown seeds.

The seed and other parts of the plant contain several active principles including aesculin (6,7-dihydroxycoumarin-6-glucoside) and aescin (Fig. 2), the latter being a blend of saponins. The drug is generally provided by the ripe seed, as described in German and Spanish pharmacopeias, while Portuguese pharmacopeia also includes the bark. The pharmaceutical importance of this species is to be found in β-aescin, which is widely used in the prevention and treatment of various peripheral vascular disorders.

*Aesculus hippocastanum* and other species belonging to the same genus may be poisonous. Esculoside is considered to be the toxic substance.

### 1.2 Propagation and Seed Collection

Propagation is generally by seed, which fall in autumn, remain dormant over winter, and germinate in spring, at least in Liguria (Italy). In the first year plants have either two or four leaves. The vegetative growth of such plants returns the following year. The plant may also be propagated by cutting; in this case the specimens have the same morphological and physiological characteristics as the parent plant.

The suitable time for collection of the drug coincides, according to the literature, with the period when the fruit falls in autumn. However, the season

[1] Istituto Botanico Hanbury, Università di Genova, Italy

Biotechnology in Agriculture and Forestry, Vol. 28
Medicinal and Aromatic Plants VII (ed. by Y.P.S. Bajaj)

**Fig. 1.** *Aesculus hippocastanum* tree. (Photo by P. Profumo 1993)

when the seed has its maximum content of aescin is the month of January; the content declines both before and during seed germination. (Profumo et al. 1987a)

In any case, the seeds, which are the sole industrial source of aescin, can be gathered and used for saponin extraction over a limited period of the year. At present, the demand for pure aescin in the world market is approximately 40–50 tons a year (data supplied by Inverni Della Beffa, Milan, Italy).

## 1.3 Secondary Metabolites and Their Importance

The horse chestnut was imported to central Europe in the 16th century, and in 1651 introduced into France. However, its therapeutic properties, which had long been known in popular medicine, did not gain official recognition until Artault de Vevey published his chemical studies between 1896 and 1909 (Benigni et al. 1964). This author showed the efficacy of the alcohol extract of horse chestnut in the treatment of venous diseases.

These results have been confirmed by subsequent research, both pharmacological and chemical, intended to identify all the possible therapeutic uses and related action mechanisms in the extracts obtained from this plant, and to single out the active principles contained within them. Now it is known that the active principle of major applied interest is aescin; the most important therapeutic activity of horse chestnut extract may be attributed to aescin. By virtue of its pharmacological characteristics and its limited side effects, aescin is now widely used medicinally. The effects of the various saponins constituting aescin (krypto-aescin and β-aescin in particular) are described as being similar in quality, but different in intensity of action.

The anti-edematose properties of aescin are evident at the cell level, which makes it efficacious in the treatment of localized edema, but not in cases of generalized edema.

Administration may be oral, intravenous (in the form of sodium aescinate), or local. It has been administered intravenously in the prevention and treatment of post-operative edema. The maximum dose for intravenous administration in adults under such conditions is reported as 20 mg a day. Acute renal failure has been recorded in patients given higher doses, when used in conjunction with other nephrotoxic drugs (Reynolds 1989).

The mechanism of action of aescin has yet to be clarified. Mussgnung (1969) reached the conclusion that the effect of aescin is due to its transformation into an anti-inflammatory glucoactive substance and, when the adrenal glands are intact, to an increased excretion of glucocorticoids which inhibit inflammation and prevent the formation of new edema. The disappearance of the edema might also be due to a surface-active effect on the cell membrane and on the connective tissue. Such an effect leads to the normalization of the exchange of extra-cellular liquids. It is noted that too great an interdependence between anti-edomatose action and increase in the level of corticosterone is excluded (Hampel et al. 1970).

Vogel and his team (1970) hypothesized that aescin has a "waterproofing" effect on the vessels and that it reduces the number and/or diameter of the small pores of the capillary wall through which the water system functions.

Recent studies (Martin et al. 1990) confirm that the active principles of *A. hippocastanum* act as a tonic on the veins, cause the blood to fluidify, and stimulate vitamin P activity. Besides all this, there is an increase in capillary resistance and a decrease in capillary permeability.

What is more, aescin is clearly anti-inflammatory, while esculoside regulates microcirculation. The increase in the tone of the veins is linked to the synthesis of $PGA_2$ at the vessel level.

Lastly, the cosmetic use of horse chestnut extracts, of pure aescin, and of the cholesterol/aescin combination should be mentioned. Such use, described in detail by Proserpio et al. (1980), is becoming increasingly important in both quantity and quality. Various formulations (such as creams, gels, or lotions) are employed, particularly in the treatment of couperose, the prevention and treatment of sun-rash, and as adjuvants in the treatment of cellulitis.

Chemically, aescin is a blend of triterpene glycosides deriving from pro-toescigenin and barringtogenol C. The two principal components, both deriving from protoescigenin, are β-aescin and kryptoescin (Fig. 2), the blend of which is defined as α-aescin.

In the literature, a number of methods have been reported for qualitative and/or quantitative analysis of *Aesculus* saponins. According to many pharma-copeias, they can be identified by thin layer chromatography (TLC) and quantitized by spectrophotometry. In recent times, some sensitive and reproduc-ible procedures have been described for β-aescin evaluation, by means of reverse-phase HPLC (e.g., Wagner et al. 1985; Pietta and Mauri 1989).

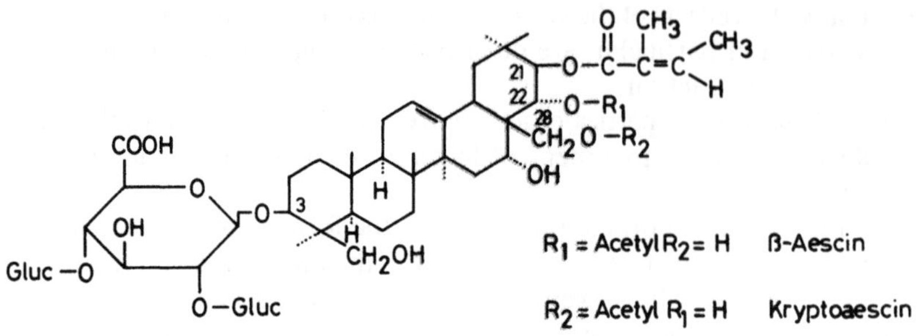


**Fig. 2.** Chemical structure of aescin

## 2 In Vitro Culture Studies

### 2.1 Review of the Tissue Culture in Horse Chestnut (see also Radojević 1991)

Trippi (1963) obtained abundant callus from fragments of stems, using Heller's agar medium supplemented with 3% glucose and various phytoregulators. The best results were obtained from adult branches, supplementing the medium with NAA and kinetin.

In 1978, Radojević used isolated anthers at various stages of development as basal explants, and obtained the first haploid somatic embryos in agarized MS (Murashige and Skoog 1962) medium supplemented with 2,4-D and kinetin; the embryoids developed to plantlets when transferred into medium lacking phytoregulators. The same author examined the ultrastructure of haploid embryos deriving from microspore culture (Radojević et al. 1980), and noted differences in the quantities of some organelles at various developmental stages of the embryoids. Anther explants gave best results when they were excised during the uninuclear microspore stage (Radojević 1980).

Later, Radojević (1988) used immature zygotic embryos and obtained embryogenic and nonembryogenic calli, embryoids, and subsequent plants. The agarized MS medium was supplemented with sucrose, 2,4-D, kin, CH and proline; the reduction (from 3 to 1 mg/l) of 2,4-D concentration produced an increase in both callus mass and number of embryoids. The somatic embryos developed into plantlets when transferred into liquid MS medium supplemented with glutamine, IAA (or IBA), and $GA_3$; the plantlet formation was inhibited by cytokinins. The author found the chromosome number of the new plants to be $2n = 40$.

In 1989, Jörgensen obtained embryogenic callus from staminal filaments of horse chestnut on woody plant medium (WPM) supplemented with BAP and 2,4-D. The callus produced embryoids when transferred to the same medium without 2,4-D.

The results mentioned above show that Radojević's finding (1980) is confirmed in the vast majority of cases; she maintains that auxin is necessary for the establishment of tissue cultures, the induction of somatic embryos, and the

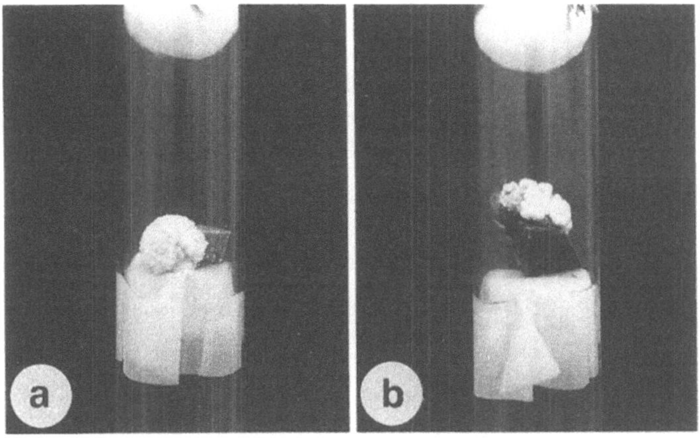


**Fig. 3a, b.** Callus grown from a cotyledonary fragment. **a** Treated with 2,4-D. **b** Treated with $GA_3$ + NAA. (Profumo et al. 1976)

maintenance of embryogenic potential. Recently, somatic embryos were also cryopreserved (Jekkel et al. 1992).

## 2.2 Establishment of Tissue Culture

Since 1976 we have carried out a number of experiments to verify the effects of many different experimental conditions on *Aesculus* tissues cultured in vitro, and to investigate the likelihood of obtaining mainly parenchymatous tissues that might produce aescin. The explants employed were cotyledonary fragments, after the fruit had fallen off the tree in autumn. The basal liquid medium contained mineral salts, glucose or sucrose, vitamins, and cysteine, plus – in different groups of flasks – coconut milk, yeast extract, 2,4-D, NAA, $GA_3$, and kin.

In these experiments (Profumo et al. 1976, 1980) callogenesis increased only when the basal medium was supplemented with 2,4-D (Fig. 3a, b) or NAA. Vascular clusters revealed their importance in callus production. The new proliferations showed intense rooting and also contained starch grains which were morphologically different from those present in the cotyledons from which they derived (Fig. 4a, b). Such experiments did not lead to interesting results as far as aescin content is concerned, but later we obtained significant results by varying media composition, extraction procedure, and the method of aescin determination.

The best culture conditions were found some years later, by culturing explants of primary leaves (Dameri et al. 1986; Profumo et al. 1986, 1987a), cotyledonary fragments (Profumo et al. 1990), or stem fragments (Gastaldo et al. in press). In these experiments the explants were put onto an agarized MS medium supplemented with sucrose, kin, NAA and 2,4-D, at pH 5.5. The whitish, compact callus thus obtained was cut into pieces, and each piece was subcultured on the

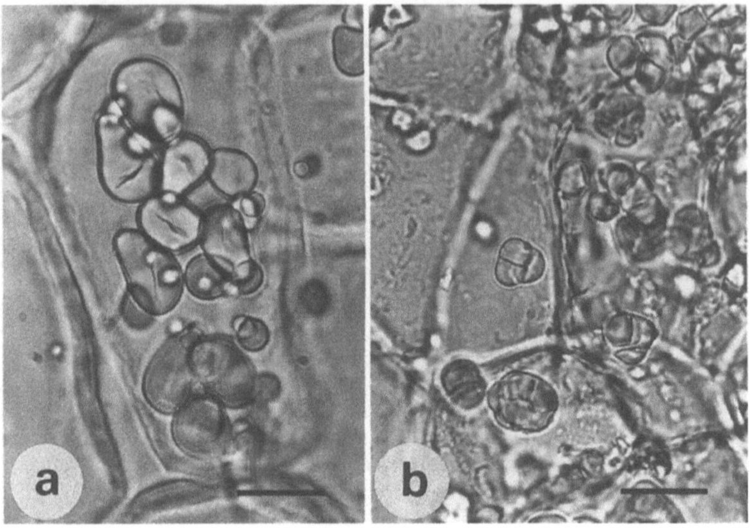

Fig. 4. a Simple starch granules in the cotyledonary tissue; b compound starch granules in the callus. *Bar* = 20 μm. (Profumo et al. 1976)

same solid medium. The initial callus gave rise to a new, yellowish, friable, embryogenic callus, from which embryoids were obtained (Fig. 5a, b). The cultures were kept in a growth room at 25 ± 1 °C and were exposed to 3000 lx from daylight fluorescent tubes under light/dark cycles of 12/12 h.

### 2.3 Extraction of Medicinal Compounds

First, an ultrastructural study (Profumo et al. 1987a) of embryogenic callus obtained from primary leaf explants showed that vacuoles sometimes contained a dark peripheral precipitate (Fig. 6) similar to that regarded by Radojević et al. (1980) as aescin in cells of androgenic embryos of horse chestnut.

Later, the aescin content was tested in calli and embryoids deriving from explants taken from primary leaves (Profumo et al. 1991) of zygotic embryos in vitro or from cotyledon explants (Profumo et al. 1992). The culture conditions were as already described above.

Aescin extraction and determination were performed according to Wagner et al. (1985) with some modification, as described in Profumo et al. (1991). In short, fine powdered tissue samples (1 g fr. wt.) were extracted overnight with 60% MeOH (25 ml) at room temperature. The extracts were filtered and adjusted to 30 ml with 60% MeOH in calibrated flasks. Then, 6 ml of each solution was diluted to 12 ml with water and adsorbed on Sep-Pak C-18 cartridges (Waters, Milford, MA, USA). Aescin was eluted from the resin with pure MeOH (4 ml) and evaluated by HPLC analysis. In early research regarding the content of aescin in calli and embryoids deriving from primary leaves,

**Fig. 5. a** Embryogenic callus from leaf explants treated with kin + NAA + 2,4-D. (Dameri et al. 1986). **b** Embryoids obtained from embryogenic callus

Profumo et al. (1991) obtained the results which appear in Fig. 7a–d. In this research, the precursor callus, the embryogenic callus, and the embryoids deriving from them were considered. The material placed in culture consisted of primary leaves deriving from the germination of the seed. Table 1 shows the aescin content (as a percentage of dry wt.) in precursor callus ($E_1$), in embryogenetic callus ($E_2$), and in embryoids. Aescin content in ripe cotyledons in vivo is used as the control value.

The difference between aescin concentration in precursor callus and in cotyledons in vivo is not significant. On the contrary, in embryogenetic callus, aescin content during the more active phase of proliferation and differentiation

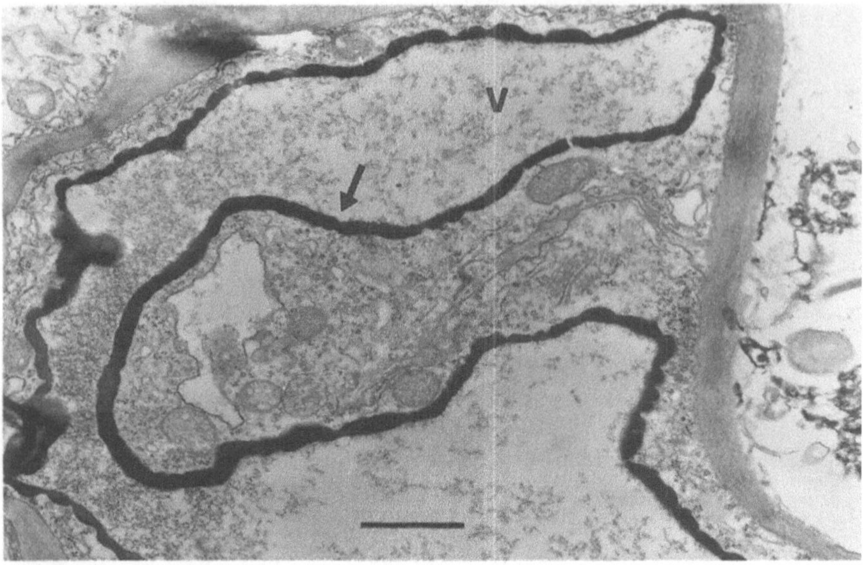


**Fig. 6.** Electron-dense vacuolar precipitate (*arrow*) in a cell of embryogenic callus; *bar* = 1 μm. *V* Vacuole. (Profumo et al. 1987)

**Table 1.** Aescin content in calli and embryoids from leaf explants of *Aesculus hippocastanum* L. cultured in vitro; ripe cotyledons in vivo are used as control. (Profumo et al. 1991)

| Sample | Dry wt. (% fr. wt.) | Aescin content/dry wt. (%) |
|---|---|---|
| Ripe cotyledons | 56.7 | 10.7 ± 0.5 |
| $E_1$ callus | 5.9 | 12.2 ± 1.9 |
| $E_2$ callus after 20 days | 4.3 | 62.2 ± 7.3 |
| $E_2$ callus after 70 days | 3.4 | 50.0 ± 2.9 |
| $E_2$ callus after 2 years | 3.4 | 31.7 ± 1.8 |
| Embryoids | 5.2 | 69.5 ± 4.6 |

Data are mean ± SEM values; the number of observations is six for each sample.

▶

**Fig. 7a–e.** High pressure liquid chromatograms of aescin in standard solution (25 mg/ml) (**a**), in ripe cotyledons (**b**), in precursor callus (**c**), in embryoids (**d**) from leaf explants, and in callus precursor from cotyledonary explants (**e**)

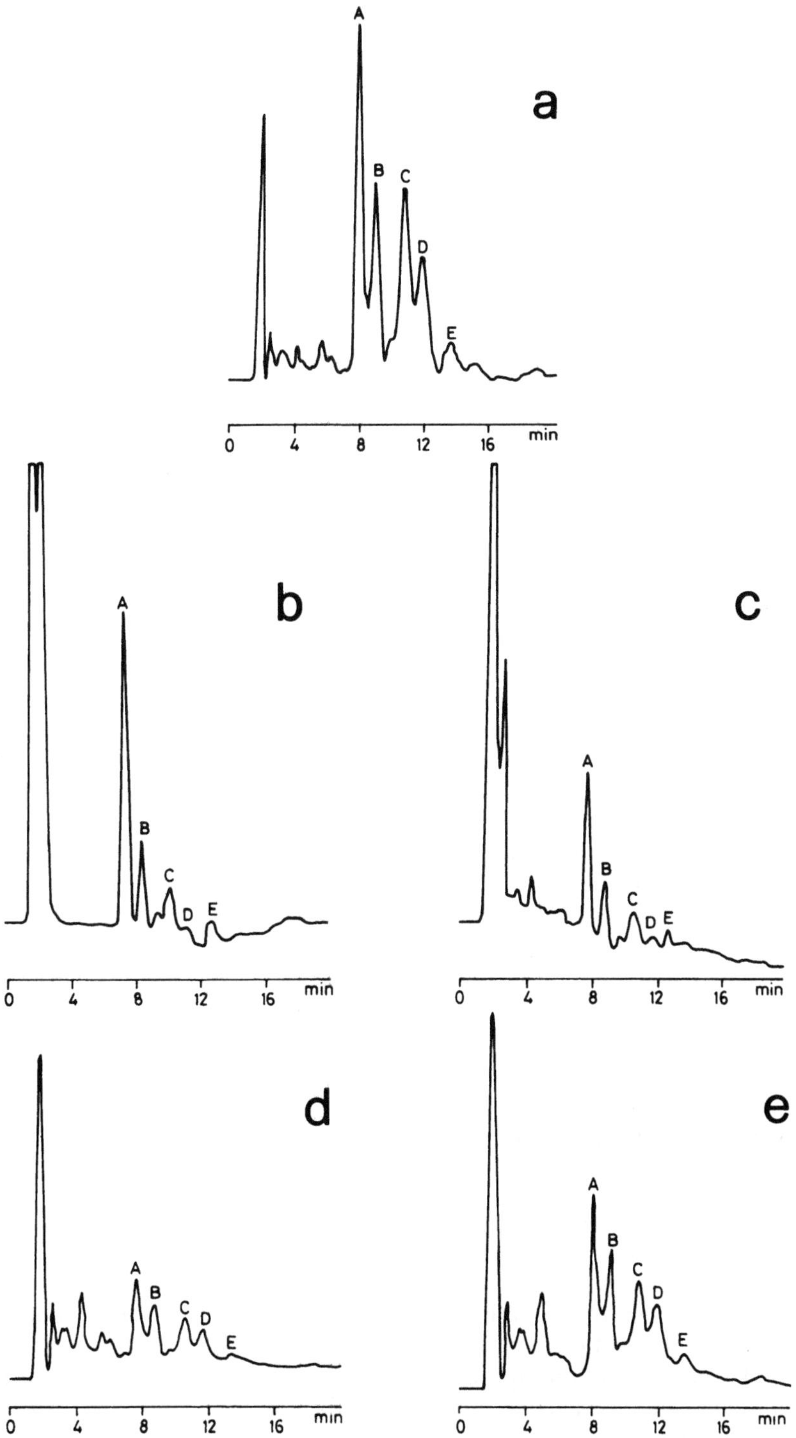
a
A
B
C
D
E
0 4 8 12 16 min
b
A
B
C
D E
0 4 8 12 16 min
c
A
B
C
D E
0 4 8 12 16 min
d
A
B
C
D
E
0 4 8 12 16 min
e
A
B
C
D
E
0 4 8 12 16 min

**Table 2.** Aescin content in calli from cotyledonary explants of *Aesculus hippocastanum* L. cultured in vitro in agarized MS medium: influence of low concentrations (0.5 mg/l) of phytoregulators. (Profumo et al. 1992)

| Culture media | Aescin content/dry wt. (%) |
|---|---|
| MS | 23.3 ± 5.1 |
| MS + 2,4-D | 47.8 ± 4.0 |
| MS + $GA_3$ | 52.9 ± 5.4 |
| MS + NAA | 53.7 ± 5.7 |

Data are mean ± SEM; the number of observations is ten for each type of sample. Dry wt./fr. wt. percentage was 8.65 ± 0.8. MS, control medium.

(20 days) is considerably higher than in control. After 70 days without subculture in fresh medium, the aescin amount decreased but remained higher than that in the control.

When embryogenic callus is transferred onto fresh medium every 25 days over a 2-year period, its aescin level is maintained at about three times that of the control. Finally, aescin is about seven times more concentrated in embryoids than in ripe seed cotyledons in vivo.

Using the same techniques as for the previous work, Profumo et al. (1992) carried out further research in order to evaluate the content of aescin in calli obtained by cultivating cotyledonary fragments in vitro. Analyses were considered on samples obtained with the use of various phytoregulators on the MS culture medium (Table 2). In all cases, the aescin content proved to be greater than that found when the medium contained no phytoregulators. The diagram obtained from analysis with HPLC (Fig. 7e) shows that the aescin concentration in calli obtained from cotyledons, even without the addition of phytoregulators, is higher than that present in the seed of horse chestnut in vivo.

## 3 Summary and Conclusions

Research carried out on explants of *Aesculus hippocastanum*, derived from both primary leaves or cotyledons, show that in vitro culture is invaluable in the production of aescin. In fact, the calli and embryoids obtained in vitro may have a higher content of aescin than is extracted from the seed – the part of the plant employed to yield the active principle.

There is a relationship between embryogenic potential and aescin production. The calli have a different active principle content, depending on whether or not they are embryogenic.

# 4 Protocol

To obtain embryogenetic callus of *Aesculus hippocastanum*, either primary leaf or cotyledon explants are used. Primary leaf fragments were excised from seedlings grown from sterile zygotic embryos in vitro on an agar medium supplemented with 3% sucrose. For callus and embryoid formation, the leaf or cotyledon explants were placed on MS medium supplemented with kin, NAA, 2,4-D (2 mg/l of each) at pH 5.5. The calli were transferred and subcultured once more on the same medium and many of them became embryogenetic and gave rise to embryoids. The material was kept at 25 ± 1 °C and exposed to 3000 lx by daylight fluorescent tubes under a 12/12-h light–dark regime.

For aescin determination, the samples were extracted with 60% MeOH. The extracts were filtered and adsorbed on Sep-Pak C-18 cartridges. Aescin was eluted from the resin with pure MeOH and evaluated by HPLC according to Wagner et al. (1985) with some modifications.

*Acknowledgments.* The authors are grateful to Dr. R.M. Dameri for her cooperation and to Mrs. S. Carli for technical assistance.

# References


Benigni R, Capra C, Cattorini PE (1964) Piante Medicinali, Chimica Farmacologia e Terapia, vol 2. Inverni & Della Beffa, Milano

Dameri RM, Caffaro L, Gastaldo P, Profumo P (1986) Callus formation and embryogenesis with leaf explants of *Aesculus hippocastanum* L. J Plant Physiol 126:93–96

Gastaldo P, Carli S, Profumo P, Somatic embryogenesis from stem explants of *Aesculus hippocastanum* L. Plant, Cell, Tissue and Organ Culture (in press)

Hampel H, Hofrichter G, Liehn HD, Schlemmer W (1970) Zur Pharmakologie der Aescin-Isomere unter besonderer Berücksichtigung von Aescin. Arzneim Forsch 20:209

Jekkel ZS, Kiss J. Kiss E, Gyulai G, Heszky LE (1992) Freeze preservation of somatic embryos of recalcitrant seed species: horse chestnut (*Aesculus hippocastanum* L.). Proc XIIIth EUCARPIA Congr Angers, France, July 6–11, 1992

Jörgensen J (1989) Somatic embryogenesis in *Aesculus hippocastanum* L. by culture of filament callus. J Plant Physiol 135:240–241

Martín MJ, Alarcón C, Motilva V (1990) Effets de l'escine et de l'aesculine sur l'excrétion rénale d'eau et d'électrolytes chez le rat. Ann Pharm Fr 48(6):306–311

Murashige T, Skoog F (1962) A revised medium for rapid growth and bioassays with tobacco tissue cultures. Physiol Plant 15:473–497

Mussgnug G (1969) Zur Biochemie der Entzündung und Oedembildung sowie der Pharmakokinetik des Aescin. Arzneim Forsch 19:1588

Pietta P, Mauri P (1989) High-performance liquid chromatographic analysis of β-escin. J Chromatogr 478:259–263

Profumo P, Dameri RM, Cremona Orsino I (1976) Frammenti cotiledonari di *Aesculus hippocastanum* L. coltivati in vitro. Primi dati sul comportamento dell'amido e dell'escina. G Bot Ital 110 (1–2):155–171

Profumo P, Dameri Orsino RM, Modenesi P (1980) Aescin content in callus from explants of *Aesculus hippocastanum* L. cotyledons grown in vitro. G Bot Ital 114:25–28

Profumo P, Gastaldo P, Dameri RM, Caffaro L (1986) Histological study of calli and embryoids from leaf explants of *Aesculus hippocastanum* L. J Plant Physiol 126:97–103

Profumo P, Gastaldo P, Martinucci R (1987a) Variations in aescin content in *Aesculus hippocastanum* seeds during the year. Fitoterapia 58(3):184–186

Profumo P, Gastaldo P, Rascio N (1987b) Ultrastructural study of different types of callus from leaf explants of *Aesculus hippocastanum*. Protoplasma 138:89–97

Profumo P, Gastaldo P, Caviglia AM, Dameri RM (1990) Somatic embryogenesis from cotyledonary explants of *Aesculus hippocastanum* L. Acta Embryol Morphol Exp New Ser 11(2):101–106

Profumo P, Caviglia AM, Gastaldo P, Dameri RM (1991) Aescin content in embryogenic callus and in embryoids from leaf explants of *Aesculus hippocastanum*. Planta Med 57:50–52

Profumo P, Caviglia AM, Gastaldo P (1992) Formation of aescin glucosides by callus tissue from cotyledonary explants of *Aesculus hippocastanum* L. Plant Sci 85:161–164

Proserpio G, Gatti S, Genesi P (1980) Cosmetic uses of horse chestnut (*Aesculus hippocastanum*) extracts, of escin and of the cholesterol/escin complex. Fitoterapia 51(2):113–128

Radojević L (1978) In vitro induction of androgenic plantlets in *Aesculus hippocastanum*. Protoplasma 96:369–374

Radojević L (1980) Embryogenèse somatique et androgenèse chez certaines espèces ligneuses. Bull Soc Bot Fr Actual Bot 127(3/4):99–107

Radojević L (1988) Plant regeneration of *Aesculus hippocastanum* L. (horse chestnut) through somatic embryogenesis. J Plant Physiol 132:322–326

Radojević L (1991) Horse chestnut (*Aesculus* spp.). In: Bajaj YPS (ed) Biotechnology in agriculture and forestry, vol 16. Trees III. Springer, Berlin Heidelberg New York, pp 111–141

Radojević L, Zylberberg L, Kovoor J (1980) Etude Ultrastructurale des Embryons Androgenetiques d'*Aesculus hippocastanum* L. Z Pflanzenphysiol 98:255–261

Reynolds JEF (ed) (1989) Martindale. The extra pharmacopoeia, 29th edn The Pharmaceutical Press, London

Trippi VS (1963) Studies on ontogeny and senility in plants. II. Seasonal variation in proliferative capacity in vitro of tissues from branches from juvenile and adult zones of *Aesculus hippocastanum* and *Castanea vulgaris*. Phyton 20(2):146–152

Tutin TG, Heywood VH, Burges NA, Moore DM, Valentine DH, Walters SM, Webb DA (1968) Flora Europaea, vol 2. University Press, Cambridge

Vogel G, Marek ML, Oertner R (1970) Untersuchungen zum Mechanismus der therapeutischen und toxischen Wirkung des Rosskastanien-Saponins Aescin. Arzneim Forsch 20:699

Wagner H, Reger H, Bauer R (1985) Saponinhaltige Drogen und Fertigarzneimittel. Dtsch Apoth Zg 125:1513–1518

# II *Althaea officinalis* L. (Marshmallow): In Vitro Culture and the Production of Biologically Active Compounds

I. Ionkova[1] and A. W. Alfermann[2]

## 1 General Account

### 1.1 Habit and Distribution of the Plant

The genus *Althaea* belongs to the family Malvaceae and includes 12 species (Trease and Evans 1978), which are located mainly in Europe, with the exception of the Scandinavian countries, and the Near East (western and north Asia). They are cultivated mainly in Germany, France, Belgium, Spain, Italy, Hungary, and Russia, and have been introduced in North and South America (Uphof 1968).

The most important species of the genus is *Althaea officinalis* L. (marshmallow) (Fig. 1), densely gray-pubescent perennial up to 1.5–2 m, with stellate hairs. Leaves triangular-ovate, acute, crenate-serrate, undivided or palmately 3-5-lobed, often somewhat plicate. Flowers solitary or clustered in axillary inflorescences shorter than the substanding leaf. Epicalix segments linear-lanceolate. Sepals ovate, acute, curved over the fruit. Petals, 15–20 mm, very pale, lilac-pink, rarely deeper pink. Anthers are purplish red. Mericarps more or less densely covered with stellate hairs. The chromosome number is 2n = 42 (Tutin et al. 1968). The plant has a woody rootstock from which numerous roots arise, up to 30 cm in length (Pahlow 1979; Widmaier 1988).

The roots (Radix Althaea naturalis and mundata); the leaves (Folia Althaeae), and the flowers (Flores Althaeae) are used in medicine.

### 1.2 Marshmallow Root

The drug Radix Althaea is mainly collected from cultivated plants at least 2 years old. The roots are dug up in the autumn, cut away from the cork, and dried, either whole or after slicing (mundata). The drug occurs in whitish, fibrous pieces about 15–20 cm long and 1–2 mm in diameter, or in small transverse slices. Odor, slight; taste, sweetish and mucilaginous. A transverse section shows a bark about 1–2 mm thick which is separated by a grayish, sinuate cambium

[1] Faculty of Pharmacy, Medical Academy, Dunav Str. 2, 1000 Sofia, Bulgaria
[2] Institut fur Entwicklungs- und Molekularbiologie der Pflanzen, Heinrich-Heine-Universität Düsseldorf, Universitätsstr. 1, 40225 Düsseldorf, Germany

Biotechnology in Agriculture and Forestry, Vol. 28
Medicinal and Aromatic Plants VII (ed. by Y.P.S. Bajaj)

**Fig. 1.** *Althaea officinalis* plant

from the white radiate wood. The section shows numerous mucilage cells, the content of which is colored a deep yellow by solution of sodium hydroxide.

### 1.3 Medically Important Compounds

On the basis of the organ-specific distribution of the compounds, the following drugs are used:

Radix Althaea contains from 5–10% (Hegnauer 1969; Hausel and Haas 1984; Widmaier 1988; Hausel 1991) to 20–35% (Trease and Evans 1978; Koleva et al. 1990) of mucilage. The amount being season-dependent (Evans 1989); 30–38% starch, 11% pectin, 5–10% sugars, 2% betain, lecithin, phytosterin, and asparagine. The latter, which is the amide of aspartic (aminosuccinic) acid has no therapeutic value, 2% tannins, 1.5% fatty acids, 7% minerals, flavonoids, the coumarin scopoletin, and phenolic acids.

Folia Althaea contains 5–10% mucilage, flavonoids, coumarin-scopoletin, and 0.02% essential oils.

Flores Althaea contains 5–8% mucilage, 0.02% essential oil, asparagin, sugars, and flavonoids.

Several groups of compounds of *Althaea* plants, proved to be of medical importance, are described here.

### *1.3.1 Mucilages*

Different plant species vary enormously in their soil and nutritive requirements, and this aspect has received considerable attention in medicinal plants. It is known that the content of polysaccharides and their composition in the plant is affected by several factors, for example, the locality where it is grown, the degree of the development stage, time of harvest, etc.

Variation in particle size results in different soils, ranging from clay via sand to gravel. Particle size is one factor influencing water-holding capacity, and *Althaea officinalis*, which produces mucilage as a water-retaining material, contains less mucilage when grown on soil with a high moisture content (Evans 1989).

The production of mucilage coincides with the fermentative disintegration of starch (Franz 1966). The muci-layer is deposited on the cell wall and the whole cell is filled with mucilage. All starch-containing cells of *A. officinalis* have the capacity to produce mucilage.

Based on the chemical property of the mucilage of *Althaea* plants, von Friedrichs (1919) found its components to be glucose and xylose. In 1946, Beauquesne (cited in Franz 1966) reported that is was made up of uronic acid, methylpentoses, and hexoses. Franz (1966) reported that crude mucilage of the roots contains glucan, an arabinogalactan, and an acidic polysaccharide – galacturonorhamnan. He observed that the neutral fraction was composed of 21% glucose, 52% galactose, and 27% arabinose. For the acidic polysaccharide, Franz reported a composition of 58% galacturonic acid, 39% rhamnose, 3% galactose, and trace amounts of glucose. He isolated α-(D-galactopyranosyl-uronic acid)-(1-2)-L-rhamnose as a partial hydrolysate of the acidic polysaccharide.

Recently, several reports have appeared on structural studies of mucilaginous polysaccharides isolated from *A. officinalis* plants, as well as from other related species and varieties.

Tomoda et al. (1977) reported on the isolation and properties of mucous polysaccharide, named *Althaea* mucilage O. It is composed of L-rhamnose: D-galactose: D-galacturonic acid: D-glucoronic acid in the molar ratio of 3:2:3:3, and the molecular weight of its ammonium salt is 34000. Later, the same group reported partial acid hydrolysis of *Althaea* mucilage O, and the isolation of five oligosaccharides (Tomoda et al. 1980). Analysis of their components, as well as reduction and methylation, and partial degradation studies showed that these oligosaccharides are O-α-(D-galactopyranosyl uronic acid)-(1-2)-L-rhamnopyranose, O-β-(D-glucopyranosyluronic acid)-(1-3)-O-α-(D-galactopyranosyl-uronic acid)-(1-2)-L-rhamnopyranose, and the hexasaccharide, the nonasaccharide, and the dodesaccharide composed of a repeating unit having the structure of the trisaccharide through position 4 of the D-galacturnoic acid residue (Fig. 2).

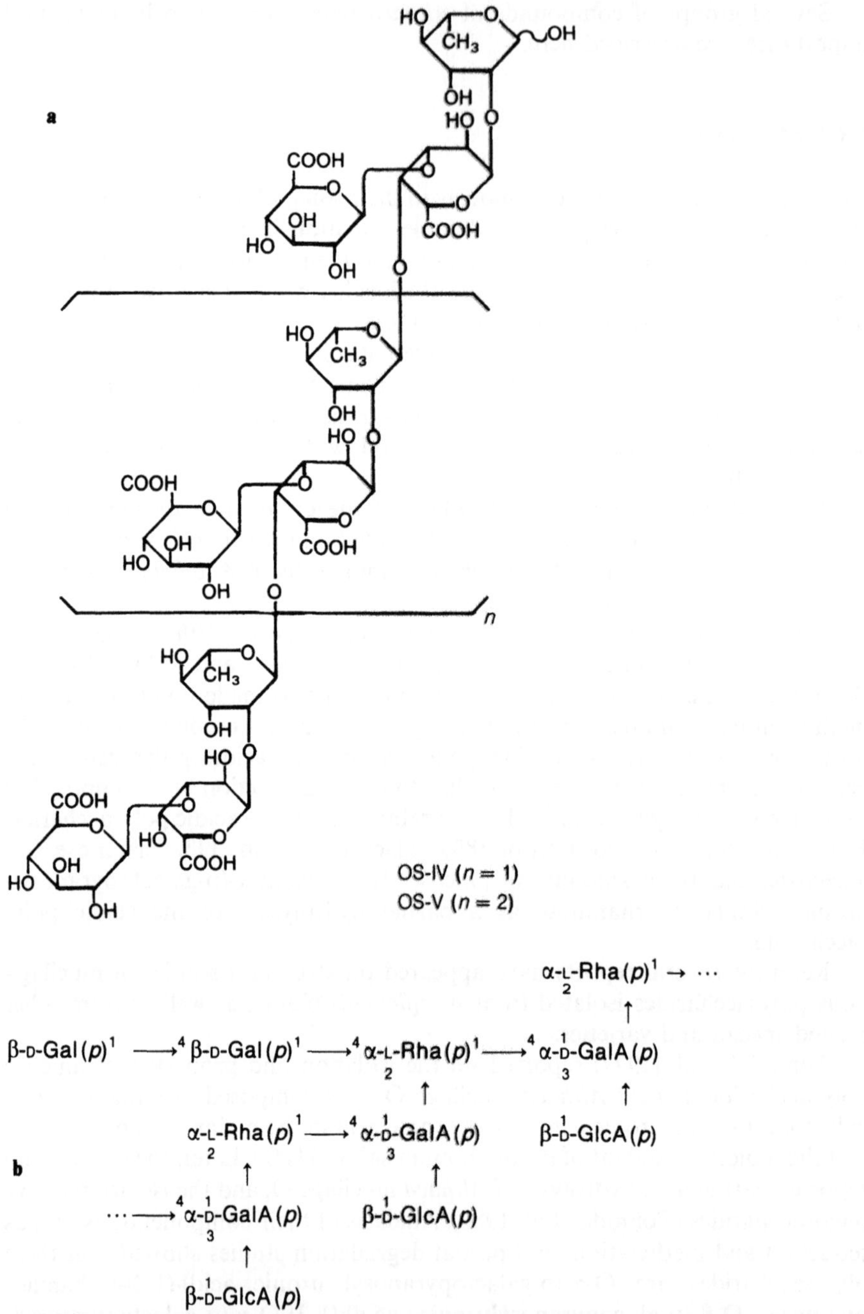


**Fig. 2. a** Structures of Oligosaccharides IV and V, obtained from the partial hydrolysate of the mucilage from the roots of *Althaea officinalis* L. (*OS* Oligosaccharide). **b** A possible structural fragment of *Althaea*-mucilage O

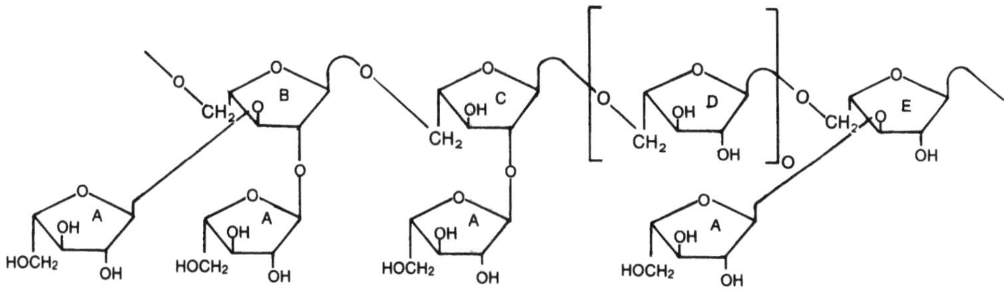

**Fig. 3.** L-Arabinan from the roots of *Althaea officinalis* L.

Capek et al. (1983) reported isolation of a water-soluble L-arabinan from the roots of the marshmallow with a highly branched structure of ~90-α-L-arabinofuranosyl residues variously linked by (1–5), (1–3), and (1–2) bonds (Fig. 3). Later, they isolated three D-glucans (Capek et al. 1984) from *A. officinalis*, var. robusta which differed in molecular weight. The polysaccharide chains are composed of α-D-glucopyranose units linked by (1–6) glycosidic bonds almost exclusively. The polymers had essentially the same structural features as water-soluble, low-molecular D-glucan, isolated from the leaves of this plant (Kardosova et al. 1983).

Blaschek and Franz (1986) and Madaus (1989) reported mucilage from the roots of *A. officinalis*, derived from extracts of different origin. They found a neutral fraction, mainly composed of glucose, and two acid fractions, both of which differ in increasing amounts of uronic acids and rhamnose. The sugar composition of the combined acid polysaccharides was determined as 24% galacturonic acid, 22% glucoronic acid, 35% rhamnose, 8% arabinose, 1% xylose, 2% manose, 5% galactose, and 4% glucose.

Despite the different variety of the plants used and different cultivation conditions, the basic structural features of the *Althaea* polysaccharides are similar to those of the mucilage described by Tomoda et al. (1977). Some minor differences are evident in the degree of branching. For example, in the acidic heteropolysaccharide isolated from the roots of the variety *Robusta*, all L-rhamnose units from branching points, as indicated by the absence of any di-O-methyl derivative of L-rhamnose in the methylation products of the polysaccharide, whereas one third of the L-rhamnose units were unbranched in the material (Capek et al. 1987). From the roots of the variety Russalka three homogeneous polysaccharide fractions were isolated. AO1, containing glucose, xylose, and arabinose; AO2 containing xylose, arabinose, rhamnose, and glucoronic acid; and AO3 with xylose, galactose, arabinose, and glucoronic acid (Koleva et al. 1990).

The component unit having the repeating structure (1–4)-O-β-(D-glucopyranosyluronic acid)-(1–3)-O-α-(D-galactopyranosyluronic acid)-α-(1-2)-O-α-L-rhamnopyranose is common to mucilages from roots and leaves of *A. officinalis* (Tomoda et al. 1981).

*Althaea rosea* Cavailles (hollyhock) has also been used as a crude drug with emollient and demulcent proprieties. Salikov et al. (1978) reported that the crude polysaccharides of the roots were made up of arabinose, xylose, galactose, glucose, rhamnose, galacturonic acid, and glucoronic acid. Later, Tomoda et al. (1983) isolated mucilage from the roots of the same plants, named *Althaea* mucilage R. The polysaccharide moiety was composed of D-galactose, D-glucose, L-rhamnose, D-galacturonic acid, and D-glucoronic acid in the molar ratio of 2:1:6:4:6.

The mucilages from the roots of *A. officinalis* have branches composed of 4-O-β-D-galactopyranosyl D-galactopyranose at position 4 of a part of the L-rhamnose units in the main chain. In contrast, the presence of the side chains composed of both D-galactose and D-glucose is another characteristic of the structure of the mucilages from *Althaea rosea* (Tomoda et al. 1983).

The common structural unit of the polysaccharides from the *Althaea* genus was identified as O-β-(D-glucopyranosyluronic acid)-(1-3)-O-(α-D-galactopyranosyluronic acid)-(1-2)-L-rhamnopyranose (Capek et al. 1987).

### *1.3.2 Flavonoids*

Flavonoids are found in all organs of the *Althaea* plant (Gudej 1989; Ninov et al. 1992); (Table 1). The flowers of *A. officinalis* L., *A. armeniaca* Ten., *A. canabina* L., *A. narbonensis* Pourr., and *A. bronssonetiifolia* Iljin., contain qualitatively identical sets of flavonoids (Table 2). The presence of kaempferol-3-O-β-D-(6″-p-coumaroyl)-glucoside (tiliroside), kaempferol-3-O-glucoside (astragalin), quercetin-3-O-glucoside (isoquercitrin), dihydrokaempferol-4′-O-glucoside, naringenin-4′-O-glucoside, and trace amounts of hypolaetin-8-O-gentiobioside and hypolaetin-8-O-glucoside has been confirmed (Gudej and Bieganowska 1990a). The dominating compounds in the flowers are tiliroside and dichidrokaempferol-4′-O-glucoside.

**Table 1.** Flavonoids of *Althaea officinalis* L. plants

| Compounds | Plant part | Reference |
|---|---|---|
| 1. Tiliroside (kaempferol-3-O-D-(6″-p-coumaroyl)-glucoside | F, L | Gudej (1989) |
| 2. Astragalin (kaempferol-3-O-glucoside) | F, R | Gudej (1989)<br>Ninov et al. (1992) |
| 3. Isoquercitrin (quercetin-3-O-glucoside) | F, R | Gudej (1989)<br>Ninov et al. (1992) |
| 4. Dihydrokamepferol-4′-O-glucoside | F | Gudej (1989) |
| 5. Naringenin-4′-O-glucoside | F | Gudej (1989) |
| 6. Hypolaetin-8-O-gentiobioside | F, L | Gudej (1989) |
| 7. Hypolaetin-8-O-glucoside | F, R | Gudej and Biegan (1990); Gudej (1991) |
| 8. Quercetin | L, R | Ionkova (1992c) |
| 9. Kaempferol | L, R | Ionkova (1992c) |
| 10. Naringenin | L, R | Ninov et al. (1992) |
| 11. Isoscutellarein-4′-methyl-ether-8-O-D-glucoside-2″-$SO_3K$ | R | Gudje and Dzido (1991) |

F = flower; L = leaf; R = root

**Table 2.** Results of the quantitative determination of the flavonoglycosides in leaves and flowers in various species of *Althaea* (results in % of dry material). (Gudej and Bieganowska 1990a)

| Substance | Content in material, % | | | | | | | | | |
|---|---|---|---|---|---|---|---|---|---|---|
| | *A. officinalis* | | *A. armeniaca* | | *A. cannabina* | | *A. narbonensis* | | *A. broussonetiifolia* | |
| | L | F | L | F | L | F | L | F | L | F |
| Dihydrokaempferol 4′-O-glucoside | – | 0.869 | – | 0.854 | – | 1.934 | – | 1.044 | – | 0.839 |
| Isoquercitrin | 0.005 | 0.099 | 0.005 | 0.203 | 0.068 | 0.258 | 0.012 | 0.189 | 0.006 | 0.202 |
| Hypolaetin 8-O-genitiobioside | 0.382 | – | 0.055 | – | – | – | 0.068 | – | 0.086 | – |
| Astragalin | 0.053 | 0.248 | 0.005 | 0.187 | 0.034 | 0.124 | 0.015 | 0.177 | 0.013 | 0.196 |
| Tiliroside | 0.172 | 0.197 | 0.042 | 0.297 | 0.054 | 0.224 | 0.056 | 0.328 | 0.055 | 0.325 |
| Total | 0.617 | 1.413 | 0.107 | 1.541 | 0.156 | 2.540 | 0.151 | 1.738 | 0.160 | 1.562 |

L = leaves. F = Flower.

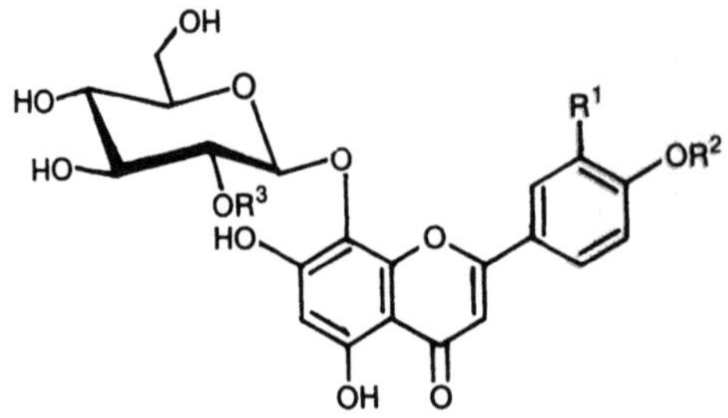

1 $R^1$ = OH, $R^2$ = H, $R^3$ = H
2 $R^1$ = H, $R^2$ = $CH_3$, $R^3$ = $SO_3^{\ominus}K^{\oplus}$

**Fig. 4.** Flavonoids from the roots of *Althaea officinalis* L. *1* Hypolaetin-8-O-β-D-glucoside; *2* 4′-methylether-8-O-β-D-glucoside-2″-SO3K

In the leaves, both qualitative and quantitative differences have been observed in the flavonoid content in *A. officinalis* L. and *A. armeniaca* Ten.. On the contrary, the leaf flavonoids of *A. cannabina* L. show differences in this aspect. The dominating flavonoids in the leaves of *A. officinalis* L., *A. armeniaca* Ten., *A. narbonensis* Pourr. and *A. bronssonetiifolia* Iljin, are tiliroside and hypolaetin-8-O-gentiobioside, while in the leaves of *A. cannabina* L. tiliroside and quercetin-3-O-glucoside are the dominating compounds.

The flavonoid aglycons quercetin, kaempferol, and naringenin were found in leaves of *A. officinalis* L., var. Russalka (Ionkova 1992c); four flavonoids have been isolated from the roots, hypolaetin-8-O-β-D-glucoside, the new flavonoid isoscutellarein-4′-methylether-8-O-β-D-glucoside-2″-$SO_3K$ (Fig. 4), kaempferol-3-O-glucoside (astragalin), and quercetin-3-O-glucoside (isoquercitrin) (Gudej 1991; Ionkova 1992c). The last two are isolated from the roots of variety Russalka. Three flavonoid aglycons, kaempferol, quercetin, and naringenin, have been established in the roots of variety Russalka by high performance liquid chromatography (Ninov et al. 1992).

The flavonoid content is much higher in the flowers than in leaves and roots and varied from 1.41% (*A. officinalis*) to 2.64% (*A. cannabina*) for the first case, from 0.11% (*A. armeniaca*) to 0.62% (*A. officinalis*) in the second case, and from 0.12 to 0.24% in the root (*A. officinalis*).

Flowers of all species of *Althaea* and leaves of *A. officinalis* are useful as sources of flavonoids.

### *1.3.3 Coumarins and Phenolic Acids*

Phenolic acids protocatechinic, p-hydroxyphenylacetic, p-hydroxybenzoic, vanillic, syringic, salicylic, cafeic, p-coumaric, clorogenic, ferulic, sinapic, and the only coumarin-scopoletin were found in the flowers, leaves, and roots of five species of the genus *Althaea*: *A. officinalis, A. armeniaca* Ten., *A. cannabina* L., *A. narbonensis* Pourr., and *A. bronssonetiifolia* Iljin. The content of phenolic acids in the flowers of the investigated plants is much higher than in the leaves and roots (Gudej and Bieganowska 1990b; Gudej 1991; Ionkova 1992c; Ninov et al. 1992). The content of scopoletin in the leaves and in the roots of *A. officinalis*, var. Russalka is similar (Ionkova 1992c).

### 1.4 Pharmacological Effects of *Althaea* Compounds

*Althaea officinalis* L. has been known as a medicinal plant since ancient times, when its roots, leaves, and flowers were used (Kohler 1887) Althaea radix is a well-known antiphlogistic, emollient, and demulcent (Reuter et al. 1988: Frohne and Tenssen 1979). The mucous material isolated from the roots of the marshmallow may be used as a crude drug in treatment of catarrh of the respiratory system, gastritis, ulcus ventriculi, various inflammations of the nasal and oral cavities, kidney pain, and tonsillitis (Kresanek and Krejca 1981).

The use of polysaccharide hydrocolloides in antitussive preparations such as cough drops, cough syrup, and cough lozenges is well documented (Franz 1989). As a consequence of inflammatory events of the mucous membranes in the respiratory tract, the naturally occurring mucin layer is disturbed in its function. After medication with polysaccharide hydrocolloids, the mucin effect can be restored by forming a protective layer in order to desensitize the irritated peripheroceptors. Local inflammations are diminished by this effect, and as a consequence, the dry cough is reduced (Franz 1989).

According to recent literature, some of the flavonoids found in *Althaea* show pronounced antiinflammatory activity, e.g., tiliroside, astragalin, and kaempferol (Della Loggia et al. 1989), hypolaetin-8-O-glucoside and its aglycone (Villar et al. 1984, 1987; Alcaraz et al. 1988). They may be a supporting factor for the anti-inflammatory effect of the mucilage present in the investigated materials (Gudej and Bieganowska 1990a).

It has been known for some time that, in animal experiments, a high polysaccharide diet reduces the blood level of both cholesterol and glucose. The mucilages of *Althaea* showed considerable anticomplementary activity and remarkable hypoglycemic activity (Tomoda et al. 1986, 1990).

Several herbal drugs including *Althaea* are at present considered as being potent in the area of immunestimulants. The polysaccharides were postulated to have an unspecific immune-modulation effect. Those containing uronic acid in their structure have better effect (Wagner et al. 1985). The mixture of polysaccharides from the variety Russalka of *Althaea officinalis* L. has a strong immunostimulating effect to the humoral immune response and phagocytic activity (Marinova et al. 1985; Koleva et al. 1990).

## 2 Biotechnological Approaches

### 2.1 Callus Induction

The studies on the production of biological active substances in in vitro cultures of *Althaea* started with our report on callus induction and transformed root

**Table 3.** Growth rate of callus obtained from different organs of *Althaea officinalis* var. Russalka. The callus was cultured for 4 weeks at 26 °C in constant darkness (Ionkova et al. 1991)

| Medium | Phytohormones (mg/l) | | | | Growth rate | |
|---|---|---|---|---|---|---|
| | Kinetin | 2,4-D | IAA | NAA | Leaf | Petiole |
| Ga | 0.1 | 1.0 | 0 | 0 | 10.4 | 11.3 |
| Gb | 0.3 | 0 | 0 | 4.0 | 11.4 | 12.6 |
| Gc | 0.3 | 0 | 2.0 | 0 | 7.9 | 9.1 |
| Gd | 3.0 | 1.1 | 0 | 0 | 8.8 | 9.7 |

cultures of *Althaea officinalis*, var. Russalka (Ionkova et al. 1991). Cultures were set up from sterile plants, obtained from aseptically germinated seeds (Botanical Garden, Sofia, Bulgaria). In general, it has been found that the biosynthetic activity of cell cultures is lower than that of the donor plant (Ionkova 1992c). There is some evidence to suggest that the variation in biosynthetic activity observed in cultures is influenced by the intrinsic biosynthetic capacity of the explanted material. Cultures derived from high-yielding plants produce higher levels of natural products than comparable cultures from low-yielding plants (Zenk et al. 1977; Kinnersley and Dougall 1981; Holden et al. 1988).

In 1990, when we started work with cell cultures from *Althaea officinalis* L. in order to find metabolites of pharmacological relevance, we established callus from the highly productive variety named Russalka (Ilieva 1984; Koleva et al. 1990). It was proposed on the basis of agrobiological indexes – introduction and selection. The first step was to find the best explant material. Callus cells were obtained from leaf and petiole. It was found that those derived from the petiole give better results. However, calli growth was low, had a pale yellow color, often with browning. This color is usually attributed to the biosynthesis of polyphenols, which accumulate in the cell and inhibit growth, inducing degeneration of the callus line (Henry et al. 1991).

Therefore extensive studies were made to optimize the factors in the medium that affect growth, i.e., growth hormones. Murashige and Skoog (1962) solid medium of any growth regulator was the basal medium and is designated MS. In various experiments, the basal medium was supplemented with auxins: indol-3-acetic acid (IAA), 1-naphthylacetic acid (NAA), 2,4-dichlorophenoxyacetic acid (2,4-D), and cytokinin-kinetin in various combinations, which play an important role in inducing friable callus. The results are summarized in Table 3. Differences in growth are observed between the various media (Ga, Gb, Gc, Gd). Regarding phytohormones, it was found that 2,4-D strongly suppressed organogenesis. With increase in concentration of 2,4-D, the calli obtained are more hydrated. NAA and IAA are weaker than 2,4-D. When IAA is replaced by NAA or 2,4-D, the calli are completely different, with more compact texture. Best results were obtained with a 2,4-D concentration of 1 mg/l (medium Ga), or an NAA concentration 4 of mg/l (medium Gb) resulting in relatively rapid growth and friable calli. Comparison of the auxins 2,4-D and NAA showed no

**Table 4.** Culture condition and polysaccharide concentration. (Ionkova et al. 1991)

| Medium | Explant | Light/darkness | Polysaccharide |
|---|---|---|---|
| Ga | Petiole | Light (4000 lx) | + + |
| Gb | Petiole | Light (4000 lx) | + |
| Gb | Petiole | Darkness T 25 °C | + + |
| Ga | Petiole | Darkness T 25 °C | + + + |
| G(-horm) | HR | Darkness T 25 °C | + + + |

significant differences for growth yields. However, comparison concerning polysaccharide production showed that the 2,4-D in lower concentrations gave better results (Table 4). Callus cannot be induced from medium containing less than 0.1 mg/l auxin, or auxin-free medium. Addition of cytokinin increases proliferation. Comparison of different concentrations of kinetin gave the best result with 0.3 mg/l.

Light is important for growth and a number of metabolic processes (Fowler 1985; Staba 1985; Ionkova 1992c). Best growth was observed in the light series; however, from Table 4, it can also be observed that the best results concerning polysaccharide accumulation were found in the dark series. As long as *Althaea* callus cultures are kept in the dark they multiply, but the capacity to produce flavonoids decreases.

Based on these results, larger amount of calli were grown on medium Ga in dark and light, for further studies of the metabolism of natural products. Parts (3 g fresh weight) of the calli are transferred to a new medium every 30 days for repeated subculture. After about 1 year of monthly subcultivation, the callus had adapted to the constant standard condition.

### 2.2 Suspension Cultures

The establishment of cell suspension from calli of *Althaea officinalis* caused no problems because the components used for the cultures on solid agar medium are also suitable for cultures in liquid medium (Hunault 1984; Ionkova 1992). Calli obtained from petiole are used for induction of suspension. Two-week-old calli (5 g fresh weight) were inoculated in 300-ml Erlenmayer flasks containing 50 ml medium Ga without agar. The cultures were incubated on a rotary shaker (80–100 rpm) in darkness or light, at 25 °C. Reactions during the first passage to a liquid medium differed. Several of the cultures dissociated slowly. In these, the cells did not dissociate much and remained closely together in the form of rather voluminous aggregates. Once established, the suspension cultures proliferated actively and were subcultured every 10 days. The suspension was chlorophyllous. In order to maintain the viability of the culture, the cells were subcultured before the stationary phase (every 10 days).

For production of polysaccharides by callus cultures of *Althaea*, NAA (4 mg/l) and 2,4-D (1 mg/l) were found to be better auxins than IAA, therefore we preferred a low concentration (2,4-D 1 mg/l); kinetin level was 0.3 mg/l.

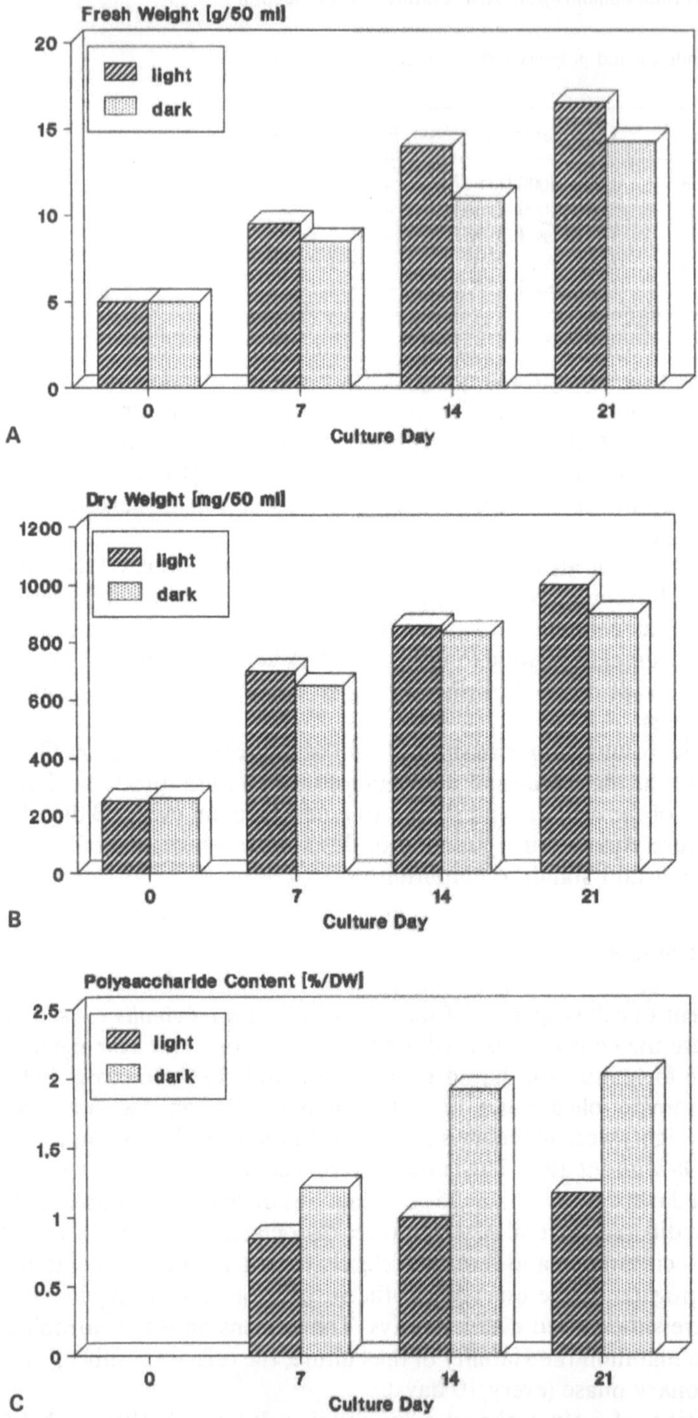


**Fig. 5A–C.** Growth and polysaccharide production in suspension cultures of *Althaea officinalis*, var. Russalka

Growth and formation of metabolites are often inversely related. This fact has been discussed in detail (Lindsey and Yeoman 1983). The accumulation of natural compounds at any stage of culture growth is the result of a dynamic balance between biosynthetic, biotransformational, and biodegradative processes. Growth of cell suspensions and polysaccharide production in *Althaea* suspension cultures are shown in Fig. 5. The maximal yield was obtained during the stationary growth phase.

The importance of light for the optimal expression of flavonoid biosynthesis in callus cells was demonstrated in Section 2.1. From the technological point of view, however, it would be more convenient if light is not required for achieving relevant product levels and could be replaced by other effectors (Berlin 1988).

The influence of darkness for polysaccharide production in suspension cultures of *A. officinalis* is given in Fig. 5. For metabolic studies, cell suspensions were used after the stabilization period of culture (fifth generation) in order to avoid the possibility of carryover of polysaccharides from parent (calli) cells.

Quantitative analysis of polysaccharides was carried out on culture media as well as on the cells, since they are thought to be excreted from the cells into the media. Metabolite yields were compared by product weight per unit weight of cells or volume of medium basis.

## 2.3 Transformed Hairy Root Cultures

The family Malvaceae is unique in the production of mucilage, which is present in many of its members. The majority of these biopolymers occur in the roots. Recently, we have begun to study the potential of hairy root cultures of *Althaea officinalis*, var. Russalka resulting from the infection of sterile plant segments with *Agrobacterium rhizogenes*. In those cases in which roots constitute a main site of synthesis in the plant, hairy root cultures are considered as an alternative system to cell suspension cultures. These hairy root cultures have metabolic features very similar to normal root, and are potentially applicable to the production of all root-derived products from dicotyledonous plants (Rhodes et al. 1987; Ionkova et al. 1989; Sign and Flores 1990).

### *2.3.1 Establishment and Maintenance of Hairy Roots*

Hairy roots of *Althaea officinalis*, var. Russalka were induced by co-cultivation of leaves or stems of in vitro-cultivated plants with the suspension of *Agrobacterium rhizogenes* strain LBA 9402, which contains the plasmid pRi 1855 and induces production of the opine agropine (Ooms et al. 1985). This plasmid shows less dependence on the internal auxin levels in the plant, with regard to its ability to produce roots, than several mannopine-producing strains (Hamill et al. 1987). Another bacterial strain is the supervirulent R 1601, which was obtained from Dr. Pythoud, Bern. The last R 1601 harbors the cosmid pTVK 291, which is responsible for the supervirulent genotype (Pythoud et al. 1987).

R 1601 confers a kanamycin resistance gene with the eukaryotic promoter and terminator of the nopaline synthesis gene.

Different strains showed various abilities to induce hairy root (HR) formation. A promoting effect of acetosyringone on the *Agrobacterium*-mediated HR initiation known for other species was observed in *Althaea* as well. The suspensions of strain LBA 9402 generated by growing bacteria in YMB medium (Ionkova 1992b), including 10 $\mu$M acetosyringon, an activator of the *vir* genes of *Agrobacterium*, produced during the wounding response of plants, was left for 2 days at 25 °C with gyratory shaking. Scratching the midrib of the leaf or the stem of the *Althaea* plant with a hypodermic needle containing a thick bacterial suspension allows inoculation with small droplets containing *Agrobacterium*, which is kept in agar medium. Approximately 14 days after inoculation HR tips appeared at the inoculated sites.

Cultures were cleared of bacteria by several passages in medium containing 500 mg/l antibiotic (Claforan Hoechst). Several segments of HR were transferred to hormone-free medium. The best growth of HR tissue was obtained by inoculating a flask of medium with three to four root tips of 2–3 cm length (approx. 0.2–0.3 g fresh weight). In our experiments, HR started to grow with a minimal lag period.

### *2.3.2 Stability and Confirmation of Transformation*

We have found that hairy root tissue from a number of plant species (*Datura, Hyoscyamus, Astragalus, Nicandra, Althaea*) grows rapidly in simple media without phytohormones (Ionkova et al. 1989, 1990b, 1991; Ionkova 1990, 1991, 1992a, b.)

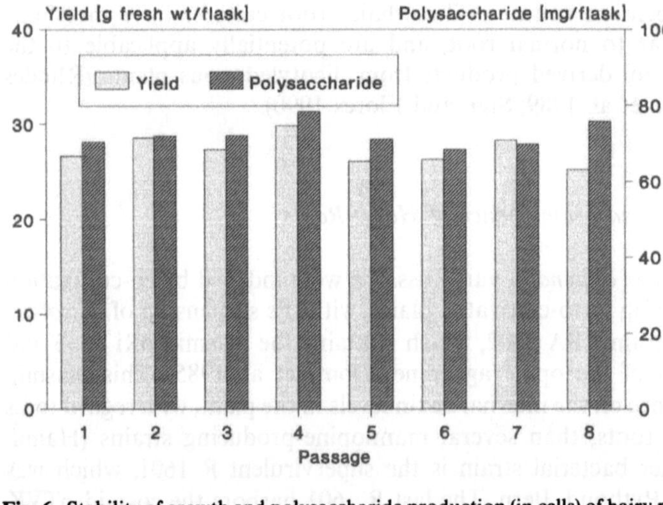


**Fig. 6.** Stability of growth and polysaccharide production (in cells) of hairy root clone R3 of *Althaea officinalis*, var. Russalka

The major advantage of hairy root cultures of *Althaea officinalis* over suspension cultures is their inherent genetic stability. The chromosome number of HR is that of the parent plant (Hamill et al. 1987; Aird et al. 1988). This stability is reflected in *Althaea* cultures in both growth rate and considerable biosynthetic capacity (Ionkova 1992c). The HR cultures were stable in their growth rate and polysaccharide production over a period of more than 2 years in culture (Fig. 6).

Opine biosynthesis is a firm indication that these roots are indeed transformed. To prove the genetic transformation, the opine mannopine/agropine was identified in several HR clones. Preparation of extracts from plant material and HR clone, opine analysis by high voltage paper electrophoresis and TLC, and detection with alkaline silver nitrate reagent were carried out as described earlier (Ionkova et al. 1990b; Ionkova 1991). The opines were identified in the extracts of the freshly isolated hairy roots, whereas normal tissues were not.

### *2.3.3 Growth Kinetics of Hairy Roots*

Hairy roots of *Althaea officinalis*, var. Russalka have a profusion of root hairs, a high degree of lateral branching, and show no geotropism (Fig. 7), resulting in high growth, rates; with about 30 g/100 ml fresh weight in a 4 weeks' growth period, average fresh weight increased 29–30-fold.

The effects of sucrose concentration and absence of $NH_4NO_3$ in the medium on the growth and yield of polysaccharides in *Althaea* HR were also examined. The response of product formation to different carbohydrates supplied in the medium appears to indicate that sucrose may be the best general carbon source. It was found that sucrose in a concentration of 2% is most suitable for maximal

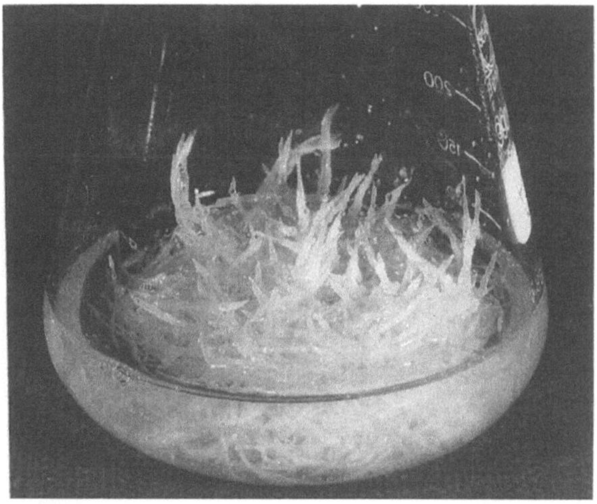

**Fig. 7.** Hairy loots – clone R3 of *Althaea officinalis*, var. Russalka

biomass yield. Low and high concentration of carbohydrate source inhibits growth of hairy roots (Fig. 8).

The most apparent effect was achieved by using a medium without $NH_4NO_3$, which resulted in better growth of the roots as well as in higher total polysaccharide accumulation (Fig. 9). Quantitative differences in tissue growth in the presence and absence of $NH_4NO_3$ have been demonstrated (Cravent et al. 1972; Ionkova 1992c). The HR clone, grown with and without $NH_4NO_3$, is morphologically different. Furthermore, the transformed tissues secreted a large amount of polysaccharides into the liquid medium without $NH_4NO_3$ (Table 5) (Ionkova, 1992c).

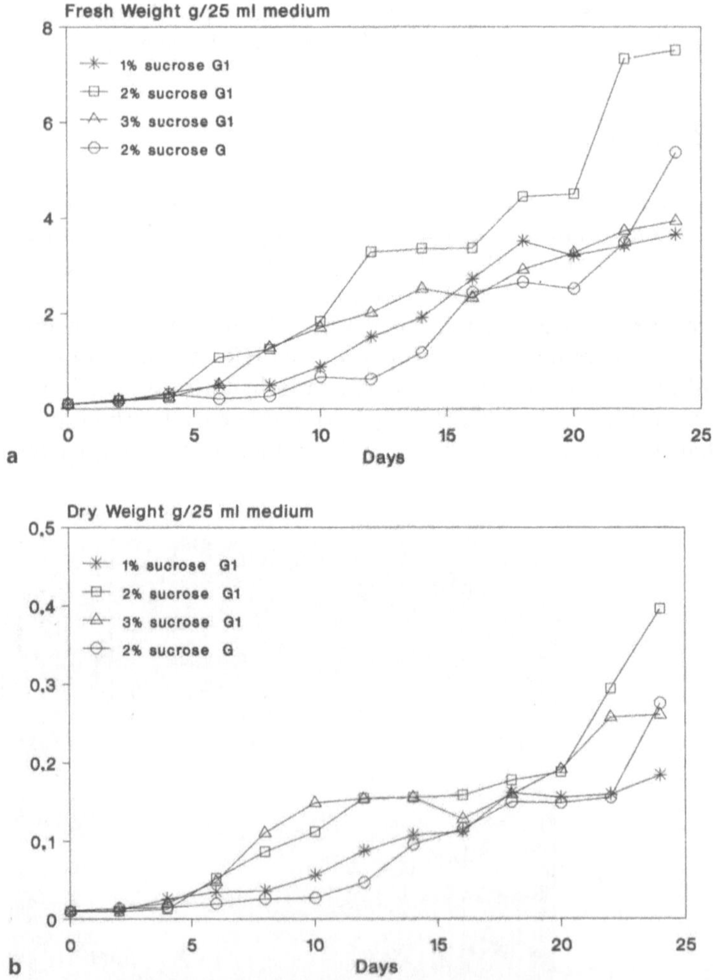


**Fig. 8a–d.** Characterization of a HR culture of *Althaea officinalis*, var Russalka. Culture period of 24 days in medium G and G1 with 1–3% sucrose: **a** Fresh weight (g/25 ml medium). **b** Dry weight g/25 ml medium). **c** pH. **d** Sugar concentration (%) measured by refractive index

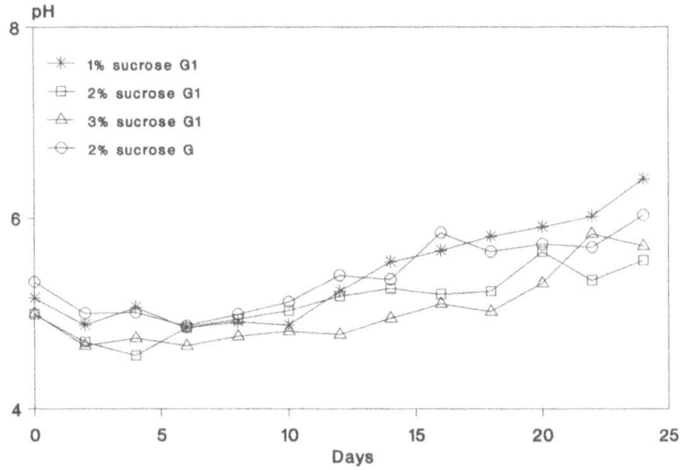


c

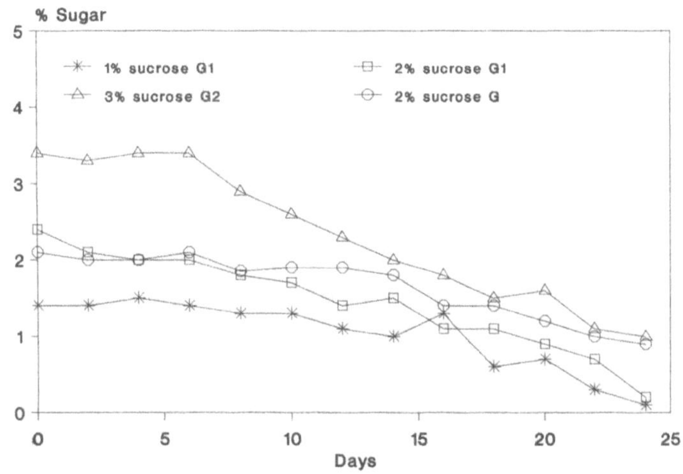


d

**Fig. 8c, d**

**Table 5.** Comparative analysis of the genetically transformed roots of *Althaea officinalis*, var. Russalka and intact plant roots. (Ionkova 1992c)

| Kind of root | Color of root mass | Consistency and kind | Polysaccharides (mg/ml) | |
|---|---|---|---|---|
| | | | In medium | In roots |
| 1. HR (G1) | Whitish yellow | Thin, long | 2.13 | 0.58 |
| 2. HR (G) | Gray-beige | Thick, juicy | 0.28 | 0.55 |
| 3. Root of plant | Gray | Thick, bast fiber-like | | |

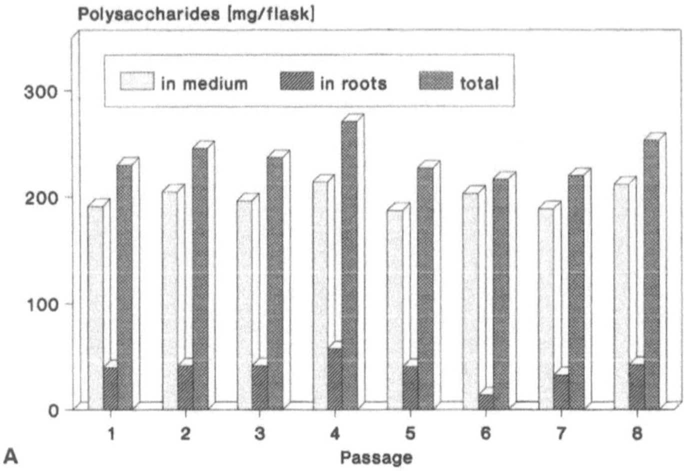


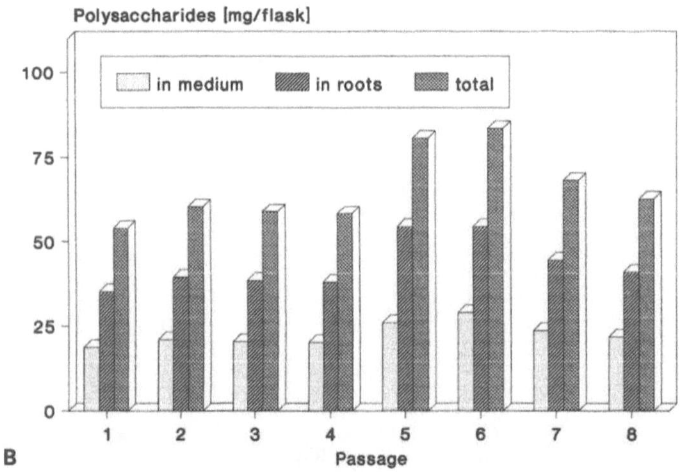


**Fig. 9A,B.** Polysaccharide production in HR of *Althaea officinalis*, var. Russalka. **A** Medium MS without $NH_4NO_3$ (G1). **B** Medium MS (G)

# 3 Production of Biologically Active Compounds from Cell and Tissue Cultures

## 3.1 Polysaccharides

Increasingly, cell cultures are now being used as a tool to elucidate carbohydrate metabolism. Cell cultures of *Althaea officinalis* produce significant amounts of polysaccharides. Macromolecules are secreted into the liquid media by a

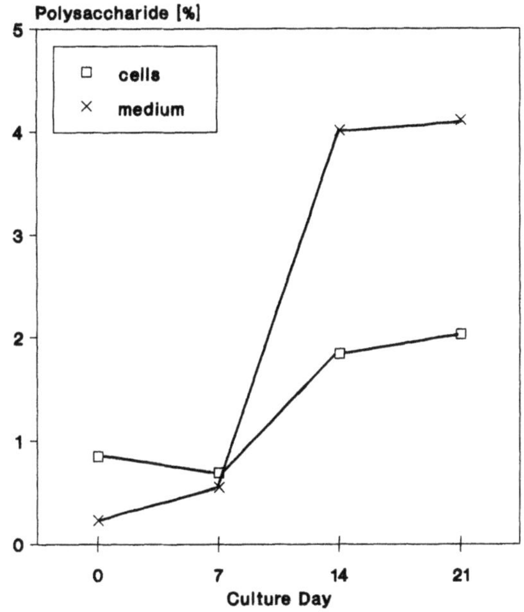


**Fig. 10.** Time course of changes in polysacccharide content during the growth of *Althaea officinalis*, var. Russalka suspension cultures (% dry wt. of cells)

nondifferentiated suspension and hairy root cultures. Figure 10 shows the time course of secretion in these cultures. These data show that the concentration of the polysaccharides changes significantly during cultivation. The concentration changed from 0.012 mg/ml in newly inoculated cultures to 0.74 mg/ml in older cultures.

The data of Table 6 include the sugar composition of the macromolecules. The composition of the suspension and callus polysaccharides is similar. Both polysaccharides contain the same array of sugars – they included arabinose, xylose, galactose, glucose, and traces of rhamnose in callus polysaccharides, with a predominance of arabinose and glucose. This preliminary evidence, obtained by chromatography using the method of Madaus (1989), indicates that the polysaccharides are heterogeneous. The amount of monosaccharides is similar in all cases. However, the polysaccharides are similar in composition to the monosaccharides of undifferentiated cells, and very different from the composition of those in the transformed and intact roots of the plant. The polysaccharides of the intact roots yield a predominance of rhamnose with uronic acids, while the undifferentiated cells possess a predominance of arabinose, without rhamnose and uronic acids. The same monosaccharides are found in the hydrolysates of tobacco cells growing in suspension (Olson et al. 1969) and in sycamore cultures (Becker et al. 1964).

The marshmallow cells grow rapidly in large amounts, and by continuous transfer, the population in the logarithmic phase of growth is more uniform than cells of plant tissues. Furthermore, the supply of metabolites is readily controlled, and every cell has direct access to the external medium containing these

**Table 6.** Sugar composition of the polysaccharides from *Althaea officinalis*, var. Russalka cultivated in vitro. (Ionkova unpubl.)

| Cells/medium | Neutral monosaccharide composition | | | | | | |
|---|---|---|---|---|---|---|---|
| | | Rh | Gl | Ga | Xy | Ar | Gluc.ac |
| Callus | In cells (dark) | ND | + + + | + + | + + | + + + | ND |
| | In cells (light) | ND | + + + | + + | + + | + + + | ND |
| Suspension | In cells | ND | + + + | + + | + + | + + + | ND |
| | In medium | ND | + + + | + + | + | + + + | ND |
| HR | In cells | tr | + + + | + + + | + + | + + + | + + |
| (medium G1) | In medium | tr | + + + | + + + | + | + + + + | + |
| HR | In cells | tr | + + + | + + + | + + | + + + | + + |
| (medium G) | In medium | tr | + + + | + + | + + | + + + + | + |

Rh = rhamnose. G1 = glucose. Ga = galactose. Xy = xylose. Ar = arabinose. Gluc.ac. = glucoronic acid. ND = Not detected. tr = trace.

metabolites. When desirable, the cells can be grown on different media. Marshmallow cells produce extracellular polysaccharides which can be isolated from the medium by simple ethanol precipitation. Media recovered from cultures in which these cells have been growing were quite viscous. The increase in the viscosity of the medium in a growing culture was due not only to changes in the concentration of the macromolecules in the medium, but also to changes in their composition and possibly an increase in molecular weight.

Although secretion is often desirable, its detection is not always simple. The presence of an extracellular component can be the result of true secretion or simply lysis. Lysis is undesirable, since expensive biomass is lost and the release of the intracellular material is not controlled as is the case when cells are harvested directly for an intracellular product. Payne et al. (1992) reported that lysis can be detected if the fraction of the desired product which is extracellular is the same as the extracellular fraction of a normally intracellular component.

In a continuous system, the actual production and release of a product can be distinguished from the simple loss of intracellular material by lysis. If the total amount of a product recovered at the end of an experiment (extracellular plus intracellular) is the same as the total product in the inoculum, it is reasonable to presume that lysis was responsible for the extracellular portion. On the other hand, if the total amount that is extracellular is more than the amount in the inoculum and also the amount expected to be made by any growth and lysis, then secretion would be probable (payne et al. 1992). Our hypothesis is that in suspension cultures of *Althaea*, probable secretion is observed in the logarithmic phase, in the absence of visible cell debris. The polysaccharides released in the late stationary phase are the result of cell lysis.

Differences in the sugar composition of polysaccharides between cell cultures and intact plants have also been reported (Stoddart et al. 1967; Wagner et al. 1988). We suggest that the difference in the sugar composition of polysaccharides in cell cultures and *Althaea* plants might represent the difference between organized and unorganized cells.

Auxin has been reported to have some effect on sugar composition of polysaccharides (Baker and Ray 1965a; Nishitani et al. 1979). Rubery and Northcote (1970) suggested that 2,4-D acts in facilitating the transfer of arabinose from UDP-arabinose to the pectic arabinogalactan. However, no difference could be found in the sugar composition of polysaccharides in cultured cells from *Althaea* grown with different auxins in culture medium. No difference in the sugar composition was detected between cells cultured on agar medium and liquid medium, or between cells in light and dark.

It is generally assumed that hairy root cultures resemble the roots of the plant from which they were derived in respect to their metabolite synthetic capacity (Flores and Filner 1985; Hamill et al. 1987). We have found HR cultures of *Althaea* to be stable in their biosynthetic capacity, in a slightly lower amount than those in 2-year-old roots of plants grown in the fields. These differences in the production of total polysaccharides between normal and transformed roots can be attributed to the different stages of development in the transformed roots. It was found that *Althaea* HR secreted a considerable amount of polysaccharides into the medium without $NH_4NO_3$, and remained stable (Fig. 9). The total concentration changed from 0.53–0.83 mg/ml (normal medium) to 2.16–2.71 mg/ml (medium without $NH_4NO_3$) in older cultures. A comparison of the relative composition of polysaccharides in medium and hairy root cells shows it to be qualitatively similar, but no monosaccharides uniform with those in the mother plant were found. The uronic acid was exclusively glucuronic acid. Traces of rhamnose were found; the amount of arabinose, xylose, galactose, and glucose is equivalent. In intact plant material, rhamnose is predominant. A comparison of the composition of the external and cellular polysaccharides in marshmallow HR seems to indicate that the external polysaccharides are synthesized in a normal manner within the cell and then excreted into the medium.

It is known that *Agrobacterium rhizogenes* produces intracellular and extracellular β-D-(1-2)-glucan with considerable viscosity; however, it is composed entirely of glucose (Zorrequieta and Ugalde 1986). From our investigation it is clear that *Althaea* HR culture mucilages are composed of several types of neutral monosaccharides.

In a number of species studied earlier (*Datura*, *Astragalus*), accumulation of active substances in HR is approximately growth-related, i.e., the product content per unit biomass stays roughly constant. Polysaccharide biosynthesis in HR of *Althaea*, however, continues after the growth of the cultures ceases. Such nongrowth-related production may be important in developing commercial strains. This may be due to the starch degradations in the stationary phase. The endogenous formation of starch has been followed in cultured cells. The biosynthesis and biodegradation of starch and biosynthesis of polysaccharides in carbohydrate metabolism are related. Thorpe and Meier (1972) and Sasaki and Kainuma (1982) studied starch metabolism. The condition favoring the breakdown of starch is darkness. Thus the significance of darkness for polysaccharide formation in *Althaea* cell cultures, found in our experiments, can be explained.

## 3.2 Flavonoids Profile In Vitro

Samples for determination of flavonoid content were prepared in the same way as in previous reports (Ionkova 1991, 1992c). The analysis of flavonoids was carried out by high pressure liquid chromatography (HPLC) (Ionkova 1991) and reversed phase Hypersyl C 18 column.

The flavonoids were isolated by preparative and column chromatography to be identified by direct comparison with their respective authentic specimens (TLC, HPLC).

All cultures (callus, suspension, hairy roots) produced flavonoids. Table 7 shows the results obtained from a study of the levels of flavonoids in different cultures. The aglycons identified included quercetin, kaempferol, and naringenin. Quercetin and kaempferol were identified in the cultures in both free and bound forms (quercetin-3-glucoside, kaempferol-3-glucoside). Except in suspension cultures, naringenin in free form could be detected only in callus and transformed root cultures of *Althaea officinalis*. In the suspension cultures in darkness, only trace amounts of flavonoids were observed and were not analyzed further.

Maximum contents of quercetin (82 ppm) and kaempferol (10 ppm) in free form were observed in the suspension cultures in light. It was found that dark-grown cell cultures of *Althaea* form few flavonoid glycosides. When light-grown callus cultures were exposed to continuous dark, increase in quantity of aglycons and decrease in glycoside quantity was observed. Hahlbrock and Wellmann (1970) found that the biosynthesis of flavonoid glycosides is light-induced. We suggest that the decrease of the flavonoid content in *Althaea* callus transferred into darkness may be due to a decrease in the activities of the enzymes involved in flavonoid biosynthesis and/or an increase of degrading enzymes. Although many flavonoids do occur in tissue culture, the types and quantities produced are restricted compared with the *Althaea*-intact plant (Table 7).

## 3.3 Coumarins and Phenolcarbonic Acids

Scopoletin has been isolated from callus, suspension, and transformed root cultures of *Althaea officinalis*, var. Russalka (Ionkova et al. 1991; Ionkova 1992c). Furthermore, in a comparative study it has been found that the amount produced is markedly influenced by the differentiation of cells. The amount of the coumarin present in HR cultures differs from that present in conventional cultures. Differentiation and tissue organization are not the only factors which influence the synthesis. It is perhaps significant that in intact plants, coumarins, when present, are frequently distributed throughout the tissues and do not appear to be associated with specialized cell types (Butcher 1977). In intact plants, scopoletin is found in the vacuoles, while in tissue culture of *Althaea* it is in the vacuoles, but is also excreted into the liquid medium. Environmental factors, particularly light or medium constituents, can influence production.

**Table 7.** Flavonoid and coumarin content in *Althaea officinalis.* var. Russalka – intact plants and in vitro cultures. (Ionkova 1992c)

| Compouds | HPLC min | Quantity (μg/g dry wt.) | | | | | |
|---|---|---|---|---|---|---|---|
| | | Callus (light) | Callus (dark) | Suspension (light) | HR | Intact L | Plant R |
| 1. Isoquercitrin | 9.84 | 100 ± 3.6 (4.1) | 10.2 ± 0.12 (1.3) | 151.6 ± 3.78 (2.8) | 74.7 ± 1.86 (2.8) | 270 ± 7 (2.9) | 151 ± 3.77 (2.8) |
| 2. Scopoletin | 10.38 | 270 ± 11.8 (4.9) | 120 ± 3 (2.8) | 231.9 ± 6.4 (3.1) | 312 ± 8.63 (3.1) | 326 ± 7.85 (2.7) | 329 ± 8.5 (3.1) |
| 3. Astragalin | 11.09 | 13 ± 0.24 (2.1) | 9.6 ± 0.13 (1.5) | – | 97 ± 1.65 (1.9) | 170 ± 4.7 (3.1) | 103 ± 2.8 (3.1) |
| 4. Quercetin | 15.34 | 1.5 ± 0.057 (1.3) | 5.1 ± 0.06 (1.3) | 818.1 ± 29.9 (4.1) | 13 ± 0.14 (1.2) | 2317 ± 100 (4.9) | 413 ± 12.9 (3.5) |
| 5. Naringenin | 16.82 | 8.2 ± 2.6 (3.5) | 3.1 ± 0.03 (1.2) | – | 1.6 ± 0.02 (1.3) | 38 ± 7.05 (3.1) | 2 ± 0.02 (1.1) |
| 6. Kaempferol | 17.80 | 70 ± 1.68 (2.7) | 5.7 ± 0.06 (1.2) | 104.1 ± 2.97 (3.2) | 1.7 ± 0.17 (1.1) | 26.3 ± 0.59 (2.5) | 445 ± 8.3 (2.1) |

n = 10 (mean ± SE), L = leaves, R = root

Free phenolcarbonic acids – caffeic, p-coumaric, and chlorogenic acids – have been detected in *Althaea* cultures (Ionkova et al. 1991; Ionkova 1992c). High levels of p-coumaric acid were found in callus grown in medium with low concentration of auxin (2,4-D 1 mg/l) in comparison to high levels (NAA 4 mg/l).

It has been shown that p-coumaric and caffeic acids can serve as precursors of chlorogenic acid (Gamborg 1967).

## 4 Summary and Conclusion

The experiments on cultured tissue of *Althaea officinalis* were performed from the following two viewpoints:

1. The standardization problems concerning the chemical composition of *Althaea* polysaccharides and biological activity of plant extracts, due to varying plant origin (Wagner et al. 1988), make the production of polysaccharides by cell cultures of *Althaea* plants desirable.
2. Plant tissue culture technique can be successfully employed to study the influence of various physical, nutritional, and hormonal factors on growth and production of polysaccharides, flavonoids, coumarins, and phenolic acids under controlled conditions.

Callus induction, suspension cultures, and transformed root cultures were examined under various culture conditions. The results show that they are capable of producing polysaccharides in relatively large amounts. 2,4-D or NAA in the presence of kinetin promotes callus induction and growth. The best medium for both growth and polysaccharide production was Ga medium supplemented with 1 mg/l 2,4-D and 0.3 mg/l kinetin. Suspension cultures were established from *Althaea* callus in Ga medium without agar.

Physical factors such as light influenced polysaccharide and flavonoid production. Total darkness, which promoted maximum polysaccharides, did not favor flavonoid synthesis.

Among the different phytohormones used, both 2,4-D and NAA promoted maximum growth, but 2,4-D is favored for polysaccharide production.

The expression of hairy root disease in *Althaea* explants involves a stable integration of the Ri plasmid into the cell genome, reflecting stable production and very fast growth. The transformed roots produce polysaccharides to a level comparable with that of intact plant roots. The composition of monosaccharides in hairy roots was similar to that of the mother plant. Such cultures are amenable to genetic manipulation, raising the possibility of increasing the product synthesizing capacity of hairy roots grown in vitro. Coupled with the inherent biochemical and genetic stability of such tissues, these methods may facilitate the biotechnological exploitation of *Althaea* hairy roots as a source of valuable products.

From the pharmaceutical point of view, the most important discovery has been that *Althaea* polysaccharides in tissue cultures can be excreted in liquid medium; furthermore, that this process may be influenced by changes in its content. This fact offers possibilities for continuous culturing, once the difficulties associated with accumulation of polysaccharides in the tissue are overcome. The easier isolation of polysaccharides from the medium in comparison with the classic method of extraction from plant material is an important advantage.

Recently, the interest of international pharmaceutical industries has been directed more and more to high molecular compounds such as polysaccharides, with distinctive biological activities, as forerunners of a new therapeutic era (Ulbrich et al. 1988). We believe that cell cultures of *Althaea* plants as a source of biologically active polysaccharides can play a role in this respect.

## 5 Protocol

Media and culture conditions for the establishment of cultures and production of biological active compounds from *Althaea officinalis* L., var. Russalka.

1. *Explants.* A convenient method to obtain cultures is to use plantlets grown aseptically in MS medium with 3% sucrose as explant source at 25 ± 2 °C (darkness or light).
2. *Callus.* Medium – basal MS salts, vitamins, 3% sucrose, 0.9% agar, containing 2,4-D 1 g/l (alternatively NAA 4 g/l), and kinetin 0.3 g/l. Subculture at 4–5-week intervals, depending on the growth rate of the lines.
3. *Cell Suspension.* Choose a fast-growing callus, transfer the tissue into 50 ml of the same medium without agar, and put on a rotary shaker (80 rpm), 25 ± 2 °C, darkness/light. Subculture the suspension every 10 days if you need to maintain suspension before the stationary phase of growth.
4. *Hairy Roots (HR).* Cocultivation of leaves or stems of sterile *Althaea* plants with a suspension of *Agrobacterium rhizogenes* (LBA 9402, R 1601). Scratch with a needle the hypodermis in the midrib of the leaf or the stem, and let it drop the dense bacterial suspension. After 2 weeks, root tips appear at the wounded sites. Clear the cultures from bacteria by several passages in medium, containing 500 mg/l Claforan (Hoechst). Transfer several segments of HR of 2–3 cm length to hormone-free liquid medium (MS salts without $NH_4NO_3$, 2% sucrose), 25 ± 2 °C, darkness, rotary shaker (80–100 rpm). Fresh biomass increased 29–30-fold in 4 weeks' growth period.
5. *Agrobacterium rhizogenes.* Bacterial suspension, YMB liquid medium for 2 days, at 25 ± 2 °C, incubated on a horizontal rotary shaker (80 rpm). *Agrobacterium* is kept in YMB-agar medium.

The above conditions are optimal for both growth and production of polysaccharides (darkness), flavonoids, coumarins, phenolcarbonic acids (light) in *Althaea officinalis* L. cultures.

## References

Aird ELH, Hamill JD, Rhodes MJC (1988) Cytogenetics analysis of hairy root cultures from a number of plant species transformed by *Agrobacterium rhizogenes.* Plant Cell Tissue Organ Cult 15(1):47–58

Alcaraz MJ, Moroney M, Hoult JRS (1989) Effects of hypolaetin-8-O-glucoside and its aglycone in in vitro test for anti-inflammatory agents. Planta Med 55:107
Anderson JW, Beardall J (1991) Molecular activities of plant cells. Sci Publications, Oxford, pp 275–282
Baker DB, Ray PM (1965a) Direct and indirect effects of auxin on cell wall synthesis in oat coleoptile tissue. Plant Physiol 40:345–352
Baker DB, Ray PM (1965b) Relation between effects of auxin on cell wall synthesis and cell elongation. Plant Physiol 40:360–368
Becker GE, Hui PA, Albersheim P (1964) Synthesis of extracellular polysaccharides by suspension of *Acer pseudoplatanus* cells. Plant Physiol 39:913–920
Bergmann L, Grosse W, Koth P (1976) Influences of ammonium and nitrate on N-metabolism, malate accumulation and malic enzyme activity in suspension cultures of *Nicotiana tabacum*. Z Pflanzenphysiol 80:60–64
Berlin J (1988) Formation of secondary metabolites in cultured plant cell and its impact on pharmacy. In: Bajaj YPS (ed) Biotechnology in agriculture and forestry 4. Medicinal and aromatic plants I. Springer, Berlin Heidelberg New York, pp 37–59
Blaschek W, Franz G (1986) A convenient method for the quantitative determination of mucilage polysaccharides in Althaea Radix. Planta Med 52:537
Bonner J (1961) On the mechanics of auxin-induced growth, In: Iowa State (ed) Plant growth regulators. University Press, Ames, pp 307–326
Burke D, Kaufman P, McNeil M, Albersheim P (1974) The structure of plant cell walls. Plant Physiol 54:109–115
Butcher DN (1977) Secondary products in tissue culture. In: Reinert J, Bajaj YPS (eds) Applied and fundamental aspects of plant cell, tissue, and organ culture. Springer Berlin Heidelberg New York, pp 668–716
Capek P, Toman R, Kardosova A, Rosik J (1983) Polysaccharides from the roots of the marshmallow (*Althaea officinalis* L.): structure of an arabinan. Carbohydr Res 117:133–140
Capek P, Toman R, Rosik J, Kardosova A (1984) Polysaccharides from the roots of *Althaea officinalis* L.: structural features of D-glucans. Collect Czech Chem Commun 49:2674–2679
Capek P, Rosik J, Kardosova A, Toman R (1987) Polysaccharides from the roots of the marshmallow (*Althaea officinalis* L., var. Robusta): Structural features of an acidic polysaccharide. Carbohydr Res 164:443–452
Cravent GH, Mott RL, Steward FC (1972) Solute accumulation in plant cells. IV. Effects of ammonium ions on growth and solute content. Ann Bot (Lond) 36:897–901
Della Loggia R, Del Negro P, Bianchi P, Romussi G, Tubaro A (1989) Topical anti-inflammatory activity of some flavonoids from *Quercus ilex* leaves. Planta Med 55:109
Dzido TH, Soczewinski E, Gudej J (1991) Computer aided optimization of high-performance liquid chromatographic analysis of flavonoids from some species of the genus *Althaea*. J Chromatogr 550:71–76
Ebert K (1982) Arznei- und Gewürzpflanzen. Wissenschaftliche Verlaggesellschaft, Stuttgart, pp 78–79
Evans WC (1989) Pharmacognosy, 13th edn. Baillière Tindall, London, 355 pp
Flores HE, Filner P (1985) Metabolic relationships of putrescine, GABA and alkaloids in cell and root cultures of Solanaceae. In: Neumann KH, Barz W, Reinhard E (eds) Primary and secondary metabolism in plant cell cultures. Springer, Berlin Heidelberg New York, pp 174–185
Fowler MW (1985) Plant cell culture – future perspectives, In: Neumann KH, Barz W, Reinhard E (eds) Primary and secondary metabolism of plant cell cultures. Springer, Berlin Heidelberg New York, pp 362–368
Franz G (1966) Die Schleimpolysaccharide von *Althaea officinalis* und *Malva silvestris*. Planta Med 14:90–110
Franz G (1989) Polysaccharides in pharmacy: current applications and future concept. Planta Med 55:493–497
Franz G (1991) Polysaccharides. Springer, Berlin Heidelberg New York
Von Friedrichs O (1919) Über einige Inhaltsstoffe der Altheewurzel. Arch Pharm (Weinheim) 257:288–298
Frohne D, Jensen U (1979) Systematik des Pflanzenreichs. Gustav Fischer, Stuttgart, 138 pp

Gamborg OL (1967) Aromatic metabolism in plants V. The biosynthesis of chlorogenic acid and lignin in potato cell cultures. Can J Biochem 45:1451–1457
Gudej J (1989) Determination of flavonoids in leaves, flowers and roots of *Althaea officinalis* L. Farm Pol 46:153–155
Gudej J (1991) Flavonoids, phenolic acids and coumarins from the roots of *Althaea officinalis* L. Planta Med 57:284–285
Gudej J, Bieganowska ML (1990a) Chromatographic investigations of flavonoid compounds in the leaves and flowers of some species of the genus *Althaea*. Chromatographia 30 (5/6):333–336
Gudej J, Bieganowska ML (1990b) Chromatographic investigation of phenolic acids and coumarins in the leaves and flowers of some species of the genus *Althaea*. J Liq Chromatogr 13(20):4081–4092
Gudej J, Dzido TH (1991) Quantitative determination of flavonoid glycosides in leaves and flowers from some species of *Althaea* genus using HPLC Technique. Acta Pol Pharm – Drug Res 48(3–4):59–62
Guhman H (1924) Variation in the root system of the common everlast (*Gnaphalium polycephalum*). Ohio J Sci 24:199–208
Hahlbrock K, Wellmann E (1970) Light-induced flavone biosynthesis and activity of phenylalanine-ammonia-lyase and UDP-apiose synthetase in cell suspension cultures of *Petroselinum hortense*. Planta Med 94:236–239
Hamill JD, Parr AJ, Rhodes MJC, Robins RJ, Walton NJ (1987) New routes to plant secondary products. Bio/Technology 5:800–804
Hartwich C (1891) Über die Schleimzellen von *Althaea officinalis* L. Pharm Zentralhalle:586–588
Hausel R (1991) Phytopharmaka. Springer, Berlin Heidelberg New York, 100 pp
Hausel R, Haas H (1984) Therapie mit Phytopharmaka. Springer, Berlin Heidelberg New York, pp 102–103
Hegnauer R (1969) Chemotaxonomie der Pflanzen Vol 5. Birkhäuser, Basel pp 29–46
Henry M, Edy AM, Desmarest P, Manoir J (1991) *Glycyrrhiza glabra* (licorice): Cell culture, regeneration, and the production of glycyrrhizin, In: Bajaj YPS (ed) Biotechnology in agriculture and forestry, vol 15. Medicinal and aromatic plants III. Springer, Berlin Heidelberg New York, pp 270–282
Holden PR, Holden MA, Yeoman MM (1988) Variation in the secondary metabolism of cultured plant cells. In: Robins RJ, Rhodes MJC (eds) Manipulating secondary metabolism in culture. Cambridge University Press, Cambridge, pp 15–29
Holzl J, Baucher E (1965) Bau und Eigenschaften der organischen Naturstoffe. Springer, Berlin Heidelberg New York
Hoppe AH (1975) Drogenkunde Band 1. de Gruyter, Berlin, pp 62–64
Hunault G (1984) In vitro culture of fennel tissues (*Foeniculum vulgare* Miller) from cell suspension to mature plant. Sci Hortic 22:55–65
Ilieva St, Peneva C, Koleva M, Ninov St, Topalova V (1984) *Althaea officinalis* L., var. Russalka. Bulg. Autorship Cert. N 65 404/3.5.1984
Ionkova I (1990) Transformed plant cultures – sources of secondary natural products. Pharmacy (Sofia) 4:61–66
Ionkova I (1991) Production of saponins by conventional and transformed root cultures of *Astragalus hamosus* (Fabaceae). Probl Pharm Pharmacol (Sofia) V:31–35
Ionkova I (1992a) Alkaloid production of *Hyoscyamus reticulatus* plant and transformed root cultures. Biotechnology (Sofia) 2:50–52
Ionkova I (1992b) Potential of transformed root cultures from *Hyoscyamus aureus* for the biosynthesis of tropane alkaloids. Biotechnology (Sofia) 3:41–43
Ionkova I (1992c) Alternative sources of biological active substances from *Althaea officinalis* L., var. *Russalka*. CR Acad Bulg Sci 9:137–141
Ionkova I (1992d) Polysaccharides from *Althaea officinalis* L., var. *Russalka* – intact plant and suspension cultures. Biotechnology 4 (in press)
Ionkova I, Witte L, Alfermann AW (1989) Production of alkaloids by transformed root cultures of *Datura innoxia*. Planta Med 55:229–230
Ionkova I, Stefanova-Gateva B, Zvetanova V (1990a) Biosynthese de substance medicinales de plantes medicinales par technologie in vitro. In: Tacher T (ed) XXI[e] Semaine Medicinale Balkanique, 2–5 Sept, Varna, Bulgaria. Sim, Sofia, 142

Ionkova I, Zvetanova V, Alfermann AW (1990b) Production of tropane alkaloids by hairy root cultures of *Hyoscyamus albus* L. Solanaceae cultivated in vitro. Pharmacy (Sofia) XL, 5:13–17
Ionkova I, Ninov St, Kolev D, Alfermann AW (1991) Secondary constituents of in vitro cell and transformed root cultures of *Althaea officinalis* L., var. Russalka, Malvaceae. Planta Med 57:A41–42
Jaretzky R, Ulbrich H (1934) Die intraplasmatischen Vorgänge bei der Schleimbildung in den Samen von *Althaea officinalis* L. Arch Pharm (Weinheim) 272:796–811
Kamada H, Okamura N, Satake M, Harada H, Shimomura K (1986) Alkaloid production by hairy root cultures in *Atropa belladonna.* Plant Cell Rep 5:239–242
Kardosova A, Rosik J, Toman R, Capek P (1983) Glucan isolated from leaves of *Althaea officinalis* L. Collect Czech Chem Commun 48:2082–2087
Kato K, Wattanabe F, Eda S (1977) Interior chains of glucuronomannan from extracellular polysaccharides of suspension-cultured tobacco cells. Agric Biol Chem 41:539–543
Kinnersley AM, Dougall DK (1981) Correlation between the nicotine content of tobacco plants and callus cultures. Planta 149:205–206
Kohler FE (1887 reprint 1990) Medizinal-Pflanzen Atlas, repr "Libri rari", Band 2. Eugen Kohler, Hannover, pp 10–13
Koleva M, Kolev D (1985) Polysaccharides. Nauka, Sofia
Koleva M, Ninov S, Daskalov V, Ilieva S (1986) Polyphenol compounds from the leaves of *Althaea officinalis* L., var. Russalka. Pharmacy (Sofia) 3:15
Koleva M, Ninov S, Siegel U, Zinsmeister H, Kolev D (1990) Isolation of polysaccharides from the roots of *Althaea officinalis*, var. Russalka CR Acad Bulg Sci 43 (N8):55–56
Kresanek J, Krejca J (1981) Les plantes médicinales. Slovart, Bratislava, Cercle d'Art, Paris, pp 27, 34, 46
Lindsey K, Yeoman MM (1983) The relationship between growth rate, differentiation and alkaloid accumulation in cell cultures. J Exp Bot 34:1055–1065
Madaus A (1989) Analytische Untersuchung der wertbestimmenden hochmolekularen Schleimpolysaccharide von *Althaea officinalis* L. Dissertation, Regensburg Univ, Germany
Mantell SH, Smith H (1983) Cultural factors that influence secondary metabolite accumulations in plant cell and tissue cultures, In: Mantell SH, Smith H (eds) Plant biotechnology. Cambridge University Press, Cambridge, pp 75–108
Marinova S, Stoichkov J, Koleva M, Ninov S (1985) Plant-immunostimulant. Bulg authorship cert N 38651, A61 K35/78
Murashige T, Skoog F (1962) A revised medium for rapid growth and bioassays with tobacco tissue cultures. Physiol Plant 15:473–497
Nakamura T (1974) Production of agar. Japan Patent (Kokai) 74-101561
Neuman D, Krauss G, Heike M, Corder D (1983) Indole alkaloid formation and storage in cell suspension cultures of *Catharanthus roseus.* Planta med 48:20–23
Ninov S, Ionkova I, Kolev D (1992) Constituents from roots of *Althaea officinalis* L., var. Russalka, Malvaceae. Fitoterapia (It) LXIII-N5: pp 474
Nishitani K, Shibaoka H, Masuda Y (1979) Growth and cell wall changes in azuki bean epocotyls II. Changes in wall polysaccharides during auxin-induced growth of excised segments. Plant Cell Physiol 20:463–472
Olson AC, Evans J, Frederick DP, Jansen E (1969) Plant suspension culture media macromolecules – pectic substances, protein, and peroxidase. Plant Physiol 44:1594–1600
Ooms G, Karp A, Burrell M, Twell D, Roberts J (1985) Genetic modification of potato development using *RiT*-DNA. Theor Appl Genet 70:440–446
Pahlow M (1979) Das große Buch der Heilpflanzen. Crafe und Unser, München, pp 112–115
Parr A, Hamil JD (1987) Relationship between biosynthetic capacities of *Agrobacterium rhizogenes*-transformed hairy roots and intact uninfected plants of *Nicotiana.* Phytochemistry 26:3241–3245
Payne G, Bringi V, Prince C, Shuler M (1992) Questions and strategies for productivity improvements. In: Plant cell and tissue culture in liquid systems. Oxford University Press, New York, pp 329–335
Pythoud F, Sinkar P, Nester E, Gordon M (1987) Increased virulence of *Agrobacterium rhizogenes* conferred by the *vir* region of pTiBo 542: application to genetic engineering of poplar. Biotechnology 5:1323–1327
Reuter HD, Deininger R, Schulz V (1988) Phytotherapie. Hippokrates, Stuttgart, 206 pp

Rhodes MJC, Robins RJ, Hamill JD, Parr AJ, Waltton NJ (1987) Secondary product formation, using *Agrobacterium rhizogenes* – transformed cultures. IAPTC Newslett 53:2–15
Roberts MF (1988) Medicinal products through plant biotechnology. In: Robins RJ, Rhodes MJC (eds) Manipulating secondary metabolism in cultures. Cambridge University Press, Cambridge, pp 201–216
Rose D, Martin S (1975) Effect of ammonium on growth of plant cells (*Ipomoea* sp.) in suspension cultures. Can J Bot 53:1942–1949
Rubery PH, Northcote DH (1970) The effect of auxin (2,4-dichlorophenoxyacetic acid) on the cultured sycamore cell wall polysaccharides in cultured sycamore cells. Biochim Biophys Acta 222:95–108
Salikow SA, Ismailov ZF, Rakhimov D, Gudynshina O (1978) Some biologically speciality and dynamic of *Althaea rosea* L. polysaccharides. Rastit Resur 14: 375–378
Sasaki T, Kainuma K (1982) Regulation of starch synthesis and external polysaccharide synthesis by gibberllic acid in cultured sweet potato cells. In: Fujiwara A (ed) Plant tissue culture 1982, Maruzen, Tokyo, pp 255–256
Sign MW, Flores HE (1990) The biosynthetic potential of plant roots. BioEssay, 12 (N1):7–13
Staba EJ (1985) Milestones in plant tissue culture systems for production of secondary products. J Nat Prod 48:203–209
Stachel SE, Messeus E, Van Montagu M, Zambryski P (1986) Identification of signal molecules produced by wounded plant cells that activate T-DNA transfer in *Agrobacterium tumefaciens*. Nature 318:624–629
Stoddart RW, Barrett AJ, Northcote DH (1967) Pectic polysaccharides of growing plant tissues. Biochem J 102:194–207
Takashi Isa (1991) Mucilage production by hairy root cultures of *Astragalus gummifer*, In: Komanine A (ed) Plant cell culture in Japan CMC pp 99–103
Thorpe TA, Meier D (1972) Starch metabolism, respiration and shoot formation in tobacco callus cultures. Physiol Plant 27:365–368
Tomodo M, Kaneko S, Ebashi M, Nagakura T (1977) Plant mucilages XVI. Isolation and characterization of a mucous polysaccharide "*Althaea*-mucilage-O". Chem Pharm Bull 25(6): 1357–1362
Tomoda M, Saton N, Shimada K (1980) Plant mucilages XXIV. The structural features of "*Althaea*-mucilage-O", a representative mucous polysaccharide from the roots of *Althaea officinalis*. Chem Pharm Bull 28(3):824–830
Tomoda M, Shimizu N, Suzuki H, Takasu T (1981) Plant mucilages XXVIII. Isolation and characterisation of a mucilage "*Althaea*-mucilage OL" from the leaves of *Althaea officinalis* L. Chem Pharm Bull 29:2277–2282
Tomoda M, Shimada K, Shimizu N (1983) Plant mucilages, XXXII. A representative mucilage, "*Althaea*-mucilage-O", from the roots of *Althaea rosea*. Chem Pharm Bull 31(8):2677–2684
Tomoda M, Shimizu M, Oshima Y, Takahashi M, Murakami M, Hikino H (1986) Hypoglycemic activity of twenty plant mucilages and three modified products. Planta Med 53:8–12
Tomoda M, Shimizu N, Gouda R, Kanari M, Yamada H, Hikino H (1990) Anti-complementary and hypoglycemic activities of the glycans from the seeds of *Malva vericulata*. Planta Med 56:169–170
Trease GE, Evans WC (1978) Pharmacognosy. Baillière Tindall, London, pp 116–117
Tutin TG, Heywood VH, Burges NA, Moore DM, Valentine DH, Walters SM, Weeb DA (eds) Flora Europea. Cambridge Univ Press, Cambridge, 253 pp
Ulbrich B, Osthoff H, Wiesner W (1988) Aspect of screening plant cell cultures for new pharmacologically active compounds. In: Pais MSM et al. (eds) Plant cell biotechnology. Springer, Berlin Heidelberg New York, pp 461–473
Uphof JCTH (1968) Dictionary of economic plants. Cramer Lehre, pp 28–29
Vigneau C (1988) Plantes médicinales. Masson, Paris, 112 pp
Villar A, Gasco MA, Alcaraz MJ (1984) Anti-inflamatory and anti-ulcer proprieties of hypolaetin-8-glucoside, a novel plant flavonoid. J Pharm Pharmacol 36:820–823
Villar A, Gasco MA, Alcaraz (1987) Some aspects of the inhibitory activity of hypolaetin-8-glucoside in acute inflammation. J Pharm Pharmacol 39:502–507
Wagner H, Proksch A, Riess-Mauer I, Vollmar A, Odenthal S, Stuppner H, Jurcic H, Le Turdu Mm, Fang JN (1985) Immunostimulierend wirkende Polysaccharide (Heteroglykane) aus höheren Pflanzen. Arzneim Forsch/Drug Res 35(II) N7:1069–1075

Wagner H, Stuppner H, Schafer W, Zenk M (1988) Immunologically active polysaccharides of *Echinaceae purpurea* cell cultures. Phytochemistry 27(N1): 119–126
Widmaier W (1988) Pflanzenheikunde, Band 2. WBV, Biologisch-Medizinische Verlaggesellschaft, Schorndorf, pp 48–51
Wilder BM, Albersheim P (1973) The structure of plant cell walls, IV. A structural comparison of the wall hemicellulose of cell suspension cultures of sycamore (*Acer pseudoplatanus*) and red kideny bean (*Phaseolus vulgaris*). Plant Physiol 51: 889–893
Wydra K, Rudolph K (1992) Analysis of toxic extracellular polysaccharides. In: Linskens H, Jackson J (eds) Plant toxin analysis. Springer Berlin Heidelberg New York, pp 113–183
Zenk MH, Deus B (1982) Natural products synthesis by plant cell cultures. In: Fujiwara A (ed) Plant tissue culture 1982, Maruzen, Tokyo, pp 391–394
Zenk MH, El-shagi H, Arens H, Stockigt J, Weiler E, Deus B (1977) Formation of the indole alkaloids serpentine and ajmaline in cell suspension culture of *Catharanthus roseus*. In: Barz W, Reinhard E, Zenk HM (eds) Plant tissue culture and its biotechnological application. Springer, Berlin Heidelberg New York, pp 27–43
Zorrefuieta A, Ugalde RA (1986) Formation in *Rhizobium* and *Agrobacterium* spp. of a 235-kilodalton protein intermediate in β-D-(1-2)-glucan synthesis. J Bacteriol 167(3): (3) 947–952

# III *Baptisia tinctoria* (L.) R. Brown: Micropropagation, In Vitro Culture and Production in Direction of Pharmaceutically Used Root Biomass

G. MEVENKAMP[1], R. LIEBEREI[1], and G. HARNISCHFEGER[2]

## 1 General Account

### 1.1 Botany, Distribution, and Importance

*Baptisia tinctoria* (L.) R. Brown (synonyms: *Sophora tinctoria* L., *Polydaria tinctoria* Michaux Willd., commonly known as rattle bush, horsefly weed, indigo weed, yellow indigo, or yellow clover broom) and other members of the genus *Baptisia* were traditional medicinal plants for the American natives (Hussey 1974; Millspaugh 1887). Leaves of *B. tinctoria* were also used as a plant-derived dye (Gr. bapto, Lat. tingere: dye). It is a member of the family of Fabaceae (Leguminosae) and forms bushy shrubs up to 1 m high with woody perennial rhizomes and roots and annual aboveground parts (Fig. 1). The round stems are usually erect, often widely branched, glabrous, occasionally slightly pubescent and yellowish green. The subsessile leaves are terminately compound, with subsessile cuneate, obovate leaflets of 1 to 1.5 cm length, and are bluish green in color.

The yellow flowers form loose terminal axillary racemes. The floral bracts are lanceolate-setaceous to ovate acuminate, the pedicels are 4 to 5 mm long, the calyx tube length is 3 to 4 mm. The corolla is papilionaceous, the upper part (standard) is about 1 cm long, the wings and the keel about 1.2 to 1.3 cm. The keel is curving upwards. The ten equally formed stamens are not united into a tube at the base.

The fruit is unilocular, unicarpellary, containing up to five seeds. Mature pods are up to 1 cm long, and 0.8 cm broad, subglobose to ovoid, strongly rounded at the base and the summit. The kidney-shaped yellow to dark brown seeds are 1 to 3 mm long, and 1 to 1.5 mm broad.

The genus *Baptisia* comprises more than 20 species. Interspecies hybridization is often seen, and has been proven by morphological analysis (Larisey 1940) as well as by identification and comparison of alkaloid and flavonoid patterns (Alston and Turner 1963; McHale and Alston 1964).

*Baptisia tinctoria* is native to the eastern parts of North America and Southern Canada (Ontario). It is a general part of the natural flora and was

[1] Institute of Applied Botany, University of Hamburg, Marseiller Straße 7, 20355 Hamburg, Germany
[2] Fa. Schaper and Brümmer, Bahnhofstraße 35, 38259 Salzgitter, Germany

Biotechnology in Agriculture and Forestry, Vol. 28
Medicinal and Aromatic Plants VII (ed. by Y.P.S. Bajaj)

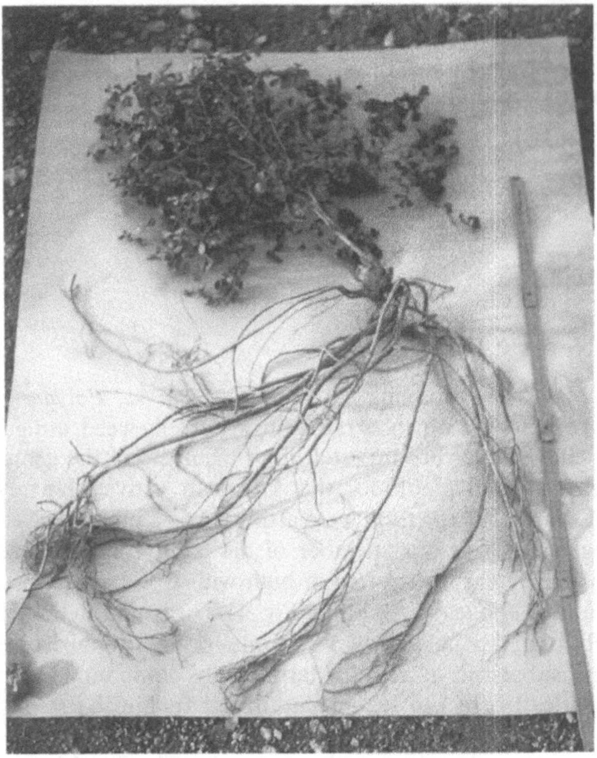

**Fig. 1.** *Baptisia tinctoria* grown from in vitro culture. The plant belongs to clone BM 9 and is 3 years old (1 year under greenhouse conditions and 2 years in the field)

described by European botanists as early as 1762 (Claytonus), and later in monographs (Meehan 1885; Larisey 1940).

The habitats of the wild, slow-growing plant are sandy grounds and natural openings with poor, well-drained slightly acidic soils. These areas were formerly kept open by animals or natural fires, but in the meantime, most of these habitats have been turned into agricultural and cultivated sites. In this way, the land suitable for *B. tinctoria* has diminished continuously. *B. tinctoria* as a slow-growing plant does not spread very fast and, due to the low germinability of its seeds (Voß 1992), reproduction is low. It is probably not regenerated naturally as rapidly as it is being eliminated by sample collection. Thus, nowadays *B. tinctoria* occurs in a few scattered locations and its populations are becoming smaller. In order to ensure a continuing ample supply of rhizomes and roots for medicinal use, it will have to be taken into culture.

The crude drug, Radix Baptisiae tinctoriae, was utilized in the traditional medicine of the American Indians for preparing liquid aqueous extracts. The extracts were applied to wounds, bruises, and as laxans, especially in septic, inflammatory processes, ulcers and abscesses, etc. (Berger 1969). The European settlers took over the use of this plant as an antiseptic against typhus and scarlet fever (Madaus 1938). In Europe, it was introduced into therapy as a homeopathic compound in the mid 19th century; later it was applied allopathically in septic processes (Madaus 1938).

Today, it is a well-known and accepted herb in phytotherapy (Harnischfeger and Stolze 1983). The aqueous-ethanolic extracts of *B. tinctoria* roots form an essential part of immune-modulating herbal products like "Esberitox". These products are used in the treatment of acute and chronic viral or bacterial infections like common wed, bronchitis, angina laryngitis, otitis, and bacterial infections of the skin.

## 1.2 Demand on the World Market and Conventional Methods for Propagation

The demand for the drug, Radix Baptisiae tinctoriae, is increasing. In 1980 the demand for the crude drug in Germany was 2000 kg/year; 10 years later this had risen to 5000 kg.

Since the rootstocks are provided by collecting wild-grown material, their diversity has to be tested. Rootstocks of *B. australis* or *B. alba* occur as falsification. Both species are used as substitute for *B. tinctoria* (List and Hörhammer 1972).

Conventional methods for propogation have never been described. There are historical references to the plant's being cultivated in South Carolina in the 18th and 19th centuries (1744 until 1890: Adrosko 1968; Buchanan 1987) as a dye producer; but no reports on the manner of cultivation or the agricultural techniques used in the past are available. Recently, studies were carried out to cultivate *B. tinctoria* starting by seedling culture and by rhizome cuttings (Ogzewalla 1987). Both methods are slow and time-consuming, and cannot be applied for mass production. Rhizomes used for vegetative propagation can be taken only from plants older than 3 years, because those of younger plants are too small and die after cutting. Furthermore, rhizomes are often attacked by soil borne fungi and bacteria, causing rhizome soft rot.

Propagation by seeds is possible, but germination in the field under natural conditions is poor, due to dormancy (Ogzewalla 1987). Less than 3% of seeds germinated under natural conditions or in laboratory experiments (Voß 1992). The best method for breaking dormancy is scarification by mechanical or chemical means, especially soaking the seeds in sulfuric acid. Detailed studies on seed dormancy and seedling variability have been carried out (Voß 1992).

## 1.3 Secondary Products in Intact Plants

### *1.3.1 Low Molecular Weight Secondary Products*

All species of *Baptisia* synthesize a remarkable spectrum of quinolizidine alkaloids (Table 1). Detailed studies on secondary compounds of this genus have been published by Cranmer and Mabry (1966) for stems and leaves, and by

**Table 1.** Secondary products identified in plants and tissue cultures of *Baptisia*

| Species | Plant part | Alkaloid | Flavonoids isoflavonoids | Reference |
|---|---|---|---|---|
| *B. tinctoria* | Roots/rhizomes | 1, 2, 3, 12, 14, 16, 17, 18 | 20 to 29 | Harnischfeger (1983) |
| *B. australis* | Aerial part | 1 to 16, 3!, 12! | | Wink et al. (1981) Wink et al. (1983) |
| | Roots/rhizomes | 1, 2, 3 | 28, 29 | Oung-Boran et al. (1969) |
| | Callus | 8! | | Wink et al. (1983) |
| | Suspension | 8! | | Wink et al. (1983) |
| *B. alba* | Whole plant | 3, 2, 12, 11, 19 | 30, 31 | Cranmer and Mabry (1966) |
| *B. leucantha* | Aerial part | 3, 2, 12, 1 | 30, 31 | Hoppe (1975) |
| *B. perfoliata* | Whole plant | 2, 3, 1, 12, 16 | | Cranmer and Mabry (1966) |
| *B. leucophaea* | Whole plant | 19, 1, 1, 8, 11, 3, 2, 16 | | Nowacki and Waller (1975) |
| *B. lecontei* | Aerial part | | 36 | Markham and Mabry (1968) |

List of compounds:
Alkaloids: 1 spartein; 2 N-methylcytisin; 3 cytisin; 4 17-oxospartein; 5 isolupanin; 6 5,6-dehydrolupanin; 7 rhombifolin; 8 lupanin; 9 tinctorin; 10 11-allylcytisin; 11 thermopsin; 12 anagyrin; 13 N-acetylcytisin; 14 N-formylcytisin; 15 13-hydroxyanagyrin; 16 baptifolin; 17 aphyllin; 18 virgilin; 19 hydroxyspartein.
Flavonoids/isoflavonoids: 20 maackiain; 21 formononetin; 22 genistein; 23 pseudobaptigenin; 24 daidzein; 25 trifolirhizin = ( − )-maackiain-3-O-β-D glucoside; 26 formononetin-7-O-β-D rutinosid; 27 pseudobaptigenin-7-O-β-D rutinoside; 28 apigenin-7-O-β-D glycoside; 29 luteolin-7-O-β-D glycoside; 30 kämpferol; 31 quercetin; 32 orobol-7-O-β-D glycoside; 33 genistein-7-O-β-D rutinoside; 34 lecontin; 35 ( + )-fustin-3-O-β-D glucoside.

Guder (1981 cit. in Harnischfeger and Stolze 1983) for the rhizomes (crude drug). Cytisin (Fig. 2), anagyrin, methylcytisin, and 1,3-hydroxyspartein are present in stems (0.45% of dry wt.) and in leaves (0.08% of dry wt.). Guder (1981) reported N-methylcytisin, cytisin (Fig. 2), and anagyrin as main alkaloids, and spartein, acetoxyanagyrin, and N-formylcystein as minor compounds, furthermore thermopsin, aphyllin, virgilin, and baptifolin (Fig. 2) have been detected. Wink et al. (1981, 1983) investigated the content of quinolizidine alkaloids in several species of the Fabaceae and compared the alkaloid pattern of intact plants with the pattern of undifferentiated cell suspension cultures. Main alkaloids from intact plants of *B. australis* were from the pyridone type, such as anagyrine, cytisine, and n-methylcytisine. In cell suspensions only lupanine was accumulated in remarkable amounts. Wink et al. (1983) concluded that in undifferentiated cell suspensions only the basic metabolic pathways up to the formation of lupanine is expressed, whereas differentiated plants possess an additional set of enzymes to form the typical alkaloid pattern.

Flavone- and isoflavonederivatives have been studied intensively (Farkas et al. 1959, 1963; Alston et al. 1965; Wagner et al. 1969; Table 1). Several quercetin-glycosides occur in inflorescences (Alston et al. 1965); from leaves and stems, various isoflavones (luteolin, orobol, genistein, sectorigenin, liochanin A, pseu-

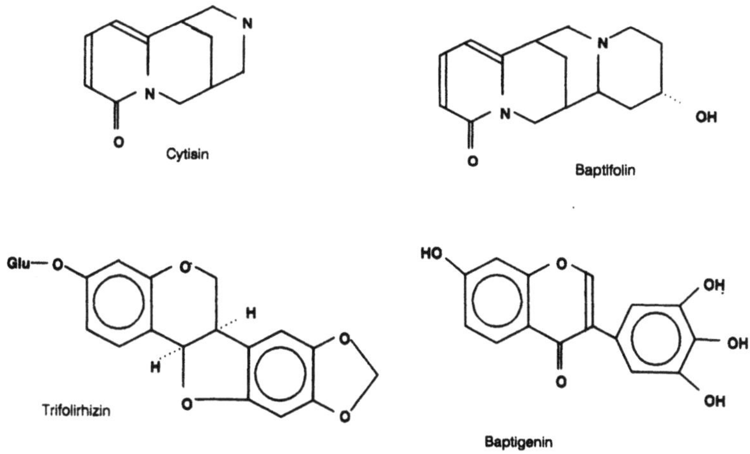


**Fig. 2.** Structures of characteristic compounds from *Baptisia tinctoria* according to Harnischfeger and Stolze (1983). Among the pharmaceutically active compounds of *B. tinctoria* are cytisin, the maackiain glucoside trifolirhizin, and glycoproteins. Formulas are drawn according to Harnischfeger and Stolze (1983)

dobaptigenin) and their corresponding 7-*O*-rhamnoglucosides have been isolated. In addition, the coumarin derivative scopoletin was identified (Markham et al. 1970). Pseudobaptisin of the crude drug was shown to be 7-β-rutinosyl-pseudobaptigenin after chemical synthesis (Wagner et al. 1969). Guder (1981) isolated the flavonoids maackiain, baptigenin (Fig. 2), and formononetin and the related glucosides from the crude drug.

### *1.3.2 Polysaccharides and Glycoproteins*

Pharmacological studies using aqueous alcoholic extracts focused attention on the immunostimulating capacity of *B. tinctoria.* Beuscher and Kopanski (1985) isolated polysaccharides from roots of *B. tinctoria,* and enriched several fractions by column chromatography (on Sephadex and Sephacryl). An immunostimulating effect was proved for at least three fractions, manifesting itself as an increased DNA synthesis in mouse lymphocytes, as well as an increased number of hemolytic antibodies produced by the same cells.

In addition to the active polysaccharides, Beuscher et al. (1989) extracted immunologically active glycoproteins from the roots of *B. tinctoria.* The glycoproteins stimulated lymphocyte DNA synthesis, antibody production and host resistance in general (Beuscher et al. 1990). These proteins were identified as arabinogalactan proteins (Beuscher et al. 1988).

Chromatographically purified fractions of aqueous-ethanolic extracts were analyzed electrophoretically (SDS-PAGE). Four distinct glycoproteins were identified in the gel. The isolated active glycoproteins were acid proteins with a pI from 4.7 to 5.3. The molecular weight was estimated as 14 to 58 kDa. The protein contained mainly the amino acids aspartic acid, glutamic acid, threonine, and alanine. Eight to 27% of the total molecular weight consisted of carbohydrates.

## 2 In Vitro Culture Studies

### 2.1 Callus and Cell Culture

Callus was successfully initiated from roots, hypocotyls, stems, leaves, and shoot tips; however, best results were obtained when young leaves or shoot tips were used as starting material. Callus was induced from explants cultivated on MS basal medium (Murashige and Skoog 1962) or a modified ammonia-reduced medium ($NH_4NO_3$ partially substituted by $KNO_3$), both supported with sucrose, 6-benzylaminopurine (BAP), and indole-3-acetic acid (IAA), and both solidified with 0.9% agar (Oxoid, Bacteriological Agar, No. 1). High concentrations of phytohormones were needed. Best results were obtained with 3 mg/l

**Fig. 3.** Multiple shoot formation of *B. tinctoria*, clone BM 9, 6 weeks after transfer of a single node segment onto propagation medium

IAA and 3 mg/l BAP together with MS basal medium and 3.5% sucrose, or together with the modified medium. Callus tissue grew vigorously at 23 °C in the dark. It was light brown to yellowish-brownish and was 2 to 3 cm in diameter after 8 weeks of incubation. Callus growing under the same conditions but with additional illumination for 14 h per day was much smaller, and callus tissue turned dark brown.

Plant organs or whole plants could be regenerated from the callus tissue by changing the phytohormone combination. Roots were formed when BAP was removed from the medium in the presence of 2 or 3 mg/l IAA (Fig. 4). Sprouts were developed when IAA was removed from the medium in the presence of 0.5 mg/l up to 3 mg/l BAP (Fig. 3). The sprouts obtained with more than 1.5 mg/l BAP became thinner when more BAP was applied. Whole plants regenerated when low concentrations of auxin and cytokinin were added to the medium. Good results were obtained with 0.5 mg/l BAP together with 1 mg/l or 2 mg/l IAA. The growing plantlets showed normal development of sprouts and leaf morphology. Ten clones of *B. tinctoria* were investigated for callus induction. The conditions recommended above must be checked in detail for every new clone, because growth behavior differed slightly from clone to clone.

## 2.2 Micropropagation

Micropropagation was carried out as multiple shoot propagation using axillary sprouts as starting material. The sprouts were cut in single node segments and

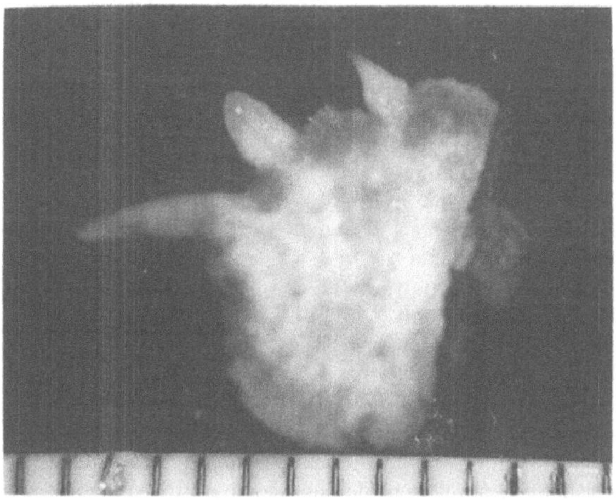

**Fig. 4.** Callus of *B. tinctoria* clone BM 9 with root primordia 2 weeks after transfer onto rooting medium enriched with 3 mg/l IAA

transferred to the propagation medium. The MS formula for macronutrients and micronutrients gave good results when BAP was added. High concentrations of BAP from 1.5 to 3 mg/l led to reduced sprout growth. The resulting sprouts were very thin and short. BAP concentrations below 1 mg/l induced multiple sprouts with normal morphology (Fig. 3).

The number of sprouts could be increased by changing the content of sucrose. The propagation rate increased from three segments per explant with 2% sucrose up to eight segments per explant with 4.5% sucrose. Sprouts obtained on the medium with 4.5% sucrose were thick, but revealed symptoms of vitrification and were therefore not suitable for further treatment. Glycine (1 mg/l) and inositol (100 mg/l) added to the medium increased the number of sprouts per explant significantly.

The cultures were incubated at 23 °C with 14 h illumination per day. Cultures for multiple sprout formation must be transferred to fresh medium 8 weeks after the beginning of cultivation.

Ten clones of *B. tinctoria* with different characteristics (Table 2) were successfully cultivated in vitro. Two clones were selected with regard to their resistance to powdery mildew. Two clones grew vigorously under field conditions but were susceptible to the disease. Two clones were highly suitable for in vitro culture and propagation, and at least four clones showed especially high medicinal quality in root material. From this stock of ten clones, at least five could be recommended for field cultivation in northern Germany.

**Table 2.** List of the cultivated clones of *Baptisia tinctoria*. Summary of results of our experiments carried out from 1990 to 1992

| Clone | Origin | Propagation rate[a] | Root formation %[b] | Field growth | Disease-resistance[c] | Recommended for the field |
|---|---|---|---|---|---|---|
| BM 1 | Seedling Hamburg | 4–7 | 96.3 | + + | r | + |
| BM 2 | Seedling Hamburg | 4–6 | 87.2 | + + | r | + |
| BM 3 | Field Hamburg | 2–4 | 88.8 | + + | r | + |
| BM 4 | Field Hamburg | 4–6 | 98.1 | + + | hs | – |
| BM 8 | KWS-Planta | 4–6 | 94.2 | + + | ws | + |
| BM 9 | KWS-Planta | 4–6 | 94.2 | + + | ws | + |
| BJ 1 | Jasper County | 1–2 | 45.0 | – | r | – |
| BJ 2 | Jasper County | 2–4 | 80.2 | – | ws | – |
| BP 1 | Portage County | 4–6 | 98.1 | + + | hs | – |
| BP 5 | Portage County | 4–7 | 94.2 | + + | hs | – |

Clones BM 1 to BM 9 originated from seedlings grown in Germany, the clones BJ 1/2 and BP 1/5 originated from wild-growing plants from natural habitats in Ohio, USA. Only plants showing resistance to powdery mildew and good growth in the field were recommended.

[a] Expressed as segments/explant × 8 weeks.

[b] Relative amount of rooted segments on rooting medium.

[c] Disease: powdery mildew; r: resistant, ws: weak susceptible, hs: high susceptible.

## 2.3 Root Induction

Roots could easily be induced in excised sprout segments. Good results were obtained on half-strength MS macronutrients and normal strength micronutrients with 2 mg/l IAA and 3% sucrose (Fig. 5). The medium was solidified with 0.9% agar (Oxoid Bacteriological Agar, No. 1). Within 3 weeks' cultivation of segments at 23 °C and 14 h illumination per day, more than 90% of the transferred shoots of eight clones tested showed root induction. Two clones revealed decreased root formation. Forty or 60% of the segments, respectively, rooted when transferred onto the medium described above. The use of indole-3-butyric acid instead of IAA led to a drastic reduction in the number of rooted plantlets in good rooting clones as well as in poor rooting clones compared to the effect of IAA.

Increased sucrose content of the rooting medium led to better results. The amount of rooted plantlets increased from 30% with 2% sucrose to 60% with 4% sucrose in the medium. An effective method to induce roots in poorly rooting clones was the sequential application of naphthaleneacetic acid (NAA) and IAA (Fig. 6). The shoot segments were incubated on the rooting medium with 2 mg/l NAA under standard conditions for 1 week. After this period, the shoots were transferred to the rooting medium with 2 mg/l IAA for another 2 weeks. This modified rooting procedure increased the amount of rooted sprouts from 40% up to more than 70%.

**Fig. 5.** Excised shoot of *B. tinctoria* clone BM 8 with young roots 2 weeks after transfer onto the rooting medium enriched with 2 mg/l IAA

**Fig. 6.** Rooted plantlets of *B. tinctoria* clone BM 9. Root formation was induced with sequential treatment of the shoots with NAA and IAA (see Sect. 2.3)

# 3 Protocol

## 3.1 Induction and Maintenance of Calli

The explants are transferred on MS medium containing 3 mg/l IAA, 3 mg/l BAP, 3.5% sucrose, 0.9% agar, and enriched with 1 mg/l glycine and 100 mg/l inositol. The cultures are incubated at 23 °C in the dark. Sufficient callus appears on the explants 2 weeks after onset of cultivation. This callus tissue can be harvested and transferred to fresh callus medium. It must be subcultured every 2 months. Callus growth continues with unchanged vigor over several subcultures.

## 3.2 Multiple Sprout Formation

Starting material for multiple sprout formation is sterile seedlings or sterilized sprouts derived from field or greenhouse culture. The shoots are cut in single-node segments 1 mm below every node. These segments are transferred to the multiple shoot proliferation medium containing macronutrients according to MS. The medium is enriched with 3.5% sucrose, 1 mg/l glycine, and 100 ppm inositol; 0.6 mg/l BAP is added to induce growth of axillary shoots. One hundred ml medium is filled into the glass vessels applied for in vitro culture of *B. tinctoria*. The vessels are 8 cm in diameter and 5 cm in height, and are closed with a glass lid. A felt ring is inserted between vessel and lid to enable a limited gas exchange between the environment and the interior of culture vessels.

The cultures are kept at 23 °C with 14 h illumination (Philips TLW 25W, distance between lamps and culture vessels 50 cm) per day.

New shoots can be harvested 8 weeks after the beginning of culture. The sprouts can be divided again into single node explants and handled as described above. Four to eight new segments can be obtained from each explant under the described conditions. The number of segments differs with each cultivated clone (Table 2).

### 3.3 Root Formation

For rooting, the segments from multiple shoot proliferation are transferred to rooting medium containing half-strength MS macronutrients, micronutrients in normal strength, and 2 mg/l IAA. The medium is enriched with 3% sucrose; 60 ml medium are filled into each culture vessel (see Sect. 3.2).

The cultures are incubated at 23 °C with 14 h illumination (see Sect. 3.2) per day. Under these conditions, first roots appear 10 days after culture initiation on rooting medium. Three weeks after induction of rooting has started, the plantlets are suitable for transfer to ex vitro conditions.

### 3.4 Ex Vitro Transfer

The rooted plantlets can be transferred to the greenhouse 3 weeks after the beginning of rooting culture. The plantlets are pricked out in a dish containing substrate in a 5-cm-thick layer. About 90% survival of the transplanted plantlets can be obtained by using a mixture of clay-peat-sand-lecatone (15:35:50:9; v/v). The dish is covered with a cap to maintain high air humidity. This protection procedure is needed for about 24 h after transplantation.

Three weeks later, the plantlets can be transferred to pots (8 cm in diameter). The same substrate as described above is used for this step. The culture in pots lasts another 4 to 6 weeks for plants to develop a vigorous root system. After this, plants are vigorous enough to be grown in the field.

The plants can be harvested 4 years after in vitro production. The root growth of transplanted *B. tinctoria* plants differs with the clones and the climatic conditions, but is significantly enhanced by the presence of rhizobial nodules.

## 4 Conclusions

Mass propagation and maintenance of genetic stability of *B. tinctoria* by means of in vitro culture is possible. The cultivars (clones) in in vitro propagation differ considerably in vigor of growth and multiplication potential. Within more than 12 passages of in vitro cultivation no visible changes in plant behavior have been seen. Most clones of *B. tinctoria* can be easily transferred into nonsterile conditions (Mevenkamp et al. 1991).

Root formation of in vitro segments varies strongly with the clones used. The rooting procedure is under study to establish suitable modifications.

The medicinal quality is not correlated to alkaloid content but is due to glycoproteins in the roots, and must be tested for various genotypes in culture. The variability of glycoproteins in quality as well as in quantity in roots of different developmental stages and clones of *B. tinctoria* is under study.

## References

Adrosko RJ (1968) Natural dyes in the United States. Smithsonian Institution Press, Washington DC

Alston RE, Turner BL (1959) Application of paper chromatography to systematics: recombination of parental biochemical components in a *Baptisia* hybrid population. Nature 184:285–286

Alston RE, Turner BL (1963) Natural hybridization among four species of *Baptisia* (Leguminosae). Am J Bot 50:159–173
Alston RE, Turner BL (1969) Natural hybridization among four species of *Baptisia* (Leguminosae). Am J Bot 50:159–173
Alston RE, Rösler H, Naifeh K, Mabry TJ (1965) Hybrid compounds in natural interspecific hybrids. Proc Natl Acad Sci USA 54:1458–1465
Anonymous (1919) Fabaceae 33rd Annu Report of the Bureau of American Ethnology to the Secretary of the Smithsonian Institution 1911–1912. Washington Government Printing Office, Washington
Berger F (1969) Handbuch der Drogenkunde V. Radices: Radix Baptisiae tinctoriae. Wilhelm Mandrich, Wien
Beuscher N, Kopanski L (1985) Stimulation der Immunantwort durch Inhaltsstoffe aus *Baptisia tinctoria*. Planta Med 5:357–472
Beuscher N, Kopanski L, Ernwein C (1988) Modulation of immune response by polymeric substances from *Baptisia tinctoria* and *Echinacea angustifolia*. Adv Biosci 68:329–336
Beuscher N, Scheit K-H, Bodinet C, Kopanski L (1989) Immunologisch aktive Glycoproteine aus *Baptisia tinctoria*. Planta Med 55:358–363
Beuscher N, Scheit KH, Bodinet C, Egerl D (1990) Modulation of host resistance by polymeric substances from *Baptisia tinctoria* and *Echinacea purpurea*. In: Masihi KN, Lange W (eds) Immunotherapeutic prospects of infectious diseases. Springer, Berlin Heidelberg New York, pp 59–63
Buchanan R (1987) A weaver's garden. Interweaver Press, Loveland Co
Claytonus DDJ (1762) Flora virginica. Lugdumi batavorum
Cranmer MF, Mabry TJ (1966) The lupine alkaloids of the genus *Baptisia* (Leguminosae). Phytochemistry 5:1133–1138
Farkas L, Szantho VV (1959) Untersuchung der Glycoside von *Baptisia tinctoria* L. I. Eine vereinfachte Synthese des Pseudobaptigenins. Acta Chim Acad Sci Hung vol 19:217–220
Farkas L, Varady J, Gottsegen A (1963) Untersuchung der Glykoside der *Baptisia tinctoria* R. Br., Synthese des 7.3′. 4′. 5′ Tetrahydroxy-isoflavons (Baptigenin). Chem Ber 96:1865–1867
Guder HJ (1981) Isolierung von Wirkstoffen aus Radix Baptisiae. Diplomarbeit, Universität Göttingen
Harnischfeger G, Stolze H (1983) Bewährte Pflanzendrogen in Wissenschaft und Medizin. Notamed, Bad Homburg
Hoppe HA (1975) Handbuch der Drogenkunde. de Gruyter, Berlin
Hussey JS (1974) Some useful plants of early New England. Econ Bot 28:311–330
Larisey MM (1940) A monograph of the genus *Baptisia*. Ann Mo Bot Gard 27:119–195
List PH, Hörhammer L (1972) Hagers Handbuch der pharmazeutischen Chemie, vol III. Springer, Berlin Heidelberg New York
Mabry TJ, Kagau J, Rösler H (1965) *Baptisia* flavonoids: nuclear magnetic resonance analysis. Phytochemistry 4:487–493
Madaus G (1938) Lehrbuch der biologischen Heilmittel. Thieme, Leipzig
Markham KR, Mabry TJ (1968) The structure and stereochemistry of two new dihydroflavonol glycosides. Tetrahedron 24:823–827
Markham KR, Mabry TJ, Swift WT (1970) Distribution of flavonoids in the genus *Baptisia* (Leguminosae). Phytochemistry 9:2359–2364
McHale J, Alston RE (1964) Utilization of chemical patterns in the analysis of hybridization between *Baptisia leucantha* and *B. sphaerocarpa*. Evolution 18:304–311
Meehan T (1885) *Baptisia tinctoria*. Meehans' Monthly 5:81–82
Mevenkamp G, Lieberei R, Ogzewlla CD, Harnischfeger G (1991) In vitro propagation of *Baptisia tinctoria*. Planta Med 57 (Suppl) 2:A 31
Millspaugh CF (1887) American medicinal plants. Boericke and Tafel, New York
Murashige T, Skoog F (1962) A revised medium for rapid growth and bioassays with tobacco tissue cultures. Physiol Plant 15:473–497
Nowacki EK, Waller GR (1975) Comparative biosynthesis of quinolizidine alkaloids from DL-lysine in five species of Leguminosae. Phytochemistry 14:155–159
Ogzewalla CD (1987) Cultivation of *Baptisia tinctoria*. Report for Schaper and Brümmer, Salzgitter

Oung-Boran P, Lebreton P, Netien G (1969) Contribution à l'étude biochimique et pharmacologique de *Baptisia australis*. Planta Med 17:301–318

Voß K (1992) Beeinflussung der Keimung von *Baptisia tinctoria* durch interne Faktoren des Samens. Diplomarbeit an der Naturwissenschaftlichen Fakultät, Universität Göttingen

Wagner H, Hörhammer W, Budweg W, Major A, Farkas L (1969) Synthese des Pseudobaptisins. Chem Ber 102:3006–3008

Wink M, Hartmann T, Witte L, Schiebel HM (1981) The alkaloid pattern of cell suspension cultures and differentiated plants of *Baptisia australis* and their biogenetic implications. J Nat Prod. 44 (1):14–20

Wink M, Witte L, Hartmann T, Theuring C, Volz V (1983) Accumulation of quinolizidine alkaloids in plants and cell suspension cultures: genera *Lupinus*, *Cytisus*, *Baptisia*, *Genista*, *Laburnum* and *Sophora*. Planta Med 48:253–257

# IV *Berberis* Species: In Vitro Culture and the Production of Protoberberine and Other Alkaloids

C. DELIU[1], M. TAMAS[2], and D. GHIRAN[2]

## 1 Introduction

### 1.1 Distribution and Importance

The Berberidaceae family comprises 14–16 genera (Heywood 1978). The most representative genus, *Berberis*, includes over 100 species distributed in Europe, Siberia, China, Afghanistan, India, Nepal, and North and South America (Goina 1929; Heywood 1978). *Berberis parvifolia* Sprague, the subject of our studies, originated from China (West Sechuan) (Krussmann 1960), and is cultivated in Romania, in the Cluj Botanical Garden.

The plants belonging to the genus *Berberis* are shrubs with palar roots; they vegetate in light, dry places, as bushes on the fringe of forests and agricultural fields. They grow slowly, reaching maturity in 5–7 years, when the bark of stem and root can be separated to be used as raw material for drugs or to extract alkaloids. The leaves are intermediary host for the phytopathogenic fungus *Puccinia graminis*, for which reason the phytosanitary laws in Romania ban cultivation of *B. vulgaris*. This species (Fig. 1) is widely distributed in Europe and Siberia.

*Berberis* species have been known and used since ancient times, extracts from these plants being recommended in eye diseases. Starting with the 17th century, the infusion of fruits is recommended in the treatment of hemorrhages, dysentery, and fever induced by inflammation of the biliary ducts (Goina 1929).

In traditional medicine in Romania, the root and stem bark of *Berberis vulgaris* (barberry) was used in the treatment of hepatitis, kidney diseases, and gingivitis. The fruit pulp is used against chest pains, and the juice is drunken to alleviate feverish states. Extracts of the wood and root bark are used as dye for wool and leather (Butura 1979). Goina et al. (1979a, b) prepared two drugs from *B. vulgaris* (Biloton mitte and Biloton forte), which show good results in the treatment of the biliary dyspeptic syndrome. Since 1983, the Faculty of Pharmacy in Cluj has produced the drug called Tonobil, a fluid extract of *B. vulgaris* stem bark, used in dyskinesia and biliary inflammation. This extract seems also to have beneficial effects in hepatic cirrhosis (Bordas et al. 1989). The firm

[1] Institute of Biological Research, 3400 Cluj-Napoca, Romania
[2] Medicinal and Pharmacy University, Faculty of Pharmacy, 3400 Cluj-Napoca, Romania

Biotechnology in Agriculture and Forestry, Vol. 28
Medicinal and Aromatic Plants VII (ed. by Y.P.S. Bajaj)

**Fig. 1.** *Berberis vulgaris* branches with leaves and fruits

Mann-Pharma (Germany) manufactures an eye lotion (Berberil) recommended in the treatment of conjuctivitis.

## 1.2 Alkaloids

Among the alkaloids obtained from *Berberis*, berberine has been most studied from a pharmacodynamic point of view. It is well known that this alkaloid has anti-inflammatory, antimicrobial, antileukemic, and antineoplastic effects (Kondo 1976; Otsuka et al. 1981). Together with berberine, berbamine and oxyacanthine also have a cardiac depressive action; they stimulate the smooth muscles of the intestine and uterus, being a vasodilator as well. In small quantities, they stimulate cardiac activity by their action on the coronary vessels. Berberine has a strong bactericidal effect on Gram-positive germs, while oxyacanthine has a strong effect on Gram-negative ones (Pitea 1970; Pitea and Margineanu 1972).

In all the 20 *Berberis* species studied, several alkaloids were identified, of which berberine (Fig. 2) is present in all species and is considered as an alkaloid specific to the genus *Berberis* (Hegnauer 1964). The alkaloids found in *Berberis* species belong to the isoquinoline type and may be classified into three groups: (1) quaternary bases of protoberberine alkaloids including berberine, palmatine (which are nonphenolic), and jatrorrhizine and columbamine (phenolic type); (2) tertiary bases or bisbenzylisoquinoline alkaloids, including berbamine, oxyacanthine, and isotetrandrine; and (3) the aporphinic type including vulracine and bervulcine.

As far as the alkaloid content in *B. vulgaris* is concerned, the first data were presented by Büchner (quoted by Pitea 1970), who estimated that the root bark contains 9.40% berberine, 3.90% tertiary bases, 2.03% phenolic bases (jatrorrhizine and columbamine), and 0.028% palmatine. According to Goina (1929), the root bark contains 5–6% berberine and 4–5% oxyacanthine, while the stem bark has a 2–2.5% berberine and 1.5–1.8% oxyacanthine, the total alkaloid content being 12.28% in the root bark and 5.54% in the stem bark. According to Pitea (1970), the total content in the stem bark is 7.96%, and 13.55% in the root bark, of which berberine represents 6.60%, tertiary bases 7.50%, and reduced bases 2%.

Some other *Berberis* species grown in the Cluj Botanical Garden have the following total alkaloid content: *B. koreana* Palib. – 6.78% in the stem bark, 11% in the root bark; *B. francisciferdinandi* C. Schneid. – 4.06% in the stem bark, 12.24% in the root bark; *B. silvataroucana* C. Schneid. – 4.04% in the stem bark, 9.41% in the root bark (Petcu 1959); *B. parvifolia* Sprague – 12.68% in the root bark, 3.76% in the stem bark (Tamas unpubl.).

## 2 In Vitro Culture Studies

### 2.1 Review of Tissue Culture/Biotechnology Studies

The possibility of obtaining protoberberine alkaloids by plant cell cultures has attracted the attention of many researchers. The most studied plants are species of *Thalictrum* (see Ikuta and Itokawa 1988) and *Coptis* (see Ikuta and Itokawa 1989). Reinhard (1967) was the first to describe the synthesis of these compounds by callus cultures of *Berberis vulgaris*. Tissue and cell cultures from 36 other *Berberis* species were induced by Hinz and Zenk (1981). From such cell suspensions, cultured in Erlenmeyer flasks, the isolation of highly jatrorrhizine-producing strains of *B. stolonifera* (Hinz and Zenk 1981) and *B. wilsoniae* (Rothenberger 1982) was successfully carried out. Similar results were obtained with *B. parvifolia* cell cultures (Deliu et al. 1989). The biosynthesis of protoberberine alkaloids was studied in *B. wilsoniae* cell cultures in bioreactors of various sizes and types as well (Rothenberger 1982; Breuling et al. 1985; Breuling 1986) and the maximum alkaloid yield obtained was 3 g/l.

A detailed knowledge of the enzymatic pathways of the formation of secondary metabolites is an essential requirement for applying rational strategies leading to increased cell culture productivity. By using cell-free extracts from *Berberis* species as well as cultures of species from several other genera, it became possible to isolate and characterize all the enzymes involved in berberine biosynthesis (Zenk 1985; Zenk et al. 1988; Bauer et al. 1992). Two enzymes, berberine bridge enzyme and (S)-tetrahydroprotoberine oxidase, are located exclusively in smooth vesicles of 1 to 1 μm. Within these vesicles, the formation of quaternary alkaloids takes place, jatrorrhizine being the main product (Amann et al. 1986). Although in almost all *Berberis* species the major alkaloid in the root bark is berberine (Petcu 1959; Rueffer 1985), the main alkaloid in tissue and cell cultures is jatrorrhizine. The explanation of this phenomenon was given by Beecher and Kelleher (1983) and Rueffer et al. (1983), who proved that the precursor of jatrorrhizine is berberine, in which, through an unknown mechanism, the methylenedioxy group is opened.

Comparative studies concerning the ratio between amounts of several alkaloids in the intact plant and the cell cultures of *B. wilsoniae* were also carried out (Rueffer 1985).

Although the biosynthesis of quaternary alkaloids is well known in many respects, only few data are available on tertiary alkaloids. Cassels et al. (1987) succeeded in isolating *B. stolonifera* strains with a high content of these bisbenzylisoquinoline alkaloids.

## 2.2 Callus Culture/Cell Suspensions

Callus tissues from *Berberis parvifolia* Sprague were first derived in 1984 from root, hypocotyl, and cotyledon of a 16-day-old seedling grown in the dark at 25 °C on B5 medium (Gamborg et al. 1968) with NAA (2 mg/l) and BA (0.2 mg/l), and subcultured at 5-week intervals (Deliu unpubl.). These parts of the seedling synthesize different amounts of alkaloids (Table 1). The same differences are also noticed among calli derived from them. All callus tissues synthesize more alkaloids than the seedling organs from which they derived. However, the quantities of protoberberine compounds are much lower in calli than in the root bark of the intact *B. parvifolia* plant.

In order to obtain cell suspensions, yellow, friable calli derived from the seedling root were transferred in the nutrient liquid B5 medium in Erlenmeyer flasks agitated on a rotary shaker. The effect of several basal media on cell growth and alkaloid production was investigated. Optimal conditions for

**Table 1.** Alkaloid content of callus tissues, 16-day seedling, and 7-year root of original *Berberis parvifolia* plant. (Deliu unpubl)

| Tissue | Growth[a] | | Alkaloid content (% dry wt.) |
|---|---|---|---|
| | mg fr. wt./flask | mg dry wt./flask | |
| Callus from | | | |
| Root | 592 | 49 | 3.46 |
| Hypocotyl | 384 | 28 | 2.95 |
| Cotyledon | 477 | 38 | 2.51 |
| Seedling | | | |
| Root | – | – | 2.21 |
| Hypocotyl | – | – | 1.84 |
| Cotyledon | – | – | 1.63 |
| 7-year root (in vivo) | – | – | 7.51 |

[a] Size of inoculum: 150 mg fresh wt. per flask.

growth and synthesis of secondary compounds were found on B5 basal medium. After 6 months of continuous subcultures, the cell suspension reached, on the 14th day of culture, the maximum productivity of 0.7 g total alkaloids/l (4.7% of dry wt.) (Deliu et al. 1985). The main alkaloid, as in callus tissues, is jatrorrhizine. Remarkable results were obtained with cells of *B. stolonifera* (Hinz and Zenk 1981) cultivated in LS medium (Linsmaier and Skoog 1965) and cells of *B. wilsoniae* (Breuling 1986) for which the optimal medium was MS (Murashige and Skoog 1962).

One of the medium constituents which could significantly influence cell culture productivity is the source of carbon. Of all carbon sources tested (sucrose, fructose, glucose, maltose, and starch), sucrose in 3% concentration had the best effect on cell growth and alkaloid synthesis in *B. parvifolia* cell cultures (Deliu unpubl.). Sucrose was found to be effective also in other *Berberis* species (Hinz and Zenk 1981; Breuling 1986).

Many of the studies, carried out mainly on *Coptis japonica* (Fukui et al. 1982) and *Thalictrum minus* (Nakagawa et al. 1986), made it evident that, at least for in vitro cultivation, there was a link between the metabolism of protoberberine alkaloids and the type and level of phytohormones in the nutrient medium. However, for *B. stolonifera* and other *Berberis* species cell cultures, Hinz and Zenk (1981) reported that no increase of productivity was observed as a result of using several types of auxins and cytokinins. However, the transfer of *B. parvifolia* cell in medium with a phytohormone content differing from the initial one led to modification in alkaloid production (Table 2). Thus, 2,4-D stimulated cell growth but inhibited the synthesis of protoberberine alkaloids, whereas IAPr inhibited both processes. Only NAA influenced the productivity of cell cultures positively (Deliu unpubl.).

Among all cytokinins tested (benzylaminopurine, kinetin, and 2-isopentenyladenine) used together with NAA, BA (1.3 $\mu$m) in combination with NAA (8 $\mu$m) was best for both cell growth and accumulation of alkaloids. This

**Table 2.** Influence of auxins in combination with BA (0.2 mg/l) on cell growth and protoberberine alkaloid accumulation in *Berberis parvifolia* cell suspensions cultivated in B5 medium. (Deliu unpubl.)

| Auxin[a] | Cell yield[b] (g/l dry wt.) | Alkaloid content (% of dry wt.) | | | |
|---|---|---|---|---|---|
| | | J | B | P | Total |
| 2,4-D | 17.52 | 2.05 | 0.24 | 0.32 | 2.61 |
| NAA | 15.19 | 3.25 | 0.68 | 0.58 | 4.45 |
| IAB | 14.73 | 2.93 | 0.45 | 0.41 | 3.79 |
| IAPr | 8.98 | 2.33 | 0.42 | 0.38 | 3.13 |

*Abbreviations*: B5 = Gamborg et al. (1968); 2,4-D = 2,4-dichlorophenoxy acetic acid; NAA = naphthaleneacetic acid; IAB = indolebutyric acid; IAPr = indolepropionic acid; J = jatrorrhizine; B = berberine; P = palmatine.

[a] Auxin concentration: 2 mg/l.

[b] Inoculum size: 1.87 g/l dry wt.

combination was used in all subsequent subcultures. From the qualitative point of view, neither auxins nor cytokinins change the initial alkaloidal spectrum of cell cultures (Deliu unpubl.). In other *Berberis* species, various phytohormone combinations were used successfully, such as NAA ($10^{-6}$ M) with 2,4-D ($10^{-6}$ M) (Amann et al. 1986) or IAA (0.2 mg/l) with 2,4-D (0.1 mg/l) and kinetin (2 mg/l) in *B. wilsoniae* (Breuling 1986).

## 2.3 Extraction and Structure of Alkaloids

Alkaloids (Fig. 2) have been extracted with MeOH from *B. parvifolia* dry biomass of seedlings, tissue and cell cultures, and root bark of mature plants. Berberine and jatrorrhizine have been identified by UV, IR, MS, TLC, and with authentic samples of berberine chloride and jatrorrhizine chloride, respectively. Palmatine, columbamine, berbamine, and oxyacanthine were identified by TLC with authentic samples. Quantitative analysis of the total protoberberine alkaloids was performed by using an original technique (Tamas et al. 1993), according to which the absorbance of alcoholic extract is measured at 350 nm in a VSU-2P (K. Zeiss, Jena) apparatus. The alkaloid content is calculated after the formula:

$$\text{alkaloid} = \frac{E_s}{E^{1\%}_{1\,\text{cm}}} \times \frac{2500}{s},$$

where $E_s$ = sample absorbance; $E^{1\%}_{1\,\text{cm}}$ = molar extinction of berberine at 350 nm (580); s = dry biomass (g).

For quantitative determination of berberine, columbamine, jatrorrhizine, and palmatine, the alkaloids were separated by TLC after the method indicated by Amann et al. (1986), by applying a linear band of alcoholic extract upon the chromatographic plate. After locating and delimiting (in UV) the bands corresponding to each alkaloid, the chromatographic substratum of the bands is detached and extracted with hot MeOH. Afterwards, the extract is centrifuged, filtered, and absorbance is determined at 350 nm.

## 2.4 Metabolic Characteristics of *Berberis* Cell Cultures

One of the common features of *Berberis* cell suspensions cultivated in the batch system is that during the cell growth cycle the accumulation of protoberberine

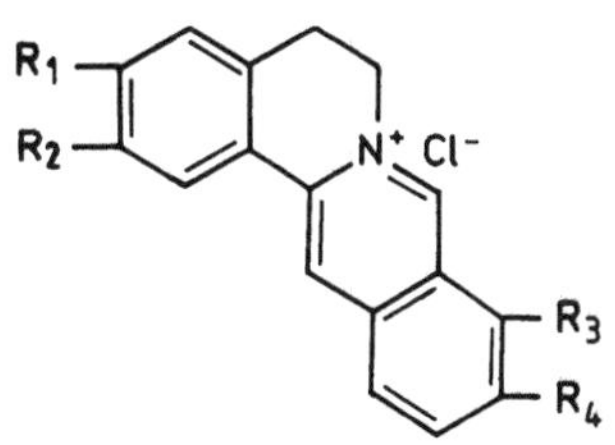


| | $R_1$ | $R_2$ | $R_3$ | $R_4$ |
|---|---|---|---|---|
| Berberline | O | O | OMe | OMe |
| | $CH_2$ | | | |
| Plamatine | OMe | OMe | OMe | OMe |
| Jatrorrhizine | OH | OMe | OMe | OMe |
| Columbamine | OMe | OH | OMe | OMe |

**Fig. 2.** Protoberberine alkaloids from the *B. parvifolia* plant, tissue, and cell cultures

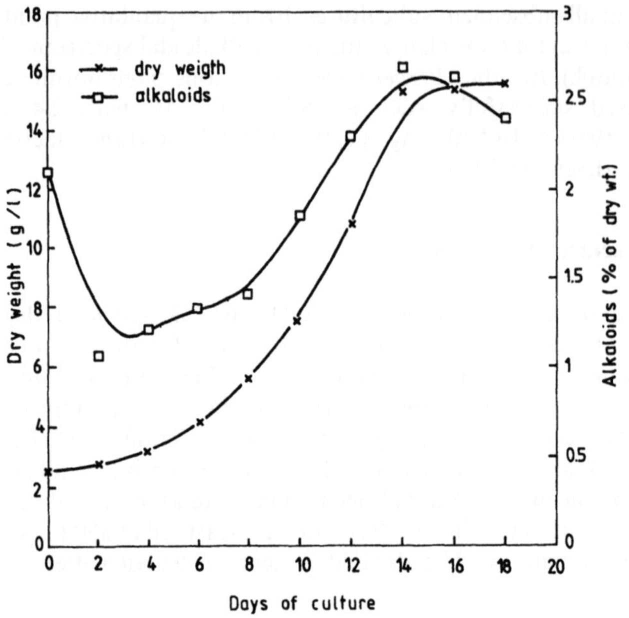


**Fig. 3.** Time course of cell growth and protoberberine alkaloid accumulation in *B. parvifolia* cell suspensions. Mean of three repetitions. (Deliu et al. 1993a)

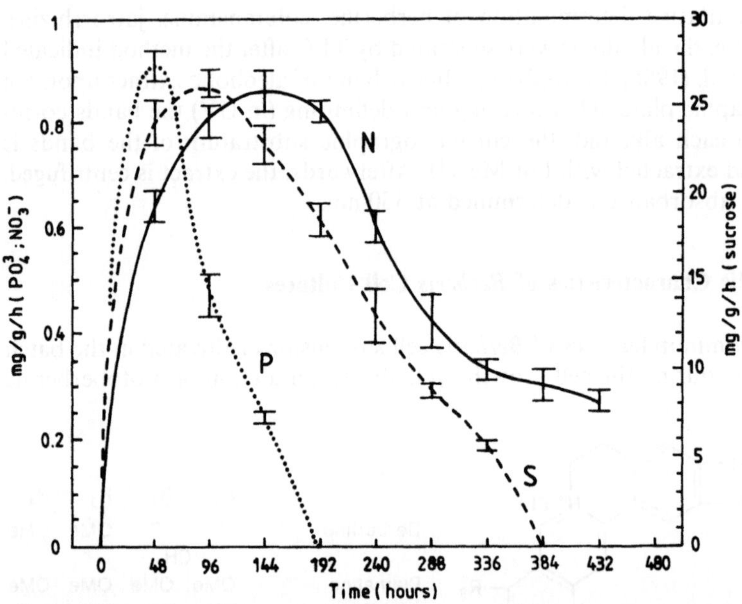


**Fig. 4.** Phosphate (*P*), sucrose (*S*), and nitrate (*N*) uptake (mg/g dry wt./h) from the culture medium by *B. parvifolia* cell suspensions. Each *bar* represents mean ± SD of three repetitions. (Deliu et al. 1993a)

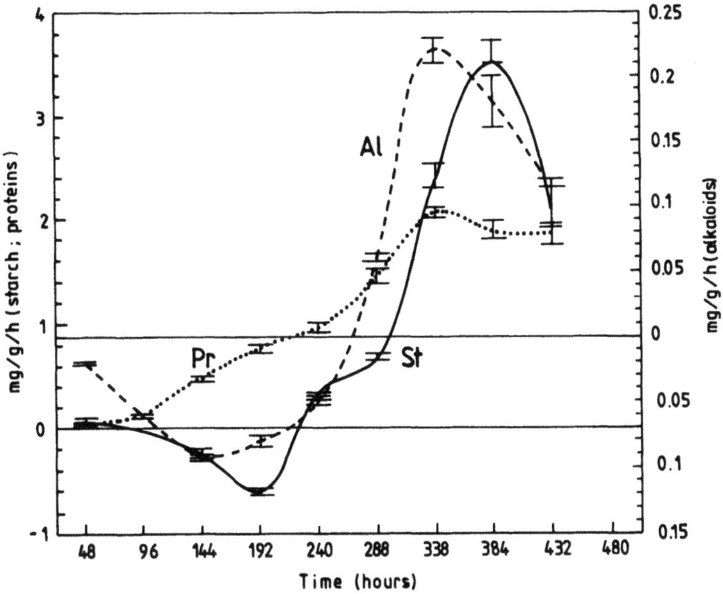


**Fig. 5.** Accumulation of proteins (*Pr*), starch (*St*), and alkaloids (*Al*) (mg/g dry wt./h) in *B. parvifolia* cell suspensions. Each *bar* represents mean ± SD of three repetitions. (Deliu et al. 1993a)

alkaloids follows a sigmoidal curve almost parallel with that of cell growth (Hinz and Zenk 1981; Deliu et al. 1985, 1993a; Breuling 1986). The synthesis of these compounds reaches maximum value when the culture enters the stationary phase (Fig. 3).

Another peculiarity of the *B. parvifolia* cell cultures is related to the uptake of the principal nutrients from the culture medium (Fig. 4). The first nutrient consumed is $PO_4^{3-}$, followed by sucrose, while $NO_3^-$ is to be found in the medium even after 18 days of culture. If the maximum intensity of consumption of these nutrients occurs in the lag phase (Fig. 4), the biosynthesis of some primary metabolites (proteins, starch) reaches the highest levels in the stationary phase (Fig. 5). Among the different metabolic processes, the existence of some direct correlations is to be noticed, such as the correlation between phosphate consumption and alkaloid accumulation (Figs. 4, 5). Thus, from the moment when the $PO_4^{3-}$ is depleted, the synthesis of secondary compounds is induced. A negative ratio between accumulation of alkaloids and starch, and a parallelism between synthesis of proteins and alkaloids has also been recorded (Fig. 5).

## 2.5 Effect of Inorganic Medium Constituents on Productivity

Hinz and Zenk (1981) reported that the variation in concentration of the main inorganic constituents of the LS basal medium had no effect on jatrorrhizine production in several *Berberis* cell cultures. However, our results clearly indicated that the productivity of *B. parvifolia* cell suspensions cultivated on B5

**Table 3.** The effect of inorganic macronutrients on cell growth and alkaloid production in *Berberis parvifolia* cell cultures. Mean ± SD of three repetitions. (Deliu unpubl.)

| Variant | Concentration (mM)[a] $NO_3^-$ | $NH_4^+$ | $PO_4^{3-}$ | Growth[b] (g/l dry wt.) | Alkaloid content (% of dry wt.) |
|---|---|---|---|---|---|
| Control | 25 | 1.0 | 1.1 | 16.02 ± 0.81 | 3.54 ± 0.23 |
| N1 | 30 | 1.0 | 1.1 | 17.06 ± 0.60 | 4.09 ± 0.19 |
| N2 | 25 | 0.0 | 1.1 | 13.24 ± 0.32 | 3.60 ± 0.14 |
| N3 | 25 | 0.5 | 1.1 | 15.38 ± 0.48 | 3.83 ± 0.09 |
| N4 | 25 | 3.0 | 1.1 | 12.60 ± 0.61 | 3.15 ± 0.25 |
| N5 | 25 | 16.0 | 1.1 | 8.81 ± 0.31 | 2.37 ± 0.11 |
| P1 | 25 | 1.0 | 0.37 | 9.79 ± 0.28 | 5.52 ± 0.20 |
| P2 | 25 | 1.0 | 0.55 | 13.17 ± 0.55 | 5.04 ± 0.14 |
| P3 | 25 | 1.0 | 2.20 | 17.60 ± 0.40 | 2.84 ± 0.13 |

[a] Basal medium: B5 with NAA and BA.
[b] Inoculum size: 2.2 g dry wt./l.

medium was negatively or positively modified, depending on the changes in phosphate, nitrate, and ammonium concentration (Table 3). Among these, $PO_4^{3-}$ had the most evident effect. As a result of the decrease in phosphate concentration to 1/3 and 1/2, the synthesis of alkaloids increased by over 50 and 40% respectively, and concomitantly the cell growth decreased. In fact, this aspect confirms one of the characteristic features of *B. parvifolia* cell suspensions concerning the inverse proportion between phosphate level and alkaloid biosynthesis. The change in the ratio between $NO_3^-$ and $NH_4^+$, through increase of ammonium ion concentration, had a negative effect on the productivity of *B. parvifolia* cell cultures (Deliu unpubl.).

Besides the action of macronutrients, the influence of inorganic micronutrients was also tested. By increasing the concentration of $CuSO_4$ and $Fe^{2+}$ (as Fe citrate), the productivity of cell cultures increased by 50 and 40% respectively (Deliu unpubl.).

## 2.6 Elicitation

Exposure of cell culture to the action of some elicitors such as fungal homogenates and fractions in order to induce or increase biosynthesis of secondary metabolites has been widely described in the literature (Eilert 1988). Elicitation of *B. parvifolia* cell cultures with preparations obtained from various parasitic or saprophytic fungi (*Torula, Fusarium, Rhodotorula, Saccharomyces*) did not always result in stimulating the accumulation of protoberberine alkaloids (Deliu unpubl.). However, by using a homogenate obtained from the fungus pathogenic *Botrytis* (Fig. 6), changes were produced in the metabolic processes depending on the concentration of elicitor and on the time of its introduction into flasks with *Berberis* cell suspension (Deliu et al. 1991). The optimal variant in which the productivity of cultures significantly increased, namely from 0.5 g alkaloids/l to 0.8 g/l, was that in which 3 ml of *Botrytis* homogenate were added to 100 ml of culture medium 18 h before harvesting the *B. parvifolia* cells. In this variant,

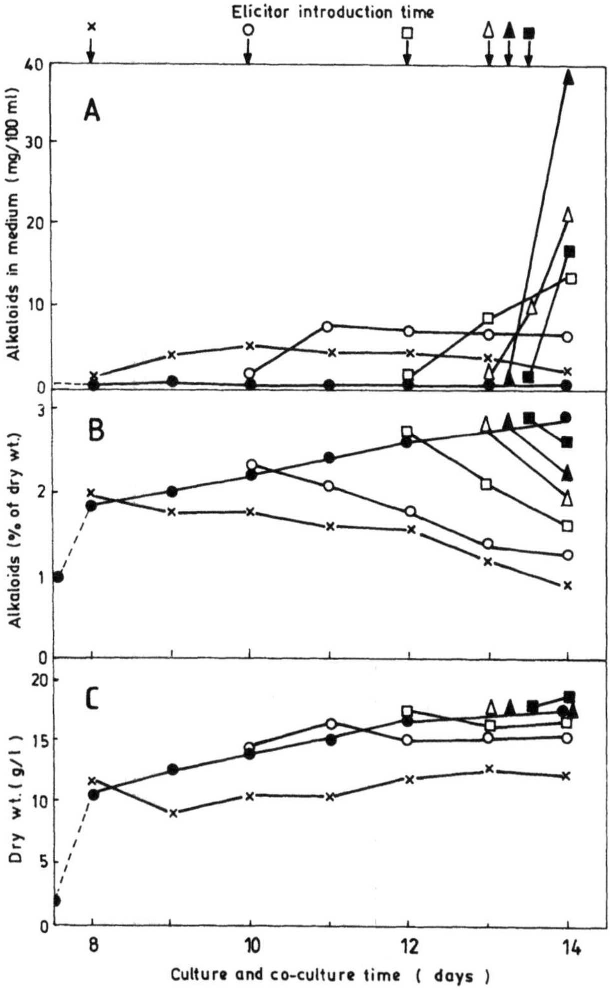


**Fig. 6A–C.** Influence of *Botrytis* preparation (3 ml/100 ml medium) on the release of the alkaloids from the *B. parvifolia* cells cultured in G medium (**A**) and accumulation of these compounds in cells (**B**) and growth of suspensions (**C**) as a function of incubation time. The *arrows* (marked with the symbols corresponding to the respective curves) indicate the time when the elicitor was added to the cell suspension. Untreated cells (—●—); cells treated after: 192 h of culture (— × —), 240 h (—○—), 288 h (—□—), 312 h (—△—), 318 h (—▲—), and 324 h (—■—). (Deliu et al. 1991)

release of alkaloids into the medium was mainly stimulated without loss of viability in *Berberis* cells.

## 2.7 Selection for Variant Cell Lines

The selection procedure for cell cultures which synthesize protoberberine alkaloids is easier due to their yellow color and fluorescence in UV light. By cloning small cell aggregates from *B. parvifolia* cell suspensions and selecting

**Table 4.** Yield of cell mass and alkaloid production in *Berberis parvifolia* cell lines isolated by cloning. (Deliu et al. 1989)

| Line no. | Cell yield[a] (g dry wt./l) | Alkaloid content (mg/g dry wt.) | | |
|---|---|---|---|---|
| | | Total | Berberine | Jatrorrhizine |
| 44 | 16.95 | 16.4 | 1.01 | 9.59 |
| 59 | 16.98 | 29.2 | 8.27 | 16.55 |
| 77 | 14.21 | 49.0 | 20.21 | 16.52 |
| 100 | 15.80 | 21.8 | 4.72 | 9.45 |
| 114 | 16.84 | 23.6 | 2.95 | 8.85 |
| 149 | 17.64 | 30.5 | 3.27 | 19.60 |
| 163 | 17.18 | 45.1 | 19.02 | 16.16 |
| 164 | 16.20 | 42.1 | 21.17 | 15.88 |
| 165 | 17.50 | 47.4 | 19.08 | 19.08 |
| 177 | 17.33 | 58.3 | 23.90 | 23.90 |
| 10-F1[b] | 16.86 | 28.7 | 3.21 | 18.66 |
| 15-F2[b] | 17.25 | 36.1 | 1.94 | 31.24 |
| 9-T1[c] | 16.63 | 34.0 | 5.44 | 21.76 |
| 13-T2[c] | 16.20 | 30.0 | 1.78 | 26.70 |
| 14-T4[c] | 15.60 | 30.0 | 1.12 | 28.00 |
| NL[d] | 17.92 | 31.0 | 3.79 | 22.56 |

[a] Inoculum size: 2.31 g dry wt./l.
[b] Cell lines isolated on media containing L-phenylalanine in concentration of 100 mg/l (F1) and 200 mg/l (F2).
[c] Cell lines isolated on media containing L-tyrosine in concentration of 100 mg/l (T1); 200 mg/l (T2) and 400 mg/l (T4).
[d] Nonselected cell line.

according to color, a series of cell lines were obtained (Table 4). Unlike the original nonselected line biosynthesizing about 3% alkaloids (dry wt.), with jatrorrhizine as the major alkaloid, one of the selected lines yielded about 6% alkaloids (dry wt.) and the ratio of jatrorrhizine/berberine was 1:1 (Deliu et al. 1989). After publication of these data, we succeeded in isolating, after repeated cloning, a cell line 164-4, with a production of 1.69 g protobereberine alkaloids/l (8.75% of dry wt.). This quantity is higher than the amount that accumulates in the root bark of the mature plant. In this cell line berberine is the main compound. If the entire aerial part of a mature (7-year-old) plant is destroyed for collecting the stem bark, the quantity obtained will be about 300–400 g dry weight with a content of 1.5% berberine (dry wt.). The same amount of berberine could be obtained in 14 days by cultivating the cell line 164-4 in 7 l of medium.

From the viewpoint of stability, a marked decrease in alkaloid synthesis is noticed in the unselected and 44 cell lines (Table 5). Some of the most stable lines, such as lines 165 and 177, were selected on second cloning.

## 2.8 Comparison Between Differentiated Plants and Cell Cultures

Comparing the ratio between the amount of three alkaloids (jatrorrhizine:berberine:palmatine) in cell cultures, seedling, and root rind

**Table 5.** Stability of protoberberine alkaloid productivity in *Berberis parvifolia* cell lines obtained by cloning and subcultured unselectively. The cells were transferred every 14 days into fresh B5 medium with NAA and BA. (Deliu unpubl.)

| Cell line | Alkaloid content (% of dry wt.) | | Alkaloid production (g/l) | |
|---|---|---|---|---|
| | 4 th[a] | 48 th[a] | 4 th[a] | 48 th[a] |
| NL[b] | 5.00 | 3.48 | 0.705 | 0.630 |
| 44 | 1.64 | 0.82 | 0.278 | 0.149 |
| 163 | 4.51 | 3.50 | 0.775 | 0.580 |
| 165 | 4.74 | 4.11 | 0.830 | 0.710 |
| 177 | 5.83 | 5.19 | 1.010 | 0.850 |
| 14-T4[c] | 3.00 | 1.89 | 0.468 | 0.321 |

[a] Subculture number.
[b] Nonselected cell line.
[c] Cell line isolated on a medium containing tyrosine (400 mg/l).

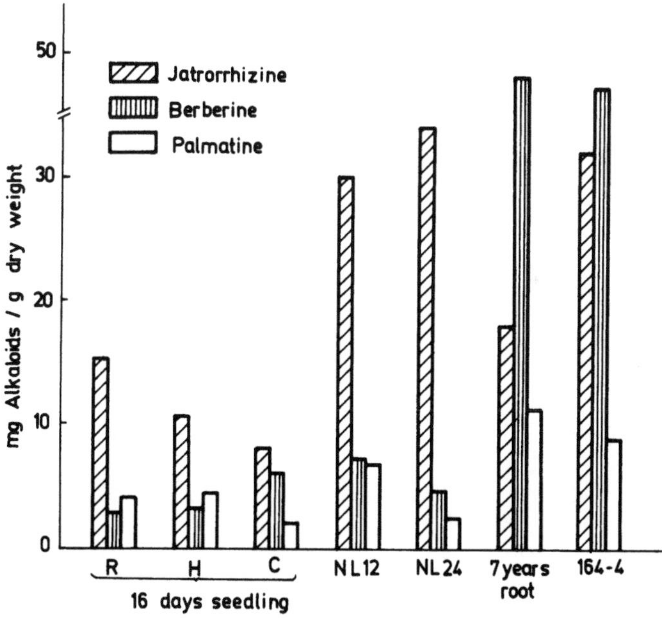


**Fig. 7.** Comparison between the alkaloid production of the cell culture, seedling, and root bark of a 7-year-old *B. parvifolia* plant. *R* Root; *H* hypocotyl; *C* cotyledon; *NL12* nonselected cell line after 12 subcultures; *NL24* nonselected cell line after 24 subcultures; *164-4* cell line isolated after repeated cloning. (Deliu unpubl.)

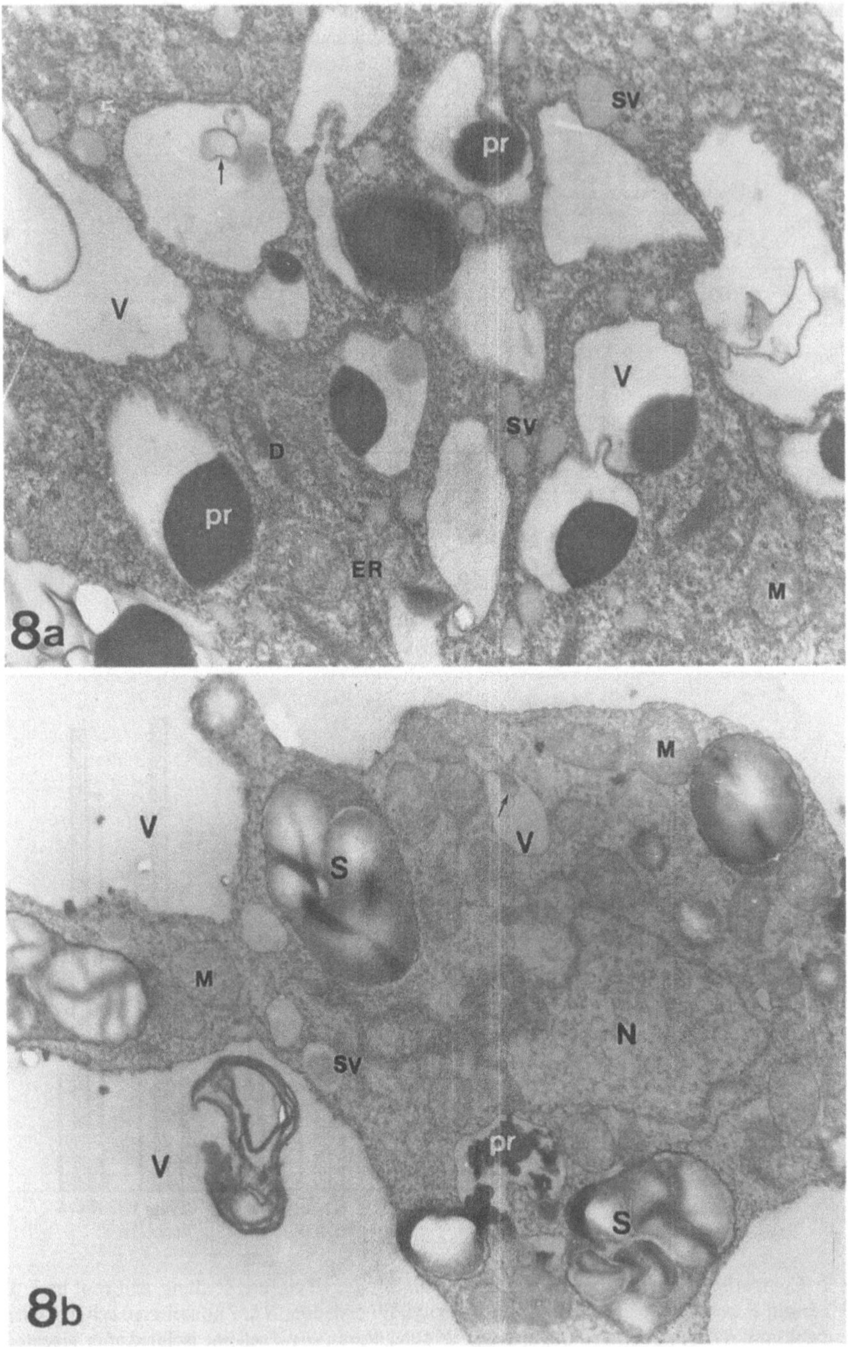


**Fig. 8a, b.** Electron micrographs of *B. parvifolia* meristematic root cell (**a**) and the cell cultivated in B5 medium (**b**). (Deliu et al. 1993b). *Arrow* Small vesicles inside small vacuoles; *SV* small vesicles in the cytoplasm; *pr* electron-dense precipitate; *ER* rough endoplasmic reticulum; *D* dictyosome; *M* mitochondrion; *N* nucleus; *S* starch; *V* vacuole some similarities or differences were noticed (Fig. 7). There were some differences among the three organs of the 16-day-old seedling. The ratio of the three alkaloids was 1:0.19:0.26 in the root, 1:0.29:0.41 in the hypocotyl, and 1:0.76:0.24 in the cotyledon. However, for the calli as well for cell cultures, irrespective of the nature of the explant from which they were induced, after 12 successive subcultures the ratio was 1:0.24:0.22. This proves that the *B. parvifolia* cells cultivated in vitro preserved the alkaloid pattern of the tissue for only a certain period of time, because after 24 subcultures the ratio was of 1:0.12:0.06. The alkaloidal spectrum undergoes a series of major changes during the ontogeny of *Berberis* plants, as the differences are great when comparing, for example, the root of the seedling with the root of the mature plant. Although the alkaloid pattern of the majority of *Berberis* cell cultures is mostly similar to that of the seedling, the 164-4 cell line of *B. parvifolia* shows an alkaloidal pattern which is close to that of the mature root.

Based on studies concerning intracellular location of the synthesis of protoberberine alkaloids (Amann et al. 1986), several biochemical and electron microscopic experiments were performed on *B. parvifolia* seedling and cell cultures (Deliu et al. 1993b). It was found that the biosynthesis of alkaloids takes place even at the level of the apical meristem of the root. Within the meristematic cells there are some vesicular formations of 1–1 μm in diameter. They are located in the cytoplasm and inside small vacuoles (Fig. 8a). Some of these vesicles release their content into small vacuoles which show the presence of electron-dense osmiophilic precipitates (alkaloids). Figure 8b shows the same formations also in *B. parvifolia* cells cultivated in vitro. Qualitative analysis by TLC attests the presence of three protoberberine alkaloids (jatrorrhizine, palmatine, and berberine) in the meristematic cells.

## 3 Conclusions

The cell suspension cultures of *B. parvifolia* and other *Berberis* species are characterized by fast growth accompanied by a relatively intense biosynthesis of protoberberine alkaloids. This peculiarity is similar to that of the meristematic cells of the root apex, where the synthesis of secondary metabolites takes place, even if alkaloid accumulation is lower due to the absence of the central vacuole.

*Berberis* plants grow slowly and, in addition, they are intermediary hosts for the pathogenic fungus *Puccinia graminis*. Isolation of highly productive cell lines whose main alkaloid is berberine opens the possibility of cultivating *Berberis* cell suspensions on a large scale.

Variation in the concentration of inorganic macro- and micronutrients or use of elicitors make it possible to increase alkaloid production.

# 4 Protocol

Sterile seedlings of *Berberis parvifolia* Sprague were grown from surface-sterilized seeds on 4% agar medium with 1/2 MS salts in a growth chamber under 16-h light at 25 °C.

In order to choose the most suitable explants, several 16-day-old seedlings with yellow-orange roots (the color of protoberberine alkaloids) were selected. These roots contain great amounts of protoberberine alkaloids. Segments from the bottom of the roots were inoculated on solid medium (Gamborg et al. 1968) with NAA (2 mg/l) and BA (0.2 mg/l). The flasks were kept in the dark at 25 °C.

In order to obtain the cell cultures, 0.5 g fragments from the very friable calli formed from the roots were transferred to different liquid media, in 200-ml Erlenmeyer flasks (20 ml medium per flask). The flasks were placed on a rotary shaker (100 rpm) at 25 °C, in the dark. The best medium for cell growth and for synthesis of the protoberberine alkaloids was that of Gamborg et al. (1968) inorganic nutrients supplemented with sucrose (30 g/l), myo-inositol (100 mg/l), thiamine (5 mg/l), nicotinic acid (1 mg/l), pyridoxine (1 mg/l), NAA (1.5 mg/l), and BA (0.3 mg/l).

The cell suspension cultures were subcultured every 14 days in the same type of flask with a dilution ratio 1:4.

*Acknowledgment.* The authors thank I. Török (Cluj) for computer assistance.

# References

Amann M, Wanner G, Zenk MH (1986) Intracellular compartimentation of two enzymes of berberine biosynthesis in plant cell cultures. Planta 167:310–320

Bauer W, Stadler R, Zenk MH (1992) Peroxidase catalyzed dimerization and demethylation of protoberberine alkaloids. Bot Acta 105:370–374

Beecher CWW, Kelleher WJ (1983) The incorporation of berberine into jatrorrhizine. Tetrahedron Lett 24:469–472

Bordas E, Petcu P, Andronescu E, Zeic A (1989) Date preliminare privind efectul medicamentului "Tonobil" în hepatita cronica agresiva cirogena la sobolanii albi. Farmacia (Bucuresti) 37:33–38

Breuling M (1986) Fermentation von Zellkulturen von *Berberis wilsoniae* Hemsl & Wils. Diss Thesis, Univ Tübingen

Breuling M, Alfermann AW, Reinhard E (1985) Cultivation of cell cultures of *Berberis wilsoniae* in 20-l airlift bioreactors. Plant Cell Rep 4:220–223

Butura V (ed) (1979) Enciclopedie de Etnobotanica Româneasca. St Encicl, Bucuresti

Cassels KB, Breitmaier E, Zenk MH (1987) Bisbenzylisoquinoline alkaloids in *Berberis* cell cultures. Phytochemistry 26:1005–1008

Deliu C, Tamas M, Pitea M, Nichiforescu E, Tesio B (1985) Continutul în alcaloizi la *Berberis parvifolia* Sprague cultivata in vitro. In: Anghel I, Brezeanu A (eds) Al III-lea Simpozion national de culturi de celule si tesuturi vegetale, Tip Univ Bucuresti, Bucuresti, pp 249–253

Deliu C, Tamas M, Ghiran D, Munteanu-Deliu C (1989) Alkaloid productivity in various *Berberis parvifolia* Sprague cell lines. Contrib Bot (Cluj-Napoca), pp 171–177

Deliu C, Ghiran D, Tamas M, Rakosy-Tican L, Cristea V (1991) Release of alkaloids in the liquid medium of *Berberis parvifolia* cell suspension cultures after treatment with *Botrytis* sp. preparation (elicitor). In: Cachita-Cosma D (ed) In vitro explant cultures – present and perspective. Brasov, Cluj-Napoca, pp 122–125

Deliu C, Nicoara A, Munteanu-Deliu C, Bercea V, Tirnoveanu D, Keul M (1993a) Some aspects of the metabolism of *Berberis parvifolia* cell suspensions. Studia (Cluj) (in press)

Deliu C, Craciun C, Craciun V, Tamas M (1993b) Ultrastructural and biochemical study of *Berberis parvifolia* root meristem and cell cultures. Plant Sci (in press)

Eilert U (1988) Elicitation: methodology and aspects of application. In: Constabel F, Vasil IK (eds) Cell culture and somatic cell genetics, vol 4. Academic Press, New York, pp 153–196

Fukui H, Nakagawa K, Tsuda S, Tabata M (1982) Production of isoquinoline alkaloids by cell suspension cultures. In: Fujiwara A (ed) Plant tissue culture 1982. Maruzen, Tokyo, pp 313–314

Gamborg OL, Miller RA, Ojima K (1968) Nutrient requirements of suspension cultures of soybean root cells. Exp Cell Res 50: 151–158

Goina T (1929) Contributii la studiul farmacognostic al plantei *Berberis vulgaris* L. Diss Thesis, Cluj Univ

Goina T, Petcu P, Pitea M (1979a) Medicament destinat tratamentului afectiunilor hepatobiliare (Biloton mitte). Rom patent 77674

Goina T, Petcu P, Pitea M (1979b) Medicament destinat tratamentului afectiunilor hepatobiliare (Biloton forte). Rom patent 77675

Hegnauer R (ed) (1964) Chemotaxonomie der Pflanzen, Band III. Birkhäuser, Stuttgart

Heywood VH (ed) (1978) Flowering plants of the world. Oxford University Press, Oxford

Hinz H, Zenk MH (1981) Production of protoberberine alkaloids by cell suspension cultures of *Berberis* species. Naturwissenschaften 67: 620–621

Ikuta A, Itokawa H (1988) Berberine: production through plant (*Thalictrum* spp.) cell cultures. In: Bajaj YPS (ed) Biotechnology in agriculture and forestry, vol 4. Medicinal and aromatic plants I. Springer, Berlin Heidelberg New York, pp 282–293

Ikuta A, Itokawa H (1989) *Coptis*: In vitro regeneration of plants and the production of berberine. In: Bajaj YPS (ed) Biotechnology in agriculture and forestry, vol 7. Medicinal and aromatic plants II. Springer, Berlin Heidelberg New York, pp 110–128

Kondo Y (1976) Organic and biological aspects of berberine alkaloids. Heterocycles 4: 197–219

Krussmann G (ed) (1960) Handbuch der Laubgehölze, Band I. Paul Parey, Berlin

Linsmaier EM, Skoog F (1965) Organic growth factor requirements of tobacco tissue cultures. Physiol Plant 18: 100–127

Murashige T, Skoog F (1962) A revised medium for rapid growth and bioassay with tobacco tissue cultures. Physiol Plant 15: 473–497

Nakagawa K, Fukui H, Tabata M (1986) Hormonal regulation of berberine production in cell suspension cultures of *Thalictrum minus*. Plant Cell Rep 5: 69–71

Otsuka H, Fujimura H, Sawada T, Goto M (1981) Studies on anti-inflamatory agents. II Anti-inflamatory constituents from rhizome of *Coptis japonica* Makino. Yakugaka Zasshi 101: 883–890

Petcu P (1959) Cercetari asupra cîtorva specii de *Berberis* aclimatizate la noi. Diss Thesis, Cluj Univ Med Pharm

Pitea M (1970) Cercetari asupra unor alcaloizi din planta *Berberis vulgaris* L. Diss Thesis, Cluj Inst Med Pharm

Pitea M, Margineanu C (1972) Contributii la studiul relatiilor dintre structura chimica si activitatea antimicrobiana a berberinei. Clujul Med 45: 465–469

Reinhard E (1967) Probleme der Produktion von Arzneistoffen durch pflanzliche Gewebekulturen. Dtsch Apoth Ztg 107: 1201–1027

Rothenberger S (1982) Untersuchungen zur Bildung von Protoberberine-Alkaloiden in Zellkulturen von *Berberis wilsoniae* Hemsl & Wills. Diss Thesis, Univ Tübingen

Rueffer M (1985) The production of isoquinoline alkaloids by plant cell cultures. In: Phillipson D, Roberts MF, Zenk MH (eds) The chemistry and biology of isoquinoline alkaloids. Springer, Berlin Heidelberg New York, pp 265–280

Rueffer M, Ekundayo O, Nagakura N, Zenk MH (1983) Biosynthesis of the protoberberine alkaloid jatrorrhizine. Tetrahedron Lett 24: 2643–2644

Tamas M, Fagarasan E, Bojita M, Deliu C (1993) Metode de analiza a alcaloizilor protoberberinici din culturi celulare in vitro de *Berberis parvifolia* Sprague. Clujul Med (in press)

Zenk MH (1985) Enzymology of benzylisoquinoline alkaloid formation. In: Phillipson D, Roberts MF, Zenk MH (eds) The chemistry and biology of isoquinoline alkaloids. Springer, Berlin Heidelberg New York, pp 240–256

Zenk MH, Rueffer M, Kutchan TM, Galneder E (1988) Biotechnological approaches to the production of isoquinoline alkaloids. In: Yamada Y (ed) Application of plant cell and tissue culture, Ciba – Foundation Symposium 137. Wiley, Chichester, pp 213–227

# V *Beta vulgaris* L. (Sugar Beet): In Vitro Culture and the Production of Glucosidases

Y. Yamasaki and H. Konno[1]

## 1 Introduction

Sugar beet (*Beta vulgaris* L.) grows wild on the eastern coast of the Mediterranean Sea and in Central Asia. About 40% of the demand for sugar in the world is supplied by sugar beet (Takahashi and Hoshikawa 1989). It is cultivated in the cold regions of the globe and in the temperate zone, where the climate is not suitable for cultivating sugarcane. The main producing countries are the former USSR, the USA, France, Poland, Germany, and Italy. About 85% of world production is from Europe (Takahashi and Hoshikawa 1989).

Numerous studies have been carried out to elucidate the sucrose storage mechanism in sugar-beet roots. In sugarcane, it has been suggested that invertase plays an important role during sucrose accumulation (Hatch et al. 1963). Sucrose translocated to the stalk enters the free space, where it is hydrolyzed by a cell wall acid invertase. The resulting hexoses are actively accumulated into the storage parenchyma and resynthesized to sucrose phosphate by sucrose phosphate synthetase. In sugar beet, however, sucrose synthetase plays a pivotal role during sucrose accumulation (Pavlinova and Prasolova 1970, 1972; Pavlinova 1971; Prasolova et al. 1976). Giaquinta (1979) has shown that the onset of storage is accompanied by the appearance of sucrose synthetase activity. Doney's group has shown that sugar beet root cell size is negatively correlated with sucrose concentration (Doney et al. 1981; Doney and Theurer 1983). High-sucrose genotypes have small cells, and low-sucrose genotypes have large cells. Therefore, genetic selection methods that increase root yield decrease sucrose content, and those that increase sucrose content decrease yield (Powers 1957; Bergen 1967; Oldemeyer 1975). Stein and Willenbrink (1976) have shown a correlation between sucrose accumulation and energy charge during beet development.

In recent years, it has become clear that sugar beet contains interesting α-glucosidases. The α-glucosidases (EC 3.2.1.20) have been studied extensively in animals, higher plants, yeasts, and molds. Most of these enzymes readily hydrolyze soluble starch, liberating glucose. The enzyme from *Mucor javanicus* (Yamasaki et al. 1973) hydrolyzes soluble starch at the same rate as maltose, but the enzymes from *Aspergillus* (Tsujisaka and Fukumoto 1963; Yamasaki et al.

[1] Research Institute for Bioresources, Okayama University, Kurashiki 710, Japan

Biotechnology in Agriculture and Forestry, Vol. 28
Medicinal and Aromatic Plants VII (ed. by Y.P.S. Bajaj)

1977) and some bacteria (Yamasaki and Suzuki 1974; Guffanti and Corpe 1976) hydrolyze soluble starch very weakly or not at all. The enzymes from higher plants hydrolyze soluble starch more weakly than does the enzyme from *M. javanicus*. The in-vivo significance of glucose production from soluble starch in plant tissues by α-glucosidase is not yet clear, although some authors (Swain and Dekker 1966; Nomura et al. 1969) have suggested that the enzyme forms a part of the nonphosphorolytic pathway for the breakdown of starch, and is functioning in seed germination by hydrolyzing the oligosaccharides produced by α- and β-amylases.

The α-glucosidases from sugar-beet seeds hydrolyze soluble starch at a faster rate than maltose (Chiba et al. 1978; Yamasaki and Suzuki 1980), as if they were glucoamylases. This strong soluble-starch-hydrolyzing activity suggests that, like glucoamylase, the enzymes may even hydrolyze starch to glucose without the preceding action of α- and β-amylases in the plant tissue. On the other hand, another type of α-glucosidase, which hydrolyzes soluble starch only very weakly, was also isolated from sugar-beet seeds at the same time (Yamasaki and Suzuki 1980). The two enzymes from sugar-beet seeds exhibit very different properties, particularly in their soluble-starch-hydrolyzing activity. They may be located in a different part of the plant tissue and play different roles there. Therefore, sugar-beet seeds offer an excellent system with which to examine the physiological function of α-glucosidase. Plant tissue cultures are often useful in studying the various physiological phenomena of higher plants. In this chapter, the purification and properties of the α-glucosidases produced by suspension-cultured sugar-beet cells are described.

## 2 In Vitro Approaches

In vitro culture studies on sugar beet have been reviewed earlier in this Series (see Atanassov 1986; Hall et al. 1994).

### 2.1 Establishment of Tissue Culture

The stem pieces of sugar beet (cv. Tsukisappu) after washing with water, are treated with 0.1% benzalkonium chloride for 3 min, 80% ethyl alcohol for 10 min, and 1% sodium hypochlorite for 30 min.

After washing two times with sterilized water, the stem is bored by a cork borer (diameter; 3–5 mm) to obtain a cylindrical part. This part is cut into 1–5-mm-thick disks. The disks are explanted on basal MS medium (Murashige and Skoog 1962) containing 3% sucrose as the carbon source and 4.52 μM 2,4-dichlorophenoxyacetic acid as growth regulator. The induced callus is carried through more than 30 transfers and then inoculated in liquid MS medium. The sugar-beet cells are maintained in darkness for many generations before being used for experiments. Suspension cells (about 0.7 g wet wt.) are inoculated into 125 ml of the medium in 500-ml flasks and the flasks are agitated on a reciprocal shaker at a speed of 88 strokes $min^{-1}$ at 25 °C for 2 weeks in the dark.

Masuda et al. (1988a) have extracted various glycosidases from cell walls of sugar-beet cells grown in suspension culture by different extraction methods, and suggested that the cell-wall glycosidases are associated with different polysaccharides. They also isolated acid and alkaline invertases, and characterized them (Masuda et al. 1988b). Moreover, several studies have been carried out to characterize invertases from tissue cultures of sugar beet (Angelova et al. 1974; Kursanov et al. 1976; Mohammad and Collin 1979; Dubinina et al. 1982; Kudryavtseva et al. 1982). Some of these have shown the difference between intact plants and tissue cultures on the pattern of invertase activity. Mohammad and Collin (1979) have reported that the pattern of invertase activity in callus is unlike that in intact plants. Kudryavtseva et al. (1982) also reported that tissue culture induced from the stored sugar-beet root, which lacks invertase activity, possesses both acid and alkaline intracellular invertases and cell-wall acid invertase. Moreover, Dubinina et al. (1982) reported that the invertase inhibitor isolated from root tissue culture is less active than that from the stored root.

## 2.2 Isolation of Intracellular α-Glucosidase

When sugar-beet cells are grown in MS medium with shaking, they produce some α-glucosidases in the cells (Yamasaki and Konno 1989a). A fraction of them can be extracted from the cells with buffer, and most of them with sodium chloride (Fig. 1). The former shows only negligible activity in comparison with the latter (Yamasaki and Konno 1989a). Three forms of α-glucosidase (Fig. 2) are extracted from the cells with sodium chloride (Yamasaki and Konno 1989a). One of them (G-I) appears and is a main enzyme in the early stage of culture, decreasing as the cells grow. Another (G-II) is negligible at the early stage of culture, but increases markedly as the cells grow. G-II is a main enzyme at the period when the cells are harvested. The third (G-III) appears at all stages of culture constantly and is a minor enzyme. G-II hydrolyzes soluble starch more rapidly than maltose, and so is like glucoamylase. G-I and G-III hydrolyze soluble starch only very weakly. On the other hand, the starch content in the cell

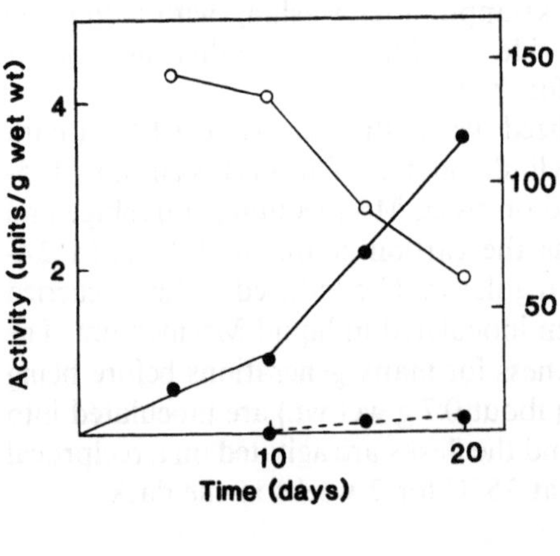


**Fig. 1.** Changes in the α-glucosidase activity and the starch content in cultured sugar-beet cells. Sugar-beet cells grown in MS liquid medium were harvested at periodic intervals. ●---● α-Glucosidase activity of fraction extracted with buffer; ●——● α-glucosidase activity of fraction with sodium chloride; ○——○ starch content

decreases as G-II increases, which suggests that G-II is involved in the metabolism of starch. G-II can be isolated by the following procedures (Yamasaki and Konno 1991a). The cells are harvested by filtration, then suspended in 25 mM acetate buffer, pH 4.5, and disrupted by sonication (20 kHz, model 5202 PZT, Ohtake Works, Tokyo) for 15 min at 0 °C. Complete disruption of the cells is confirmed by light microscopy. The homogenate is centrifuged at 8000 *g* for 10 min. The precipitate is washed, and suspended in 25 mM acetate buffer, pH 4.5, containing 2 M NaCl and stirred overnight. The suspension is centrifuged at 8000 *g* for 10 min and the supernatant collected. The supernatant is brought to

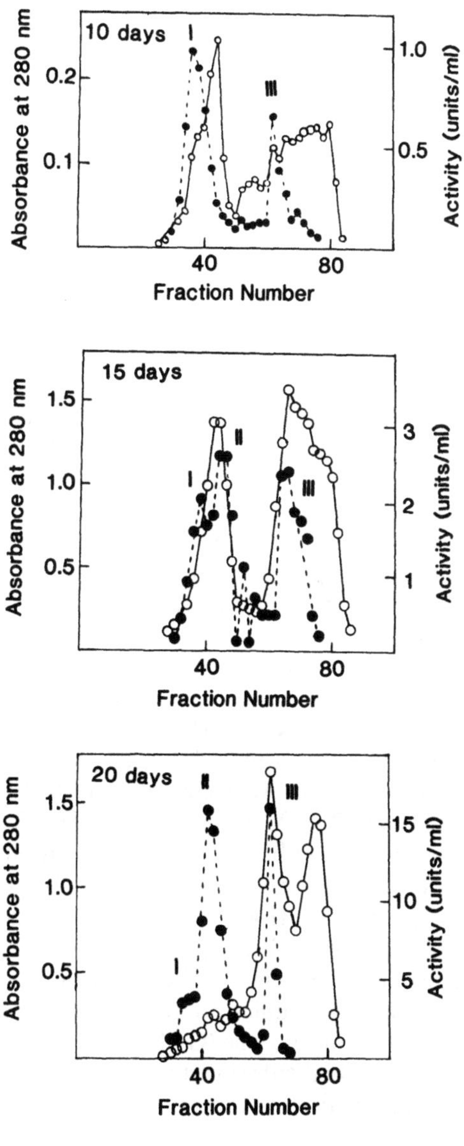


**Fig. 2.** Gel chromatography on Sephacryl S-200 HR. Fraction extracted with sodium chloride was harvested at periodic intervals and loaded on a Sephacryl S-200 HR column (bed volume = 236.5 ml; bed height = 93 cm). The column was eluted with 20 mM acetate buffer, pH 4.5, containing 150 mM sodium chloride. The eluate was collected in 2-ml fractions. ○——○ $A_{280}$; ●---● α-glucosidase activity; *I* G-I; *II* G-II; *III* G-III

**Table 1.** Summary of purification of α-glucosidase from suspension-cultured sugar-beet cells

| Purification step | Total activity | Yield | Specific activity | Purification |
|---|---|---|---|---|
| | Units | % | Units/mg | |
| Ammonium sulfate ppt. | 293 | 100 | 2 | 1 |
| Sephacryl S-200 HR chromatography | 148 | 50 | 5 | 2.5 |
| CM-cellulose chromatography | 132 | 45 | 13 | 6.6 |
| Mono Q HR 5/5 chromatography | 117 | 40 | 21 | 10.3 |
| Preparative PAGE | 26 | 9 | 151 | 75.6 |

0.9 saturation with $(NH_4)_2SO_4$. The precipitate is collected by centrifugation and dissolved in 25 mM acetate buffer, pH 4.5. The enzyme solution is divided into three fractions by gel chromatography on Sephacryl S-200 HR. G-II, having the highest α-glucosidase activity, is collected, and brought to 0.9 saturation with $(NH_4)_2SO_4$. The precipitate is collected by centrifugation and dissolved in 25 mM acetate buffer, pH 4.5, and dialyzed overnight against the buffer. The dialysate is put on a CM-cellulose column (1.6 × 10 cm) equilibrated with 20 mM acetate buffer, pH 4.5. The column is first eluted with the same buffer to wash off the unadsorbed protein, and then with the same buffer containing 300 mM sodium chloride. The eluate with α-glucosidase activity is concentrated using an Amicon ultrafiltration device (PM-10 membrane; Amicon Co.). The concentrate is dialyzed overnight against deionized water. After adding 1 M Tris-HCl buffer, pH 8.0, to 20 mM, the dialysate is chromatographed on a column (0.5 cm × 5 cm) of Mono Q HR 5/5 equilibrated with 20 mM Tris-HCl buffer, pH 8. After washing the column with the buffer, a linear gradient of sodium chloride (0–200 mM) in the same buffer is applied. As the column is too small to load the whole sample; only a part of the sample was used for one chromatography. The chromatography is automatically repeated using the LKB HPLC system with the Superloop at 4 °C. The eluate with α-glucosidase activity is concentrated using an Amicon ultrafiltration device. The concentrate is dialyzed overnight against 20 mM acetate buffer, pH 4.5. The concentrate is used for preparative disk gel electrophoresis, pH 4.0 (Reisfeld et al. 1962).

The running column (1.5 × 20 cm) is composed of 16 ml of separating gel and 4 ml of concentrating gel. Electrophoresis is carried out at a constant current of 42 mA for 10 h. After electrophoresis, the gel is cut into 0.3-cm sections and crushed in 20 mM acetate buffer, pH 4.5. The active fractions are pooled. After removing insoluble materials by filtration, the enzyme solution is concentrated by ultrafiltration with Amicon PM-10 and dialyzed overnight against 20 mM acetate buffer, pH 4.5. This procedure leads to a 76-fold purification with 8.8% recovery (Table 1). The purified enzyme is shown to be homogeneous by PAGE. The molecular weight of the enzyme is calculated to be 53 000 from SDS-PAGE. The value is similar to that from gel filtration with Sephacryl S-200 HR. Therefore, the enzyme (G-II) does not contain quaternary structure. Maltohexaose is the best substrate for the enzyme (Table 2). However, the enzyme readily hydrolyzes also maltopentaose and maltoheptaose, indicating little or no differ-

**Table 2.** The kinetic parameters of α-glucosidase for various substrates

| Substrate | $K_m$ | $V_{max}$ | $V_{max}/K_m$ ratio |
|---|---|---|---|
| | mM | μ mol/min/mg protein | |
| Maltose | 3.47 | 3.53 | 1.0 |
| Maltotetraose | 1.43 | 7.84 | 5.5 |
| Maltopentaose | 0.64 | 7.06 | 11.0 |
| Maltohexaose | 0.35 | 5.88 | 16.8 |
| Maltoheptaose | 0.38 | 8.30 | 21.8 |
| Soluble starch | 0.61 (g/l) | 2.80 (mg/min/mg protein) | 4.6 |

ence in the substrate specificities of maltopentaose, maltohexaose, and maltoheptaose. The result shows that the enzyme has higher substrate specificity for malto-oligosaccharide than maltose. Therefore, the α-glucosidase from sugar-beet cells may hydrolyze starch to glucose without the preceding action of β-amylase in the plant tissue, although some authors (Swain and Dekker 1966; Nomura et al. 1969) have suggested that α-glucosidase forms a part of the nonphosphorolytic pathway for the breakdown of starch, and functions in seed germination by hydrolyzing the oligosaccharides produced by α- and β-amylases.

To be able to understand the mode of action of the α-glucosidase from sugar-beet cells, the enzyme is incubated with amylose under standard conditions, and the product formed is determined by gas liquid chromatography. As a result, the α-glucosidase from sugar-beet cells hydrolyzes amylose, liberating α-glucose as the other α-glucosidases (Yamasaki and Suzuki 1979). Therefore, the enzyme from sugar-beet cells is an α-glucosidase, although it may be regarded as a glucoamylase from its high soluble-starch-hydrolyzing activity.

### 2.3 Isolation of Extracellular α-Glucosidase

Sugar-beet cells secreted two α-glucosidases into the culture medium when grown in MS medium with shaking (Yamasaki and Konno 1989b). The α-glucosidases have been studied in animals, yeasts, and molds extensively. However, there are few reports on extracellular α-glucosidases from plants. This section deals with the purification and properties of extracellular α-glucosidase from suspension-cultured sugar-beet cells (Yamasaki and Konno 1991b).

Sugar-beet cells (about 0.7 g wet wt.) are inoculated into 125 ml of the MS medium in 500-ml flasks and the flasks are agitated on a reciprocal shaker at a speed of 88 strokes $min^{-1}$ at 25 °C for 20 days in the dark. The culture supernatant was collected by filtration, and concentrated by an Ultrafiltration System (model 1XX-42 OLC; Millipore Ltd.). The concentrate is brought to 0.9 saturation with ammonium sulfate. The precipitate is collected by centrifugation and dissolved in 50 mM acetate buffer, pH 4.5. The enzyme solution is divided into two fractions (N-I and N-II) by gel chromatography on Sephacryl S-200 HR. The major fraction of α-glucosidase (N-I) is collected and brought to 0.9 saturation with ammonium sulfate. The precipitate is collected by centrifugation and dissolved in 50 mM acetate buffer, pH 4.5. The solution is dialyzed overnight against 20 mM acetate buffer, pH 4.5. The dialysate is put on a CM-cellulose column (1.6 × 10 cm) equilibrated with 20 mM acetate buffer, pH 4.5. The column is first washed with the same buffer to remove the unadsorbed protein, and then α-glucosidase is eluted with the same buffer containing 300 mM sodium chloride. Active fractions are concentrated using an Amicon ultrafiltration device (PM-10 membrane; Amicon Co.) followed by dialysis overnight against deionized water. One M Tris-HCl buffer, pH 8, is added to the dialysate to make the concentration of 20 mM, and the enzyme solution is chromatographed on a column (0.5 cm × 5 cm) of Mono Q HR 5/5 equilibrated with 20 mM Tris-HCl buffer, pH 8. After the column is washed with the buffer, it is eluted with a linear gradient of sodium chloride (0–200 mM) in the same buffer. Active fractions are concentrated using an Amicon ultrafiltration device, and dialyzed overnight against 20 mM acetate buffer, pH 4.5. The dialysate is electrophoresed on a preparative disk gel, pH 4.0, as described by Reisfeld et al. (1962). After electophoresis, the enzyme was extracted from gel slices. The extract is filtered to remove insoluble materials, and concentrated by ultrafiltration with Amicon PM-10 followed by dialysis overnight against 20 mM acetate buffer, pH 4.5. The purified enzyme is homogeneous by polyacrylamide gel electrophoresis. The molecular weight of the enzyme is calculated to be 54000 from SDS-polyacrylamide gel electrophoresis, and 55000 from gel filtration with Superose 12 HR 10/30. Therefore, the extracellular enzyme does not contain quaternary structure, as in the intracellular enzyme (G-II).

The enzyme hydrolyzes malto-oligosaccharide more rapidly than maltose (Table 3); it also readily hydrolyzes soluble starch. The results show that the substrate specificity of this enzyme is similar to α-glucosidases (Chiba et al. 1978; Yamasaki and Suzuki 1980) from sugar-beet seeds. The products formed from

**Table 3.** Substrate specificity of α-glucosidase

| Substrate | Relative rate of hydrolysis | $K_m$ value |
|---|---|---|
| | % | mM |
| Maltose | 100 | 2.310 |
| Maltotriose | 125.2 | 1.367 |
| Maltotetraose | 125.3 | 1.154 |
| Maltopentaose | 223.5 | 0.670 |
| Maltohexaose | 215.0 | 0.612 |
| Maltoheptaose | 250.9 | 0.788 |
| Amylose EX-I[a] | 179.1 | |
| Soluble starch | 126.5 | |
| Phenyl α-glucoside | 11.8 | |
| Sucrose | 0 | |

[a] Degree of polymerization (Dpn.): about 18.

amylose with the enzyme are determined by gas liquid chromatography. As a result, the α-glucosidase from sugar-beet cells hydrolyzes amylose to liberate α-glucose like other α-glucosidases (Yamasaki and Suzuki 1980).

## 2.4 Effects of Tunicamycin on α-Glucosidase Secretion from Cultured Sugar-Beet Suspension Cells

In a previous section, the presence of two α-glucosidases in suspension-cultured sugar-beet cells was described. The secreted enzyme protein is proposed to possess a signal for secretion into the culture medium. In cultured peanut suspension cells, a large proportion of peroxidase is rapidly secreted during the culture, but inhibition of glycosylation results in inhibition of the enzyme secretion (Ravi et al. 1986). Moreover, there are some reports suggesting a role for carbohydrate in invertase secretion (Gascon et al. 1968; Prentice and Robbins 1976; Schwaiger and Tanner 1979). The degree of glycosylation may affect the destination of enzyme proteins out of or within cells. Tunicamycin is an inhibitor of protein glycosylation (Tkacz and Lampen 1975) and has been used in vivo to produce under- or nonglycosylated proteins in plants (Schwaiger

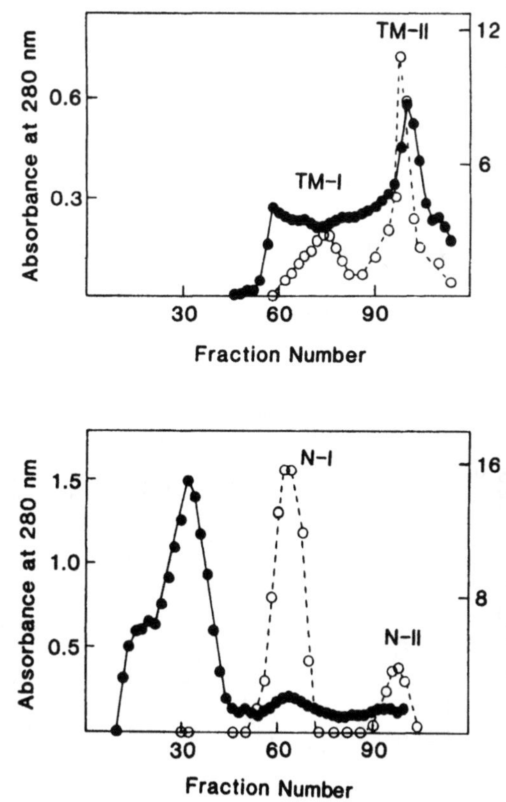


**Fig. 3.** Gel chromatography on Sephacryl S-200 HR. Chromatography was carried out as in Fig. 2. *Upper figure* filtrate cultured with tunicamycin; *lower figure* filtrate cultured without tunicamycin; ●——● $A_{280}$; ○---○ α-glucosidase activity

and Tanner 1979; Quader 1984; Ravi et al. 1986). Thus, it is of interest to study whether tunicamycin has an effect on the secretion of α-glucosidase from plant cells.

Sugar-beet cells (about 0.7 g wet wt.) are inoculated into 125 ml of the MS medium in 500-ml flasks, and the flasks are agitated on a reciprocal shaker at a speed of 88 strokes $min^{-1}$ at 25 °C for 20 days in the dark. Cells cultured for 2 weeks are washed once with an equal volume of fresh medium to remove extracellular α-glucosidase. The washed cells are resuspended into an equal volume of fresh medium containing 5 μg/ml of tunicamycin, and cultured for 5 days. The culture supernatant is collected by filtration, and concentrated by an Ultrafiltration System (model MC-4; Bioengineering Co., Ltd., Tokyo, Japan). The concentrate is brought to 0.9 saturation with ammonium sulfate. The precipitate is collected by centrifugation and dissolved in 50 mM acetate buffer, pH 4.5. The enzyme solution is divided into two fractions (TM-I and TM-II) by gel chromatography on Sephacryl S-200 HR (Yamasaki and Konno 1989b). α-Glucosidase activity of TM-II is about twice as high as that of TM-I (Fig. 2). In the culture without tunicamycin, also, two active fractions (N-I and N-II) are obtained by the gel filtration of the culture supernatant (Yamasaki and Konno

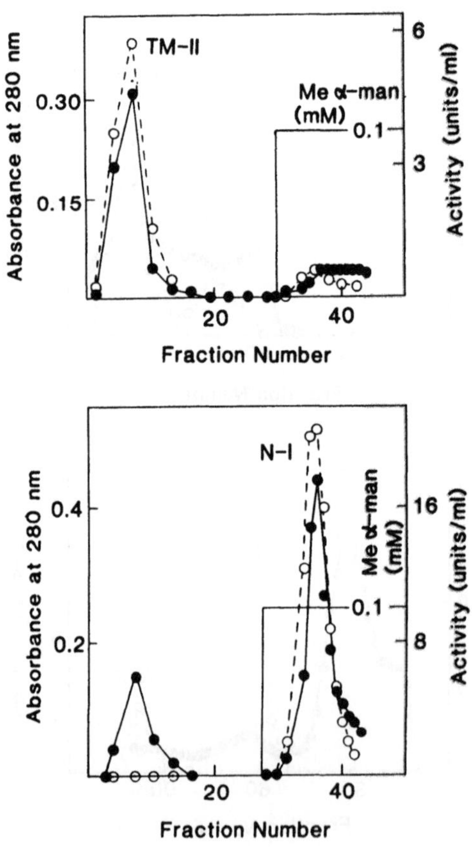


**Fig. 4.** Chromatography of TM-II (*Upper figure*) and N-I (*lower figure*) on Con A-Sepharose. TM-II and N-I were put separately on a Con A-Sepharose column (bed volume = 10 ml; bed height = 5 cm) equilibrated with 100 mM acetate buffer, pH 6, containing 1 mM $CaCl_2$, 1 mM $MgCl_2$, and 1 mM $MnCl_2$. The column was first eluted with the same buffer to wash off unadsorbed protein, and then with 100 mM methyl α-mannoside in the same buffer. ●——● $A_{280}$; ○---○ α-glucosidase activity

1989b). α-Glucosidase activity of N-I is about three times as high as that of N-II (Fig. 3). TM-II and N-II are eluted from the column at the same position and the enzyme activity of TM-II is as high as that of N-II, showing that tunicamycin has no inhibitory effect on the secretion of N-II. On the other hand, TM-I and N-I seem to differ slightly from each other in the profile of elution from the column, and the enzyme activity of TM-I is much lower than that of N-I. The results suggest that tunicamycin inhibits glycosylation of N-I, which causes inhibition of the secretion of N-I, and that N-I is an N-linked glycoprotein. This is supported by Con A-Sepharose column chromatography. Con A specifically binds with α-D-mannopyranose and α-D-glucopyranose residues in glycoconjugates (Goldstein 1976). Therefore, the specific binding to Con A has been used to examine whether a protein may be a glycoprotein (Faye and Berjonneau 1979; Chatfield and Armstrong 1988) and to stain glycoproteins in a gel after electrophoresis (Wood and Sarinana 1975; Avigad 1978). N-I is adsorbed on Con A-Sepharose and eluted from it by 100 mM of methyl α-mannoside, whereas TM-II is not adsorbed on Con A-Sepharose (Fig. 4). TM-I is also adsorbed on Con A-Sepharose.

In normal culture from 15 to 20 days, N-I is mainly secreted in the culture medium and tunicamycin inhibits the secretion by 81%. As Ferro-Novick et al. (1984) have suggested, oligosaccharides of N-I may be important in facilitating proper folding of the polypeptide during membrane penetration.

## 3 Summary and Conclusions

It has been suggested that α-glucosidase forms a part of the nonphosphorolytic pathway for the breakdown of starch, and functions in seed germination by hydrolyzing the oligosaccharides produced by α- and β-amylases. However, α-glucosidases from higher plants readily hydrolyze soluble starch, liberating glucose. In higher plants, an α-glucosidase that hydrolyzes soluble starch weakly is found only in sugar-beet seeds. Sugar-beet seeds contain two α-glucosidases which exhibit very different properties, particularly in soluble-starch-hydrolyzing activity. Therefore, sugar beet offers an excellent system with which to examine the physiological function of α-glucosidase. Four α-glucosidases have been fractionated from sugar-beet cells. One of them (G-II), which is a main enzyme of four α-glucosidases, hydrolyzes soluble starch at a faster rate than maltose, is not detected during the phase of starch accumulation in cells, and undergoes a marked increase with the decrease in starch content. Therefore, a possible role for the enzyme is in the metabolism of starch in the cell. The molecular weight of the enzyme is calculated to be 53 000 and the enzyme does not contain quaternary structure. On the other hand, two α-glucosidases (N-I and N-II) are secreted into the culture medium when sugar-beet cells are grown in MS medium with shaking. α-Glucosidase activity of N-I is about three times as high as that of N-II. However, α-glucosidase activity of N-I, which inhibited the glycosylation by tunicamycin, is much lower than that of N-II. The secreted enzyme protein is proposed to possess a signal for secretion into the culture medium. In the two α-glucosidases from sugar-beet cells, the degree of glycosylation affects the destination of the enzyme proteins out of or within the cells.

## References


Angelova AA, Atanassov AI, Stambolova MA, Nikolov TK (1974) Study of invertase activity and sugar content in a sugar beet tissue culture grown on a medium with chloramphenicol. Fiziol Rast (Mosc) 21:1021–1024

Atanassov AI (1986) Sugar beet (*Beta vulgaris* L.) In: Bajaj YPS (ed) Biotechnology in agriculture and forestry, vol 2, Crops I. Springer, Berlin Heidelberg New York, pp 462–470

Avigad G (1978) Lectin-bound alcohol and lactic dehydrogenases as a reagent for the visual detection of glycoproteins. Anal biochem 86:443–449

Bergen P (1967) Dry matter of the petiole as an index for the selection of sugar beet plants. J Am Soc Sugar Beet Technol 14:396–399

Chatfield JM, Armstrong DJ (1988) Cytokinin oxidase from *Phaseolus vulgaris* callus cultures. Affinity for Concanavalin A. Plant Physiol 88:245–247

Chiba S, Inomata S, Matsui H, Shimomura T (1978) Purification and properties of an α-glucosidase (glucoamylase) in sugar beet seed. Agric Biol Chem 42:241–245

Doney DL, Theurer JC (1983) Genetics of cell size and sucrose concentration in sugarbeet. Crop Sci 23:904–907

Doney DL, Wyse RE, Theurer JC (1981) The relationship between cell size, yield, and sucrose concentration of the sugarbeet root. Can J Plant Sci 61:447–453

Dubinina IM, Burakhanova EA, Kudryavtseva LF (1982) Regulation of invertase activity in sugar beet tissue cultures. Fiziol Rast (Mosc) 29:1188–1194

Faye L, Berjonneau C (1979) Evidence for the glycoprotein nature of radish β-fructosidase. Biochimie 61:51–59

Ferro-Novick S, Hansen W, Schaure I (1984) Genes required for completion of import of proteins into the endoplasmic reticulum in yeast. J Cell Biol 98:44–53

Gascon S, Neumann NP, Lampen JO (1968) Comparative study of the properties of the purified internal and external invertase from yeast. J Biol Chem 243:1573–1577

Giaquinta RT (1979) Sucrose translocation and storage in the sugar beet. Plant Physiol 63:828–832

Goldstein IJ (1976) Carbohydrate binding specificity of Concanavalin A. In: Bittiger J, Schnebli HP (eds) Concanavalin A as a tool. Wiley, London, pp 55–65

Guffanti AA, Corpe WA (1976) Partial purification and characterization of α-glucosidase from *Pseudomonas fluorescens* W. Arch Microbiol 107:269–276

Hall RD, Pedersen C, Krens FA (1994) Regeneration of plants from protoplasts of *Beta vulgaris* (Sugarbeet). In: Bajaj YPS (ed) Biotechnology in agriculture and forestry, vol 29. Plant protoplasts and genetic engineering V. Springer Berlin Heidelberg New York, pp 14–35

Hatch MD, Sacher JM, Glasziou KT (1963) Sugar accumulation cycle in sugarcane. I. Studies in enzymes of the cycle. Plant Physiol 38:338–343

Kudryavtseva LF, Burakhanova EA, Dubinina IM (1982) Invertase in differentiated and dedifferentiated tissues of the sugar beet. Fiziol Rast (Mosc) 29:868–875

Kursanov AL, Dubinina IM, Kudryavtseva LF, Burakhanova EA (1976) Some aspects of carbohydrate metabolism in cultures of dedifferentiated tissues of the sugar beet. Fiziol Rast (Mosc) 23:1119–1127

Masuda H, Komiyama S, Sugawara S (1988a) Extraction of enzymes from cell walls of sugar beet cells grown in suspension culture. Plant Cell Physiol 29:623–627

Masuda H, Takahashi T, Sugawara S (1988b) Acid and alkaline invertases in suspension cultures of sugar beet cells. Plant Physiol 86:312–317

Mohammad AMS, Collin HA (1979) Growth and invertase activity of sugar beet callus. New Phytol 82:293–300

Murashige T, Skoog F (1962) A revised medium for rapid growth and bioassays with tobacco tissue cultures. Physiol Plant 15:473–497

Nomura T, Kono Y, Akazawa T (1969) Enzymic mechanism of starch breakdown in germinating rice seeds. II. Scutellum as the site of sucrose synthesis. Plant Physiol 44:765–769

Oldemeyer RK (1975) Introgression hybridization as a breeding method in *Beta vulgaris*. J Am Soc Sugar Beet Technol 18:269–273

Pavlinova OA (1971) Sucrose metabolism in the sugarbeet root. Fiziol Rast (Mosc) 18:722–730

Pavlinova OA, Prasolova MF (1970) Sucrose-synthesizing enzymes of the sugar beet. Fiziol Rast (Mosc) 17:295–301

Pavlinova OA, Prasolova MF (1972) Physiological role of sucrose synthetase in sugarbeet roots. Fiziol Rast (Mosc) 19:920–925

Powers L (1957) Identification of genetically superior individuals and the prediction of genetic gains in sugar beet breeding programs. J Am Soc Sugar Beet Technol 9:408–432

Prasolova M, Mambetkulov S, Pavlinova OA, Pechanov VA (1976) Action of inorganic ions of sucrose synthetase of sugarbeet roots. Fiziol Rast (Mosc) 23:292–299

Prentice N, Robbins GS (1976) Composition of invertases from germinated barley. Cereal Chem 53:874–880

Quader H (1984) Tunicamycin prevents cellulose microfibril formation in *Oocystis solitaria*. Plant Physiol 75:534–538

Ravi K, Hu C, Reddi PS, van Huystee RB (1986) Effect of tunicamycin on peroxidase release by cultured peanut suspension cells. J Exp Bot 37:1708–1715

Reisfeld RA, Lewis UJ, Williams DE (1962) Disc electrophoresis of basic proteins and peptides on polyacrylamide gels. Nature (Lond) 195:281–283

Schwaiger H, Tanner W (1979) Effects of gibberellic acid and of tunicamycin on glycosyl-transferase activities and on α-amylase secretion in barley. J Biochem 102:375–381

Stein M, Willenbrink J (1976) On accumulation of sucrose in the growing sugar beet. Z Pflanzenphysiol 79:310–322

Swain RR, Dekker EE (1966) Seed germination studies II. Pathways for degradation in germinating pea seedlings. Biochim Biophys Acta 122:87–100

Takahashi B, Hoshikawa K (1989) *Beta* L. In: Hotta M, Ogata K, Nitta A, Hosikawa K, Yanagi M, Yamasaki K (eds) Useful plants of the world. Heibonsha, Tokyo, pp 148–149

Tkacz JS, Lampen JO (1975) Tunicamycin inhibition of polyisoprenyl N-acetylglucosaminyl pyrophosphate formation in calf-liver microsomes. Biochem Biophys Res Commun 65:248–257

Tsujisaka Y, Fukumoto J (1963) Studies on maltose splitting enzymes of moulds. I. Crystallization of transglucosidase of *Aspergillus niger*. Nippon Nogeikagaku Kaishi 37:668–672

Wood JG, Sarinana FO (1975) The staining of sciatic nerve glycoproteins on polyacrylamide gels with concanavalin A-peroxidase. Anal Biochem 69:320–322

Yamasaki Y, Konno H (1989a) α-Glucosidases of suspension-cultured sugar-beet cells. Phytochemistry 28:2583–2585

Yamasaki Y, Konno H (1989b) Effects of tunicamycin on α-glucosidase secretion from cultured sugar-beet suspension cells. Agric Biol Chem 53:2499–2500

Yamasaki Y, Konno H (1991a) Purification and properties of α-glucosidase from suspension-cultured sugar-beet cells. Phytochemistry 30:2861–2863

Yamasaki Y, Konno H (1991b) Extracellular α-glucosidase from suspension-cultured sugar-beet cells. Agric Biol Chem 55:1675–1676

Yamasaki Y, Suzuki Y (1974) Purification and properties of α-glucosidase from *Bacillus cereus*. Agric Biol Chem 38:443–454

Yamasaki Y, Suzuki Y (1979) Purification and properties of three forms of α-glucosidase from germinated green gram (*Phaseolus vidissimus Ten.*) Agric Biol Chem 43:481–489

Yamasaki Y, Suzuki Y (1980) Two forms of α-glucosidase from sugar-beet seeds. Planta 148:354–361

Yamasaki Y, Miyake T, Suzuki Y (1973) Properties of crystalline α-glucosidase from *Mucor javanicus*. Agric Biol Chem 37:251–259

Yamasaki Y, Suzuki Y, Ozawa J (1977) Three forms of α-glucosidase and a glucoamylase from *Aspergillus awamori*. Agric Biol Chem 41:2149–2161

# VI *Bowiea volubilis* Harv. ex Hook.f. (Sea Onion): In Vitro Culture and the Production of Cardiac Glycosides

J. F. Finnie, F. E. Drewes, and J. van Staden[1]

## 1 General Account

When *Bowiea volubilis* Harv. ex Hook.f., a member of the family Hyacinthaceae, (also known as the sea/climbing onion) was first described, it was stated in Curtis's *Botanical Magazine* that "though possessing little beauty, this is certainly one of the most curious plants ever introduced into Europe" (Dyer 1941). More recently, *B. volubilis* has attracted attention as a source of cardiac glycosides. The plant has long been known among tribes of southern Africa as a strong medicinal herb. The strong toxic properties of this plant have been the cause of many deaths due to overdoses administered by herbalists. Jaretsky (Watt and Breyer-Brandwijk 1962) suggested that the toxic properties of the bulb were due to cardiac glycosides similar to digoxin and digitoxin. Katz (1950, 1953a, b, 1954, 1957a, b) identified the cardiac glycosides and named them bovoside A, B, C, D, and E.

### 1.1 Morphology and Distribution

There is uncertainty as to the actual number (one to three) of *Bowiea* species (Phillips 1951; Riley 1963; Dyer 1976; Dahlgren et al. 1985). The three described species of *Bowiea* include the common form *B. volubilis* Harv. ex Hook.f. (Dahlgren et al. 1985), *B. kilimandscharica* Mildbraed, and *B. gariepensis* v Jaarsveld (Van Jaarsveld 1983). There is debate as to whether the latter two are synonym and sub-species respectively (Bruyns and Vosa 1987). Dahlgren et al. (1985) indicated that *Bowiea* is pronouncedly peripheral in the Hyacinthaceae and may merit the rank of a separate family.

*Bowiea* is a geophyte with a depressed, globose bulb up to 15 cm in diameter, which consists of thickened fleshy scales, the outer layers becoming green when exposed to sunlight. There are one to a few leaves which are rudimentary, approximately 2–4 cm long, filiform, and which wither and fall early during growth. The "stem" (inflorescence) is annual in nature and leafless, scrambling, and copiously and intricately branched. The branches are reflexed and spreading, succulent, green in color, and function as the assimilatory organ. The

[1] NU Research Unit for Plant Growth and Development, Department of Botany, University of Natal, Pietermaritzburg 3200, Republic of South Africa

Biotechnology in Agriculture and Forestry, Vol. 28
Medicinal and Aromatic Plants VII (ed. by Y.P.S. Bajaj)

pedicels are curved, with those towards the base being mostly sterile. There are numerous green, diurnal flowers with six spreading segments that are fused at the base; these segments are persistent and are reflexed after fruit set. There are six stamens and the anthers are introrse. The ovary is conical and trigonous with numerous axile ovules; the style is terete with a clavate three-lobed stigma. Seeds are few, oblong, and flattened with shiny black testas, contained within a

**Fig. 1.** *Bowiea volubilis* Harv. ex Hook.f.

loculicidal ovoid capsule (Dyer 1941, 1976, Phillips 1951; Dahlgren et al. 1985; Fig. 1).

*Bowiea*'s distribution is fairly wide in southern Africa, with localities from Namibia through the Cape Province, Natal Transvaal and north into Zimbabwe, Zambia, Tanzania, and central Africa.

## 1.2 Medicinal Importance

The bulbs of *Bowiea* are highly sought after by the indigenous populations for use as a poison and a medicine, and for their magical properties (Table 1). *Bowiea volubilis* has long been used for treating a variety of ailments, and African herbalists have probably used it for centuries. Unfortunately, accidental (and intended) overdoses, and perhaps local differences in plant toxicity, have resulted in fatalities. The symptoms of overdose are vomiting, salivation, palpitations, and cramps followed by death within minutes or hours, depending on the amount ingested. Walker (1914) recorded cases of *Bowiea* poisoning with the postmortem results showing congestion and signs of irritation of the gastrointestinal tracts, stoppage of the heart in diastole, and congestion of the liver and other organs. Van der Riet (De Almeida 1930) was supposedly the first to extract a poisonous alkaloid from the bulb, and reported that it gave (with hydrochloric acid) a color reaction characteristic of veratrine. Walker (1914) isolated resinous bodies from the bulb. Lewis (Meltzer 1928) extracted a highly toxic impure alkaloid from the bulb. De Almeida (1930), after extracting and subjecting the extract to pharmacological experiments, suggested that the active principle was colchicine. A great deal of work has been conducted on the bulb, its extracts, and

**Table 1.** Herbal, medicinal, and magical uses of *Bowiea*

| Uses | Reference |
|---|---|
| Poison | Steyn (1934); Watt and Breyer-Brandwijk (1962); Palmer (1985); Gelfand et al. (1985) |
| Purgative, Use against dropsy, and cure for barrenness | Steyn (1934); Watt and Breyer-Brandwijk (1962); Palmer (1985) |
| Cure for sore eyes | Watt and Breyer-Brandwijk (1962); Riley (1963); Palmer (1985) |
| Remedy for ascites | Watt and Breyer-Brandwijk (1962) |
| To facilitate delivery, and to procure abortion | Gerstner (1939) |
| Cure for venereal disease, arrow poison | Gerstner (1939); Moll and Moll (1989) |
| Cure for skin diseases, acne pimples | Watt and Breyer-Brandwijk (1962); Riley (1963) |
| "Makes the enemy flee from them" | Watt and Breyer-Brandwijk (1962) |
| Love potion | Watt and Breyer-Brandwijk (1962); Palmer (1985) |
| Repels muggers | Palmer (1985) |
| Protects huts against lightning and hail | Riley (1963) |

its effects on a number of laboratory animals to determine the minimum lethal dose (Meltzer 1928; Table 2).

Jaretsky (Watt and Breyer-Brandwijk 1962) first suggested that cardiac glycosides were probably responsible for the action of the bulb, with the two glycosides being probably of the scillaren type. Research workers have confirmed the digitalis action of the *Bowiea* extract and reported that the glycoside of *Bowiea* is slightly more cumulative than digoxin, but not nearly as much as digitoxin (Watt and Breyer-Brandwijk 1962). Katz (1950, 1953a, b, 1954, 1957a, b) did extensive work on the chemistry of *Bowiea* (Fig. 2), from which he isolated a number of highly active cardiac glycosides of the scillaren type. A number of researchers subsequently studied the constituents of *Bowiea* (Table 3). Jaretsky and Scheermesser (1938) registered a German patent for a glycoside preparation for human cardiac use.

It has been reported that the cardiac glycoside action of the bulb is estimated to be 30 times more intense than that of *Digitalis*, and that of the flower, 60 times. It is evident that the range and toxicity of substances in *Bowiea* make it a valuable source of cardiac glycosides and a fascinating experimental plant.

**Table 2.** Toxicity of *Bowiea* extracts. (Watt and Breyer-Brandwijk 1962)

| Plant part | No. of fatal doses per gram |
|---|---|
| Flower | 110 000 |
| Bulb | 16 000 |
| Stem and leaves | 18 000 |
| Seed | Inactive |
| Other reports | |
| Plant/substance | |
| | 2 000 000 |
| | 1 600 000 |
| Bowieatoxine A | 1 000 000–1 500 000 |
| Bowieatoxine B | 3 600 000–4 500 000 |

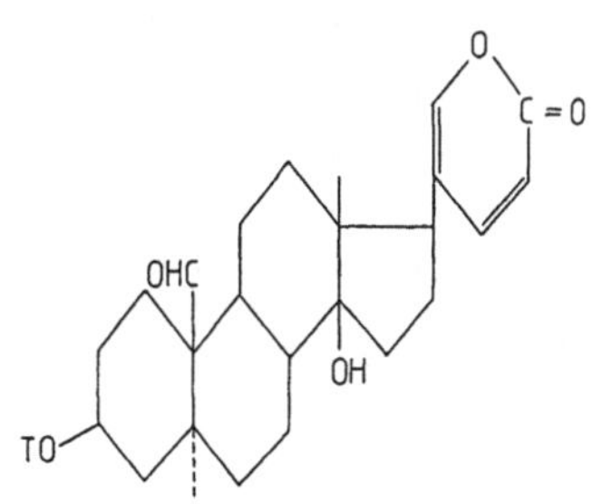


**Fig. 2.** Proposed structure of bovoside A α-L-thevetoside. *T* α-L-thevetosyl. (Katz 1953b)

**Table 3.** List of compounds isolated from *Bowiea*

| Compound | Chemical formula |
|---|---|
| Bovoside A | $C_{31}H_{44}O_9$ |
| Bovoside B | $C_{31}H_{44}O_{12}$ or $C_{31}H_{42}O_{12}$ |
| Bovoside C | $C_{31}H_{44}O_{11}$ or $C_{31}H_{42}O_{11}$ |
| Bovoside D | $C_{31}H_{44}O_{10}$ |
| Bovoside E | |
| *Bowiea* substance G | Aglycon of bovoside A |
| Bovogenin | Aglycon of bovoside E |
| Bovieatoxin A | |
| Bovieatoxin B | |
| Bowieatoxine A | |
| Bowieatoxine B | |
| Bowieatoxigenin | |
| Bovoruboside | $C_{31}H_{42}O_{11}$ |
| Bovocryptoside | $C_{31}H_{42-44}O_{11}$ |
| Bovoxanthotoxin | |
| Bovoerythrotoxin | |
| Bovocyanotoxin | |
| Bovocolatoxin | |
| Kilimandscharogenin A | $C_{24}H_{34}O_5$ |
| Kilimandscharogenin B | $C_{24}H_{34}O_9$ |
| Sinistrin | |
| Nabogenin (mixture comprising primarily scilliglaucosidin) | |
| Hellebrigenin | |

## 1.3 Collection and Conventional Propagation of *Bowiea*

Traditional means of *Bowiea* propagation include seed production and germination (with associated variability), the production and separating of offsets, and scaling (a slow propagation alternative).

*Bowiea* bulbs are collected from the Natal/KwaZulu area in South Africa and are sold in urban herbal medicine shops for high prices (approximately ten times more than most medicinal plants; Cunningham 1986). This is coupled with their scarcity in the Natal/KwaZulu area. The popularity of medicinal plants amongst rural and urban Zulu populations ensures that exploitation and future conservation concerns are likely to increase in direct proportion to population growth in these areas. (Wainwright et al. 1977; Ellis 1986; Cook et al. 1988). Law enforcement as the only management strategy has failed to control overutilization of indigenous medicinal plants, and illegal gathering takes place in state forests and conservation areas. An alternative management strategy is to provide a legal source of supply through cultivation to satisfy the demand of both the urban herb trader and the rural herbalist. In vitro culture provides an alternative means for mass production of these endangered plants (Cook et al. 1988).

## 2 In Vitro Culture Studies

### 2.1 Micropropagation

*Bowiea* tissue cultures have been successfully established (Jha and Sen 1985; Cook et al. 1988; Van Staden et al. 1991). Elmer-English (a coauthor in the last-mentioned paper) has done extensive work on explant source optimization, which is summarized here.

Explant sterilization varies depending on the explant (Table 4). The sterilization is normally preceded by a complex series of explant preparation steps. The bulbs were removed from the sterilized soil in which they were growing, all outer damaged areas of the bulb scales and roots were removed, and the bulb thoroughly washed in running water. The bulbs were cut into the desired orientation, and immersed in 70% ethanol (plus Tween 20 as a surfactant) for 5 min, and then subjected to a further immersion in 1% Benlate (50% A.I. benomyl) for 5 min. The bulbs were then subjected to a mercuric chloride soak, followed by standard sterile technique (Table 4). Cook et al. (1988) recommend the use of a preconditioning step for mother plant material, as suggested by Debergh and Maene (1981).

All the reports quote the use of MS (Murashige and Skoog 1962) as the standard basal medium. Jha and Sen (1985) supplemented the standard organic components with folic acid (0.4 mg/l), biotin (0.05 mg/l), riboflavin (0.05 mg/l), calcium pantothenate (1 mg/l), p-aminobenzoic acid (0.02 mg/l), casein hydrolysate (100 mg/l), and coconut milk (15% v/v). Agar was used at either 0.6 or 0.8%. The supplementation of the media with growth hormone was variable, with 2,4-dichlorophenoxyacetic acid (2,4-D) or naphthalene acetic acid (NAA) as the source of auxin, and benzyladenine (BA) as the source of cytokinin. (Table 5 and Fig. 3).

Environmental conditions for growth varied from 22–25 °C, a maximum relative humidity of 60%, and a light intensity of 3200 lx with a 16:8-h light: dark period (Jha and Sen 1985); 25 °C, a light intensity of 50 $\mu$mol m$^2$/s with an 18:6-h light:dark period (Cook et al. 1988); and 25 °C in continuous light at an intensity of 35 $\mu$mol m$^2$/s (Van Staden et al. 1991).

**Table 4.** Summary of the work done on in vitro culture of *Bowiea* (sterilization of *Bowiea* explants)

| Sterilant/ explant | Time of immersion (min) | Survival (%) | Reference |
|---|---|---|---|
| 0.1% $HgCl_2$ Inflorescence | 8–18 | | Jha and Sen (1985) |
| 0.1% $HgCl_2$ Bulb | 10 | 60.8 | Cook et al. (1988) |
| 0.1% $HgCl_2$ Bulb | 10 | 85 | Van Staden et al. (1991) |

**Table 5.** Percent production of callus (c) and shoots (s) on *Bowiea* bulb explants. (Cook et al. 1988)

| Time (Days) | NAA (mg/l) | BA (mg/l) 0 | 1 | 2 | 10 |
|---|---|---|---|---|---|
| 60 | 0 | 0 | 0 | 0 | 0 |
| | 1 | 0 | 50S | 0 | 0 |
| | 2 | 0 | 33C | 0 | 0 |
| | 10 | 11C | 11C | 25C | 42C |
| 120 | 0 | 0 | 10S | 20S | 0 |
| | 1 | 0 | 72S | 33S | 0 |
| | 2 | 0 | 53S | 19S | 0 |
| | 10 | 15C | 21C | 30C | 53C |

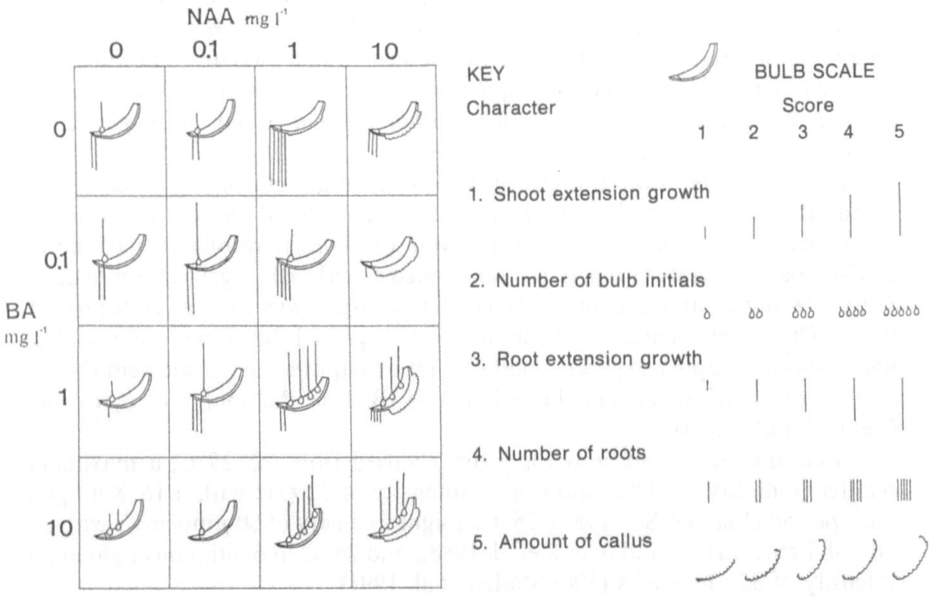


**Fig. 3.** Diagrammatic representation of morphogenesis on twin bulb scale explants of *B. volubilis* with 16 different combinations of NAA and BA

Segments of inflorescence on 1 mg/l 2,4-D and coconut milk proliferated within 1 week, and callus was formed within 6–8 weeks on 80% of the explants. On transfer to 0.5 mg/l 2,4-D and 100 mg/l casein hydrolysate, green nodular callus developed and rhizogenesis was observed. Reduction of 2,4-D concentrations (0.2 mg/l) resulted in shoot formation, and these primordia developed into complete bulbous plants on hormone-free media; these in vitro-produced bulbs were used as secondary explants (Jha and Sen 1985). Bulbs were either split in two or used as individual scales. On basal tissue in the presence of BA (2 and

**Fig. 4.** Shoot development from the base plate of twin scale explants of *B. volubilis* 21 days after being placed on a medium containing 1 mg/l BA only

5 mg/l) the basal tissue of the split bulbs produced daughter shoots. In the presence of BA (2 and 5 mg/l) and 2,4-D (0.05 mg/l), small white globular shoot primordia developed on the adaxial surface of both the outer and inner scales.

Van Staden et al. (1991) observed that the explants started greening after 7 days in culture. This was followed by the development of either shoots or roots from the base plate region of the bulb scale. The degree of morphogenesis is dependent on the plant growth regulator combination. If single bulb scale explants and pieces of bulb scale rings were used, adventitious buds developed over most of the adaxial surface (Figs. 4 and 5).

To maximize bulblet production, a system of shake culture was developed (Jha and Sen 1985). Using a hormone-free liquid MS medium shaken at 180 rpm, massive plantlet regeneration occurred. Growing neo-formed (5–8 shoots) bulblets (produced on induction medium) in agitated liquid culture resulted in the proliferation of more than 100 daughter bulblets, while in comparison, 10–15 clumps were produced in static liquid culture, and only 8–10 bulblets on solid medium.

To enhance rooting, clusters of 5–8 shoots were transferred from a shoot induction medium to hormone-free, half-strength MS medium (both solid and liquid). Under these conditions efficient rooting (80–90%) was obtained (Jha and Sen 1985).

For transfer to the soil, Jha and Sen (1985) maintained their plants for 4–6 weeks in half-strength macro- and microsalts of MS medium with 0.5% sucrose. Under aseptic conditions, these plantlets were transferred from 24 to 28–30 °C

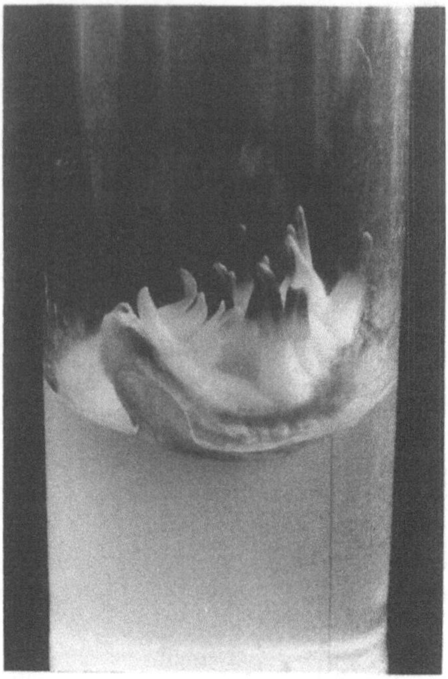

**Fig. 5.** Adventitious bud development on the adaxial surface of a *B. volubilis* bulb scale cultured on a medium containing 1 mg/l NAA and 10 mg/l BA

for a further 4 weeks. The plantlets were then washed in tap water and directly transferred to potting soil, with an 80% survival rate. Cook et al. (1988) transferred culture-produced plantlets directly into a sterile sand : peat (2 : 1) mixture and kept them under mist for 2 weeks, resulting in 90% survival. Over a further 2-week period the bulbs were "hardened off" by reducing the amount of watering. Bulbs with a diameter of 2–3 cm be produced within 1 year.

Using each scale produced in vitro for subsequent shake culture, 400–600 bulblets could be produced in 16–20 weeks (Jha and Sen 1985). The technique recommended by Cook et al. (1988) resulted in the formation of 52.5 $\pm$ 13.6 daughter bulblets from each bulb over a 3-month period. Elmer-English (unpubl.) reported that, using a system of "bulb scale rings", up to 30 adventitious buds could be formed per ring. Fewer buds were formed on "rings" taken from the inner scales than from the outer or middle scales. The recommended hormone combination consisted of 1 mg/l NAA and 10 mg/l BA (Fig. 6).

It is significant to note that the formation of buds was restricted to the adaxial surface (Figs. 3 and 4). Van Staden et al. (1991) conducted an anatomical investigation to determine the reasons for such a phenomenon. Bulb scales that had turned green had abundant chloroplasts on the abaxial side of the scale. More stomata were observed on the abaxial side of the scale, together with a much thicker cuticle. Small globular bodies were present only in the epidermal layers of the adaxial surface. These bodies did not stain for starch, and disappeared in the presence of alcohol. After 2 days in culture, it was possible to observe signs of mitotic division in the adaxial epidermal cells, and after 7 days, there was evidence of well-established adventitious bud meristemoids (Fig. 7).

**Fig. 6.** Adventitious bud formation on the adaxial surface of a 1/4 piece of a bulb scale ring of *B. volubilis* cultured on a medium containing 1 mg/l NAA and 10 mg/l BA

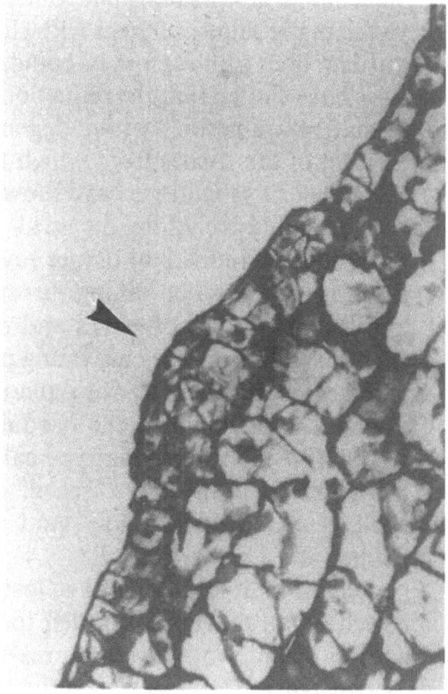

**Fig. 7.** Transverse section through the adaxial epidermis of an explant from a piece of a bulb scale ring of *B. volubilis* 7 days after being placed on a medium containing 1 mg/l NAA and 10 mg/l BA. *Arrowhead* indicates the initial protrusion of the bud on the epidermal surface

It remains unclear why only the adaxial epidermis of *Bowiea* bulb scales has the capacity to become meristematic. The presence of the observed differences on the two surfaces is unlikely to have any affect on meristematic activity, and mitotic activity apparently occurred at random despite the fact that all cells contained the same contents. Although it cannot be unequivocally determined whether the origin of the meristemoid is uni- or multicellular, circumstantial evidence points to the fact that cell division originates in a single epidermal cell and the meristemoid develops from this point.

### 2.2 In Vitro Production of Cardiac Glycosides

Cardioactive glycosides appear in several plant families which mostly are not related to each other; the majority of the research into cardiac glycosides revolves around the usage of *Digitalis* and its manipulation for enhanced secondary metabolite formation. A comprehensive review on the production of cardenolides and other secondary products in *Digitalis* spp. was published (Rücker 1988). *Digitalis* glycosides belong to the cardenolide type, and in their stereo ring system an unsaturated five-membered lactone ring is substituted at position 17. Initial studies on cardioactive glycosides in a number of *Digitalis* species showed that, although various cardenolides could be identified in the callus, the yield was very small in comparison with those obtained from intact plants. During the course of cultivation the cardenolides further decreased and disappeared completely (Rücker 1988).

Extensive comparisons between the cardenolide content in the callus and in the original tissue were undertaken with *D. purpurea* over a wide range of organ tissues, and in all cases this phenomena was verified. Root cardenolides could be detected in the callus; whereas with shoots, the calli from cotyledons were devoid of cardenolides although they could be identified in the original tissue. Other studies have shown that the reduction of individual cardenolides differs strongly over cultivation periods with a dependence upon environmental influences and the origin of the tissues from which the callus was derived.

Detailed investigations have shown that all green organs of the plant are rich in cardenolides including the seed coat (Lückner and Diettrich 1989). The investigation of individual tissues revealed that the mesophyll cells and similar types of chloroplast containing "green" cells contained most of the compounds.

*D. lanata* cultures of leaves and roots preserved their capacity for forming cardenolides. During the cultivating periods, the digoxin content in the root and in the leaf cultures rose with increasing age, and the cardenolide pattern changed. $GA_3$ stimulated the production of digoxin, whereas BA inhibited it (Rücker, 1988). Parenchyma-type callus cells derived from tissues treated with 2,4-D from *D. lanata* were free of cardenolides, but were, however, able to transform added cardenolides, and were shown to be stored in the vacuoles (Lückner and Diettrich 1989).

In callus cultures which have lost their capacity to synthesize cardenolides, renewed organ differentiation led to the renewed synthesis of cardenolides. In the callus of embryogenic structures of *D. lanata* and shoot differentiating callus of *D. pupurea*, the influence of cultivation conditions was studied. A considerable increase in cardenolide synthesis could be brought about by growth regulators, sugars, applied nitrogen, calcium, phosphate, and by the influence of light.

Research by Lückner and Diettrich (1989) on the formation of cardenolides in cultures of *D. lanata* was in summary thought to be regulated by two principles: at a given stage of a developmental program, the genes involved in cardenolide biosynthesis became competent for expression and a certain basal level of cardenolides is produced; and, by illumination of the competent structures, the genes become fully expressed.

There is the prospect that, although cell cultures may not produce the cardenolides of interest, they may be able to perform specific biotransformation reactions on substrates added to the medium. In numerous metabolic studies with *Digitalis* cells, distant precursors and intermediates of the cardenolide biosynthetic pathway have been used as substrates in an attempt to produce the pharmaceutically important compounds. These studies have shown only very limited success.

The data that have been collected in our laboratory is that the cardiac glycoside profiles of in vitro-produced plantlets are identical to those of in vivo-grown *B. volubilis* plants (Fig. 8). The difference in concentration of the compounds in both tissue-cultured and field-grown plants is not significant. *Bowiea* extracts do not contain colchicine. Due to the fact that bulbs are readily

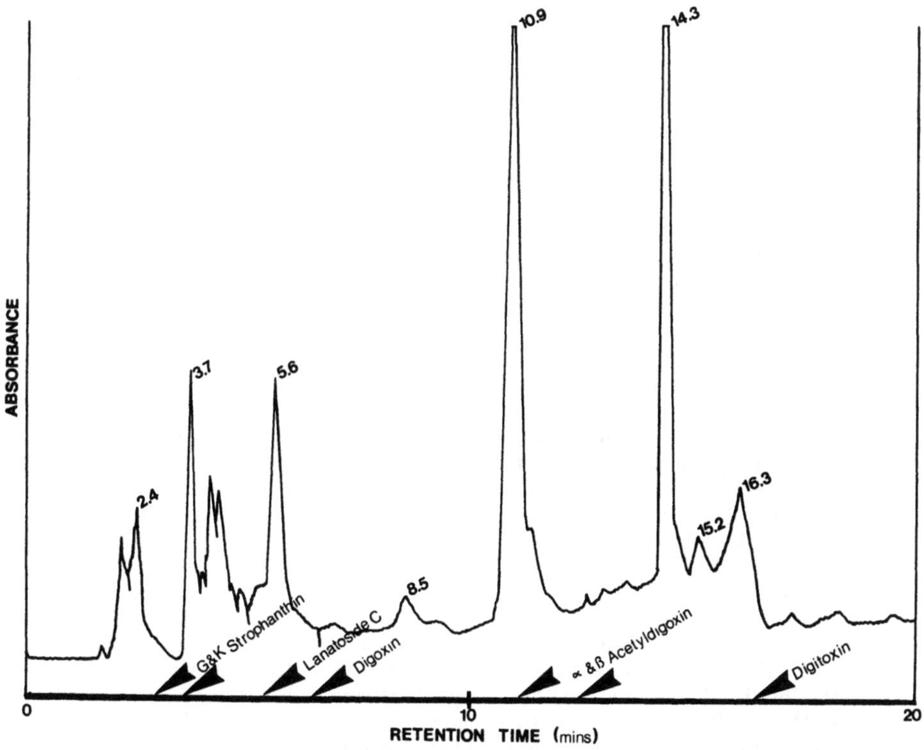


**Fig. 8.** Typical HPLC profile for a *B. volubilis* cardiac glycoside extract. (HPLC program after Kreis et al. 1986)

produced in culture, all the work has been aimed at characterizing the components in the fully formed plantlets. Current research is being performed on the isolation, quantification, and identification of the glycoside components in both in vitro-grown plantlets and callus-derived cultures.

## 3 Conclusion

In vitro culture of *Bowiea volubilis* can play an important role in the boosting of plant stocks from which larger numbers of bulbs could be produced using traditional propagation techniques (Cook et al. 1988). However, more importantly, these data indicate the potential for biotechnological manipulation of this plantlet not only from a conservation outlook, but also as a source for new and more effective compounds in the battle against heart disease.

## References

Bruyns PV, Vosa CG (1987) Taxonomic and cytological notes on *Bowiea* Hook. F. and allied genera (Liliaceae). Caryologia 40:287–298

Cook EL, Cunningham A, Van Staden J (1988) The tissue culture of an exploited Zulu medicinal plant *Bowiea volubilis*. S Afr J Bot 54:509–510

Cunningham AB (1986) An investigation of the herbal medicine trade in Natal/KwaZulu. Working Rep 8. Institute of Natural Resources, University of Natal, Pietermaritzburg

Dahlgren RMT, Clifford HT, Yeo PF (1985) The families of the monocotyledons: structure, evolution and taxonomy. Springer, Berlin Heidelberg New York

De Almeida AG (1930) On the plant poisons of the natives of Mozambique. S Afr J Sci 27:296–302

Debergh P, Maene L (1981) A scheme for commercial propagation of ornamental plants by tissue culture. Sci Hortic 14:335–345

Dyer RA (1941) *Bowiea volubilis*. Flowering Plants S Afr 21: Plate 815

Dyer RA (1976) The genera of Southern African flowering plants. Department of Agriculture and Technical Services, Pretoria

Ellis CG (1986) Medicinal plant use – a survey. Veld Flora 72:72–74

Gelfand M, Mavi S, Drummond RB, Ndemera B (1985) The traditional medical practitioner in Zimbabwe: his principles of practice and pharmacopoeia. Mambo Press, Gweru

Gerstner J (1939) A preliminary checklist of Zulu names of plants. Bantu Stud 13:49–64

Jaretsky R, Scheermesser F (1938) Glucosides from *Bowiea volubilis*. Chem Abstr 32:3555

Jha S, Sen S (1985) Regeneration and rapid multiplication of *Bowiea volubilis* Harv in tissue culture. Plant Cell Rep 4:12–14

Katz A (1950) Über die Glykoside von *Bowiea volubilis* Harvey 1. Helv Chim Acta 33:1420–1428

Katz A (1953a) Über die Glykoside von *Bowiea volubilis* Harvey 2. Helv Chim Acta 36:1344–1353

Katz A (1953b) Über die Glykoside von *Bowiea volubilis* Harvey 3. Helv Chim Acta 36:1417–1423

Katz A (1954) Über die Glykoside von *Bowiea volubilis* Harvey 4. Helv Chim Acta 37:832–836

Katz A (1957a) Über die Glykoside von *Bowiea volubilis* Harvey 9. Helv chim Acta 40:487–491

Katz A (1957b) Über die Glykoside von *Bowiea volubilis* Harvey 10. Helv Chim Acta 40:831–846

Kreis W, May U, Reinhard E (1986) UDP-glucose: digitoxin 16'-O-glucosyltransferase from suspension-cultured *Digitalis lanata* cells. Plant Cell Rep 5:442–445

Lückner M, Diettrich B (1989) The integration of cardenolide biosynthesis in the differentiation

program "somatic embryogenesis" of *Digitalis lanata*. In: Kurz WGW (ed) Primary and secondary metabolism of plant cell cultures. Springer, Berlin Heidelberg New York, pp 117–123
Meltzer E (1928) Preliminary observations on "*Bowiea volubilis*". J Med Assoc S Afr 2:6
Moll E, Moll G (1989) Poisonous plants. Struik, Cape Town
Murashige T, Skoog F (1962) A revised medium for rapid growth and bioassay with tobacco tissue cultures. Physiol Plant 15:473–497
Palmer E (1985) The South African herbal. Tafelberg, Cape Town
Phillips EP (1951) The genera of South African flowering plants. Cape Times, Cape Town
Riley HP (1963) Families of flowering plants of Southern Africa. University of Kentucky Press, Lexington
Rücker W (1988) *Digitalis* spp.: In vitro culture, regeneration, and the production of cardenolides and other secondary products. In: Bajaj YPS (ed) Biotechnology in agriculture and forestry Vol 4. Medicinal and aromatic plants I. Springer, Berlin Heidelberg New York, pp 388–418
Steyn DG (1934) The toxicology of plants in South Africa, together with a consideration of poisonous foodstuffs and fungi. Central News Agency, Johannesburg
Van Jaarsveld EJ (1983) *Bowiea gariepensis* new species: a new *Bowiea* species (Liliaceae) from the northwest Cape (South Africa). J S Afr Bot 49:343–346
Van Staden J, Elmer-English CW, Finnie JF (1991) Physiological and anatomical aspects related to in vitro shoot initiation in *Bowiea volubilis*. S Afr J Bot 57:352–355
Wainwright J, Schonland MM, Candy HA (1977) Toxicity of *Callilepis laureola*. S Afr Med J 52:313–315
Walker PH (1914) *Bowiea volubilis* and "Mlambo". S Afr Med Rec 12:141–142
Watt JM, Breyer-Brandwijk MG (1962) The medicinal and poisonous plants of Southern and Eastern Africa. Livingstone, Edinburgh

# VII *Camptotheca acuminata* Decne: In Vitro Culture and the Production of Camptothecin

A. J. van Hengel[1], R. M. Buitelaar[2], and H. J. Wichers[2]

## 1 Introduction

### 1.1 *Camptotheca acuminata* Decne

*Camptotheca acuminata* belongs to the Nyssaceae family. This is a small family of trees and shrubs, including the ornamental handkerchief and tupelo trees, and it is represented by three genera, *Nyssa, Davidia*, and *Camptotheca. Camptotheca* is represented by a single species, *Camptotheca acuminata* (Fig. 1), which grows only in east Tibet and West China. The plant is cultivated as an ornamental, but the greatest commercial value lies in its production of camptothecin. This alkaloid is also reported to be found in *Nothapodytes foetida* (Wight) Sleumer and *Ophiorrhiza mungos* L. (Hutchinson et al. 1979). *Camptotheca acuminata* also contains alkaloids that are related to camptothecin, such as 10-hydroxycamptothecin, 18-hydroxycamptothecin, 10-methoxycamptothecin and 9-methoxycamptothecin (Erdelmeier et al. 1986; Poehland et al. 1989).

### 1.2 Importance of Camptothecin

The strongly cytotoxic quinoline alkaloid camptothecin was first isolated by Wall et al. (1966). Its structure was elucidated as part of a screening program for antitumor agents. In China, camptothecin is widely used in the treatment of cancer. Camptothecin inhibits the growth of a wide range of tumors including murine leukemia L1210, Walker 256 rat carcinosarcoma L5178Y, K1964 and P388 leukemia, Novikoff hepatoma, mastcell P815 sarcoma, and reticulum cell sarcoma (A-RCS) (Loh and Ahmed 1990). Brief phase I clinical trials in the early 1970s, however, failed because of excessive toxicity. Renewed interest in camptothecin as a potential clinical antitumor drug has come from the identification of its molecular target and the elucidation of its mechanism of action. Camptothecin is a strong inhibitor of both DNA and RNA synthesis. This inhibition is caused by the interaction of camptothecin with its intracellular target

[1] Department of Molecular Biology, Agricultural University, Dreijenlaan 3, 6703 HA Wageningen, The Netherlands
[2] Agrotechnological Research Institute ATO-DLO, P.O. Box 17, 6700 AA Wageningen, The Netherlands

Biotechnology in Agriculture and Forestry, Vol. 28
Medicinal and Aromatic Plants VII (ed. by Y.P.S. Bajaj)

**Fig. 1.** *Camptotheca acuminata* Decne. (Graf 1992)

mammalian DNA topoisomerase I (Hsiang and Liu 1988). Topoisomerase I breaks one DNA strand, passes the unbroken strand through the break, and reseals the break as a part of the replication process. Camptothecin is the only known topoisomerase I inhibitor. It acts in the form of a reversible drug-enzyme-DNA complex. It stops the replication forks and thus produces DNA single-strand breaks and DNA-protein cross-links (Covey et al. 1989; Hsiang et al. 1989). The A-ring of camptothecin is suggested to bind to the enzyme-DNA complex at specific DNA sites (Jaxel et al. 1989; Herzberg et al. 1990). Phase II studies have been conducted to evaluate the antitumor effect and toxicity of CPT-11, a camptothecin derivate, on leukemia and lymphoma. The results indicate that this new topoisomerase I inhibitor shows considerable antitumor effects on refractory leukemia and lymphoma with acceptable side effects (Ohno et al. 1990).

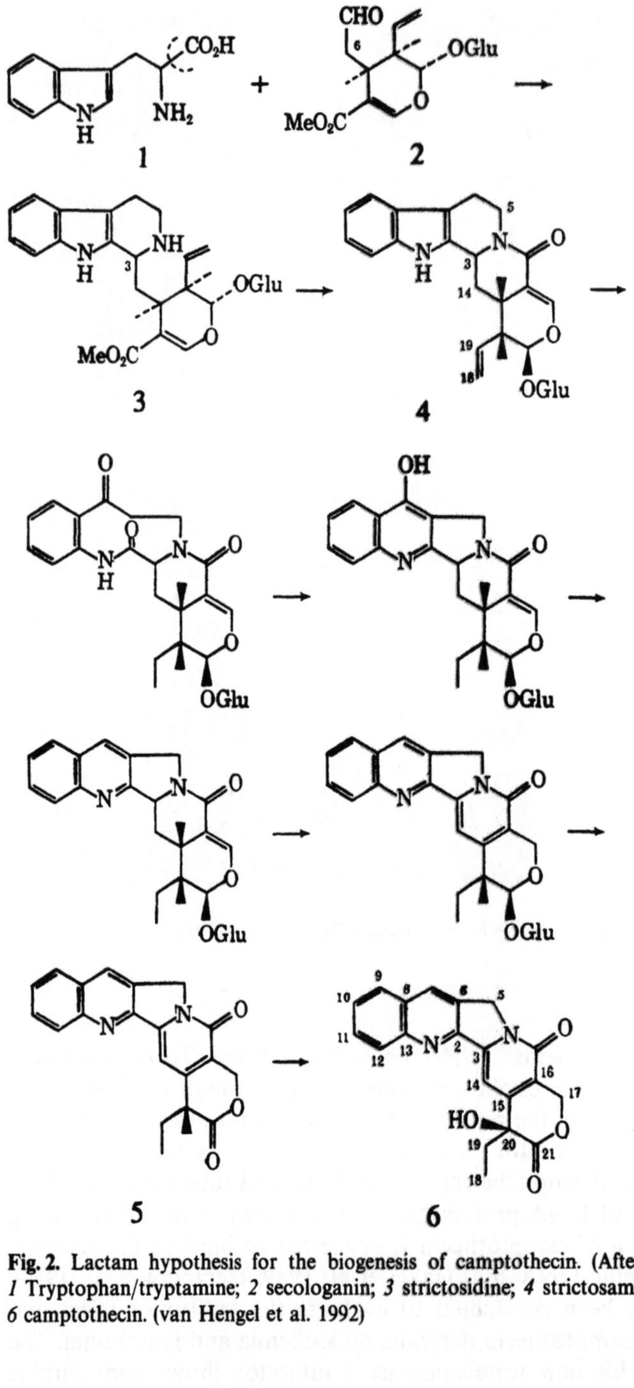


**Fig. 2.** Lactam hypothesis for the biogenesis of camptothecin. (After Hutchinson et al. 1974). *1* Tryptophan/tryptamine; *2* secologanin; *3* strictosidine; *4* strictosamide; *5* desoxycamptothecin; *6* camptothecin. (van Hengel et al. 1992)

Camptothecin has also been reported to be a naturally occurring plant growth regulator (Buta and Worley 1976). Camptothecin treatment inhibits sprouting and reduces weight loss of potato tubers (Wang et al. 1980a). The proposed mechanism of action is a camptothecin-induced xylem disruption. Storage of radishes can also be improved by means of camptothecin solutions. Root growth and top sprouting as well as weight loss of partially trimmed radishes was reduced after treatment with camptothecin (Wang et al. 1980b).

### 1.3 Biosynthesis of Camptothecin

Camptothecin, being a quinoline alkaloid, is biogenetically linked to the group of indole alkaloids (Sheriha and Rapoport 1976). Precursors for the biosynthesis of camptothecin are secologanine and tryptamine, which is derived from tryptophan (Fig. 2). Secologanine is a secondary monoterpene metabolite. A formal Pictet-Spengler condensation of tryptamine and secologanine yields strictosidine. Strictosidine has also been reported to be the penultimate precursor of the highly investigated complex *Catharanthus* alkaloids occurring in *Catharanthus roseus* G. Don. (Scott 1976). By an intramolecular cyclization in strictosidine, strictosamide is formed. The latter compound, a neutral glucoside, is the penultimate precursor of camptothecin (Heckendorf and Hutchnison 1977). The transformation of strictosamide into camptothecin is possible via three basic transformations. An oxidation of ring B and C results in a keto-lactam, followed by a recyclization of ring B and C. Ring D undergoes an oxidation. Ring E loses its glucose group and is also oxidated to form camptothecin.

This biogenetic hypothesis has been proposed independently by Cordell (1974) and Hutchinson et al. (1979). The proof for this route of biosynthesis was delivered by NMR and labeling experiments (Sheriha and Rapoport 1976). After feeding of the radioactive precursors tryptophan, tryptamide, and mevalonate to *Camptotheca acuminata* plants, label was incorporated into camptothecin (respectively 1.9, 0.02, and 0.2% of the produced camptothecin contained incorporated label). Feeding of the labeled precursors took place by uptake by roots, or by injection in the stems. Injection resulted in a higher incorporation of label, because uptake by roots results in a longer transport route and root metabolism possibly degrades the labeled precursors. NMR spectroscopic analyses have shown that the first steps in the synthesis of camptothecin are highly stereospecific. For example, the enzyme strictosidine synthetase forms only the C-3 S epimere of strictosidine. Later in the synthesis there seems to be almost no stereospecificity.

## 2 In Vitro Culture Studies

Sakato and Misawa (1974) studied the growth of *Camptotheca acuminata* cell cultures, whereas our group (van Hengel et al. 1992) studied, besides growth, also camptothecin production by several cell lines of *Camptotheca acuminata.*

Japanese patents (from which only the abstracts were accessible) report claims on the propagation of *C. acuminata* and *Nothapodytes foetida* by means of adventitious bud culture. Shoots were induced on a medium containing auxins and cytokinins. The shoots were then transferred to a medium suitable for the formation of roots. This propagation is used for the production of camptothecin in mature plants (Kanebo 1987, 1991).

## 2.1 Establishment of Callus and Cell Suspension Cultures

For the induction of callus, surface-sterilized stem parts were placed on agar with different nutrient media. Basic media were MS or B5 salt medium supplemented with B5 vitamins. The optimum levels of growth regulators such as kinetin and NAA or 2,4-D were examined by combining a range of concentrations with MS basal medium. As a result, callus was induced excellently on the medium containing 0.5 mg/l 2,4-D and 0.1 mg/l kinetin.

All calli which were introduced in liquid MS and B5 media formed homogeneous and undifferentiated cell suspension cultures. The diameter of the aggregates was usually smaller than 3 mm. The color of the cell suspension cultures could change from gray/green into red towards the end of the growth phase. The initial pH of the media was pH 6.5. During culturing, the pH showed a minimum at pH 5–5.5 before the exponential growth (Fig. 3). Once the

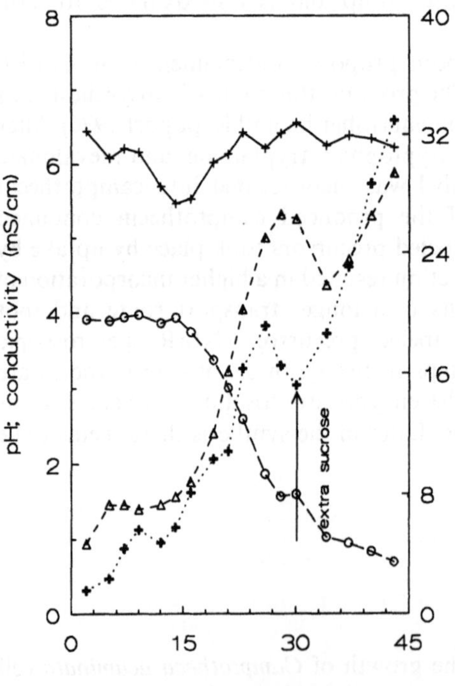


**Fig. 3.** Growth curve of *Camptotheca acuminata* cell suspension culture CAS-3 in a medium containing B5 salts and vitamins, supplemented with 40 g/l sucrose and 4 mg/l NAA. – + –, pH; – – △ – –, PCV (%); – – ○ – –, conductance (mS/cm); .. + .., CDW (g/l). (van Hengel et al. 1992)

exponential growth had started, the pH rose until 6.5 at the start of the stationary phase. In the media used, and extra addition of sucrose during the stationary phase could restore growth in cell lines. Four well-growing cell lines, initiated on three different media were used for further experiments. These cell lines were designated CAS1-4 (Table 1).

### *2.1.1 Effect of Growth Regulators on Callus and Cell Suspension Cultures*

The effect of addition of kinetin and the auxins NAA, 2,4-D, or IAA to basal MS medium on the production of biomass expressed as cell dry weight (CDW) per ml of cell suspension culture was investigated. Matrix experiments with combinations of kinetin and one auxin, both ranging in concentration from 0.1 to 10 mg/l were carried out (Table 2). Callus was induced excellently in the medium containing 0.5 mg/l 2,4-D and 0.1 mg/l kinetin.

**Table 1.** Callus initiation on different media. (van Hengel et al. 1992)

| Medium | Salts | Vitamins | Plant growth regulator (mg/l) | Carbon source (%) | Callus | Cell line code |
|---|---|---|---|---|---|---|
| A | MS | B5 | 2,4-D 0.4 | Sucrose 3 | + | CAS-3<br>CAS-4 |
| B | MS | – | NAA 3 | Sucrose 3 | – | – |
| C | MS | B5 | – | – | – | – |
| D | B5 | B5 | Coconut milk 2 | Glucose 1 fructose 1 | – | – |
| E | B5 | B5 | NAA 4 | Sucrose 4 | + | CAS-1 |
| F | B5 | B5 | 2,4-D 1 kinetin 0.5 | Sucrose 4 | + | CAS-2 |

Successful initiation after 3 weeks = +; failure = –

**Table 2.** Effect of growth regulators on the growth of callus. (Sakato and Misawa 1974)

| Auxin | (mg/l) | Kinetin (mg/l) 0.1 | 1 |
|---|---|---|---|
| NAA | 0.1 | + + | – |
| | 1 | + | ± |
| | 10 | – | – |
| 2,4-D | 0.1 | – | – |
| | 1 | + | + |
| | 10 | ± | ± |

Excellent growth after 40 days is indicated by + +, good growth by +, poor growth by ±, no growth by –

**Table 3.** Effect of gibberelin on the growth of cell suspension cultures. (Sakato and Misawa 1974)

| $GA_3$ (mg/l) | CFW (mg/ml) | CDW (mg/ml) |
|---|---|---|
| 0 | 92.4 | 10.0 |
| 0.05 | 146.8 | 13.7 |
| 0.10 | 147.3 | 12.8 |
| 1 | 153.9 | 12.5 |
| 5 | 136.9 | 12.4 |

CFW = cell fresh weight, CDW = cell dry weight. The medium used here was MS salts with addition of 30 g/l sucrose, 0.1 mg/l 2,4-D and 3 mg/l kinetin

The most vigorous growth of cell suspension cultures obtained in these ranges was in a medium containing 0.1 mg/l 2,4-D and 3 mg/l kinetin. In the latter medium the effect of gibberellin $A_3$ ($GA_3$) on the growth was investigated (Table 3). Concentrations ranging from 0 to 5 mg/l showed a remarkable stimulation of growth by $GA_3$, especially at 0.05 mg/l.

### *2.1.2 Effect of Carbon and Amino Acid nutrients on Cell Suspension Cultures*

On media supplemented with 3% sucrose, more biomass is produced than on media supplemented with other concentrations or carbon sources like glucose, fructose, or soluble starch.

The possible precursors of alkaloids, L-tryptophan, L-tyrosine, L-phenylalanine, L-histidine, and anthranilic acid were added to the media to investigate their effect on the growth of cell suspension cultures. Among these amino acids, addition of 0.115 mM of L-tryptophan and L-phenylalanine showed the most remarkable growth-promoting effect (Sakato and Misawa 1974). This reported growth-promoting effect is only relative, because growth seems to be equal to the reported growth in control media without additional amino acids. Although L-tryptophan is a precursor of camptothecin (Sheriha and Rapoport 1976), after the addition of this amino acid to the medium an increase of alkaloids in the cells could not be detected by means of TLC.

### *2.1.3 Effect of Physical Conditions on the Growth of Cell Suspension Cultures*

Sakato and Misawa (1974) also investigated the effect of initial pH of the culture medium on growth. Good growth was observed between pH 4 and 7.5, whereas the highest growth level was obtained at pH 4.3 (Fig. 4). The effects of light intensity and wavelength, however, need further investigation, as a large inter-experimental variation occurred that hampered interpretation of the results. The comparison of the produced biomass at two different shaking rates (148 and 184 rpm) also results in data with a similarly large variation. The effect of two

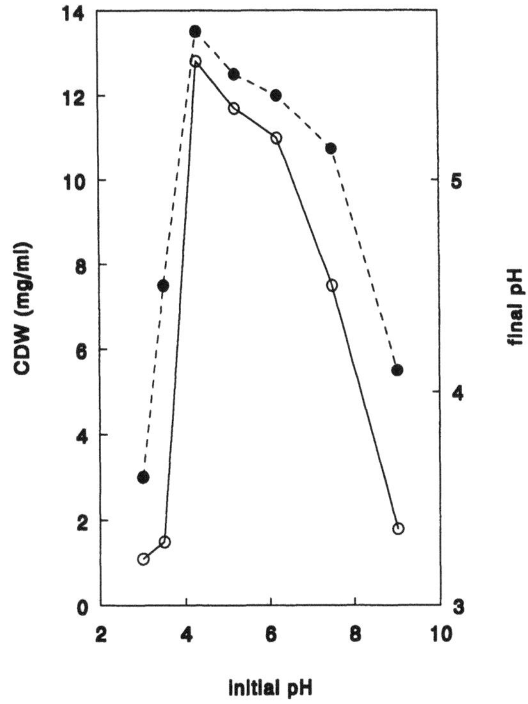


**Fig. 4.** Effect of initial pH of culture medium on growth of cell suspensions. The medium used was MS salts with addition of 30 g/l sucrose, 0.1 mg/l 2,4-D and 3 mg/l kinetin. *Open circles* represent the data on CDW and *filled circles* those on final pH of the medium. (Sakato and Misawa 1974)

different temperatures on growth was investigated. Cultures grown at a temperature of 30 °C showed better growth than cultures grown at 26 °C.

## 2.2 Production of Camptothecin

### *2.2.1 Extraction and Analysis of Camptothecin*

Extraction and analysis of camptothecin have been the subject of investigation by a number of investigators. Wall et al. (1966) reported a hot hexane-heptane extraction of dried plant material followed by an extraction with 95% ethanol. The concentrate of the ethanol extraction is partitioned between water and chloroform. Silica gel chromatography of the methanol-insoluble material from the chloroform extract, followed by crystallization from methanol-acetonitril, gave camptothecin as pale yellow needles. An isolation of camptothecin by means of gel permeation chromatography is described by Buta and Novak (1978). Overpressure layer chromatography is used by Erdelmeier et al. (1986) for the separation of *Camptotheca* alkaloids. Sakato et al. (1974) report an isolation of camptothecin from cells of *C. acuminata* suspension cultures using several ethanol and chloroform extractions. Reversed-phase high-performance liquid chromatography has proven to be very useful for detection of camptothecin (Poehland et al. 1989; Loh and Ahmed 1990). A simple extraction procedure

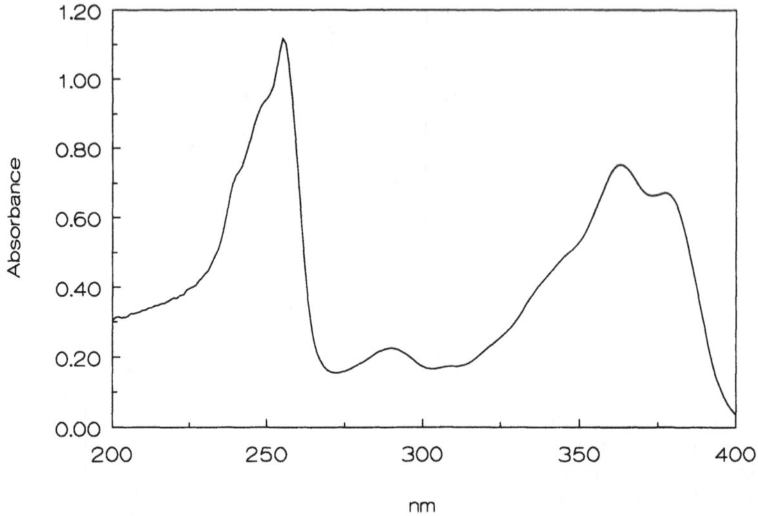


**Fig. 5.** UV-VIS absorption spectrum of camptothecin in chloroform between 200 and 400 nm. (van Hengel et al. 1992)

for HPLC analysis was developed in our laboratory (van Hengel et al. 1992). An amount of freeze-dried cells was ground in a mortar and sonificated in methanol. Water and dichloromethane were added for extraction. The dichloromethane phase, which was proved to contain the camptothecin, was recovered and evaporated to dryness. The remaining residue was dissolved in chloroform. This method is substantially faster than that of Poehland et al. (1989), and gives good results. For determination of camptothecin, its absorption spectrum was measured between 200 and 400 nm (Fig. 5). Three peaks of absorption were identified, at 256, 290, and 363 nm. The absorption maximum was found at 256 nm. This wavelength was chosen for the detection and quantification of camptothecin in the HPLC analyses. The running solvent used for TLC was a mixture of $CHCl_3$–EtOH–EtOAc-hexane (20.0:9.5:65.2:5.3). TLC analyses of extracts of plant material and camptothecin stock solutions both gave blue/violet spots ($R_f$-value 0.67), which were visible under UV light of 254 and 366 nm wavelength.

### *2.2.2 Identification of Camptothecin*

Several methods were used for the identification of camptothecin. NMR spectroscopy and X-ray analyses were first carried out by Wall et al. (1966) for the elucidation of the structure of this alkaloid. $^{13}C$ and $^{1}H$ NMR spectroscopy have been useful for the identification of camptothecin and its intermediates in biosynthesis studies (Hutchinson et al. 1979). Another suitable way of identification is mass spectrometry (Sakato et al. 1974; van Hengel et al. 1992).

In our laboratory the putative camptothecin peaks were collected after fractionation by HPLC. These samples were evaporated to dryness and brought into the mass spectrometer. The isolated compound was identified as camptothecin (MW = 348) based on the fragmentation pattern of the reference camptothecin solutions [m/z; relative intensities 348 (100%), 319 (33%), 248 (47%), 219 (44%), 109 (43%)].

### *2.2.3 Production by Cell Suspension Cultures*

The plant material was extracted and anâlyzed by TLC and HPLC for camptothecin content. The stem parts from which callus and cell suspension cultures were derived contained an amount of camptothecin, equivalent to 0.59% of the CDW of the plant material. The highest camptothecin productions by several cell lines in different media are shown in Table 4. Multiplication of CDW and the camptothecin content as a percentage of the CDW in Table 4 does not always give the maximum camptothecin content in mg/l, because the production maximum did not always occur at the same time as the maximum CDW. No camptothecin was detected in the media of cell suspension cultures. This indicates that camptothecin is accumulated intracellularly. The maximum camptothecin content in the cell suspension cultures was found to be 0.004% of the CDW. The production in cell suspension cultures is about 100 times less

**Table 4.** Highest camptothecin production, final cell density, and growth rate by several *C. acuminata* cell lines in different media

| Cell line | Medium | Final cell density CDW (g/l) | PCV (%) | Growth rate ($day^{-1}$) | Camptothecin content $10^{-3}$% of CDW | mg/l |
|---|---|---|---|---|---|---|
| CAS-1 | B5; NAA | 18.1 | 31.5 | 0.178 | 3.1 | 0.89 |
| | MS; NAA | 19.6 | 27.8 | 0.109 | 2.6 | 0.67 |
| | B5; 2,4-D | 19.8 | 36.9 | 0.206 | 0.85 | 0.090 |
| | MS; 2,4-D | 19.8 | 26.9 | 0.206 | 1.3 | 0.39 |
| CAS-2 | B5; NAA; K | 19.1 | 33.9 | 0.217 | 0.92 | 0.21 |
| | MS; NAA; K | 20.8 | 24.5 | 0.115 | 2.6 | 0.69 |
| | B5; 2,4-D; K | 17.5 | 31.3 | 0.230 | nd | nd |
| CAS-3 | B5; NAA | 19.1 | 26.6 | 0.103 | 2.1 | 0.49 |
| | MS; NAA | 17.3 | 21.4 | 0.087 | 3.6 | 0.98 |
| | B5; 2,4-D | 19.5 | 31.0 | 0.119 | 2.7 | 0.52 |
| | MS; 2,4-D | 20.9 | 26.9 | 0.106 | 1.8 | 0.53 |
| CAS-4 | B5; NAA | 21.0 | 21.8 | 0.069 | 1.9 | 0.35 |
| | MS; NAA | 7.4 | 11.0 | 0.053 | 2.4 | 0.24 |
| | B5; 2,4-D | 19.7 | 25.2 | 0.071 | 1.2 | 0.32 |
| | MS; 2,4-D | 20.0 | 27.0 | 0.085 | 1.5 | 0.29 |
| Plant stem | | | | | 590 | |

MS = MS salts and B5 vitamins; B5 = B5 salts and vitamins; NAA = 4 mg/l NAA; 2,4-D = 0.4 mg/l 2,4-D; K = 0.5 mg/l kinetin; CDW = cell dry weight; PCV = packed cell volume; nd = not detected.

than the production in the original plant material. As shown in Table 4, kinetin did not stimulate the production of camptothecin. However, quantitative and qualitative changes of the plant growth regulators might not only improve growth as described by Sakato and Misawa (1974), but also the production of camptothecin. In the media used, addition of NAA gave the highest production

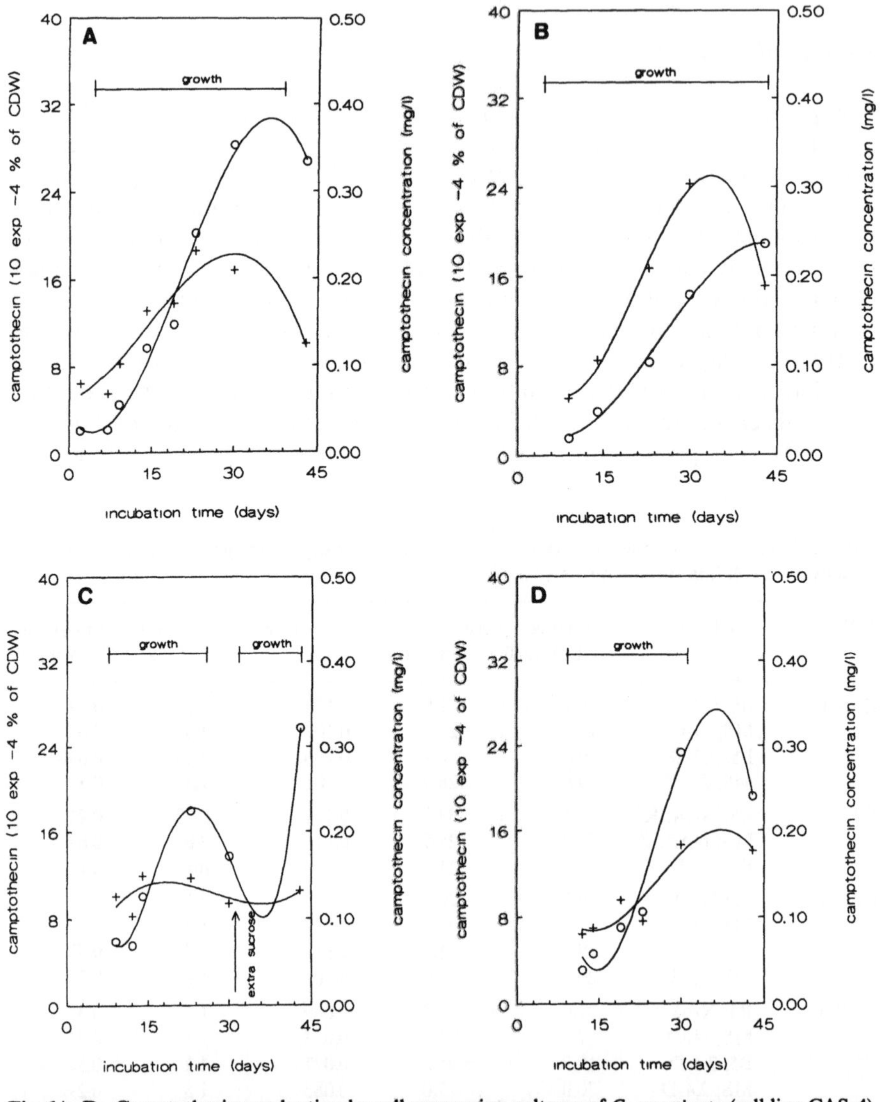


**Fig. 6A–D.** Camptothecin production by cell suspension cultures of *C. acuminata* (cell line CAS-4), grown in Erlenmeyer shake flasks, as a percentage of CDW ( + ) and as concentration in the total culture broth (O). The cell suspension cultures were grown in different media. **A** B5 salts and vitamins, 40 g/l sucrose and 4 mg/l NAA. **B** MS salts and B5 vitamins, 40 g/l sucrose and 4 mg/l NAA. **C** B5 salts and vitamins, 40 g/l sucrose and 0.4 mg/l 2,4-D. **D** MS salts and B5 vitamins, 40 g/l sucrose and 0.4 mg/l 2,4-D. (van Hengel et al. 1992)

of camptothecin, and MS salt media supplemented with B5 vitamins had an additional effect as well. The camptothecin production as percentage of the CDW in CAS-3 and CAS-4 are similar. This is not surprising, because both suspension cell cultures are derived from callus that was obtained on the same solid medium. Figure 6 shows the camptothecin production by the CAS-4 cell suspension culture in the different media tested, as mentioned in Table 4. The camptothecin content varied with time and was dependent on the growth medium used. The production of camptothecin is positively coupled to the growth of the cells. This is demonstrated by the strong correlation between camptothecin production and the increasing CDW and packed cell volume (PCV). *Camptotheca acuminata* cell suspension cultures are apparently able to degrade the camptothecin produced when growth stops and CDW declines (Fig. 6). Figure 6C shows that an extra addition of sucrose at day 30 not only leads to further growth, but also to an increase in camptothecin production. In Fig. 6B, the camptothecin production expressed in mg/l culture liquid is relatively poor, because of the slow growth rate of this cell suspension culture.

## 3 Summary

Callus and cell suspension cultures generated from *Camptotheca acuminata* stem parts were subcultured on different media. The cell suspension cultures acquired in this way were able to grow in media with different salts and growth regulators than the original callus initiation media. In these different media, differences in growth could be observed, the original media causing fastest growth. When cell growth stopped in liquid media, it could be restored by addition of extra sucrose. Qualitative and quantitative changes of plant growth regulators in the growth media showed differences in growth of callus and cell suspension cultures; however, the effect of those changes on the production of camptothecin is still unclear.

Camptothecin was detected in different parts of the original plant as well as in callus and cell suspension cultures. The camptothecin contents of callus and cell suspension cultures were approximately 100 times less than the camptothecin content in the stem parts from which they were derived. Cell suspension cultures of roots or leaves could produce more camptothecin, because roots are reported to contain 2.5 times more camptothecin than shoots (Kanebo 1991) and leaves contain 2 times more camptothecin than stems (van Hengel unpubl.).

No camptothecin was detected in the media of cell suspension cultures. This indicates that camptothecin is accumulated intracellularly. The camptothecin production in the cell suspension cultures was dependent on the media used. The original callus initiation medium does not necessarily cause the best camptothecin production, whereas growth seemed to be fastest on these original media. The camptothecin production is positively correlated to the growth period. When the CDW and PCV increase, the intracellular camptothecin content as well as the camptothecin production per liter rises. The cell suspension cultures break down camptothecin when growth stops because of sucrose limitation. The feeding of tryptophan, a precursor of camptothecin, did not increase the amount of alkaloids in the cell suspension cultures. A possible way of improving the camptothecin production is genetic transformation. A 3–22 times greater activity of strictosidine-synthase, a key enzyme in the camptothecin biosynthesis, was observed in transgenic tobacco plants (McNight et al. 1991). Overexpression of strictosidine synthase may allow manipulation of the alkaloid pathway for altering the abundance and spectrum of alkaloids produced in transgenic plants. Genetic transformation can also be used to obtain hairy root cultures. A Japanese patent reports a transformation method for *Camptotheca acuminata* tissue cultures using antibiotics to select transformed tissue (Mitsui-Toatsu-Chem 1989). Hairy root cultures and crown gall cultures can be obtained in this way. The production of camptothecin can be increased by in vitro propagation and cell suspension cultures.

# 4 Protocol

## 4.1 Culture Conditions

1. Surface sterilize stem parts of *Camptotheca acuminata* by immersing them for 4 min in a sodium hypochlorite solution (1.75 g/l). Then rinse the material five times with sterilized water, once quickly and four times for 10 min.
2. Place surface sterilized stem parts on agar plates. These plates should contain plant growth regulators such as auxins and kinetin and be solidified with 0.8% agar. A medium containing 0.5 mg/l 2,4-D and 0.1 mg/l kinetin supplemented with 3% sucrose shows excellent callus induction. The initial pH of the medium should be between pH 4 and 7.5.
3. Allow callus to initiate by culturing under a day/night regime (14/10 h) at 30 °C.
4. After a period of 6 weeks of callus growth, cell suspension cultures can be initiated by transferring the callus to the same medium without agar. Incubate the cell suspension cultures on a rotary shaker (150 rpm) under the same growing conditions.
5. Subculture the cell suspension cultures by adding 50 ml of an approximately 3-week-old cell suspension culture to 200 ml of fresh medium in a 500-ml Erlenmeyer flask.

## 4.2 Camptothecin Extraction Procedure

1. Harvest the cells of full-grown cultures by sieving.
2. Freeze-dry the cells and ground them in a mortar.
3. Add 2 ml methanol and sonificate (2 × 30 s, 20 microns peak to peak).
4. Mix vigorously for 5 min on a magnetic stirrer after addition of 18 ml water and 20 ml dichloromethane.
5. Centrifugate for 10 min at 2000 rpm, recover the dichloromethane phase, and evaporate it to dryness.
6. The remaining residue contains camptothecin and can be used for analyses or further purification.

## References

Buta JG, Novak MJ (1978) Isolation of camptothecin and 10-methoxycamptothecin from *Camptotheca acuminata* by gel permeation chromatography. Ind Eng Chem Prod Res Dev 17:160–161

Buta JG, Worley JF (1976) Camptothecin, a selective plant growth regulator. J Agric Food Chem 24:1085–1086

Cordell GA (1974) The biosynthesis of indole alkaloids. Lloydia 37:219–298

Covey JM, Jaxel C, Kohn KW, Pommier Y (1989) Protein-linked DNA strand breaks induced in mammalian cells by camptothecin, an inhibitor of topoisomerase I. Cancer Res 49:5016–5022

Erdelmeier CAJ, Erdelmeier I, Kinghorn AD, Farnsworth NR (1986) Use of Overpressure layer chromatography (OPLC) for the separation of natural products with antineoplastic activity. J Nat Prod 49:1133–1137

Graf AB (ed) (1992) Hortica color cyclopedia of garden flora and indoor plants, 1st edn. Roehrs, East Rutherford, NJ

Heckendorf AH, Hutchinson CR (1977) Biosynthesis of camptothecin. II. Confirmation that isovincoside, not vincoside, is the penultimate biosynthetic precursor of indole alkaloids. Tetrahedron Lett 48:4153–4154

Herzberg RP, Busby RW, Caranfa MJ, Holden KG, Johnson RK, Hecht SM, Kingsbury WD (1990) Irreversible trapping of the DNA-topoisomerase I covalent complex. J Biol Chem 265:19287–19295

Hsiang YH, Liu LF (1988) Identification of mammalian DNA topoisomerase I as an intracellular target of the anticancer drug camptothecin. Cancer Res 48:1722–1726

Hsiang YH, Lihou MG, Liu LF (1989) Arrest of replication forks by drug-stabilized topoisomerase I-DNA cleavable complexes as a mechanism of cell killing by camptothecin. Cancer Res 49:5077–5082

Hutchinson CR, Heckendorf AH, Daddona PE, Hagaman E, Wenkert E (1974) Biosynthesis of camptothecin. I. Definition of the overall pathway assisted by carbon-13 nuclear magnetic resonance analysis. J Am Chem Soc 96:5609–5611

Hutchinson CR, Heckendorf AH, Straughn JL, Daddona PE, Cane DE (1979) Biosynthesis of camptothecin. 3. Definition of strictosamide as the penultimate biosynthetic precursor assisted by $^{13}C$ and $^{2}H$ NMR spectroscopy. J Am Chem Soc 101:3358–3369

Jaxel C, Kohn KW, Wani MC, Wall ME, Pommier Y (1989) Structure-activity study of the actions of camptothecin derivates on mammalian topoisomerase I: Evidence for a specific receptor site and a relation to antitumor activity. Cancer Res 49:1465–1469

Kanebo (1987) Production of cytostatic substance. Japanese patent 62096088

Kanebo (1991) Propagation of plant and the production of camptothecin from the mature plant. Japanese patent 03272628

Loh JP, Ahmed AE (1990) Determination of camptothecin in biological fluids using reversed-phase high-performance liquid chromatography with fluorescence detection. J Chromatogr 530:367–376

McNight TD, Bergey DR, Burnett RJ, Nessler CL (1991) Expression of enzymatically active and correctly targeted strictosidine-synthase in transgenic tobacco plants. Planta 185:148–152

Mitsui-Toatsu-Chem (1989) Transformation of plant by controlled addition of antibiotic to plant tissue culture with bacterium. Japanese patent 01191688

Ohno R, Okada K, Masaoka T, Kuramoto A, Arima T, Yoshida Y, Ariyoshi H, Ichimaru M, Sakai Y, Oguro M, Ito Y, Morishima Y, Yokomaku S, Ota K (1990) An early phase II study of CPT-11: a new derivative of camptothecin, for the treatment of leukaemia and lymphoma. J Clin Oncol 8:1907–1912

Poehland BL, Troupe N, Carté BK, Westley JW (1989) Reversed-phase high-performance liquid chromatographic assay for camptothecin and related alkaloids. J Chromatogr 481:421–427

Sakato K, Misawa M (1974) Effects of chemical and physical conditions on growth of *Camptotheca acuminata* cell cultures. Agric Biol Chem 38:491–497

Sakato K, Tanaka H, Mukai M, Misawa M (1974) Isolation and identification of camptothecin from cells of *Camptotheca acuminata* suspension cultures. Agric Biol Chem 38:217–218

Scott AI (1976) Biogenetic type synthesis. In: Jones JB, Sih CJ, Perlman D (eds) Applications of biochemical systems in organic chemistry, part 2. Wiley, New York, pp 555–626

Sheriha GM, Rapoport H (1976) Biosynthesis of *Camptotheca acuminata* alkaloids. Phytochemistry 15:505–508

van Hengel AJ, Harkes MP, Wichers HJ, Hesselink PGM, Buitelaar RM (1992) Characterization of callus formation and camptothecin production by cell lines of *Camptotheca acuminata.* Plant Cell Tissue Organ Cult 28:11–18

Wall ME, Wani MC, Cooke CE, Palmer KH, McPhail AT, Sim GA (1966) Plant antitumor agents. I. The isolation and structure of camptothecin, a novel alkaloidal leukaemia and tumor inhibitor from *Camptotheca acuminata.* J Am Chem Soc 88:3888–3890

Wang CY, Buta JG, Moline HE, Hruschka HW (1980a) Potato sprout inhibition by camptothecin, a naturally occurring plant growth regulator. J Am Soc Hortic Sci 105:120–124

Wang CY, Buta JG, Hruschka HW (1980b) Effect of camptothecin on the storage quality of radishes. HortScience 15:72–73

# VIII *Chrysanthellum americanum*: Micropropagation and Flavonoid Production

TH. GASPAR[1]

## 1 Introduction

*Chrysanthellum americanum* (CA) is the name generally used to designate *Chrysanthellum indicum* DC subsp. *afroamericanum* B.L. Turner (Turner 1982). CA is a wild Asteraceae of tropical Africa and America. It is not cultivated, although the interest it arouses is increasing because of its numerous therapeutic properties, the most important being hepatoprotection and regulation of the lipoproteic metabolism (Lièvre et al. 1984; Lièvre 1985; Larive 1987; Guillot et al. 1988; Requéna 1989). Components that were isolated among the active constituents (Becchi et al. 1979, 1980; Honoré-Thorez 1985) are two saponins (chrysantellins a and b) and five flavonoids : a flavone : luteolin 7-O-glucoside, two flavanones : eriodictyol 7-O-glucoside and isookanin 7-O-glucoside or flavonomarein, a chalcone : okanin 4′-O-glucoside or marein, and an aurone : maritimetin 6-O-glucoside or maritimein (Fig. 1). This flavonoid association constitutes an original feature of CA that can account for its pharmacological properties.

In this chapter, the following aspects of *Chrysanthellum americanum* are discussed. (1) How to achieve a micropropagation method of this plant in order to obtain large-scale and regular production in temperate areas (Saive et al. 1988), (2) The analysis of the flavonoids produced in vitro with view to a direct use of this material for therapeutic purposes (Saive et al. 1988), (3) As it has been shown that three flavonoids behave as hydrogen donors for peroxidase in the presence of hydrogen peroxide (Darimont et al. 1989), there may be a relationship between this characteristic and the known antiradicalar activity of these flavonoids (Brasseur et al. 1987).

## 2 Material and Methods

### 2.1 Plant Material and Culture

Achenes of CA were isolated from dried whole plants provided by Dr Honoré-Thorez, Arlon, Belgium. They were surface sterilized by 20 min soaking in 5%

[1] Université de Liège, Institut de Botanique B 22, Sart Tilman, 4000 Liège, Belgium

Biotechnology in Agriculture and Forestry, Vol. 28
Medicinal and Aromatic Plants VII (ed. by Y.P.S. Bajaj)

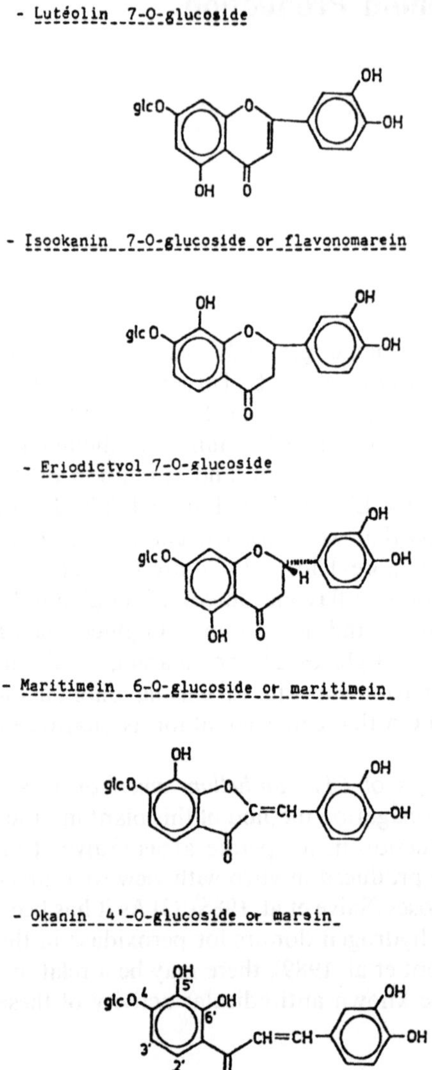


**Fig. 1.** Structure of some flavonoids of *Chrysanthellum americanum*

sodium hypochlorite supplemented by a few drops of Teepol, then washed three times for 10 min in sterile water. They were sown on Murashige and Skoog (1962) medium (MS) containing 0.8% Difco Bacto agar and 30 g/l sucrose. Primary explants consisting of 0.5 to 1.5 cm upper parts (stem cut just below cotyledons) of 15-day-old seedlings were cultivated on MS supplemented with BAP or IPA (1 mg/l) and NAA or IBA (0.1 mg/l), and adjusted to pH 5.8. Alternatively, the culture was initiated from node segments taken from greenhouse-cultivated plants, disinfected like achenes and cultivated on MS + BAP (1 mg/l) and NAA (0.1 mg/l). The growth regulators were added before autoclaving (20 min, 120°C). The cultures were performed in 500-ml cylindrical jars

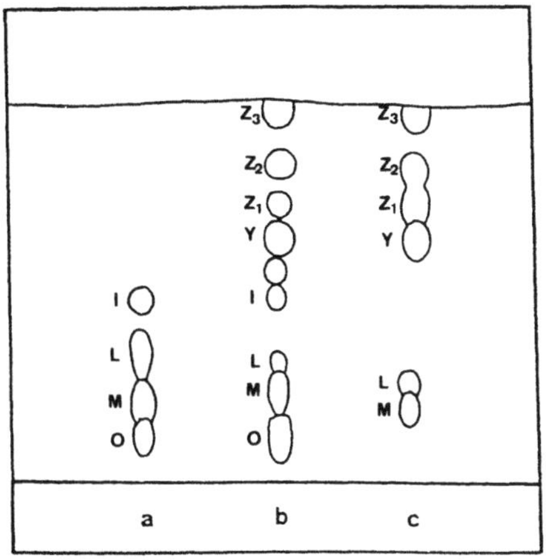


**Fig. 2.** Thin layer chromatograms of: **a** reference flavonoids, **b** extract of greenhouse-cultivated plants and **c** extract of in vitro-cultivated plants. *I* Isookanin 7-O-glu; *L* luteolin 7-O-glu; *M* maritimein; *O* = okanin 4′-O-glu

containing 100 ml medium and 12 explants, covered by a sheet of transparent polyethylene (Reynolds Films Inc.).

The cultures were maintained under 1450 lx and 16 h of fluorescent light (Sylvania Grolux). Night and day temperatures were respectively 23 and 17.5 °C. Rooted plantlets were transferred from in vitro conditions to compost (mold aerated with perlite) and grown in a shade-house at 17 °C under 16 h at 10000 lx.

## 2.2 Flavonoid Analysis

Plants were extracted with boiling water for 20 min. Flavonoids were then extracted with ethylacetate from the filtered water solution. A global analysis of the flavonoids was achieved by means of thin layer chromatography on cellulose (Fig. 2) (Polygram Cel 400 0.1 mm from Macherey & Nagel, Düren, FRG). The solvent used was tert-butanol, 3; acetic acid, 1; water, 1 (TBA). UV at 336 nm and aminoethanol diphenylborate (1% in methanol + 5% polyethyleneglycol 400) were used for visualizing the flavonoids. Further identification of the flavonoids was achieved by means of their absorption spectra after purification by two dimensional paper chromatography (Fig. 3) (Whatman 3 MM, 46 × 57 cm, solvents: TBA, then acetic acid, 1; water, 1) and spots elution by 2 ml of methanol.

## 2.3 Flavonoids as Hydrogen Donors for Peroxidase

The flavonoids marein, maritimein and flavonomarein were obtained from the society Extrasynthèse (France). Aqueous and hydroalcoholic extracts of *Chrysanthellum* were prepared by the society Iphym (France). Incubation mixtures (50 ml) at laboratory temperature ( ±20 °C) contained phosphate buffer (pH 5.8; I 0.2), $H_2O_2$ ($8.8 \times 10^{-5}$ M), horseradish peroxidase (HRP from Boehringer, 4.4

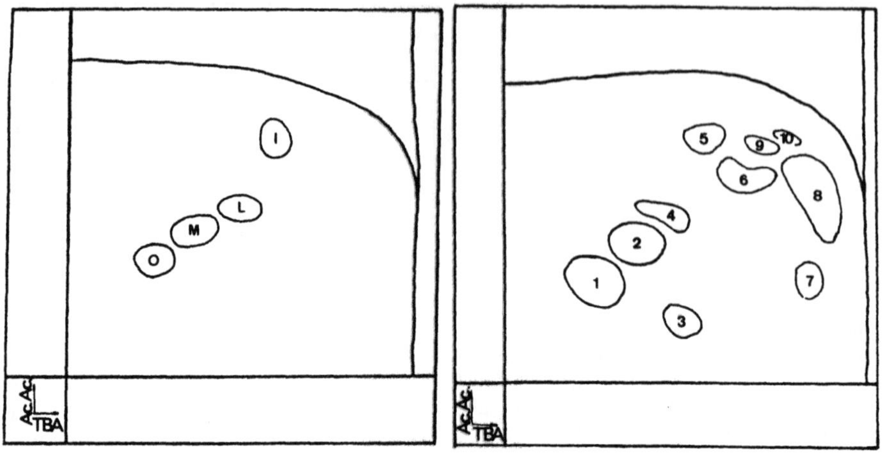


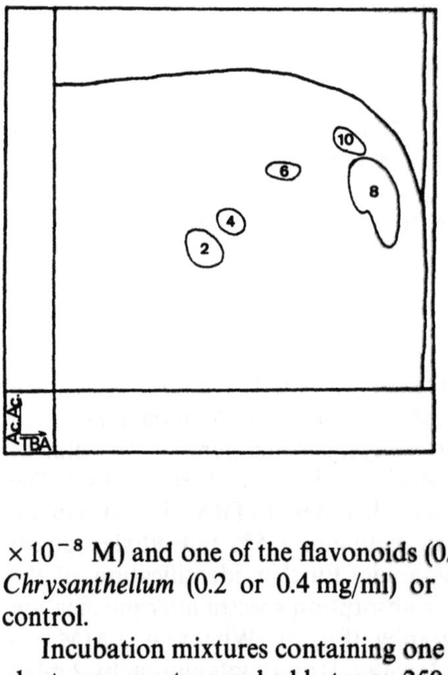


**Fig. 3.** Two-dimensional paper chromatograms of *1* reference flavonoids, *2* extract of greenhouse-cultivated plants and *3* extract of in vitro-cultivated plants. *I* Isookanin 7-O-glu; *L* luteolin 7-O-glu; *M* maritimein, *O* = okanin 4′-O-glu

$\times 10^{-8}$ M) and one of the flavonoids (0.8, 2 or $4 \times 10^{-5}$ M) or one extract from *Chrysanthellum* (0.2 or 0.4 mg/ml) or uric acid (2, 4, 6, or $12 \times 10^{-5}$ M) as control.

Incubation mixtures containing one of the flavonoids or an extract from the plant were spectrographed between 250 and 550 nm, after increasing incubation times using a Pye Unicam SP8-100 spectrophotometer. After different incubation times (between 0 and 48 min), 5 ml of the incubation mixtures were mixed with 1 ml of a reagent made of an aqueous guaiacol solution (1%) and HRP ($10^{-5}$ M). The spectrophotometric estimation (through the amount of oxidized guaiacol) of the remaining $H_2O_2$ in the incubation mixtures was made after 30 s at 420 nm. $H_2O_2$ concentration was calculated from the equation:

$$[H_2O_2] = O.D \times 8.35 \times 10^{-5}\ M.$$

The initial speed of the enzymatic reaction between HRP and the flavonoid was calculated from the curve tracing the diminution of the $H_2O_2$ concentration in function of time.

# 3 Results and Discussions

## 3.1 Micropropagation

In a first series of assays it was shown that seedling shoot tips and stem nodes from greenhouse plants as primary explants developed an average number of six side branches from axillary buds in about 1 month on MS medium without regulators. These side axes could serve as secondary explants. Axillary budding was considerably increased in the presence of a cytokinin (BAP or IPA, 1 mg/l) and an auxin (IBA, 0.1 mg/l); up to 21 apices per explant were obtained in 45 days of culture.

The shoots were easily rooted on MS medium (minerals diluted four times) using IBA or NAA (1 mg/l) as a rooting auxin, with or without a cytokinin (BAP or IPA, 0.1 mg/l). They continued their development into tufty plants on this medium, where they flowered. The greatest number of capitula, however, was obtained on MS diluted ten times in the absence of any regulators.

Acclimatation in the greenhouse needed high relative humidity (under glass) for the first 10 days. Preliminary rooting in vitro was not a prerequisite.

The rapid vegetative multiplication of CA could thus be achieved in vitro through axillary shoot development, as it was for other plants of the Asteraceae family (Wambugu and Rangan 1981; Moncousin and Gaspar 1983; Winand et al. 1986; Bajaj et al. 1992). Rooting and acclimatation posed no problem. Microplants kept either on the multiplication or the rooting medium rapidly flowered normally.

## 3.2 Flavonoid Analysis

TLC of extracts from 3-month-old (from achenes) greenhouse-cultivated plants (Fig. 2b) revealed all the flavonoids previously found (Fig. 2a), but also two additional flavonoids not yet mentioned for this plant (X and Y in Fig. 2a; Z spots are not flavonoids). The flavonoids synthesized in vitro (2-month-old rooted and flowered plants) were less numerous (Fig. 2c).

The absorption spectra obtained after two-dimensional paper chromatography (Fig. 3), were as follows (maxima, nm):

Spot 1 (381, 318, 265) = okanin 4′-O-glucoside (Mabry et al. 1980).

Spot 2 (416, 329, 274, 244) = maritimetin 6-O-glucoside (present in vitro) (Shimokoriyama 1956).

Spot 3 (416, 340, 265) probably an aurone (Markham 1982) but faint spot.

Spot 4 (248, 266, 255) = luteolin 7-O-glucoside (present in vitro).

Spot 5 (322, 283) = isookanin 7-O-glucoside (Shimokoriyama 1956).

Spot 6 composed of two flavonoids (327, 269 and 335, 268) = probably two flavones (Markham 1982).

Spot 7 (276, 390) = very faint spot and spectrum difficult to interpret.

Spot 9 (318, 277) = a flavonone (Markham 1982), probably eriodictuol 7-O-glucoside.

Spots 8 and 10 contained no flavonoids.

The CA flavonoid composition as known from the literature was recovered from greenhouse-cultivated plants. The present work, however, showed evidence for the presence of two other underterminated flavonoids. Plants cultivated in vitro did not produce all the flavonoids found in greenhouse plants. It has not been shown whether this is due to the environmental origin, i.e., the culture medium, or more simply to the physiological state of the plants. These plants, however, had developed flowers, as had those in the greenhouse. Such differences in secondary metabolites pattern between in vitro plants and wild or greenhouse ones have also been found in *Artemisia alba* (Lamproye et al. 1986).

### 3.3 Flavonoid Peroxidation by HRP

Preliminary spectrophotometric analyses had shown that the spectra of the three flavonoids examined remained unchanged (at least for 1 h) in the presence of $H_2O_2$ alone (different concentrations) or in the presence of boiled HRP. Such spectra were progressively modified in the presence of both fresh $H_2O_2$ and HRP, as illustrated in Fig. 4 for marein.

The spectra of Fig. 4 seem to indicate that the initial flavonoid underwent a first transformation before being converted to a tautomeric form, as shown by the isosbestic point. This HRP peroxidation of the flavonoids was confirmed by $H_2O_2$ consumption in function of time; it depended on the amount of flavonoid in the incubation mixture, as shown by Fig. 5.

Spectra of aqueous and hydroalcoholic extracts of *Chrysanthellum* underwent similar spectral changes in the same conditions, and $H_2O_2$ also was

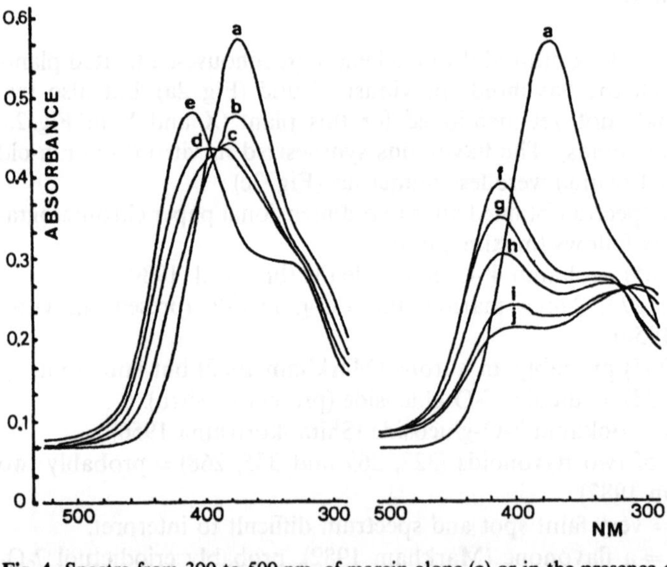


**Fig. 4.** Spectra from 300 to 500 nm, of marein alone (**a**) or in the presence of $H_2O_2$ and HRP (see Sect. 2) after 1 (*b*), 2 (*c*), 3 (*d*), 6 (*e*), 9 (*f*), 11 (*g*), 13 (*h*), 22 (*i*) and 60 (*j*) min

consumed. The initial speed of $H_2O_2$ consumption in the presence of HRP for different concentrations of the flavonoids, uric acid, and three extracts of *Chrysanthellum* is shown in Fig. 6. These results allow classification as follows for the affinity towards peroxidase: marein > maritimein > flavonomarein > uric acid. Extract no. 17 of *Chrysanthellum* was twice as active as samples nos. 20 and 26.

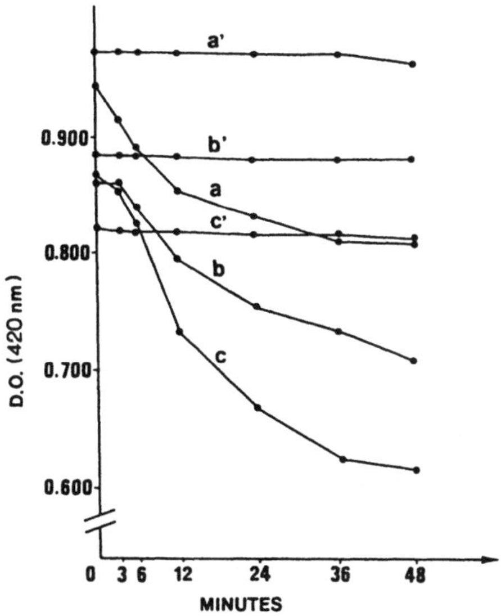


**Fig. 5.** $H_2O_2$ titration (expressed by optical density DO at 420 nm) in incubation mixtures containing marein (0.2 mg/10 ml in *a, a'*; 0.5 mg/10 ml in *b, b'*; 1 mg/10 ml in *c, c'*) in the presence of $H_2O_2$, with (*a, b, c*) or without (*a', b', c'*) HRP, after increasing incubation times (min)

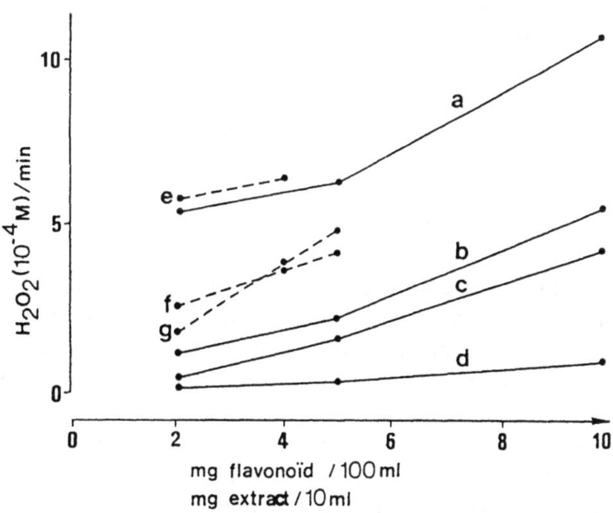


**Fig. 6.** Variation of the initial speed of $H_2O_2$ consumption in the presence of HRP and flavonoids (*a* marein; *b* maritimein; *c* flavonomarein; *d* uric acid) or extracts from *Chrysanthellum indicum* (*e, f, g*: extracts n$^{os}$ 17, 20, 26), in relation to their concentration

Flavonoids of *Chrysanthellum indicum* thus behave as hydrogen donors for peroxidase in the presence of $H_2O_2$ and oxygen. Other flavonoids had already been shown to be substrates for peroxidase (Gaspar et al. 1982), particularly the chalcone isoliquiritigenin (Rathmell and Bendall 1972). That the flavonoids behave as hydrogen donors for peroxidase does not justify their antiradicalar properties, although, in the absence of $H_2O_2$, peroxidase may use radicals to oxidize hydrogen donors (Gaspar et al. 1982; Lobarzewski et al. 1991). Peroxidase is also able to peroxidize lipids (Niinomi et al. 1987). Therefore some antilipoperoxidizing properties of *Chrysanthellum* flavonoids might be explained through a competitive inhibition of peroxidase, using or not free radicals. The previously shown antiradicalar properties of flavonoids were demonstrated on another independent basis (Brasseur et al. 1987). It must, however, be mentioned that other peroxidasic substrates such as tiron (4,5-dihydroxy-1,3-benzene disulfonic acid) are generally able to chelate radicals such as $O_2^-$ or OH (Kahn 1989). It is also interesting to note that the here established classification of flavonoids based on their affinity towards peroxidase corresponds to the classification proposed by Brasseur et al. (1987) on the basis of their antiradicalar properties.

## 4 Summary and Conclusions

An in vitro shoot proliferating cluster of *Chrysanthellum americanum* was initiated from shoot tips and nodes on MS medium. Optimal propagation by development of axillary buds was obtained using MS with BAP or IPA as a cytokinin (1 mg/l) and NAA or IBA as an auxin (0.1 ml/l). Under these conditions, average multiplication was about 14/month. Rooting was achieved on MS with mineral salts diluted four times in the presence of an auxin (1 mg/l IBA or NAA) with or without a cytokinin (0.1 mg/l BAP or IPA).

The flavonoid composition known from the literature for *C. americanum* was recovered from greenhouse-cultivated plants. The analysis showed the presence of two additional flavonoids. In vitro-cultivated plants contained a smaller number of flavonoids although they apparently were in the same physiological state (flowering) as greenhouse plants.

Marein, maritimein, and flavonomarein, three flavonoids from *C. americanum*, behave as hydrogen donors for the purified commercial horseradish peroxidase, in the presence of hydrogen peroxide. Their relative $H_2O_2$ consumption, i.e., their relative affinity towards peroxidase, has been compared with this of nonpurified extracts of *Chrysanthellum*. The relationship between these characteristics and the known antiradicalar activity of the flavonoids is discussed. Classification of the flavonoids on the basis of their affinity towards peroxidase corresponds to the previously established classification based on antiradicalar activity.

The medicinal plant *Chrysanthellum americanum* thus can be mass propagated by in vitro axillary branching and further rooting of the shoots. Such micropropagated material, at the flowering stage in vitro, is relatively rich in flavonoids, although somehow qualitatively different from greenhouse plants. Such flavonoids behave as hydrogen donors for peroxidase, a property which may contribute to their known antiradicalar activity.

*Acknowledgments.* Support and receipt of *Chrysanthellum* extracts from Dr. Guillot (Iphym) is gratefully acknowledged, as well as the help of Dr. Honoré-Thorez for the bibliography.

## References

Bajaj YPS, Sidhu MMS, Gill APS (1992) Micropropagation of *Chrysanthemum*. In: Bajaj YPS (ed) Biotechnology in agriculture and forestry, vol 20. High-Tech and micropropagation IV. Springer, Berlin Heidelberg New York, pp 69–80

Becchi M, Bruneteau M, Trouilloud M, Combier H, Sartre J, Michel G (1979) Structure of a new saponin: chrysanthelline A from *Chrysanthellum* P. Eur J Biochem 102:11–20

Becchi M, Bruneteau M, Trouilloud M, Combier H, Pontanier H, Michel G (1980) Structure de la chrysanthelline. Eur J Biochem 105:271–277

Brasseur Th, Angenot L, Pincemail J, Deby C (1987) Action antiradicalaire de flavonoïdes et d'extraits de *Chrysanthellum indicum*. Plant Méd Phytothér 21:131–137

Darimont E, Hofinger M, Guillot B, Gaspar Th (1989) Flavonoïdes de *Chrysanthellum indicum*: substrats peroxydasiques. J Pharm Belg 44:325–328

Dubernard PM (1988) Etude de l'effet du *Chrysanthellum americanum* sur les lithiases rénales. Phytotherapy 24:19–20

Gaspar Th, Penel C, Thorpe T, Greppin H (1982) Peroxidases 1970–1980. A survey of their biochemical and physiological roles in higher plants. Univ Genève – Centre Botan Publ, Genève, 324 pp

Guillot B, Despages G, Delorge F (1988) Stress et drainage en phytothérapie. Apport du *Chrysanthellum americanum*. In: Bertrand JP (ed) Les entretiens de Monaco sur les thérapeutiques du stress. Rocher, Monaco, pp 129–139

Honoré-Thorez D (1985) Description, identification et usages thérapeutiques de *Chrysanthellum americanum*: *Chrysanthellum indicum* DC Subsq *afroamericanum* BL Turner. J Pharm Belg 40:323–331

Kahn V (1989) Tiron as a substrate for horseradish peroxidase Phytochemistry 28:41–44

Lamproye A, Hofinger M, Ramaut JL, Gaspar Th (1986) *Artemisia alba*: micropropagation et huile essentielle. Meded Fac Landbouwwet Rijksuniv Gent 51/2a:523–526

Larive G (1987) De l'hypolipidémie à la protection vasculaire. Apport moderne du *Chrysanthellum americanum* Vatke. Plant Méd 25:6–10

Lièvre H (1985) Défense et illustration du drainage. Cah Biothér 87:12–16

Lièvre H, Guillot B, Raymond E (1984) *Chrysanthellum*, un hépatotrope normolipémiant et vasculotrope. J Jeune Prat 7:1–8

Lobarzewski J, Greppin H, Penel C, Gaspar Th (1991) Biochemical, molecular, and physiological aspects of plant peroxidases. Univ Genève Publ, Genève, 550 pp

Mabry T, Markham KR, Thomas M (1980) The systematic identification of flavonoids. Springer, Berlin Heidelberg New York

Markham KR (1982) Techniques of flavonoid identification. Academic Press, London

Moncousin C, Gaspar Th (1983) Peroxidase as a marker for rooting improvement of *Cynara scolymus* L. cultured in vitro. Biochem Physiol Pflanz 178:263–271

Murashige T, Skoog F (1962) A revised medium for rapid growth and bioassays with tobacco tissue cultures. Physiol Plant 15:473–497

Niinomi A, Morimoto M, Shimizu M (1987) Lipid peroxidation by the (peroxidase/$H_2O_2$/phenolic) system. Plant Cell Physiol 28:731–734

Rathmell WG, Bendall DS (1972) The peroxidase-catalysed oxidation of a chalcone and its possible physiological significance. Biochem J 127:125–132
Requéna Y (1989) Apport médical de *Chrysanthellum americanum* en gastroentérologie. Cah Biothér 103 (Suppl):95–101
Saive J, Gaspar Th, Ramaut JL, Hofinger M (1988) *Chrysanthellum americanum*: micropropagation and flavonoid production with in vitro grown plants. J Pharm Belg 43:15–18
Shimokoriyama M (1956) Antochlor pigments of *Coreopsis tinctoria*. J Am Chem Soc 79:214–220
Turner BL (1982) New species and combinations in *Chrysanthellum* (Asteraceae-Coreopsideae). Phytologia 52:291–293
Wambugu FM, Rangan TS (1981) In vitro clonal multiplication of *Pyrethrum* (*Chrysanthemum cinerariaefolium* Vis) by micropropagation. Plant Sci Lett 22:219–226
Winand C, Ramaut JL, Gaspar Th (1986) Multiplication végétative par culture in vitro d'*Arnica montana* L. J Pharm Belg 41:147–150

# IX *Citrus paradisi* Macf. (Grapefruit): In Vitro Culture and the Bioproduction of Sesquiterpenes Nootkatone, Valencene, and Other Secondary Metabolites

J. A. Del Río and A. Ortuño[1]

## 1 General Account

The genus *Citrus* (family Rutaceae), produces coumarins, flavanones, flavones, and flavonols which occur in the free form and/or as glycosides (Maier and Metzler 1967; Horowitz and Gentili 1977), in addition to an extremely complex group of compounds which are volatile and are responsible for determining the characteristic aroma of each *Citrus* variety (Ting and Rouseff 1986).

### 1.1 Distribution and Cultivation Zones

Grapefruit (*Citrus paradisi* Macf.) is an important *Citrus* crop. Its origin is unknown, although recent studies point to a natural crossing between pummelo (*Citrus grandis*) and sweet orange (*Citrus sinensis*), in which the first parental species is the dominant genotype (Barrett and Rhodes 1976). The first written description is by Hughes, who in a journey to Barbados in 1750 refers to it as "the forbidden fruit" (Webber 1943).

Annual world production of grapefruit is 4.6 million metric tons, representing 7% of total citric produce, of which 89% (4.1 million metric tons) is produced in the Northern hemisphere, principally by the United States followed by Israel, Cuba, China, Cyprus, Mexico, Tunisia, Turkey, Spain, Morocco, Italy, and Algeria. Less than 500 000 t (average of FAO figures for 1987 to 1990; FAO 1991) is produced in the Southern Hemisphere by Argentina and South Africa followed by Australia and Brazil.

### 1.2 Conventional Propagation Practices

For commercial growing *Citrus paradisi* plants are obtained by vegetative propagation, mainly by grafting. Among the most widely used rootstocks (propagated by seeds) for grapefruit grafting has been sour orange (*Citrus aurantium* L.). However, because of its susceptibility to Quick Decline, it is gradually being replaced by Troyer Citrange (a hybrid of *Poncirus trifoliata* L.

[1] Department of Plant Biology (Plant Physiology), University of Murcia, Espinardo, 30071 Murcia, Spain

Biotechnology in Agriculture and Forestry, Vol. 28
Medicinal and Aromatic Plants VII (ed. by Y.P.S. Bajaj)

Raf. and *Citrus sinensis* L. Osbeck), Carrizo Citrange (*Poncirus trifoliata* L. Raf.), and Cleopatra Mandarin (*Citrus reshni* Hort. ex Tan.).

### 1.3 Importance

*Citrus paradisi* accumulates naringin as the predominant flavanone glycoside in its fruits, leaves, and juice. To a lesser extent, it also produces narirutin, prunin, hesperidin, and neohesperidin (Fig. 1; Sinclair 1972; Albach et al. 1981; Robertson and Nisperos 1983; Albach and Wutscher 1988; Jourdan et al. 1985; Rouseff et al. 1987; Berhow and Vandercook 1989, 1991; Shaw et al. 1991).

Commercial interest in these compounds is great: pharmacologically, hesperidin has been shown to act in vascular permeability (Bruckner and Szent-Györgyi 1936; Gábor 1988) and naringin to behave as an antioxidant (Chen et al. 1990), protecting against lipid peroxidation (Salvayre et al. 1988; Guengerich and Kim 1990), and antimutagenic activity (Francis et al. 1989).

Furthermore, naringin can be used as an alternative to caffeine or quinine in tonic beverages and other nonalcoholic drinks and produces a distinctly bitter taste at low concentration (usually 1 part naringin in 50 000 parts water) (Anon 1982). Naringin and neohesperidin also have another important industrial application in that they can be chemically converted into their corresponding intensely sweet dihydrochalcones (Horowitz and Gentili 1963; Krbechek et al. 1968; Horowitz 1986; Bär et al. 1990; Borrego et al. 1991), with values relative to sucrose of 1000 for neohesperidin dihydrochalcone and 300 for naringin dihydrochalcone.

In addition, although components such as the volatile aldehydes (Keterson et al. 1971) and several acetate esters (Moshonas 1971; Wilson and Shaw 1980; Shaw and Wilson 1981) are important to the flavor and aroma of grapefruit oils, sesquiterpene nootkatone (Fig. 2) has also been suggested as a major flavor-impact compound in this species (MacLeod and Buigues 1964), and is widely used in commercial flavors and fragrances (Furia and Bellanca 1975).

Grapefruit seed oil, registered as DF-100, is also used commercially for its antimicrobial properties (Nishina et al. 1991).

### 1.4 Demand in the World Market

The spot price for grapefruit oil in 1988 was $10/lb. Its primary use is in beverage flavoring, followed by perfumery, where it is used often as a base for artificial bergamot due to natural bergamot's higher price and greater susceptibility to phototoxicity. The main sources of grapefruit oil are Israel and Belize.

Nootkatone is obtained from grapefruit oil by distillation. Its price in March 1988 was $4200/kg.

Apart from pectins, the flavonoid naringin is also obtained from grapefruit peel. Annual world consumption is approximately 100 000 kg, its price varying

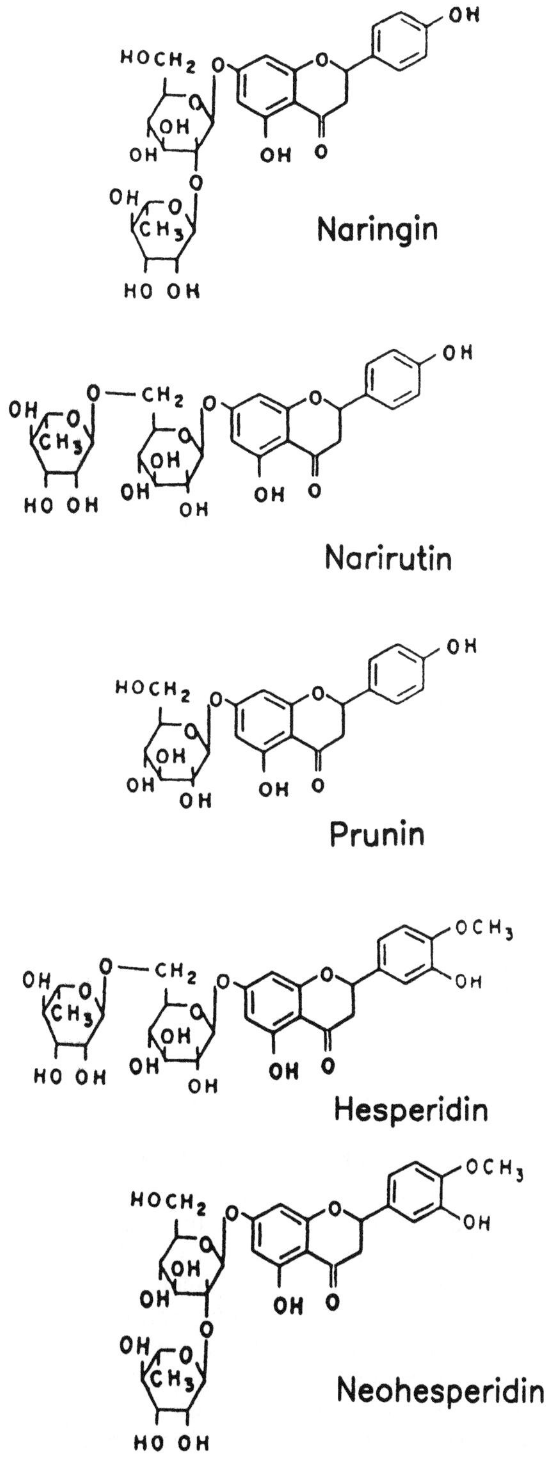


**Fig. 1.** Naringin, narirutin, prunin, hesperidin, and neohesperidin structures

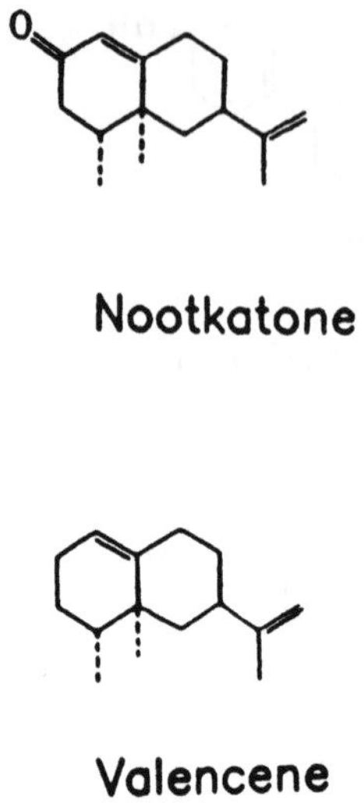


**Fig. 2.** Nootkatone and valencene structures

from \$30–40/kg depending on its purity. The principal producers are Israel, the United States, Spain, and Brazil.

Naringin, as well as being directly used in preparations, can be converted into the sweetener neohesperidin dihydrochalcone (which can also be obtained from neohesperidin) whose price (1993) is around \$500/kg with an annual consumption of approximately 3000 kg.

The antimicrobial properties of grapefruit seed oil (DF-100) mean that it is widely used in canned meats, canned vegetables, jams, juices, lacteous products, etc. Its price (1993) is around \$300/kg.

# 2 In Vitro Culture Studies

## 2.1 Review of Tissue Culture Studies

Extensive studies have been conducted on various aspects of in vitro culture of various species of *Citrus*, and recent reviews have been published on micropropagation (Navarro 1992), genetic transformation (Moore et al. 1993), somatic hybridization (Louzada and Grosser 1994), somatic embryogenesis (Kunitake and Mii 1995), and secondary metabolites (Mansell and McIntosh 1991). The ability of *Citrus* cultures to biotransform flavanones (Lewinsohn et al. 1986, 1989) and the precursors added to produce aromatic compounds (Drawert et al. 1984), and also their ability to bioproduce flavonoids (Barthe et al. 1987; Gavish et al. 1989; Mansell and McIntosh 1991) and essential oils (Drawert and Berger 1982; Del Río et al. 1991) have been described (Table 1).

**Table 1.** Studies on in vitro production of secondary metabolites in *Citrus paradisi* Macf.

| Tissue cultures established | Medium[a] (mg/l) | Compound added | Product detected | Reference |
|---|---|---|---|---|
| BIOCONVERSION | | | | |
| Cell suspension | MT | Naringenin<br>Hesperetin | Prunin<br>Hesperetin-7-<br>-O-glucoside | Lewinsohn et al. (1986) |
| Cell suspension | LS + 2,4-D (0.2) + NAA (0.18) | Hesperetin | Hesperetin-7-<br>-O-glucoside | Lewinsohn et al. (1989) |
| Cell suspension | LS + 2,4-D (0.2) + NAA (0.18) | Valencene | Nootkatone | Drawert et al. (1984) |
| BIOPRODUCTION | | | | |
| Callus from flavedo, albedo, juice vesicles, leaf, stem, and cotyledon | MT + 2,4-D (0.6) + Kin (1.3) | | Naringin (0.2–3 ppm)<br>Limonin + limonoate A-ring monolactone (0.1–4 ppm) | Barthe et al. (1987) |
| Regenerated shoots from callus | MT + BA (5) + NAA (0.1) | | Naringin (1051 ppm)<br>Limonin + limonoate A-ring monolactone (77 ppm) | Barthe et al. (1987) |
| Cell suspension | MT | | Naringin (75–100 ng/mg fr.wt.)<br>Limonin + limonoate A-ring monolactone (18–23 ng/mg fr.wt.) | Mansell and McIntosh (1991) |
| Mature embryos | MT | | Naringin (5–130 $\mu$mol/g fr.wt.)<br>Neohesperidin (5–50 $\mu$mol/g fr.wt.) | Gavish et al. (1989) |
| Callus and cell suspension | LS + 2,4-D (0.2) + NAA (0.2) | | d-Limonene ($0.14–0.47 \times 10^{-6}$% fr.wt.)<br>Valencene ($0.37–0.82 \times 10^{-6}$% fr.wt.) | Drawert and Berger (1982) |
| Callus from fruit | MT + BA (1.12) + NAA (0.93) | | Nootkatone (160 $\mu$g/100 g fr.wt.)<br>Valencene (57 $\mu$g/100 g fr.wt.) | Del Río et al. (1991) |

[a] Media: LS, Linsmaier and Skoog (1965); MT, Murashige and Tucker (1969). BA, 6-Benzylaminopurine; 2,4-D, dichlorophenoxyacetic acid; Kin, kinetin; NAA, naphthaleneacetic acid.

## 2.2 Callus Culture

### 2.2.1 *Induction and Maintenance*

Fruits of *Citrus paradisi* Macf. (cv. Star Ruby, 6–10 mm diameter) were surface sterilized by treatment with 2% (w/v) sodium hypochlorite for 10 min and then rinsed repeatedly with sterile water. For initiation of callus culture, these fruits were bisected equatorially and placed with the cut surface on Murashige and Tucker (1969) modified basal agar medium for optimal growth, as follows: 5 μM benzylaminopurine (BA), 5 μM naphthalene acetic acid (NAA), and 10% orange juice. A pH of 5.8, sucrose (30 g/l), and agar (9 g/l) were used in all media.

The callus cultures were maintained at 24°C in a 16-h light/8-h dark regime (for flavonoid assays) and in darkness (for sesquiterpene assays). In all cases, they were subcultured at 4-week intervals and fresh weight was determined each time a subculture was made.

### 2.2.2 *Microscope Study of Cells*

Different cell parameters (cell diameter and cellular density) of the in vitro growing tissue, corresponding to both the outer and the inner (0.5 and 1.5 cm below the surface, respectively) zones of 3-cm-thick calli subcultured nine times, under the conditions previously established, were measured by light microscopy using both an ocular micrometer and an integration plate coupled to a Photomicroscope II (Carl Zeiss, Oberkochen, FRG).

In addition, the cell wall thickness of the two above zones was measured using the conditions for sample preparation and techniques of morphometric analysis for electron microscopy previously described (Ortuño et al. 1990). For this, a Zeiss EM10 electron microscope (Carl Zeiss, Oberkochen, FRG) was used.

For light microscopy, semithin sections (1 μM thick) were cut on a Reichert Jung ultramicrotome and stained with 1% methylene blue, at least 15 measurements being made in each preparation. Selected areas were trimmed for ultrathin sections (50–70 nm), which were stained with uranyl acetate and lead citrate and then used for electron microscopy.

To ensure randomness, three 0.5–1-mm-thick pieces of each tissue (outer and inner zones of callus) were obtained from each cross section of the callus before fixation. Blocks containing these tissue pieces were randomly selected for sectioning prior to electron microscopy and only one ultrathin section per block and grid was obtained and then placed on formwar-coated slot grids.

For each section, 20–25 electron micrographs were obtained and in order to avoid any bias in the selection of the microscope field, the upper-right part of the holed grid was selected. The final magnification was 6800 and 15 800 times for tissues, corresponding to the outer and inner callus zones, respectively.

The analysis of histological sections, corresponding to these zones of callus, revealed apparent morphological differences in size, shape, and density: the

**Table 2.** Cellular characteristics in two zones of *Citrus paradisi* calli subcultured 9 times, grown in darkness. Data corresponding to cell diameters (a, equatorial; b, polar; μm), cell wall thickness (μm), cellular density (number of cells/$10^4$ $\mu m^2$) are given. Data correspond to mean values ± SE (n = 3). Student's t-test was used to evaluate the significance of differences in the mean values. Prior to the t-test the data were transformed into a logarithmic function. (Del Río et al. 1991)

| Callus culture | | | |
|---|---|---|---|
| Parameters | | Outer zone | Inner zone |
| Cell diameters | a | 11 ± 2 | 28 ± 3** |
| | b | 19 ± 4 | 25 ± 7* |
| Cell wall thickness | | 1.07 ± 0.18 | 0.34 ± 0.07** |
| Cellular density | | 34.25 ± 6.52 | 9.73 ± 0.77* |

*, $P < 0.05$; **, $P < 0.001$

small cells located in the outer zone had an average size of 11 × 19 μm and the medium-sized cells located in the inner zone had an average size of 28 × 25 μm (Table 2). Thus, the cells of the inner callus zone were significantly longer ($p < 0.05$) and wider ($p < 0.001$) than those of the outer callus zone. Both zones presented a typically parenchymatous shape with a significantly higher ($p < 0.05$) cellular packing in the outer zone, as can be deduced by comparing the cellular density of the two zones ( = 3.5 times greater in the outer zone) (Table 2). The reduction in cellular density observed in the inner zone of the callus was higher than might be expected from a comparison of the corresponding cell sizes of the two zones (Table 2). These results suggest that there is a greater number of intercellular spaces between the cells of the inner zone, which is in accordance with light microscope observations (data not shown).

Ultrastructural studies reveal that the cells of the outer zone have a greater cytoplasmic density (Fig. 3) and that their walls are significantly ($p < 0.001$) thicker (1.07 μm) than those of the inner zone cells (0.34 μm) (Table 2). In addition, the presence of numerous dense granules next to the cell wall (presumably osmiophilic structures) was detected in the inner zone cells oriented towards the intercellular spaces (probably oil cavities) (Fig. 3B). These were detected to a lesser degree in the outer zone cells (Fig. 3A).

## 2.3 Extraction and Structure of Secondary Compounds

### 2.3.1 *Sesquiterpenes*

Calli grown in darkness were collected and homogenized three times with n-pentane (1 g fresh weight/4 ml), adding 200 μg of lauric acid methylester as an

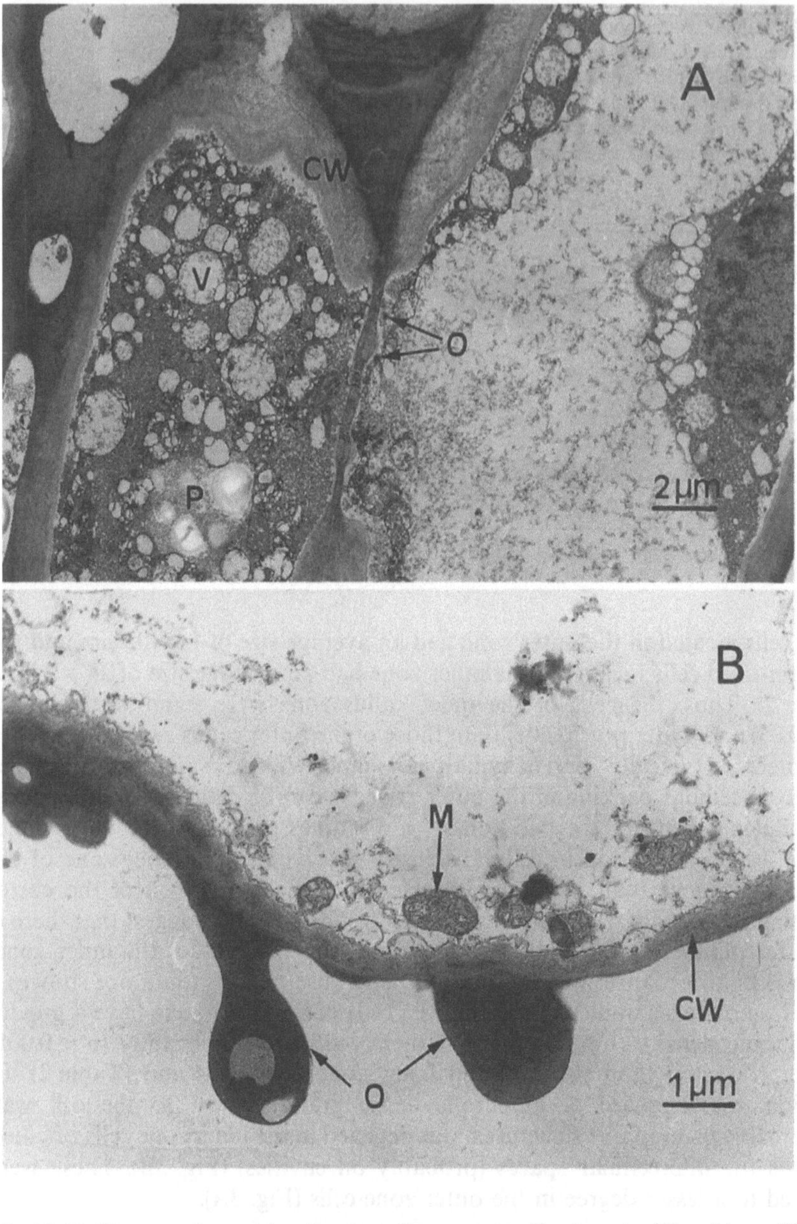


**Fig. 3A, B.** Electron micrographs of cross section corresponding to outer (**A**) and inner (**B**) zones of *Citrus paradisi* calli subcultured 9 times, grown in darkness. *CW* Cell wall; *M* mitochondria; *O* osmiophilic structures; *P* plastids; *V* vacuole. (Del Río et al. 1991)

internal standard. The homogenate was decanted and the organic phase dried with anhydrous $Na_2SO_4$. This phase was then concentrated to 0.5 ml under a nitrogen atmosphere before analysis.

The extracts were analyzed by gas chromatography with a Hewlett Packard 5890 equipped with an FID and a glass capillary column coated with Carbowax

20 M (25 m × 0.25 mm i.d., 0.2 μM film thickness). The flow rate of carrier gas was 28 ml/min He. The injection volume was 0.5 μl and the split ratio was 50/1. The injector and detector temperature was 255 °C. The following column temperature-programming sequence was used: an initial temperature of 75 °C was maintained for 8 min before being increased to 255 °C at a rate of 4 °C/min. Each peak area on the gas chromatogram was calculated automatically with a Hewlett Packard 3390A integrator. For capillary GLC-mass spectrometry a Hewlett Packard 5993 mass spectrometer was used, using a column identical to that used above. An injector and transfer line temperature of 255 °C, ionization energy 70 eV, and scan time of 1 s were used. Peak identification was confirmed by comparing the retention time and the mass spectrum with those of an authentic sample (nootkatone and valencene, see Fig. 2, from Extrasynthèse, S.A., Genay, France). Quantitative determinations were based on the known amount of added standard.

Table 3 shows the capacity of the callus cultures of three *Citrus* species (*C. paradisi*, *C. limonia*, and *C. aurantium*) to accumulate nootkatone and valencene, and the levels of these metabolites present in the exocarp of the mature fruits (Del Río et al. 1991). These results show that the callus can synthesize valencene and nootkatone in the same way as the corresponding fruits, although to a lesser extent ( ≤ 2%). There are, however, differences in the production capacity and the nature of the metabolites produced by the callus of the different species. Although both valencene and nootkatone are found in *C. paradisi* and *C. aurantium*, the callus of *C. limonia* only produces valencene. The highest levels of nootkatone are found in the callus of *C. paradisi*, while levels of valencene are similar for the three species studied.

An analysis of the formation of the two metabolites in *C. paradisi* shows that the maximum values for valencene are found in young 3-month-old callus, and that these values decrease with age. Nootkatone, on the other hand, behaves in

**Table 3.** Valencene and nootkatone levels present in mature fruits (A) and in callus cultures (B) of *Citrus* species. Data represent mean values ± SE (n = 3) of valencene and nootkatone (μg/100 g fr. wt.). Only the exocarp of fruits was used in the different assays. Calli subcultured 9 times were used; ND, not detected. (Del Río et al. 1991)

| | Valencene | |
|---|---|---|
| | (A) | (B) |
| *C. paradisi* | 4721 ± 378 | 8.0 ± 0.6 |
| *C. limonia* | 13096 ± 1171 | 9.5 ± 0.5 |
| *C. aurantium* | 299 ± 27 | 6.2 ± 0.4 |
| | **Nootkatone** | |
| | (A) | (B) |
| *C. paradisi* | 12302 ± 980 | 160.0 ± 13.6 |
| *C. limonia* | 221 ± 18 | ND |
| *C. aurantium* | 2025 ± 162 | 15.9 ± 1.3 |

the opposite way, and is not detected in 3-month-old callus, although after this age its concentration increases (Fig. 4), for which reason valencene might be considered a possible precursor of nootkatone in this species. In the calli subcultured nine times there is a significantly ($p < 0.05$) higher production of nootkatone in the inner zone ($170 \pm 8.97$ μg/100 g fr. wt.) than in the outer zone ($111 \pm 6.56$ μg/100 g fr. wt.) of the callus, in accordance with microscope observations (see Fig. 3).

These results are in accordance with those obtained when studying variations in nootkatone and valencene levels during the development of grapefruit (Del Río et al. 1992a). It was established that valencene synthesis coincides with the linear stage of growth in the fruit, the level of this sesquiterpene declining in later stages before the appearance of 2-hydroxyvalencene and then nootkatone, when the fruit enters into its maturation-senescence phase. Based on these results we have suggested that: (1) 2-hydroxyvalencene might be an intermediary in the biosynthetic pathway for nootkatone in grapefruit, generated from the valencene, in accordance with the results described for *Citrus limon* cell suspension (Drawert et al. 1984);. (2) nootkatone might be seen as an indicator of maturity in grapefruit, this view being supported by the increases observed in nootkatone levels in this species when the maturation-senescence process involved in the expression of this secondary metabolite is accelerated (Del Río et al. 1992a, 1992c; García Puig et al. 1993; Ortuño et al. 1993).

### 2.3.2 *Flavonoids*

Flavonoids were extracted from calli subcultured eight times and grown in light with dimethylsulfoxide (DMSO) in the ratio of 20 mg dry weight/ml. The corresponding extracts were filtered through a 0.45-μm nylon membrane before

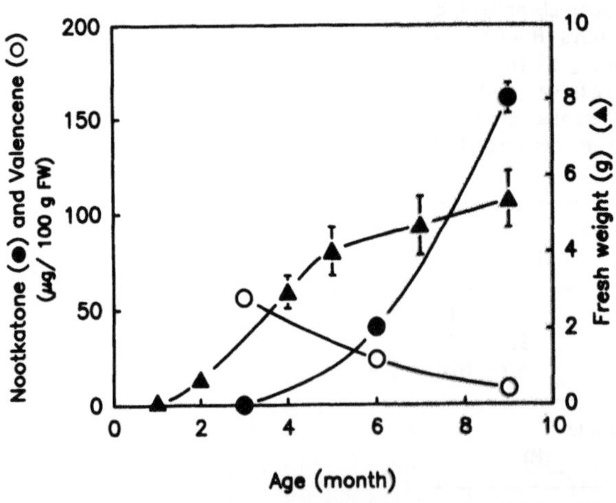


**Fig. 4.** Growth and changes in the levels of valencene and nootkatone with age of callus cultures of *Citrus paradisi*. Experiments were replicated 3 times. *Vertical bars* denote ± SE. (Del Río et al. 1991)

analysis by HPLC with a Beckman liquid chromatograph, a Model 110 B solvent-delivery module, and a System Gold Module 168 diode-array detector. Reverse phase chromatography was performed with a $C_{18}$ analytical $\mu$-Bondapak column (Waters Assoc., Milford, MA, USA) with an average particle size of 5 $\mu$m using isocratic elution with a water-methanol-acetonitrile-acetic acid (15:2:2:1, by volume) solution at a flow rate of 1 ml/min at 35 °C. The change in absorbance was monitored at 280 nm with a V/UV diode-array detector. Under these conditions the narirutin and the naringin standards (from Zoster, S.A., Murcia, Spain) showed a retention time of 9.54 and 11.39, respectively.

A representative chromatogram of an extract of these calli of *Citrus paradisi* is shown in Fig. 5. In this chromatogram compounds 1 and 2 have $R_t$'s identical to those of narirutin and naringin, respectively. The absorption spectra of these compounds, obtained by means of a V/UV diode-array detector, have two maxima at 283 and 326 nm. These data are consistent with the compounds having flavanone skeletons identical to those of narirutin and naringin. The procedure used for the isolation of compound 2 was similar to that described in a previous paper (Castillo et al. 1992) and this identity was confirmed by its $^1$H-NMR (200 MHz) spectrum (Bruker, Germany) in hexadeutero-DMSO. This $^1$H-NMR spectrum is shown in Fig. 6 and is identical to that obtained by us for naringin in immature fruits of *Citrus aurantium* (Castillo et al. 1992) and in callus cultures of *Citrus aurantium* (Del Río et al. 1992b).

The levels of narirutin and naringin present in both callus cultures and fruits of *Citrus paradisi* are given in Table 4. These calli accumulate the same major flavonoids as the fruit grown on the tree, although to a lesser extent particularly when compared with immature fruits. The levels of naringin present in the whole calli represent 1.7 and 0.2% of the levels present in mature and immature fruits, respectively; while the corresponding amounts of narirutin represent 1.3 and 0.1% of the levels present in mature and immature fruits, respectively.

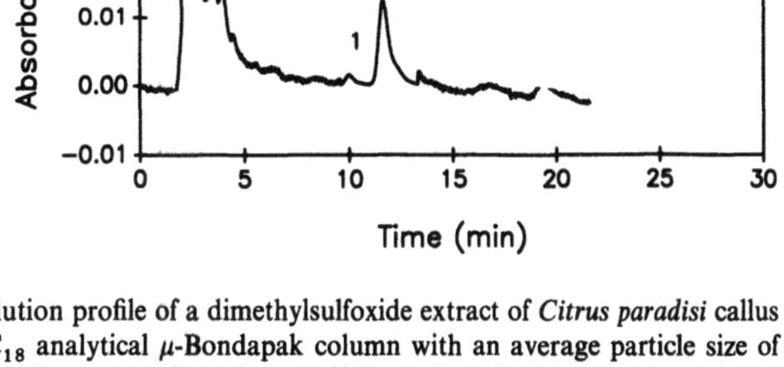


**Fig. 5.** Characteristic HPLC elution profile of a dimethylsulfoxide extract of *Citrus paradisi* callus subcultured 8 times, using a $C_{18}$ analytical $\mu$-Bondapak column with an average particle size of 5 $\mu$m. The column was eluted with water-methanol-acetonitrile-acetic acid (15:2:2:1 by volume) with a flow rate of 1 ml/min, at 35 °C. Eluent monitored at 280 nm. Identification: *1* narirutin; *2* naringin

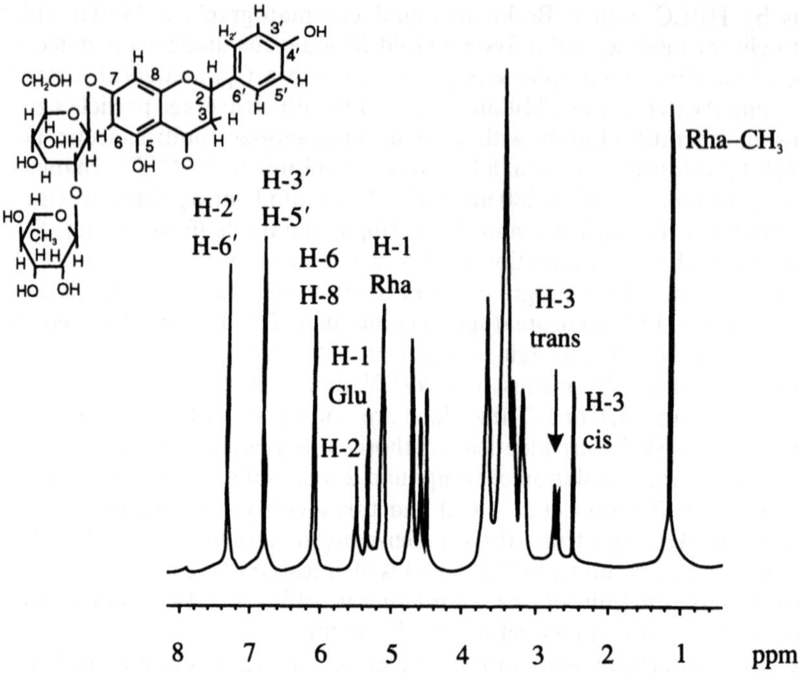


**Fig. 6.** $^1$H-NMR spectrum of compound 2 of Fig. 5 isolated by semipreparative HPLC of extracts of *Citrus paradisi* callus subcultured 8 times. The spectrum is identical to that of naringin

**Table 4.** Narirutin and naringin levels in fruits and in callus cultures of *Citrus paradisi.* Data represent mean values ± SE (n = 3) of narirutin and naringin (μmol/g dry wt.). Peel (flavedo + albedo) from fruit of 90 mm (mature fruits), complete fruit of 11 mm (immature fruits), calli subcultured 8 times, and agar medium corresponding to the last subculture were used in the different assays

| | Narirutin | | Naringin | |
|---|---|---|---|---|
| | Concentration (μmol/g dry wt.) | Total (μmol) | Concentration (μmol/g dry wt.) | Total (μmol) |
| Callus whole | 0.15 ± 0.05 | 0.12 ± 0.03 | 2.0 ± 0.41 | 1.7 ± 0.52 |
| Agar medium | – | – | – | 0.2 ± 0.02 |
| Fruit | | | | |
| Immature | 130 ± 6.2 | 91 ± 1.4 | 930 ± 13.1 | 651 ± 10.1 |
| Mature | 11 ± 0.8 | 341 ± 1.8 | 117 ± 18.6 | 3673 ± 30.0 |

The excretion of naringin into the agar medium is shown in Table 4, and represents about 10% of the quantity accumulated in the callus. In accordance with these results, the excretion of flavonoids in different *Citrus* species has been revealed at both plant level (Berhow and Vandercook 1991; Castillo et al. 1992) and cell culture level (Gavish et al. 1989; Lewinsohn et al. 1989; Mansell and McIntosh 1991), the endoplasmic reticulum perhaps being the membranous structure involved in this process as we suggested earlier (Del Río et al. 1992b).

## 3 Conclusions and Prospects

Callus cultures of various *Citrus* species are capable of synthesizing nootkatone and valencene (Table 3). Although the amounts of these metabolites produced by the callus cultures are $\leq 2\%$ of those found in the exocarps of the corresponding mature fruits, the fact that these plant cells express this secondary metabolism is considered important. In agreement with these results, other authors have described how callus cultures of various *Citrus* species accumulate small quantities of other terpenoid compounds (De Billy and Paupardin 1971; Bricout and Paupardin 1974; Paupardin 1974).

The capacity of *C. paradisi* callus cultures to accumulate nootkatone is expressed only after a certain age (Fig. 4) and this might be associated with the development of morphological differentiation in the cells of these calli, as has been suggested for monoterpenes (Paupardin 1976; Brown and Charlwood 1986). Our results show clear differences between the cellular characteristics (Fig. 3) and the production capacity of the outer and inner zones of the callus. The greater capacity for nootkatone production of the cells of the inner zone of the callus can be clearly observed. In this zone, the cell wall is thinner and even disintegrates at some points, generating lysigenous intercellular spaces and a lesser cytoplasmic density when compared with the cells of the outer zone (Fig. 3). These cellular characteristics bring to mind the differences detected by Bosabalidis and Tsekos (1982) in the peripheral cells of the oil gland and the inner secretory cells, respectively, in *C. deliciosa* fruits, suggesting that the cellular tissues present in our callus cultures have a similar integrated cellular function.

Finally, the results expressed here reveal the bioproduction of naringin, as the principal flavonoid, and narirutin in callus cultures of *Citrus paradisi*, similar to that which is found in developing fruit (Figs. 5 and 6; Table 4). In contrast to these results, other authors have suggested that *Citrus* callus culture has apparently an altered flavonoid pathway (Thorpe et al. 1971; Vandercook and Tisserat 1989). We think that these discrepancies might be related to the morphological characteristics of the respective calli, since the degree of expression of these compounds is related to certain steps of cellular growth (Del Río et al. 1992b), and is generally associated with very young tissue such as that found in immature fruits (Table 4). This is in accordance with the results obtained by us for different parts of *Citrus aurantium* (Castillo et al. 1992) and in other *Citrus* sp. (Hasegawa and Maier 1981; Jourdan et al. 1985; Berhow and Vandercook 1991).

Preliminary results suggest that the expression of both the sesquiterpenes and the flavonoids in the callus cultures can be modified by either accelerating the processes of cell differentiation or by prolonging the juvenile stages of cell growth.

*Acknowledgments.* The authors wish to thank D. A. García-Lidón and D. I. Porras, C.R.I.A.-Murcia (Spain), Da. A. Namesny, Valencia (Spain), and Dra. A. García Luis, Department of Plant Biology of the Polytechnic University of Valencia (Spain), for their generosity in providing data on grapefruit culture and production techniques. ALI 92-0524, from CICYT, Spain.

## References

Albach RF, Wutscher HK (1988) Flavanone content of whole grapefruit and juice as influenced by fruit development. J Rio Grande Valley Hortic Soc 41:89–96

Albach RF, Redman GH, Cruse RR (1981) Annual and seasonal changes in naringin concentration of Ruby Red grapefruit juice. J Agric Food Chem 29:808–811

Anon (1982) Naringin, a bitter flavouring, available in high-purity grades. Food Dev 16:29–29

Bär A, Borrego F, Benavente O, Castillo J, Del Río JA (1990) Neohesperidin dihydrochalcone: properties and applications. Food Sci Technol 23:371–376

Barrett HC, Rhodes AM (1976) A numerical taxonomic study of affinity relationship in cultivated *Citrus* and its close relatives. Syst Bot 1:105–136

Barthe GA, Jourdan PS, McIntosh CA, Mansell RL (1987) Naringin and limonin production in callus cultures and regenerated shoots from *Citrus* sp. J Plant Physiol 127:55–65

Berhow MA, Vandercook CE (1989) Biosynthesis of naringin and prunin in detached grapefruit. Phytochemistry 28:1627–1630

Berhow MA, Vandercook CE (1991) Sites of naringin biosynthesis in grapefruit seedlings. J Plant Physiol 138:176–179

Borrego F, Castillo J, Benavente-García O, Del Río JA (1991) Application potential of the citrus origin sweetener neohesperidin dihydrochalcone. Int Food Ingredients 2:23–26

Bosabalidis A, Tsekos I (1982) Ultrastructural studies on the secretory cavities of *Citrus deliciosa* Ten. II. Development of the essential oil-accumulating central space of the gland and process of active secretion. Protoplasma 112:63–70

Bricout J, Paupardin C (1974) Sur la composition de l'huile essentielle de tissus de péricarpe de Citron (*Citrus limonia* Obseck) cultivés in vitro. C R Acad Sci Paris Ser D 278:719–722

Brown JT, Charlwood BV (1986) The accumulation of essential oils by tissue cultures of *Pelargonium fragrans* (Willd.). FEBS Lett 204:117–120

Bruckner V, Szent-Györgyi A (1936) Chemical nature of citrin. Nature (Lond) 138:1057–1057

Castillo J, Benavente O, Del Río JA (1992) Naringin and neohesperidin levels during development of leaves, flower buds, and fruits of *Citrus aurantium.* Plant Physiol 99:67–73.

Chen YT, Zheng RL, Jia ZJ, Ju Y (1990) Flavonoids as superoxide scavengers and antioxidants. Free Radical Biol Med 9:19–21

De Billy F, Paupardin C (1971) Sur l'évolution des huiles essentielles dans les tissus de péricarpe de Citron (*Citrus limonia* Obseck) cultivés in vitro. C R Acad Sci Paris Ser D 273:1690–1693

Del Río JA, Ortuño A, García Puig D, Iborra JL, Sabater F (1991) Accumulation of the sesquiterpenes nootkatone and valencene by callus cultures of *Citrus paradisi*, *Citrus limonia* and *Citrus aurantium.* Plant Cell Rep 10:410–413

Del Río JA, Ortuño A, García Puig D, Porras I, García-Lidón A, Sabater F (1992a) Variations of nootkatone and valencene levels during the development of grapefruit. J Agric Food Chem 40:1488–1490

Del Río JA, Ortuño A, Marín FR, García Puig D, Sabater F (1992b) Bioproduction of neohesperidin and naringin in callus cultures of *Citrus aurantium.* Plant Cell Rep 11:592–596

Del Río JA, García Puig D, Ortuño A, Sabater F, García-Lidón A, Porras I (1992c) Sesquiterpene nootkatone as an indicator of ripening in *Citrus paradisi* Macf. Proc VII Int Soc Citricult 420–422

Drawert F, Berger RG (1982) Über die Biogenese von Aromastoffen bei Pflanzen und Früchten. XIX. Mitt: Vergleich der Biosynthese von Aromastoffen in Segment-, Kallus- und Suspensionkulturen von *Citrus* arten. Chem Mikrobiol Technol Lebensm 7:143–147

Drawert F, Berger RG, Godelmann R (1984) Regioselective biotransformation of valencene in cell suspension cultures of *Citrus* sp. Plant Cell Rep 3:37–40

FAO (1991) Citrus fruit fresh and processed. FAO, Rome.

Francis AR, Shetty TK, Bhattacharya RK (1989) Modulating effect of plant flavonoids on the mutagenicity of N-methyl-N'-nitro-N-nitrosoguanidine. Carcinogenesis 10:1953–1955

Furia TE, Bellanca N (1975) Fenarolis handbook of flavor ingredients vol 2, 3rd edn. CRC Press, Cleveland, 438 pp

Gábor M (1988) Szent-Györgyi and the bioflavonoids: new results and perspectives of pharmacological research into benzo-pyrone derivatives. In: Cody V, Middleton E Jr, Harborne JB, Beretz A (eds) Plant flavonoids in biology and medicine. II. Biochemical, cellular, and medicinal properties. Alan R Liss, New York, pp 1–15

García Puig D, Ortuño A, Sabater F, Pérez ML, Porras I, García-Lidón A, Del Río JA (1993) Effect of ethylene on sesquiterpene nootkatone production during the maturation-senescence stage in grapefruit (*Citrus paradisi* Macf.). In: Pech JC, Latché A, Balagué C (eds) Cellular and molecular aspects of the plant hormone ethylene. Kluwer Academic, Dordrecht, pp 146–147

Gavish H, Lewinsohn E, Vardi A, Fluhr R (1989) Production of flavanone-neohesperidosides in *Citrus* embryos. Plant Cell Rep 8:391–394

Guengerich FP, Kim DM (1990) In vitro inhibition of dihydropyridine oxidation and aflatoxin B1 activation in human liver microsomes by naringenin and other flavonoids. Carcinogenesis 11:2275–2279

Hasegawa S, Maier VP (1981) Some aspects of *Citrus* biochemistry and juice quality. Proc Int Soc Citricult 2:914–918

Horowitz RM (1986) Taste effects of flavonoids. In: Cody V, Middleton E Jr, Harborne JB (eds) Plant flavonoids in biology and medicine: biochemical, pharmacological, and structure-activity relationships. Alan R Liss, New York, pp 163–175

Horowitz RM, Gentili B (1963) Dyhydrochalcone derivatives and their use as sweetening agents. USA Patent 3,087,821

Horowitz RM. Gentili B (1977) Flavonoid constituents of *Citrus*. In: Nagy S, Shaw PE, Veldhuis MK (eds) *Citrus* science and technology, vol 1. AVI, Westport, pp 397–426

Jourdan PS, McIntosh CA, Mansell RL (1985) Naringin levels in citrus tissues. II. Quantitative distribution of naringin in *Citrus paradisi* Macfad. Plant Physiol 77:903–908

Keterson JW, Hendrickson R, Braddock RJ (1971) Florida citrus oils. Univ Florida, Agric Exp Stn Bull 749, Gainesville

Krbechek L, Inglett G, Holik M, Dowling B, Wagner R, Ritter R (1968) Dihydrochalcones. Synthesis of potential sweetening agents. J Agric Food Chem 16:108–112

Kunitake H, Mii M (1995) Somatic embryogenesis in *Citrus* species. In: Bajaj YPS (ed) Biotechnology in agriculture and forestry, vol 30. Somatic embryogenesis and synthetic seed I. Springer Berlin Heidelberg New York (in press)

Lewinsohn E, Berman E, Mazur Y, Gressel J (1986) Glucosylation of exogenous flavonones by grapefruit (*Citrus paradisi*) cell cultures. Phytochemistry 25:2531–2535

Lewinsohn E, Berman E, Mazur Y, Gressel J (1989) (7) Glucosylation and (1–6) rhamnosylation of exogenous flavonones by undifferentiated *Citrus* cell cultures. Plant Sci 61:23–28

Linsmaier EM, Skoog F (1965) Organic growth factor requirements of tobacco tissue cultures. Physiol Plant 18:100–127

Louzada ES, Grosser JW (1994) Somatic hybridization of *Citrus* with sexually incompatible wild relatives. In: Bajaj YPS (ed) Biotechnology in agriculture and forestry, vol 27. Somatic hybridization in crop improvement I. Springer, Berlin Heidelberg New York, pp 427–438

MacLeod WD, Buigues N (1964) Sesquiterpenes I. Nootkatone, a new grapefruit flavor constituent. J Food Sci 29:565–568

Maier VP, Metzler DM (1967) Grapefruit phenolics. II. Principal aglycons of endocarp and peel and their possible biosynthetic relationship. Phytochemistry 6:1127–1135

Mansell RL, McIntosh CA (1991) *Citrus* spp.: In vitro culture and the production of naringin and limonin. In: Bajaj YPS (ed) Biotechnology in agriculture and forestry, vol 15. Medicinal and aromatic plants III. Springer Berlin Heidelberg New York, pp 193–210

Moore GA, Jacoco CC, Neidigh JL, Lawrence SD, Cline K (1993) Transformation in *Citrus*. In: Bajaj YPS (ed) Biotechnology in agriculture and forestry, vol 23. Plant protoplasts and genetic engineering IV. Springer, Berlin Heidelberg New York, pp 194–208

Moshonas MGH (1971) Analysis of carbonyl flavor constituents from grapefruit oil. J Agric Food Chem 19:769–770

Murashige T, Tucker DPH (1969) Growth factor requirements of citrus tissue culture. In: Chapman HD (ed) Proc 1st Int Citrus Symp, vol III. Univ California, Riverside, pp 1155–1161

Navarro L (1992) Citrus shoot-tip grafting in vitro. In: Bajaj YPS (ed) Biotechnology in agriculture and forestry, vol 18. High-Tech and micropropagation II. Springer, Berlin Heidelberg New York, pp 327–338

Nishina A, Kihara H, Uchibori T, Takashi O (1991) Antimicrobial substances in "DF-100" extract of grapefruit seeds. J Antibact Antifung Agents 19:401–404

Ortuño A, Sánchez-Bravo J, Moral JR, Acosta M, Sabater F (1990) Changes in the concentration of indole-3-acetic acid during the growth of etiolated lupin hypocotyls. Physiol Plant 78:211–217

Ortuño A, García Puig D, Sabater F, Porras I, García-Lidón A, Del Río JA (1993) Influence of ethylene and ethephon on the sesquiterpene nootkatone production in *Citrus paradisi.* J Agric Food Chem 41:1566–1569

Paupardin C (1974) Sur l'evolution de l'huile essentielle dans des tissus de fruits de Citron (*Citrus limonia* Osbeck) cultivés in vitro dans diverses conditions. Rev Gén Bot 81:223–241

Paupardin C (1976) Sur la différenciation d'un tissu sécréteur et la formation d'huile essentielle par des tissus végétaux cultivés in vitro. C R Congr Natl Soc Sav Sect Sci 101:619–628

Robertson GL, Nisperos MO (1983) Changes in the chemical constituents of New Zealand grapefruit during maturation. Food Chem 11:167–174

Rouseff RL, Martin SF, Youtsey CO (1987) Quantitative survey of narirutin, naringin, hesperidin, and neohesperidin in *Citrus.* J Agric Food Chem 35:1027–1030

Salvayre R, Nègre A, Affany A, Lenoble M, Douste-Blazy L (1988) Protective effect of plant flavonoids, analogs, and vitamin E against lipid peroxidation of membranes. In: Cody V, Middleton E Jr. Harborne JB, Beretz A (eds) Plant flavonoids in biology and medicine. II. Biochemical, cellular, and medicinal properties. Alan R Liss, New York, pp 313–316

Shaw PE, Wilson CW III (1981) Importance of nootkatone to the aroma of grapefruit oil and the flavor of grapefruit juice. J Agric Food Chem 29:677–679

Shaw PE, Calkins CO, McDonald RE, Greany PD, Webb JC, Nisperos-Carriedo MO, Barros SM (1991) Changes in limonin and naringin levels in grapefruit albedo with maturity and the effects of gibberellic acid on these changes. Phytochemistry 30:3215–3219

Sinclair WB (1972) The grapefruit: its composition, physiology and products. Univ California, Div Agric Sciences

Thorpe TA, Maier VP, Hassegawa S (1971) Phenylalanine amonio-lyase activity in *Citrus* fruit tissue cultured in vitro. Phytochemistry 10:711–718

Ting SV, Rouseff RL (1986) Chemical constituents affecting quality characteristics of *Citrus* products. In: Citrus fruits and their products. Marcel Dekker, New York, pp 105–106

Vandercook CE, Tisserat B (1989) Flavonoid changes in developing lemons grown in vivo and in vitro. Phytochemistry 28:799–803

Webber HJ (1943) Cultivated varieties of citrus. In: Webber HJ, Batchelor LD (eds) The *Citrus* industry I. Univ California Press, Berkeley, pp 475–668

Wilson CW III, Shaw PE (1980) Glass capillary gas chromatography for quantitative determination of volatile constituents in cold-pressed grapefruit oil. J Agric Food Chem 28:919–922

# X *Claviceps purpurea* (Ergot): Culture and Bioproduction of Ergot Alkaloids

V. Křen[1], P. Harazim[2], and Z. Malinka[2]

## 1 Introduction

Ergot (*Claviceps purpurea*), best known as a disease of rye and some other grasses, is probably the most widely cultivated fungus and has now become an important field crop. The main reason for its importance is the presence of ergot alkaloids, extensively used in medicine. Currently, ergot alkaloids cover a large field of therapeutic uses as drugs of high potency in the treatment of uterine atonia, postpartum bleeding, migraine, orthostatic circulatory disturbances, senile cerebral insufficiency, hypertension, hyperprolactinemia, acromegaly, and parkinsonism (Berde and Schild 1978). Recently, new therapeutic uses have emerged, such as, e.g., against schizophrenia (Markstein et al. 1992), applications based on newly discovered antibacterial (Eich et al. 1985) and cytostatic effects (Eich at al. 1986), immunomodulatory (Šterzl et al. 1987) and hypolipemic activity (Golda et al. 1980; Cincotta and Meier 1989).

Of the naturally occurring ergot alkaloids only two are used in the therapy: ergotamine and ergometrine. The rest of the therapeutically important ergot compounds are semisynthetic compounds (Fig. 1).

Ergot alkaloids are traditionally obtained by extraction of ergot sclerotia artificially cultivated on cereals. The parasitic cultures are not able to produce all alkaloids, e.g., clavine, necessary for most semisynthetic drugs. Crop fluctuations and market demand have led to the development of saprophytic cultivation in production plants. Submersed cultures of various *Claviceps* strains produce relatively high yields of clavines or lysergic acid derivatives, e.g., elymoclavine and paspalic acid.

Present trends in ergot cultivation are the development of saprophytic cultivation processes and the improvement of field production such as the introduction of new hosts and ergot strains. In spite of constant effort to prepare ergot alkaloids synthetically (Ninomiya 1992), their bioproduction is still economically more competitive.

[1]Institute of Microbiology, Academy of Sciences of the Czech Republic, 142 20 Prague 4, Czech Republic
[2]Galena-Pharmaceutical Works, 747 00 Opava, Czech Republic

Biotechnology in Agriculture and Forestry, Vol. 28
Medicinal and Aromatic Plants VII (ed. by Y.P.S. Bajaj)

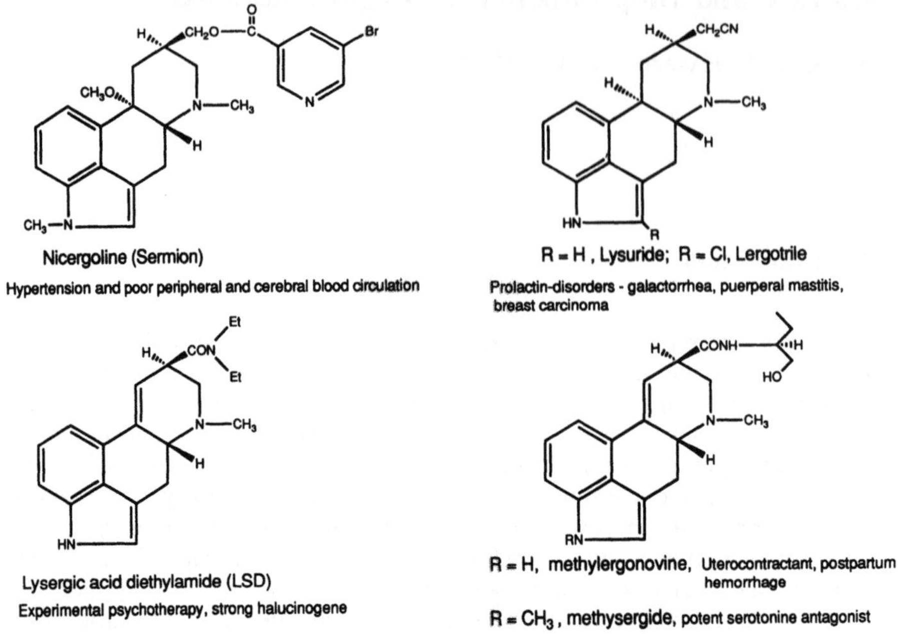


**Fig. 1.** Some therapeutically used semisynthetic ergot alkaloids

## 2 Ergot in Agriculture – Pest and Benefactor

Ergot – once dreaded as a pest and the cause of epidemic intoxications has now become a very profitable crop for farmers; however, the danger of intoxication and crop damage still persists.

### 2.1 Ergot – a Plant Disease

*Claviceps purpurea* (Fr.) TuL. was first recognized as a fungus in 1711, and its life cycle was described in 1853 by Tulasne (1853). However, the fungus was already well known in the Middle Ages as causing the last outbreaks of ergotism or "epidemic gangrene" called St. Anthony's fire (Gilman et al. 1985). Despite control of harvesting and of the use of infected rye, probably the last outbreak of ergotism was recorded in Ethiopia in 1979 (Tanner 1987), and in France in 1951 (Gabbai et al. 1951; Edwards et al. 1991).

The causal fungus *Claviceps purpurea* infects the flowers, prevents the seeds from developing, and replaces many kernels with hard, seed-like fungus bodies called sclerotia. About 5 days after infection, the sticky honeydew, containing thousands of conidia or secondary spores collects on the surface of the infected florets. Conidia are carried to healthy florets by contact, rain, and numerous insects that are attracted to the sugary honeydew. The florets are susceptible to the infection only before pollination when the stigma is open. As the crop matures, the fungus continues to grow, replacing the ovule, and eventually forming the sclerotium. The sclerotium falls off when the glumes become brittle. The sclerotia typically contain about 30% fat and 0.01 – 1% alkaloids. After dormancy, the sclerotium germinates with several stromata which have stalks and spherical heads, and which in the perithecia bear asci with sexual ascospores (Fig. 2). The ascospores are ejected when the host plant flowers, and the life cycle is concluded.

Most native grasses and all varieties of cultivated grains are susceptible to ergot (Brady 1962, Loveless 1971; Loveless and Peach 1986). Grains in which sterility occurs, such as triticale, and cross-pollinated hosts, such as rye, are more likely to become infected by the ergot fungus because their florets are open longer than those of fertile, self-pollinated hosts, thereby providing prolonged access to ergot spores (Wood and Coley-Smith 1980b, 1982). This finding is

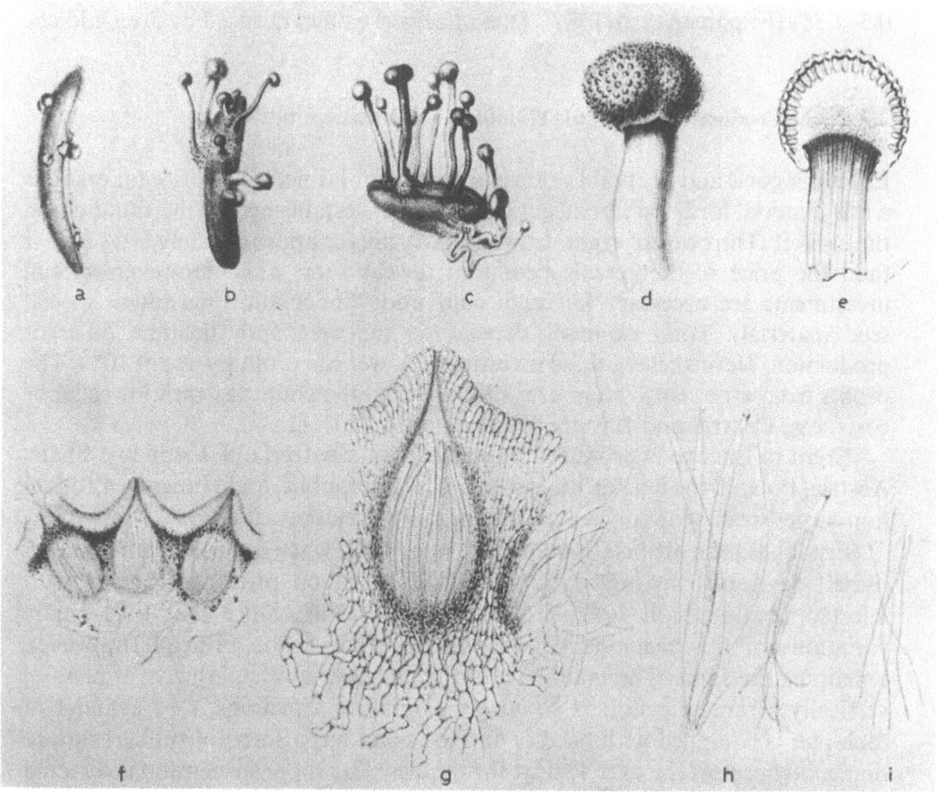


**Fig. 2a–i.** Developmental stages of ergot (Bové 1970) **a–c** Germinating sclerotium, development of stromata. **d** Head of stroma (spheridium) **e** Cross section of spheridium with perithecia (cavities). **f, g** Details of perithecia. **h** Paraphyses with ascospores **(i)**

employed in the advanced technology of parasitic ergot production (see below). Oat is seldom affected by ergot probably because the young ovary is protected by tightly enclosing floral bracts. Forage grasses are most seriously damaged by ergot when they are grown for seed production. Also subtropical and tropical grains (pearl millet) are affected by *Claviceps* (*C. fusiformis*), thus causing problems, mostly in India and other subtropical countries (Chahal et al. 1985).

The disease can be controlled by crop rotation, deep plowing, and the use of good-quality seeds or selective herbicides (Bretag 1985) or fungicides (Wood and Cooley-Smith 1980a). Ergot can be removed from the affected crop by flotation in a 20% salt solution.

Ergot alkaloids are produced by a few other organisms that, however, have a strong impact on agriculture. The fungal endophyte *Acremonium coenophialum* (Bacon 1988), associated with tall fescue grass, produces ergopeptide alkaloids (Arechavaleta et al. 1992) that sometimes cause toxicosis of cattle in America. Also some other filamentous fungi from the genera *Aspergillus and Penicillium* produce ergot alkaloids (Reshetilova and Kozlovskii 1990). Seeds of morning-glory plants *Ipomoea* spp. occurring widely as weeds among cultivated field crops such as soybean, contain a relatively high amount of ergot alkaloids 0.5–1.3% (Friedman et al. 1989). These plants are often misused by drug addicts.

### 2.2 Field Production of Ergot Alkaloids – Parasitic Cultivation

Ergot is a good and profitable alternative crop for farmers. The price for ergot as a raw material for the pharmaceutical industry is established by the situation on the market. The price of ergot grown parasitically is approximately 80% higher than the price of the cereals produced on the same area. However, special investments are necessary for ergot cultivation (inoculation machines, special seed material). Total expenses depend on the area and the time of ergot production. Nevertheless, these investments lower the profit by about 10%. The profits from ergot cultivation are stilll higher in the countries with lower labor costs, e.g., Central and Eastern Europe (up to 180%).

Ergot in Europe is produced parasitically in Switzerland, Germany, Spain, Austria, Poland, the former Yugoslavia, Czech Republic, and Hungary. Production varies from year to year depending on the market situation.

Ergot has been artificially cultivated since 1942, when A. Stoll at Sandoz Ltd. (Basel) developed an inoculation machine based on pricking the ears with infected needles (Stoll 1942; Stoll and Brack 1944). Since that time, many variations of this machine have been constructed, the principle, however, remaining the same. The main part of the machine consists usually of pairs of vertically arranged cylinders rotating in opposite directions. One cylinder of each pair is equipped with needles and the other has a smooth (rubber) surface that gently presses the ears against the needles. The spore suspension is directed from a storage tank through pipes over the needles. The inoculation of the plants is repeated after several days. The dose of conidia is usually $10^{12}$/ha.

Inoculation material (spores) is prepared either by surface or by submerged saprophytic cultivation. The first method (surface) is rather laborious but the spores are of good quality and can be stored, after drying, for a long time without loss of viability. The substrate is usually represented by sterilized rye grains soaked in a nutritive medium. The cultivation proceeds in bottles. Alternatively, cultivation in plastic bags on the surface of the liquid medium (Kybal and Sikyta 1985, 1986) can be performed. The mycelia with the spores are then collected, dried, and powdered (Harazim et al. 1988).

Submerged preparation of the inoculum is analogous to the submerged production of ergot alkaloids (see below). However, in the submerged culture, nonuniform conidia or blastospores develop, having good infectivity but lower viability (after drying). This problem was solved by the addition of 1–2% suspension of Siloxide (fine hydrophic silica gel) to the fermentation broth containing spores and mycelium. The resultant mixed suspension is then filtered, gently dried, and granulated. The resulting product is of good viability and storability, and is easy to handle (Kybal et al. 1990; Valík and Malinka, 1992).

The strains used for preparation are carefully bred with respect to total ergot alkaloid content in sclerotia, to the type of alkaloids produced, and to infectivity. Typically, industrial strains produce, e.g., α-ergokryptine, ergocristine, or ergotamine. Some strains produce directly a mixture of alkaloids necessary for a particular medicine preparation so that the separation step can be reduced (e.g., ergocornine-ergokryptine strains). Yields of the field cultivation are 700 to 1200 kg/ha; the total alkaloid content ranges from 0.4 to 1.2%.

An important factor in ergot yields is the host cereals. Rye is the most typical host for ergot in nature and also in artificial parasitic production. Triticale and wheat are used for ergot production and show good yields. An introduction of male sterile lines of some cereals, namely of rye, was a qualitative step forward in the field production of ergot. The florets remain open until the ovaries are effectively pollinated. The chances of ergot spores gaining entry to the florets and causing infection of the ovaries are thereby greatly increased (Wood and Cooley-Smith, 1980b). This is especially the case in secondary infection (after artificial inoculation) caused by honeydew that can reach the rest of the ovaries not punctured by inoculating needles. Finally, up to 100% of all ovaries in the ear are infected which increases the ergot yield dramatically. Another advantage is that the sclerotia are nearly uniform and do not fall out from the ears before harvest. The harvest is done by normal combined harvesters without any modifications, thus no hand picking or "brushing" is necessary.

Male sterile rye (genetically or cytoplasmatically inherited) is selected by the known methods (Duvick 1966) and marketed for the above purposes. In Europe a brand Hyclaro is produced by Rentschler Co., (Laupheim, Germany) (Rentschler 1972).

## 3 Saprophytic Ergot Cultivation

### 3.1 Chemistry of Ergot Alkaloids

Ergot alkaloids, like most secondary metabolites of microorganisms and plants, are produced as a family of related compounds based on the tetracyclic ergoline (partially hydrogenated indolo-[4, 3-f, g]-quinoline). In most of the naturally occurring ergot alkaloids, the nitrogen atom at position 6 is methylated and position 8 bears an additional C-atom. In most cases, ring D carries a double bond in position 8–9 or 9–10 (8-ergolenes and 9-ergolenes, respectively) (Fig. 3).

The asymmetric carbon in the 8-position of the 9-erogolenes gives rise to the two stereoisomeric ergolenes and isoergolenes. Based on the substituent at C-8, ergot alkaloids are classified as clavines and lysergic acid derivatives. For more details and the complete list of all clavines, see the review by Floss (1976).

Rarely, ring D is open at 6–7 (so-called secoclavines or chanoclavines) or cycled by oxygen. Ergolene acids are three ergolene derivatives with a carboxyl group at position 8, i.e., ( + )-lysergic acid, ( + )-isolysergic acid (9-ergolenes), and paspalic acid (8-ergolenes). The "classical" ergot alkaloids are all amide derivatives of lysergic or isolysergic acid. So-called simple lysergic acid amides are ergine, lysergic acid α-hydroxyethylamide, ergometrine, and their C-8 epimers, e.g., erginine.

The ergot peptide alkaloids consist of a lysergic acid molecule that is linked by an acid amide type of linkage to a cyclole-structured tripeptide. A well-known example of this group is ergotamine (Fig. 3). For a detailed list of these structures see the review by Kobel and Sanglier (1986).

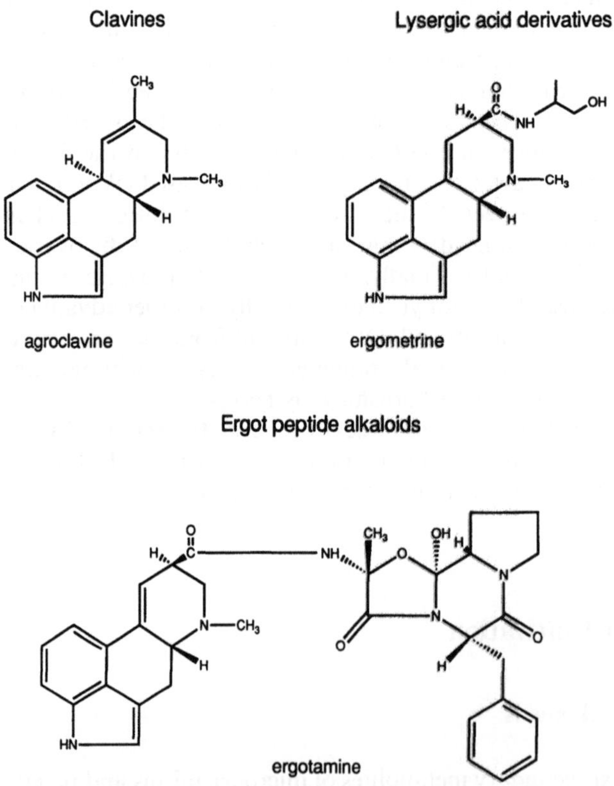


Fig. 3. Examples of the main classes of ergot alkaloids

## 3.2 Biosynthesis of Ergot Alkaloids

The main pathway of biosynthesis of ergot alkaloids stems from basic precursors, i.e., L-(tryptophan and dimethylallylidiphosphate via dimethylallyltryptophan, chanoclavine, to clavine alkaloids (Fig. 4). In the strains producing clavine alkaloids (e.g., *Claviceps fusiformis*) the pathway terminates, in the strains producing derivatives of lysergic acid, the pathway continues via paspalic and lysergic acid the pathway continues *via* paspalic and lysergic acid to their simple derivatives (e.g., ergometrine – *C. paspali*) or to ergopeptides (e.g., ergotamine – *C. purpurea*).

The first pathway-specific enzyme of the ergot alkaloid synthesis is isoprenylation of the tryptophan catalyzed by tryptophandimethylallyltransferase (EC 2.5.1.34) (Heinstein et al. 1971; Shibuya et al. 1990). Its product, 4-($\gamma$,$\gamma$-dimethylallyl)tryptophan, is N-methylated (Otsuka et al. 1980; Gröger et al. 1991). The methyl group originates from S-adenosylmethionine. The further cyclization steps leading to chanoclavine-I remain unclear, and the intermediates of these reactions are highly unstable. Chanoclavine is oxidized to chanoclavine-I-aldehyde using NADPH + $O_2$ by cytochrome P-450 and further cyclized to agroclavine.

An important regulation step in ergot alkaloid biosynthesis is the oxidation of agroclavine to elymoclavine. This reaction is catalyzed by cytochrome P-450 (found in the microsomal fraction) (Ambike et al. 1970). This enzyme (and probably some other enzymes in the pathway) is feedback-inhibited by its product, elymoclavine (Cheng et al. 1980).

Elymoclavine is oxidized by cyt p-450 to paspalic acid (Kim et al. 1983) that isomerizes (spontaneously, to lysergic acid. This acid is taken up by the multienzyme complex synthesizing the tricyclic peptide of the ergopeptines (Floss 1976).

## 3.3 Regulation of Ergot Alkaloid Biosynthesis

Well-founded knowledge of the regulation of the alkaloid synthesis and the metabolism of the producer is a prerequisite for the successful design of a biotechnological process of alkaloid production. Some products of primary metabolism serve in secondary metabolism as precursors. However, they exhibit also regulatory effects (tryptophan) or act as structural elements (lipids) influencing thus the biosynthesis of secondary metabolites in a complex way.

### *3.3.1 Carbon Sources and Saccharides*

Saccharides are the main carbon and energy source under both parasitic and saprophytic conditions. Sucrose is an important component of phloem sap of the host plants (Basset et al. 1972). The best substrates for alkaloid production are slowly metabolizable saccharides, e.g., sucrose, maltose, and polyoles (mannitol, sorbitol) (Křen et al. 1984).

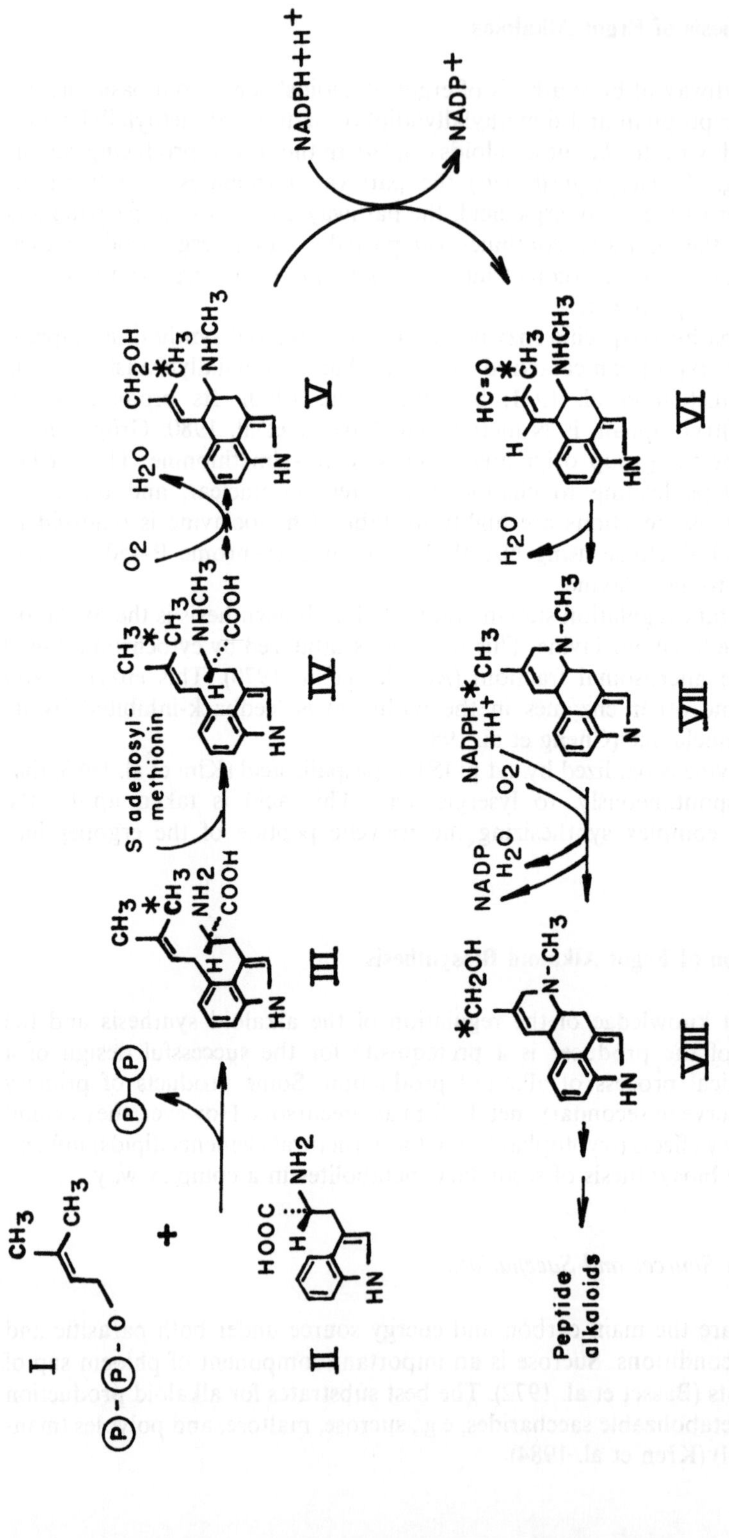


**Fig. 4.** Biosynthesis of ergot alkaloids. **I** Dimethylallylpyrophosphate. **II** Tryptophan **III** 4-Dimethylallyltryptophan **IV** N-Methyldimethylallyltryptophan **V** Chanoclavine-I **VI** Chanoclavine-I-aldehyde **VII** Agroclavine **VIII** Elymoclavine

Sucrose is metabolized fast from the very beginning of submerged *C. purpurea* cultivation (Fig. 5). The monosaccharides released are quickly utilized (Křen et al. 1984).

Glucose is not suitable as a sole carbon source because it supports mycelial growth and the alkaloid formation is virtually inhibited by glucose-catabolite repression (Křen et al. 1987a). Glucose also strongly supports extracellular polysaccharide (β-glucan) formation that increases viscosity and complicates the oxygen transport (Buck et al. 1968). However, glucose, when added in an optimized proportion, supports alkaloid production (Fig. 6).

Secondary carbon sources, i.e., intermediates of citrate or glyoxalate cycle, are essential for good saccharide utilization by saprophytically cultivated *Claviceps* strains.

*C. purpurea* growing on sucrose as the sole C-source produces lactate, and the severe acidification of the medium diminishes the activity of the Krebs cycle enzymes, e.g., malatdehydrogenase and citratsynthase (Pažoutová and Řeháček 1981). Addition of citrate or succinate to the medium keeps the citrate cycle active, ensures catabolism of pyruvate, and, by ATP production, helps to control the rate of glycolysis.

The key regulatory enzyme of glycolysis in *Claviceps* is phosophofructokinase (EC 2.7.1.11), which has rather atypical regulatory properties – feedforward regulation and hyperbolic saturation kinetics (Křen and Řeháček 1984).

The hexosomonophosphate pathway prevails in glucose breakdown during the vegetative phase of fermentation; the share of the glycolytic pathway becomes more pronounced during alkaloid synthesis (Gaberc-Porekar et al. 1990). The supply of NADPH is important for ergot alkaloid biosynthesis, e.g., for cyt P-450-catalyzed steps.

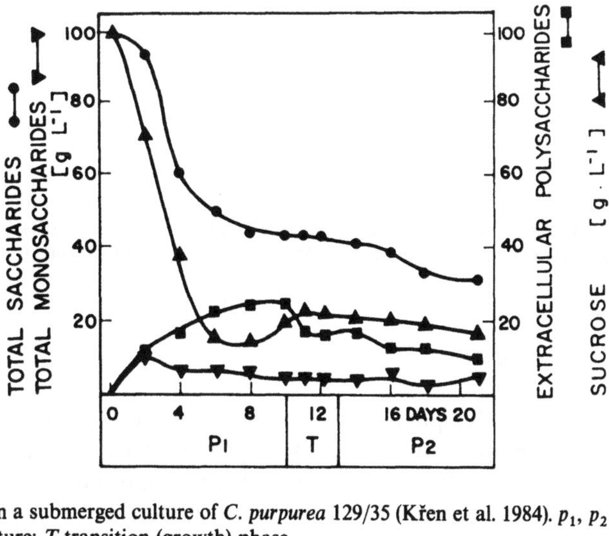


**Fig. 5.** Sucrose metabolism in a submerged culture of *C. purpurea* 129/35 (Křen et al. 1984). $p_1$, $p_2$ Production phases of the culture; *T* transition (growth) phase

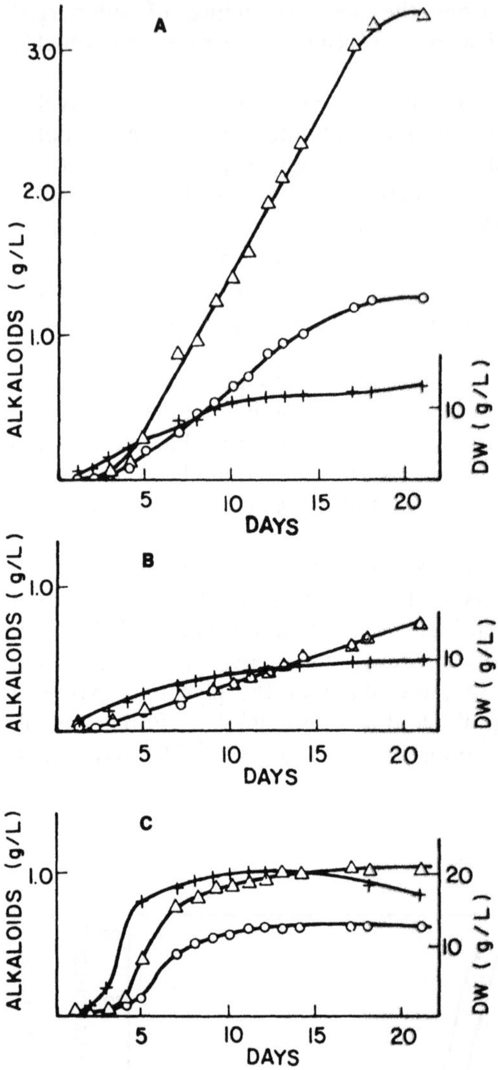


**Fig. 6A–C.** Effect of different carbon sourced on the alkaloid production by *C. purpurea* 129/35 in submerged cultivation (Flieger et al. 1988). **A** Sucrose-glucose (80 + 20 g/l) **B** Maltose–glucose (95 + 5 g/l) **C** Glucose (100 g/l). (△) Agroclavine; (○) Elymoclavine; ( + ) DW (biomass)

### *3.3.2 Phosphate*

Inorganic phosphate has unique role among the inorganic nutrients in *C. purpurea* cultivation. In most of the microbial secondary metabolites (e.g., antibiotics), a suboptimal level of phosphate has a stimulative role for their production (Martín 1977). In high alkaloid-producing *Claviceps* strains, the optimum level of phosphate in the medium ranges from 1 to 4 mM. The phosphate is taken up by the fungus during the first 2–4 days of the cultivation and is accumulated in the mycelium. Higher phosphate concentrations stimulate the vegetative growth of the mycelium and glucane production, and inhibit alkaloid biosynthesis. A drop in intracellular phosphate coincides with the cessation of mycelial growth and the onset of alkaloid production (Pažoutová and Řeháček 1984). High levels of phosphate also induce alkaloid-degrading enzymes, thus lowering the yields (Robbers et al. 1978).

### *3.3.3 Lipids*

Ergot alkaloid biosynthesis and lipid metabolism have some common regulatory points. Hydroxymethylglutaryl-coenzyme-A reductase is the key regulatory enzyme of isoprenyl unit production. The activity pattern of this enzyme in *Claviceps* (Křen et al. 1986a) reveals that the mevalonate distribution is shared at the beginning of the fermentation by sterol and alkaloid biosynthesis, and later it is used solely for alkaloid building.

Substances known to modulate lipid biosynthesis (and cell lipid composition), e.g., chlorophenoxy acids, are also able to influence ergot alkaloid biosynthesis and to amend the yields (Křen et al. 1990a). Chlorophenoxypropionate in minute doses significantly stimulates alkaloid production in *C. purpurea* strain 59 (Fig. 7).

Cell lipid composition is an important marker in *Claviceps* taxonomy and in establishing fingerprints of the industrially important strains, by using, e.g., fatty acids (Křen et al. 1985a, b) and sterols (Křen et al. 1986b).

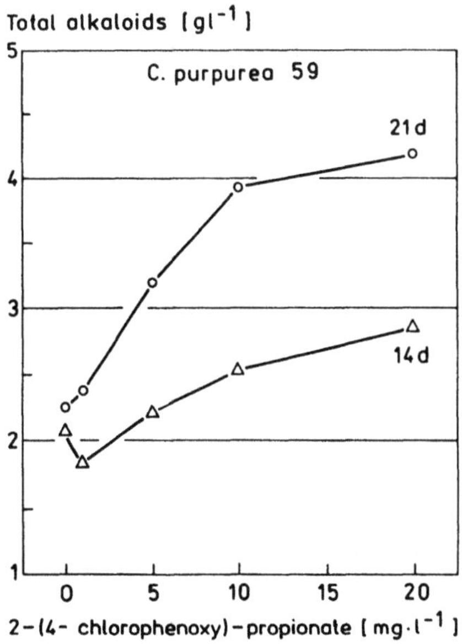


**Fig. 7.** Effect of DL-2-(4-chlorophenoxy)-propionate on ergot alkaloid production in *Claviceps purpurea* 59 on 14th (△) and 21st (○) day of cultivation

### 3.4 Genetics of *Claviceps* Fungus

*Claviceps purpurea* is a monoecious fungus. Its sexual cycle with an exchange of genetic material occurs exclusively under parasitic conditions. The second possibility of genetic material exchange is a parasexual cycle involving mitotic recombination.

Tudzynski and Esser (1982) accomplished meiotic recombination of two different mutant strains of *Claviceps purpurea* by joint parasitic cultivation on rye. Spalla and Marnati (1978) crossed two *Claviceps* auxotroph ($Ade^-$, $Phe^-$) strains differing in nutritional requirements and in alkaloid-production pattern by protoplast fusion. However, it was not certain whether the offspring were true recombinants or merely heterokaryons. Brauer and Robbers (1987) accomplished parasexual crossing (protoplast fusion) of various *Claviceps* auxotrophs. The prototrophic fusion results were submitted to segregation analysis, proving the recombination. Recently, a heterologous transformation system for *Claviceps* was established using a bleomycin-resistant gene (van Engelenburg et al. 1989) and a hygromycin-resistant gene carrying the acetamidase gene (Comino et al. 1989). This and the discovery of linear mitochondrial plasmids in *Claviceps* (Tudzynski et al. 1983, Düvell et al. 1988), have created a good methodological basis for molecular genetic manipulation applicable to industrial *Claviceps* strains.

Spores of *Claviceps* strains are quite resistant to mutagenesis, and therefore, especially for obtaining an auxotroph strain, special mutagenic methods must be used. Protoplasts were submitted to mutagenesis by chemical mutagens (NTG, EMS) with a good auxotroph yield (Keller 1983 Srikrai and Robbers 1983). This method is also suitable for nonsporulating strain improvement.

For obtaining high-yielding *Claviceps* strains, the spores are usually mutagenized by physical mutagens (UV, X-rays) or by chemical mutagens. Mutagenic treatment should not be too drastic (survival rate 2–5%) because of the danger of chromosomal aberrations and pleiotropic mutations leading to unstable strains.

Resulting mutants are tested for alkaloid production in flask experiments and the stability of the selected strains must be verified in consecutive cultivations. The increase in alkaloid production in the successive mutation steps is almost directly dependent on the number of colonies tested. Also the type of the alkaloid produced must be checked, as alkaloid-pathway mutants can be obtained (Maier et al. 1980, Křen et al. 1986d).

### 3.5 Biotechnology of Submerged Cultivation of Ergot

The prerequisite for successful, high-yielding cultivation is a high-quality strain. There exist many *Claviceps* strains (Schmauder 1982), some of them of unclear ethiology. Originally, all the strains were isolated from the parasitic sclerotia and later were mutagenized and selected. Some of the strains are deposited in public collections; however, most of the high-producing strains are in the possession of pharamaceutical companies.

The *Claviceps* strains display high variability and they often tend to degenerate. Therefore, continuous selection and standard storage techniques are an integral part of the bioprocess. The best method for strain storage is freezing in liquid nitrogen or freeze drying in a preserving solution consisting of sucrose and skimmed milk.

Cultivation conditions for *Claviceps* strains do not basically differ from cultivation of other filamentous fungi, e.g., Penicillia. A crucial step is the inoculum preparation. If the inoculum is of bad quality, the production stage usually cannot be recovered. The production stage must be incoculated by at least 10% of the inoculum. During fermentation all the key parameters must be controlled, pH being 4.5–6.2 (optimum 5.4), aeration 0.5–1.5 l of air/l of medium, mixing rate 75–200 rpm and temperature 24 °C. Some synthetic antifoam reagents diminish yields. Good results are obtained by the use of plant oils (or ergot oil from sclerotia) that are partly utilized by the fungus. The inoculum stage takes place for 6–10 days (or longer in multistage inoculum), the production stage takes place for 12–18 days. Long cultivation puts a high demand on the strict sterility of the process. Sometimes, a low amount of broad spectrum antibiotics (chloramphenicol) could be added (Křen et al. 1986c) to protect against prokaryotic contamination.

For most *Claviceps* strains synthetic media can be used. In industry, sucrose or sorbitol is used (seldom other C-sources) in concentrations of 70–250 g/l. High osmolarity supports alkaloid production, so 10–20 g/NaCl could be supplemented (Puc and Sočič 1977). As a secondary C-source, usually citrate at a concentration of 5–10 g/l is used. $NH_4^+$ salts, asparagine, or $NO_3^-$ salts serve as N-source and phosphate – see above – and other inorganic salts are supplemented. The inoculation medium differs from the production medium by its higher phosphate concentration – up to 10 mM – different N-source, some additional vitamins (biotine, aneurine), and by the addition of complex nutrients – yeast extract, corn steep, malt extract, etc.

More than 120 different media with approximately 50 components were described for *Claviceps* strain cultivation. The choice of a suitable inoculation and production medium is important because strains differ considerably in nutrient demands.

### *3.5.1 Cultivation of Ergot on Plant Tissue Cultures*

Plant tissue culture techniques provide completely controlled conditions to elucidate the growth and physiology of the cells and to study host-parasite interactions at the cellular level. There exists probably only one paper on this interesting technique in ergot research on the cultivation of *C. fusiformis* on a *Pennisetum typhoides* cell culture, (Roy and Kumar 1985), who established the methodology and studied the nutritional demands of the mixed culture with respect to growth and alkaloid production.

### *3.5.2 Alkaloid Production by Immobilized Cells*

The use of immobilized cells of the producing strain is a promising method for extending the production period of ergot cultures. *Claviceps purpurea* was immobilized in Ca-alginate by Kopp and Rehm (1983, 1984). The immobilized cells maintained alkaloid production for approximately 150 days under a semicontinuous regime. Higher matrix (alginate) concentration promoted overall production, but mostly alkaloids in a lower oxidation state were produced. The immobilized cells formed sclerotia-like structures, suggesting that immobilization (matrix) can simulate parasitic conditions.

The physiology of immobilized *C. fusiformis* during long-term semicontinuous cultivation was studied by Křen et al. (1987b) (Fig. 8). The immobilized cells maintained alkaloid production for 770 days, during which they underwent profound morphological changes – vacuolization and mitochondrial degeneration. The beads remained mechanically stable for the whole time.

Immobilization of *Claviceps* cells in other matrices (carrageenan, pectate) and their co-immobilization with oxygen-supplying systems was studied by Křen (1990). The immobilized *Claviceps* cells can be used not only for alkaloid production, but also for biotransformation.

### *3.5.3 Biotransformation of Ergot Alkaloids*

There are more than 60 papers on ergot alkaloid bioconversion, recently reviewed by Křen (1991). Ergot alkaloids were submitted to various microbial

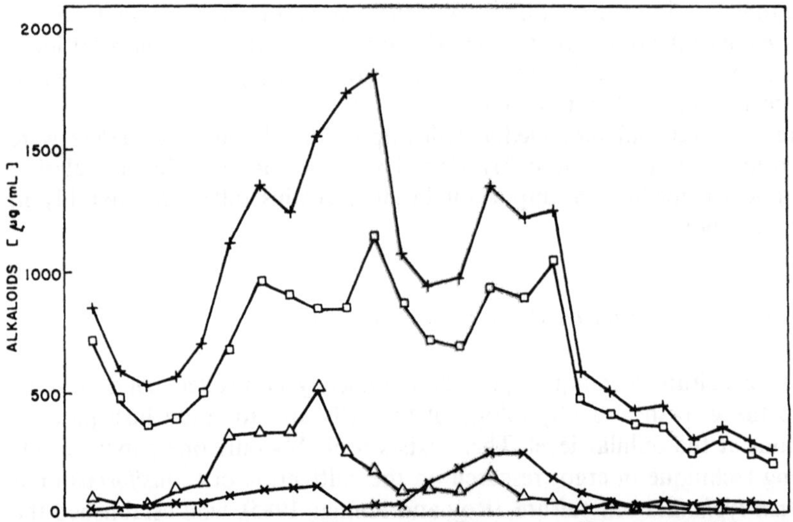


**Fig. 8.** Semicontinuous production of ergot alkaloids by Ca-alginate immobilized *Claviceps fusiformis* cells (Křen et al. 1987b). The medium was replaced every 20 days. ( + ) Total ergot alkaloids; (□) elymoclavine; (△) agroclavine; ( × ) chanoclavine

and enzymatic systems for biotransformation. Also various *Claviceps* strains were often used, e.g., for oxidations or glycosylations. Claviceps monoxygenases (cyt P-450) have high specifity for some clavine structures which cannot be oxidized to the desired product by any other known system.

A good example is the oxidation of agroclavine to elymoclavine by free or immobilized cells of *C. fusiformis* (Křen et al. 1989). Agroclavin can be produced easily in high yields – up to 10 g/l – but the elymoclavine (desired for semi synthetic drugs) production is limited by the feedback mechanism. Agroclavine, overproduced by some *C. purpurea* strains, can be consecutively oxidized by *C. fusiformis* strains having high monooxygenase activity (however, with low ergoline production).

Some *Claviceps* strains can also be used for ergot alkaloid fructosylation (Křen et al. 1990b).

## 4 Summary and Conclusions

Biotechnology of *Claviceps purpurea* (ergot) producing ergot alkaloids is discussed. Both parasitic (field production) and saprophytic cultivation of the ergot strains are described. *C. purpurea* in agriculture was considered to be dangerous pest, but has recently become an important crop. Bioproduction of ergot alkaloids by submerged fermentation employing various modern techniques such as immobilization, plant cell cultures, and biotransformation, is described.

Ergot alkaloids have a long history and, without doubt, also a promising future. The demand for this commodity is constantly growing, so that more effective technologies must be developed.

## 5 Protocol

Submerged cultivation of *Claviceps sp.* SD-58 for clavine alkaloid production.

The culture from a slant agar medium (Sabouraud) is inoculated into inoculation medium TI (60 ml/300 ml conical flasks) and aseptically cultivated on rotary shaker for 10 days; 5 ml of the mycelial suspension is transferred into the same amount of production medium CS2 and shaken for 20 days at 24 °C. Ergot alkaloid production (agroclavine, elymoclavine, chanoclavine) is 0.8–2.5 g/l. After biomass separation, the liquid is alkalized by ammonia and extracted by ether. Alkaloid can be separated by TLC ($SiO_2$, $CHCl_3$:MeOH:$NH_3$ = 8:2:0.02) and the spots visualized by Ehrlich reagent. Media: TI, (g/l): Sucrose, 100; L-asparagin, 10; $(NH_4)_2SO_4$, 10; $KH_2PO_4$, 0.25; $Ca(NO_3)_2$ 1; yeast extract, 0.1; L-cystein, 0.1; inorganic salts and trace elements, pH 5.5. $CS_2$ (g/l): sucrose, 100; citric acid, 16.8; $(NH_4)_2SO_4$, 10; $KH_2PO_4$, 0.25; $CaCl_2$, 1; inorganic salts and trace elements, pH 5.5.

## References

Ambike SH, Baxter RM, Zahid ND (1970) The relationship of cytochrome P-450 levels and alkaloid synthesis in *Claviceps purpurea*. Phytochemistry 9: 1953–1958

Arechavaleta M, Bacon CW, Plattner RD, Hoveland CS, Radclife DE (1992) Accumulation of

ergopeptide alkaloids in symbiotic tall fescue grown under deficits of soil water and nitrogen fertilizer. Appl Environ Microbiol 58:857–861
Bacon CW (1988) Procedure for isolating the endophyte from tall fescue and screening isolates for ergot alkaloids. Appl Environ Microbiol 54:2615–2618
Basset RA, Chain EB, Dickerson AG, Mantle PG (1972) Comparative metabolism of *Claviceps purpurea* in vivo and in vitro. Biochem J 127 3p–4p
Berde B, Schild HO (eds) (1978) Ergot alkaloids and related compounds. Springer, Berlin Heidelberg New York
Bové FJ (1970) The story of ergot. Karger, Basel
Brady LR (1962) Phylogenetic distribution of parasitism by *Claviceps* species. Lloydia 25:1–36
Brauer KL, Robbers JE (1987) Induced parasexual processes in *Claviceps* sp. strain SD-58. Appl Environ Microbiol 53:70–73
Bretag TW (1985) Control of ergot by a selective herbicide and stubble burning. Trans Br Mycol Soc 85:341–343
Buck KW, Chen AG, Dickerson AG, Chain EB (1968) Formation and structure of extracellular glucans produced by *Claviceps* species. J Gen Microbiol 51:337–352
Chahal SS, Rao VP, Thakur RP (1985) Variation in morphology and pathogenicity in *Claviceps fusiformis*, the causal agent of pearl millet ergot. Trans Br Mycol Soc 84:325–332
Cheng L-J, Robbers JE, Floss HG (1980) End-product regulation of ergot alkaloid formation in intact cells and protoplasts of *Claviceps* species, strain SD-58. J Nat Prod 43:329–339
Cincotta AH, Meier AH (1989) Reduction of body set stores and total plasma cholesterol and triglyceride concentrations in several species by bromocryptine treatment. Life Sci 45:2247–2254
Comino A, Kolar M, Schwab H, Sočič H (1989) Heterologous transformation of *Claviceps purpurea*. Biotechnol Lett 11:389–392
Düvell A, Hessberg-Stutzke H, Oeser B, Rogmann-Backwindel P, Tudzynski P (1988) Structural and functional analysis of mitochondrial plasmids in *Claviceps purpurea*. Mol Gen Genet 214:128–134
Duvick BDN (1966) Influence of morphology and sterility on breeding methodology. In: Frey KJ (ed) Plant breeding. Iowa State University Press, Ames, pp 85–138
Edwards RJ, Fulde GWO, McGrath MA (1991) Successful limb salvage with prostaglandin infusion: a review of ergotamine toxicity. Med J Aust 155:825–827
Eich E, Eichberg D, Schwarz G, Class F, Loos M (1985) Antimicrobial activity of clavines. Arzneim-Forsch/Drug Res 35 (II) :1760–1762
Eich E, Becker C, Sieben R, Maidhof A, Müller WEG (1986) Clavines as antitumor agents, 3: Cytostatic activity and structure/activity relationship of 1-alkyl agroclavines and 6-alkyl 6-noragroclavines. J Antibiot 39:804–812
Flieger M, Votruba J, Křen V, Pažoutová S, Rylko V, Sajdl P, Řeháček Z (1988) Physiological control and process kinetics of clavine alkaloid production by *Claviceps purpurea*. Appl Microbiol Biotechnol 29:181–185
Floss HG (1976) Biosynthesis of ergot alkaloids and related compounds. Tetrahedron 32:873–912
Friedman M, Dao L, Gumbmann MR (1989) Ergot alkaloid and chlorogenic acid content in different varieties of morning-glory (*Ipomoea* spp) seeds. J Agric Food Chem 37:708–712
Gabbai A, Lisbonne L, Pourquier H (1951) Ergot poisoning at Pont St. Esprit. Br Med J 2:650–651
Gaberc-Porekar V, Didek-Brumec M, Sočič H (1990) Carbohydrate metabolism during submerged production of ergot alkaloids. Appl Microbiol Biotechnol 34:83–86
Gilman AG, Goodman LS, Rall TW, Murad F (eds) (1985) Goodman and Gilman's The pharmacological basis of therapeutics, 7th edn. Macmillan, New York, pp 931–940
Golda V, Petr R, Cerman J, Suba P, Votruba M (1980) Studies on behaviour, brain and plasma biochemistry in normotensive, genetically hypertensive, nonobese and/or genetically hypertensive obese rats: effects of bromocryptine. Sci Pap Fac Med Charles Univ Hradec Králové 23:283–305
Gröger D, Gröger L, D'Amico D, He M-X, Floss HG (1991) Steric course of the N-methylation in the biosynthesis of ergot alkaloids by *Claviceps purpurea*. J Basic Microbiol 31:121–125
Harazim P, Valík J, Malinka Z, Kybal J (1988) A method for fermentative preparation of ergot inoculum. Czech Pat 243 009
Heinstein PF, Lee SL, Floss HG (1971) Isolation of DMAPP-trytophan dimethylallyltransferase from the ergot fungus (*Claviceps* sp.). Biochem Biophys Res Commun 44:1244–1251

Keller U (1983) Highly efficient mutagenesis of *Claviceps purpurea* by using protoplasts. Appl Environ Microbiol 46: 580–584

Kim S-U, Cho Y-J, Floss HG (1983) Conversion of elymoclavine to paspalic acid by a particulate fraction from an ergotamine-producing strain. Planta Med 48: 145–148

Kobel H, Sanglier J-J (1986) Ergot alkaloids. In: Rehm H-J, Reed G (eds) Biotechnology, vol 4. VCH, Weinheim, pp 569–609

Kopp B, Rehm H-J (1983) Alkaloid production by immobilized mycelia of *Claviceps purpurea*. Eur J Appl Microbiol Biotechnol 18: 257–263

Kopp B, Rehm H-J (1984) Semicontinuous cultivation of immobilized *Claviceps purpurea*. Appl Microbiol Biotechnol 19: 141–145

Křen V, (1990) Complex exploitation of *Claviceps* immobilized cells. In: de Bont J, Visser J, Mattiasson B, Tramper V (eds) Physiology of immobilized cells. Proc Int Symp Wageningen, 10–13.12.1989. Elsevier, Amsterdam, pp 551–556

Křen V (1991) Bioconversions of ergot alkaloids (review). Adv Biochem Eng 44: 123–144

Křen V, Řeháček Z (1984) Feedforward regulation of phosphofructokinase in a submerged culture of *Claviceps purpurea* producing clavine alkaloids. Speculations Sci Technol 7: 223–226

Křen V, Pažoutová S, Rylko V, Sajdl P, Wurst M, Řeháček Z (1984) Extracellular metabolism of sucrose in a submerged culture of *Claviceps purpurea* in relation to formation of monosaccharides and clavine alkaloids. Appl Environ Microbiol 48: 826–829

Křen V, Řezanka T, Řeháček Z (1985a) Occurrence of ricinoleic acid in submerged cultures of various *Claviceps* species. Experientia 41: 1476–1477

Křen V, Řezanka T, Sajdl P, Řeháček Z (1985b) Identification of fatty acids in submerged cultures of *Claviceps* species. FEMS Microbiol Lett 30: 359–363

Křen V, Pažoutová S, Rylko V, Řeháček Z (1986a) Saprophytic production of clavine alkaloids and activity of 3-hydroxy-3-methylglutaryl coenzyme-A reductase. Folia Microbiol 31: 282–287

Křen V, Řezanka T, Sajdl P, Řeháček Z (1986b) Identification of sterols in submerged cultures of different *Claviceps* species. Biochem Physiol Pflanz 181: 505–510

Křen V, Chomátová S, Břemek J, Pilát P, Řeháček Z (1986c) Effect of some broad-specturm antibiotic on the high-production strain *Claviceps fusiformis* W1. Biotechnol Lett 8: 327–332

Křen V, Pažoutová S, Sedmera P, Rylko V,Řeháček Z (1986d) High-production mutant *Claviceps purpurea* 59 accumulating secoclavines. FEMS Microbiol Lett 37: 31–34

Křen V, Mehta P, Rylko V, Flieger M, Kozová J, Řeháček Z (1987a) Substrate regulation of elymoclavine formation by some saccharides. Zentralbl Mikrobiol 142: 71–85

Křen V, Ludvík J, Kofroňová O, Kozová J, Řeháček Z (1987b) Physiological activity of immobilized cells of *Claviceps fusiformis* during long-term semicontinuous cultivation. Appl Microbiol Biotechnol 26: 219–226

Křen V, Břemek J, Flieger M, Kozová J, Malinka Z, Řeháček Z (1989) Bioconversion of agroclavine by free and immobilized *Claviceps fusiformis* cells. Enzyme Microb Technol 11: 685–691

Křen V, Pažoutová S, Řezanka T, Víden I, Amler E, Sajdl P (1990a) Regulation of lipid and ergot alkaloid biosynthesis in *Claviceps purpurea* by chlorophenoxy acids. Biochem Physiol Pflanz 186: 99–108

Křen V, Flieger M, Sajdl P (1990b) Glycosylation of ergot alkaloids by free and immobilized *Claviceps purpurea* cells. Appl Microbiol Biotechnol 32: 645–650

Kybal J, Sikyta B (1985) A device for cultivation of plant and animal cells. Biotechnol Lett 7: 467–470

Kybal J, Sikyta B (1986) Renaissance der Oberflächenkultur: Produktion von Mutterkornalkaloiden und Sporenbildung. Acta Biotechnol 6: 245–251

Kybal J, Nesrsta M, Strnadová K, Břemek J, Valík J, Považská H (1990) A method for production of storable preparates with high content of viable spores of filamentous fungi. Czech Pat Appl PV 01609-90 (2.4.90)

Loveless AR (1971) Conidial evidence for host restriction in *Claviceps purpurea*. Trans Br Mycol Soc 56: 419–434

Loveless AR, Peach JM (1986) Evidence from ascospores for host restriction in *Claviceps purpurea*. Trans Br Mycol Soc 86: 603–610

Maier W, Erge D, Schmidt J, Gröger D (1980) A blocked mutant of *Claviceps purpurea* accumulating chanoclavine-I-aldehyde. Experientia 36: 1353–1354

Markstein R, Seiler MP, Jaton A, Briner U (1992) Structure activity relationship and therapeutic uses of dopaminergic ergots. Neurochem Int 20:211S–214S
Martín JF (1977) Control of antibiotic synthesis by phosphate. Adv Biochem Eng 6:105–127
Ninomiya I (1992) Recent progress in the synthesis of indole alkaloids. J Nat Prod 55:541–564
Otsuka H, Quingley FR, Gröger D, Anderson JA, Floss HG (1980) In vivo and in vitro evidence for N-methylation as the second pathway-specific step in ergoline biosynthesis. Planta Med 40 :109–119
Pažoutová S, Řeháček Z (1981) The role of citrate on the oxidative metabolism of submerged cultures of *Claviceps purpurea* 129. Arch Microbiol 129:251–253
Pažoutová S, Řeháček Z (1984) Phosphate regulation of phosphatases in submerged cultures of *Claviceps purpurea* 129 producing clavine alkaloids. Appl Microbiol Biotechnol 20:389–392
Puc A, Sočič H (1977) Carbohydrate nutrition of *Claviceps purpurea* for alkaloid production related to the osmolarity of media. Eur J Appl Microbiol 4:283–287
Rentschler & Co (1972) Verfahren Zur Erzeugung von Mutterkorn auf Gramineen. Ger Pat 2 048 398
Reshetilova TA, Kozlovskii AG (1990) Biosynthesis of alkaloids by mycelial fungi (review). Prikl Biokhim Mikrobiol 26:291–306
Robbers JE, Eggert WW, Floss HG (1978) Physiological studies on ergot. Time factor influence on the inhibitory effect of phosphate and the induction effect of tryptophan on alkaloid production. Lloydia 41:120–129
Roy S, Kumar A (1985) Production of alkaloids by ergot (*Claviceps fusiformis* Lov.) on *Pennisetum typhoides* (Burm.) Stapf and Hubb. in vitro. In: Neumann KH, Bart W (eds) Primary and secondary metabolism of plant cell cultures. Springer, Berlin Heidelberg New York, pp 117–123
Schmauder H-P (1982) saprophytic production of ergot alkaloids. In: Atal CK, Kapur BM (eds) Cultivation and utilization of medicinal plants. Regional Res Lab, CSIR, Jammu-Tawi, India, pp 188–206
Shibuya M, Chou H-M, Fontoulakis M, Hassam S, Kim S-U, Kobayashi K, Otsuka H, Rogalska E, Cassady JM, Floss HG (1990) Stereochemistry of the isoprenylation of tryptophan catalyzed by 4-($\gamma,\gamma$-dimethylallyl) tryptophan synthase from *Claviceps*, the first pathway-specific enzyme in ergot alkaloid biosynthesis. J Am Chem Soc 112:297–304
Spalla C, Marnati MP (1978) Genetic aspects of the formation of ergot alkaloids. In: Hütter R, Leisinger T, Nuesch J, Wehrli W (eds) Antibiotics and other secondary metabolites. FEMS SYMP 1978, Academic Press, London, pp 219–232
Srikrai S, Robbers JE (1983) Methods for mutation and selection of the ergot fungus. Appl Environ Microbiol 45:1165–1169
Šterzl J, Řeháček Z, Cudlín J (1987) Regulation of the immune response by ergot alkaloids. Czech Med 1:90–98
Stoll A (1942) Altes und Neues über Mutterkorn. Mitt Naturforsch Ges Bern 1942:45–80
Stoll A, Brack A (1944) Zum feldmässigen Anbau von Mutterkorn. Pharm Acta Helv 19:118–123
Tanner JR (1987) St. Anthony's fire, then and now: a case report and historical review. Can J Surg 30: 291–293
Tudzynski P, Esser K (1982) Genetics of the ergot fungus *Claviceps purpurea* Part 2. Exchange of genetic material via meiotic recombination. Theor Appl Genet 61:97–100
Tudzynski P, Düvell A, Esser K (1983) Extrachromosomal genetics of *Claviceps purpurea*. I. Mitochondrial DNA and mitochondrial plasmids. Curr Genet 7:145–150
Tulasne LR (1853) Mémoire sur l'ergot des glumacées. Ann Sci Nat 20:5–56
Valík J, Malinka Z (1992) New inoculating material for parasitical production of ergot. In: Steiner W, Haltrich H (eds) Biotechnology in Central European Initiative Countries. 1st Conf, Graz, Austria, April 13, 1992. p 81
van Engelenburg F, Smit R, Goosen T, van den Broek H, Tudzynski P (1989) Transformation of *Claviceps purpurea* using a bleomycin resistance gene. Appl Microbiol Biotechnol 30:364–370
Wood G, Cooley-Smith JR (1980) The effectiveness of fungicides used against *Claviceps purpurea* attacking male-sterile barley in field trials. Ann Appl Biol 96:169–175
Wood G, Cooley-Smith JR (1982) Observation on the prevalence and incidence of ergot disease in Great Britain with special reference to open-flowering male-serile cereals. Ann Appl Biol 95:41–46
Wood G, Cooley-Smith JR (1982) Epidemiology of ergot disease (*Claviceps purpurea*) in open flowering male-sterile cereals. Ann Appl Biol 100:73–82

# XI *Coleonėma album*: In Vitro Culture and Production of Essential Oils

G. Reil[1], Z. Akkan[1], and R. G. Berger[2]

## 1 Introduction

The genus *Coleonėma* (family Rutaceae) comprises a number of popular ornamental plants (Williams 1981). *Coleonėma album* Bartl. et H. L. Wendl (Thunb.), a small shrub native to South Africa, has neither been well characterized so far nor has in vitro culture been reported previously. The flowers bloom from March to May and show a typical, disk-like thickened floral axis. Morphological peculiarities include stomata confined to the lower side and a narrow median zone of the leaf and, frequently, mucilaginous inner walls of the epidermal cells. Nuclear DNA content and ploidy level have been reported (Guerra 1984).

Although the scent of the species has long been known, neither flowers nor vegetative parts have gained industrial importance. Chemical studies on the plant were limited to limonoid bitter principles (Dreyer et al. 1972), and to several coumarins (Gray 1981). Of the closely related *C. pulchellum*, an investigation of the volatile oil has been reported (Brophy and Lassak 1986).

In this chapter are described: (1) the investigation of the volatile constituents of the aerial parts of *Coleonėma album*, (2) the accumulation of essential oil compounds in a photomixotrophic callus culture initiated from the green parts of this plant, and (3) the effect of phytohormone and light-induced cellular differentiation on the formation of secondary volatile products (Berger et al. 1990).

## 2 In Vitro Culture Studies

### 2.1 Materials and Methods

#### *2.1.1 Cell Cultures*

*Coleonėma album* plants were from the greenhouse of the Munich Botanical Garden. Leaves and stems from 7-month-old plants were sterilized with 2% w/v

[1] Institut für Lebensmitteltechnologie und Analytische Chemie der Technischen Universität München, 85350 Freising, Germany
[2] Institut für Lebensmittelchemie der Universität Hannover, 30453 Hannover, Germany

Biotechnology in Agriculture and Forestry, Vol. 28
Medicinal and Aromatic Plants VII (ed. by Y.P.S. Bajaj)

NaOCl solution for 10 min. Callus initiation using excised tissue of both organs was tested on six different media, i.e., $B_5$ medium (Gamborg et al. 1968), revised MS medium (Pech et al. 1975), LS medium (Linsmaier and Skoog 1965), SH medium (Schenk and Hildebrandt 1972), MG medium, originally used for Muscat grape cells (Ambid et al. 1983), and LH medium: micro and macronutrients as for LS medium, vitamins and growth factors as for SH medium.

Stems and leaves were equally well suited as organs for culture initiation. Best growth was obtained on SH medium. After several subcultivations, calli from both stems and leaves showed, as expected, converging characteristics in terms of macroscopic and microscopic appearance and essential oil patterns. In order to clarify the effect of carbon source and growth substances on callus growth and the accumulation of secondary products, different variants of the basic SH medium, differing in the amounts of carbohydrates and effectors, were developed: original SH medium: 3% sucrose, 0.5 mg/l 2,4-D, 2 mg/l pCPA, 0.1 mg/l kinetin; SH 1: 1% sucrose, 1% lactose, 0.1 mg/l NAA, 0.1 mg/l kinetin; SH 2: 1% sucrose, 1% lactose, 0.05 mg/l NAA, 0.2 mg/l kinetin; SH 3: 1% sucrose, 1% lactose, 0.5 mg/l NAA, 2 mg/l kinetin.

Cells were kept at 25 °C. Photomixotrophic cultures were obtained by illumination using a photoperiod of 16 h white light (6000 lx, Osram HQL/R 80 W), unless indicated otherwise. Callus cells were transferred to fresh medium every 5 to 7 weeks, and suspension cells every 4 weeks.

### *2.1.2 Preparation of Extracts for GLC Analysis*

Extracts from the green parts of the plant and from callus cultures were prepared immediately after harvesting the plant material. Callus cells were carefully separated from adhering medium, mixed with the twofold amount per weight of MeOH and 1 ml internal standard (100 μg 2-undecanone/ml) in a cooled Waring Blendor (30 s). The homogenate was centrifuged (5000 *g*, 10 min, 4 °C), the supernatant was diluted with a saturated solution of NaCl to a MeOH concentration below 40%. This solution was extracted three times with pentane/diethyl ether (2/1). The dried extract was concentrated at 40 °C using a Vigreux column (Drawert et al. 1969), and submitted to capillary gas chromatography.

Shoots and leaves of the parent plant were cut into MeOH and treated like the callus material.

### *2.1.3 Determination of Chlorophyll*

The sum of chlorophyll a and b was determined spectrophotometrically in a 80% (v/v) acetone extract according to Aron (1949).

### *2.1.4 Gas Liquid Chromatography and Coupled Mass Spectrometry*

A 25 m × 0.3 mm i.d. fused silica capillary column coated with Carbowax 20 M and a Carlo Erba gas chromatograph Fractovap 4130 were used. Conditions:

temperature program 50 °C 5 min isothermal, 50–190 °C at 2 °C/min, at 190 °C isothermal; carrier gas hydrogen, 1.6 ml/min (70 °C); injector split ratio 1:10. A 30 m × 0.5 mm i.d. WCOT capillary column coated with Carbowax 20 M and a Carlo Erba gas chromatograph Fractovap 2350 were used for sniffing analysis (Nitz et al. 1984). Conditions: temperature program 65 °C 5 min isothermal, 65–190 °C at 2 °C/min; carrier gas hydrogen, 1 ml/min (65 °C); split ratio FID: sniffing port was 1:5.

GLC-MS: a Finnigan gas chromatograph 9610, directly coupled with a mass spectrometer (Finnigan 4021, Quadrupol), was used. Spectra were recorded at an electron energy of 70 eV in combination with an Incos data system. Spectra were analyzed either by using reference compounds or by comparison with an internal NBS library (ca 36000 spectra).

#### *2.1.5 Electron Microscopy*

Heterotrophic and photomixotrophic callus cells on SH 3 medium were fixed with 2.5% glutaraldehyde in 0.1 M K-phosphate buffer pH 7.4, 1 h. The tissue was rinsed with phosphate buffer and postfixed with 1% osmium tetroxide/-phosphate buffer for 2 h. The material was serially dehydrated in ethanol and embedded in epon. Ultramicrotome sections were stained with lead citrate and viewed with the EM 10 A Zeiss electron microscope.

### 2.2 Results

#### *2.2.1 Cell Cultures*

Different media for callus initiation were examined and plant sections were explanted in the dark and under illumination. After 3 weeks, callus formation was observed on SH, MG, LS, and LH-medium. No initiation was achieved on B5 and MS medium during a period of 12 weeks. Under light conditions, green primary calli were obtained. Callus growth and chlorophyll formation were best on SH medium. All subsequent investigations were, therefore, conducted on this medium. Both the heterotrophic and photomixotrophic callus tissue grew slowly. In the dark, homogeneous smooth gray-colored tissue was formed; callus developed under light consisted of firm and nonhomogeneous aggregates with green, brownish, and white cells. By transfer to SH medium variants with higher cytokinin/auxin ratio, this heterogeneity was supported. On SH 3 medium a more friable tissue was formed.

Photomixotrophic callus growing on SH 3 medium had a chlorophyll content of 110 μg/g fresh weight (parent leaves: 315 μg/g fresh weight).

#### *2.2.2 Chemical Analysis of Volatile Compounds*

The composition and yield of volatile constituents in *C. album* cell cultures growing on different SH media variants were compared with the essential oil of the green parts of the plant. Figure 1 shows distinct qualitative and quantitative differences between extracts of the tissues from the intact plant and of photomixotrophic callus cultures on media variants SH 1 and SH 3.

*Stem and Leaves of the Plant.* GLC-MS examination of the essential oil of *C. album* led to the detection of 41 constituents, mainly monoterpenes, sesquiter-

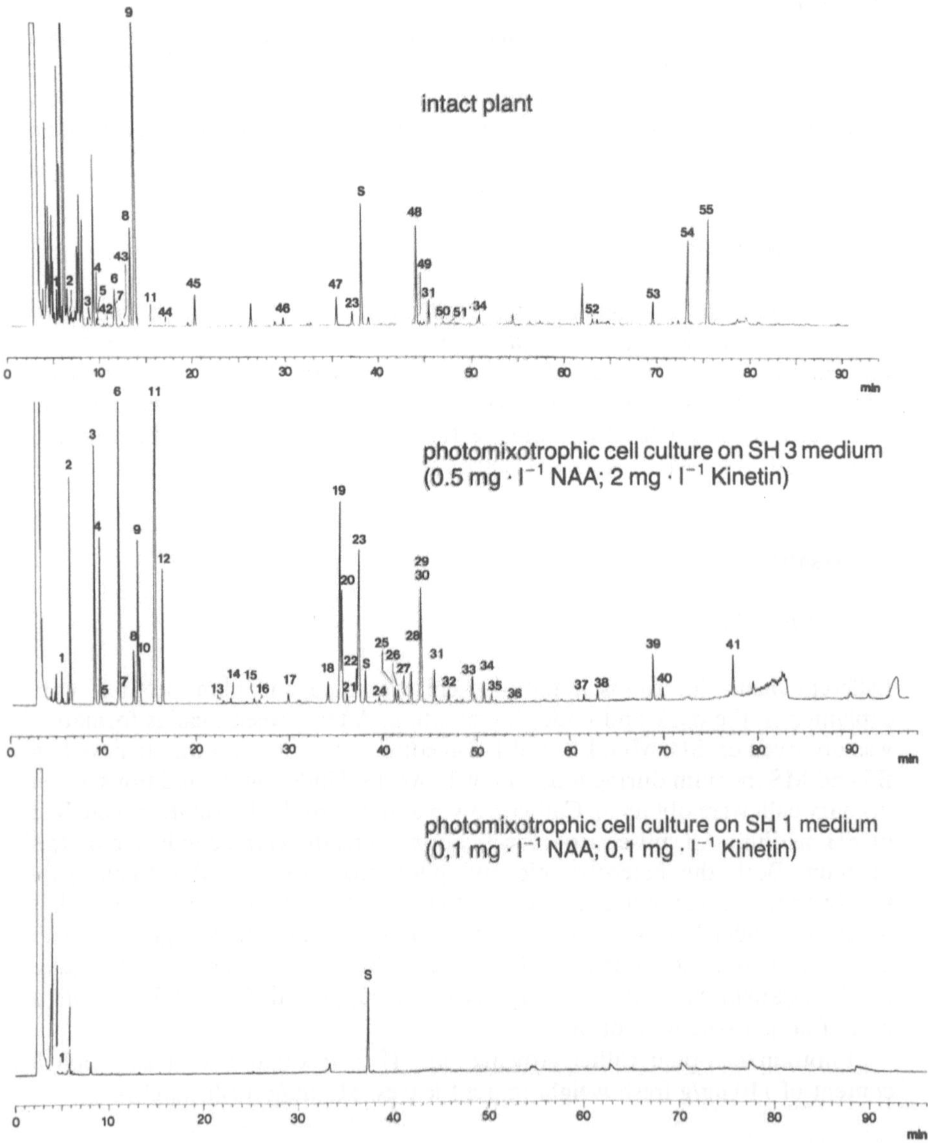


**Fig. 1.** Capillary GLC profiles of extracts from the green parts of *Coleonema album* plants and from photomixotrophic calli derived thereof. Peak # as in Tables 1 and 2; *S* internal standard; chromatographic conditions as in Section 2.1. (Berger et al. 1990)

**Table 1.** Compounds identified in the essential oil of green parts of *Coleonema album* plants. (Berger et al. 1990)

| | Compound | Quantity (mg/kg fr. wt.) | Identification method[a] |
|---|---|---|---|
| 1 | Ethyl 2-methylbutanoate | 15 | MS, R |
| 2 | α-Pinene | 139 | MS, R |
| 3 | β-Pinene | 190 | MS, R |
| 4 | Sabinene | 38 | MS, R |
| 6 | Myrcene | 241 | MS, R |
| 7 | α-Phellandrene | 3 | MS, R |
| 8 | Limonene | 38 | MS, R |
| 9 | β-Phellandrene | 126 | MS, R |
| 10 | 1,8-Cineol | 35 | MS, R |
| 11 | Z-β-Ocimene | 449 | MS, R |
| 12 | E-β-Ocimene | 99 | MS, R |
| 14 | Allo-ocimene | 5 | MS, R |
| 16 | β-Thujone | 5 | MS, R |
| 18 | Pinocamphone | 20 | MS |
| 20 | Linalool | 102 | MS, R |
| 21 | Methyl citronellate | 9 | MS, R |
| 23 | α-Guaiene | 167 | MS, R |
| 25 | Pinocarveol, $M^+$ 119 m/z 41(100), 55, 92, 91, 70, 83, 43 | 12 | MS |
| 26 | α-Humulene | 12 | MS, R |
| 28 | Verbenone, $M^+$ 150 m/z 107(100), 39(86), 80(85), 41(82), 135(76), 91(72), 79(55) | 26 | MS |
| 29 | Germacrene D | 76 | MS, R |
| 30 | α-Terpineol | 62 | MS, R |
| 31 | Bicyclogermacrene | 25 | MS, R |
| 32 | E,E-α-Farnesene | 10 | MS, R |
| 33 | Myrtenol | 25 | MS, R |
| 34 | Germacrene B | 21 | MS, R |
| 35 | Carvylacetate I | 10 | MS |
| 36 | Carvylacetate II | 2 | MS |
| 37 | Eugenol methyl ether | 8 | MS, R |
| 38 | E-Nerolidol | 11 | MS, R |
| 39 | Eugenol | 39 | MS, R |
| 40 | Isoeugenol methyl ether | 18 | MS, R |
| 41 | Isoeugenol | 86 | MS, R |

[a] Means of identification were: MS, comparison of mass spectra with MS data system; R, comparison of GC retention times with authentic reference compounds.

penes, and phenylpropanoids with Z-β-ocimene, myrcene, α- and β-pinene, β-phellandrene being the major monoterpenes (Table 1). Several compounds of the extract could not be identified by GLC-MS, and the complete structural elucidation warrants further investigation. Pinocarveol and verbenone were identified by comparison with literature mass spectral data. α-Guaiene and germacrene D were the major sesquiterpenes, and eugenol was the main phenylpropanoid compound. The total yield of essential oil was 0.24% (g/g) of the fresh weight. Major volatiles common to both *C. album* and *C. pulchellum* are α-pinene, β-pinene, myrcene, β-phellandrene, and germacrene D (Brophy and Lassak 1986).

*Cell Culture.* Heterotrophic and photomixotrophic callus cultures on media SH, SH 1, SH 2, and the heterotrophic culture on medium SH 3 contained structurally simple compounds as the main volatile constituents: primary alcohols, acetic acid, acetoin, etc. No volatile terpenes could be detected in these cultures.

First indications for oil production of the green cell culture on medium SH 3 were derived from sensory evaluations. This cell line had a spicy odor reminiscent of fresh pepper and distinctly differing from the smell of the parent plant.

GLC-MS analysis confirmed the presence of mono- and sesquiterpenes in the callus extract. The composition of the essential oil of a 48-day-old cell culture is shown in Table 2. Eleven monoterpenes and five sesquiterpenes were identified. Main constituents were the hydrocarbons β-phellandrene and δ-selinene. Sniffing analysis revealed that an unidentified compound was responsible for the characteristic peppery odor. This component, which could not be detected in the parent plant extract, and several other constituents require further structural elucidation. The total yield of essential oil was 0.007% of the fresh weight. Production of flavor compounds has been observed for now more than 4 years of continued subculturing.

**Table 2.** Compounds identified in the essential oil of photomixotrophic callus cultures of *Coleonema album*. (Berger et al. 1990)

| | Compound | Quantity (mg/kg fr. wt.) | Identification method[a] |
|---|---|---|---|
| 1 | Ethyl 2-methylbutanoate | 1.7 | MS, R |
| 2 | α-Pinene | 0.5 | MS, R |
| 3 | β-Pinene | 0.1 | MS, R |
| 4 | Sabinene | 0.2 | MS, R |
| 6 | Myrcene | 1.5 | MS, R |
| 7 | α-Phellandrene | 1.7 | MS, R |
| 8 | Limonene | 3.4 | MS, R |
| 9 | β-Phellandrene | 23.9 | MS, R |
| 11 | β-Ocimene | 0.1 | MS, R |
| 23 | α-Guaiene | 0.9 | MS, R |
| 31 | Bicyclogermacrene | 1.4 | MS, R |
| 34 | Germacrene B | 0.6 | MS, R |
| 42 | δ-3-Carene | 0.1 | MS, R |
| 43 | α-Terpinene | 0.1 | MS, R |
| 44 | p-Cymene | 0.1 | MS, R |
| 46 | δ-Elemene | 0.6 | MS, R |
| 48 | δ-Selinene | 5.7 | MS |
| 49 | δ-Guaiene | 0.1 | MS, R |

[a] Means of identification were: MS, comparison of mass spectra with MS data system; R, comparison of GC-retention times with authentic reference compounds.

### *2.2.3 Effects of Light Intensity*

Different light intensities were applied to parallel cultures. Increased illumination (from 1500 to 12 000 lx) resulted in a decreased growth index (from 4 to 1.5), and increased chlorophyll contents (from 150 to 450 μg/100 g fr.wt.) with the ratio of chlorophyll a/b approaching the ratio found in the intact plant (3/1). A concurrent increase of the essential oil constituents was monitored: starting with less than 30 mg/kg fr. wt., more than 70 mg of total volatiles were formed at 6000 lx. Incubations at 9000 and 12 000 lx, respectively, resulted in a further stimulation of terpene biosynthesis, but quantitative data suggested a saturation of light induction beyond 12 000 lx: while a continuous illumination at 12 000 lx resulted in maximum formation of chlorophylls, the terpenes showed no significant additional increase when the photoperiod was extended from 8 over 12 and 16 to 24 h.

### *2.2.4 Comparison of the Essential Oil of Green Parts of the Plant and of the Mixotrophic Cell Culture*

The pattern of essential oil constituents showed significant qualitative and quantitative differences:

- Thirteen constituents of the plant were also produced by callus cells: ethyl 2-methylbutanoate, α- and β-pinene, sabinene, myrcene, α- and β-phellandrene, limonene, Z-β-ocimene, and the sesquiterpenes α-guaiene, bicyclogermacrene, germacrene B.
- Z-β-ocimene is the main compound of the plant essential oil, whereas in cell culture β-phellandrene is the major constituent. α- and β-phellandrene are main compounds in the essential oil of pepper (Muller et al. 1968). As the aroma quality of the callus volatiles resembled that of pepper, the extract was supposed to contain other typical pepper constituents, such as the caryophyllenes. However, none of the unidentified callus oil components could be attached to known pepper constituents.
- The pinane derivatives (except α- and β-pinene), the oxygen containing monoterpenes and the following volatiles of the plant were not present in cell culture extracts: allo-ocimene, E-β-ocimene, 1,8-cineole, E-nerolidol, E,E-farnesene, α-humulene, germacrene D, the phenylpropanoids, and the unidentified compounds.
- The cell culture accumulated some volatiles which were not detectable in the plant: δ-3-carene, p-cymene, α-terpinene, and the sesquiterpenes δ-elemene, δ-guaiene, and δ-selinene.

### *2.2.5 Ultrastructural Features*

Examination of the ultrastructure of the terpene-producing callus cells revealed that they contained a heterogeneous population of plastids. Upon illumination, chloroplast differentiation occurred. In all chloroplasts, thylakoids were arranged mainly in grana (Fig. 2). Most of the chloroplasts contained significant amounts of starch. Besides these plastids, different leucoplast forms (characterized by the absence of plastoribosomes and thylakoids) were observed, some containing starch grains, some containing osmiophilic droplets (Figs. 3–5). Chloroplasts and leucoplasts were not associated in the same cell. Thus, it is supposed that there exist at least two different kinds of cell in this culture, which are characterized by different plastid forms. In heterotrophic callus cultures growing on the same medium variant, no plastids could be visualized.

## 2.3 Discussion

The biotechnological production of essential oil components used in pharmaceuticals and aroma therapy, in perfumery and in the food industry is attractive from a commercial viewpoint (Robins 1987). In recent years, tissue cultures of a variety of herbs and of other flavor sources have been established, and the production of flavor compounds has been investigated. The number of volatiles

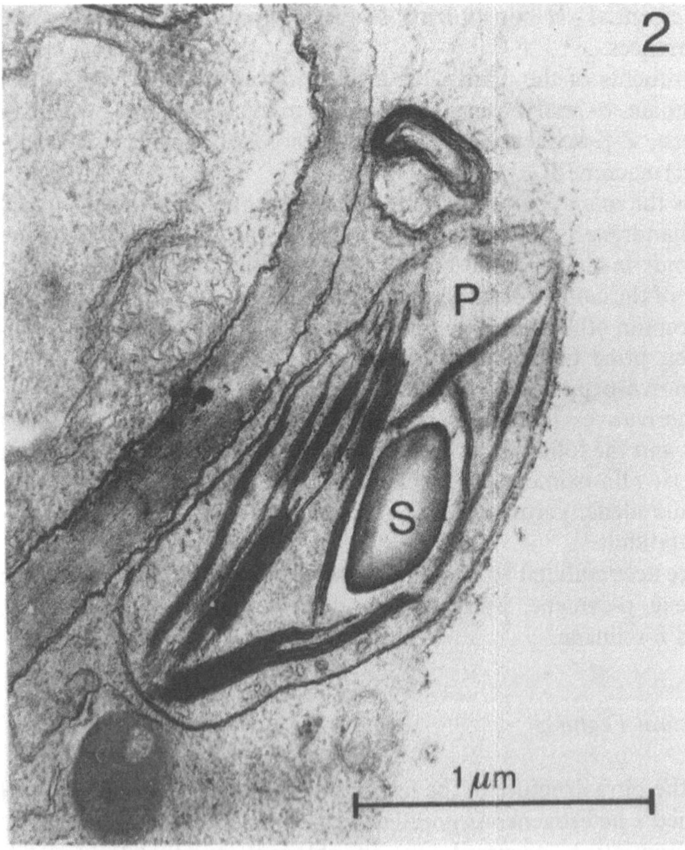


**Fig. 2**

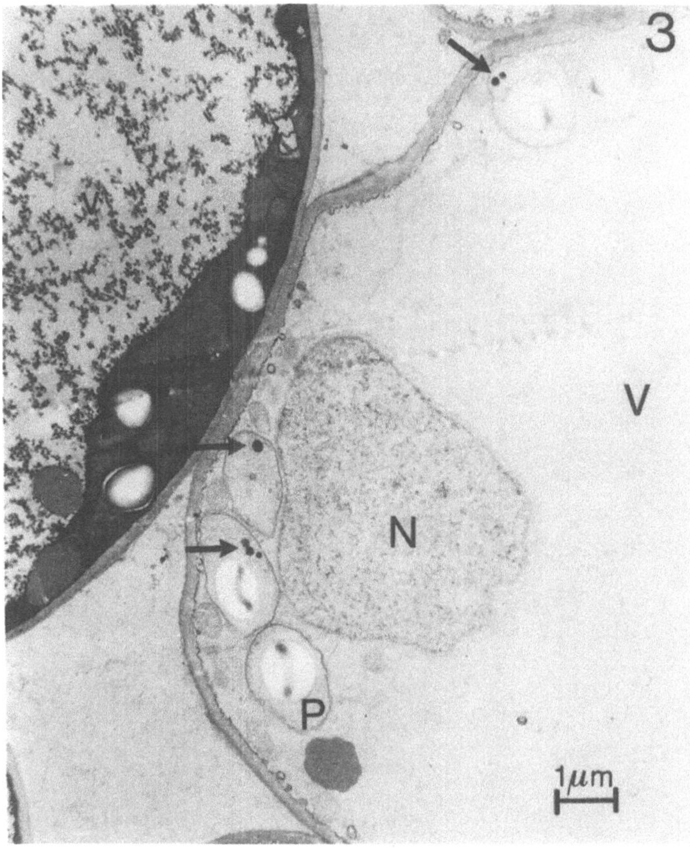


Fig. 3

identified in plant cell cultures is increasing (review in Koch-Heitzmann and Schultze 1988). However, reports on the accumulation of significant amounts of essential oil components are rarely found (e.g., Reichling et al. 1985; Lappin et al. 1987; Sugisawa et al. 1988), indicating that there are still principal obstacles in producing these secondary metabolites de novo by in vitro systems at present.

Most higher plants excrete mono- and sesquiterpenes into specialized storage sites, such as glandular hairs and resin ducts. The lack of these products in many cell cultures has been related to the low degree of cellular specialization of the culture. In tissue cultures that do not contain appropriate storage sites, essential oil components may severely disturb membrane functions (Knobloch

**Figs. 2–5.** Ultrastructure of photomixotrophic calli of *Coleonema album* grown on SH 3 medium. *N* Nucleus; *P* plastid; *S* starch grain; *V* vacuole. **Fig. 2** chloroplast. **Fig. 3** cell with leucoplasts and osmiophilic droplets (*arrows*); **Figs. 4** and **5** leucoplasts and osmiophilic droplets (*arrows*) (Berger et al. 1990)

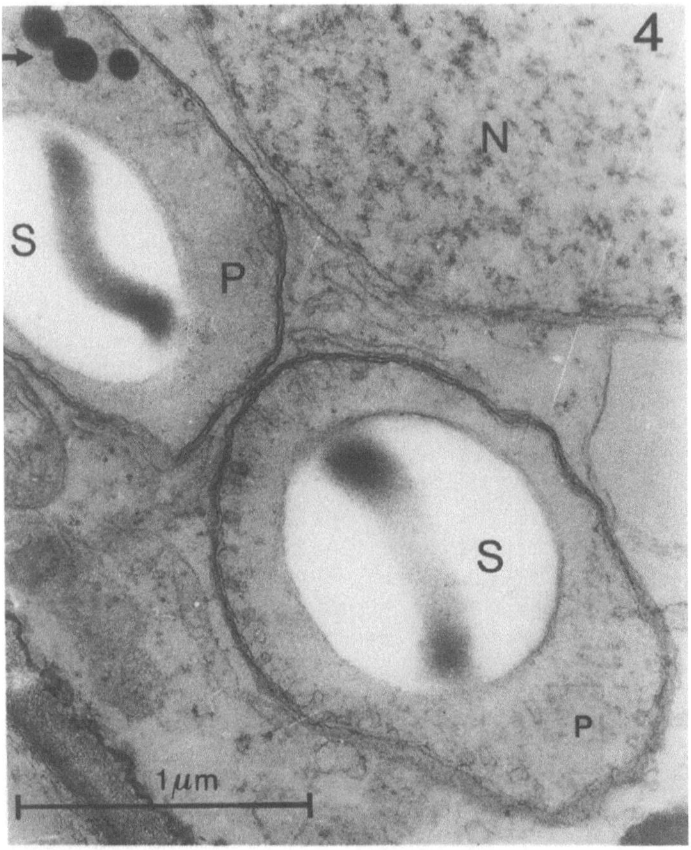


Fig. 4

et al. 1986). In cultures of *Lavandula angustifolia* and *Rosmarinus officinalis* grown under natural daylight, for example, monoterpenes were detected after regeneration of shoots only (Webb et al. 1984), indicating that the corresponding genetic information was suppressed in the callus stage.

In contrast to most of the earlier reports on in vitro cultures of essential oil plants, photomixotrophic tissue cultures of *C. album*, when cultivated under appropriate conditions, are a rich source of volatile secondary metabolites. The absolute concentrations of volatiles in the callus culture were lower than those accumulated by the plant, but the estimated productivities (time of accumulation in cell culture: 4 weeks vs. plant: 30 weeks) of both sources differed by a factor of about 4 only. In the case of *C. album*, illumination and distinct growth factor conditions were a prerequisite. It is generally assumed that some correlation exists between the degree of cell differentiation and the level of accumulation of volatiles (Banthorpe and Njar 1984; Watts et al. 1984; Brown and Charlwood 1986). The importance of chloroplast differentiation for the formation of secondary products has been demonstrated in various tissue cultures in

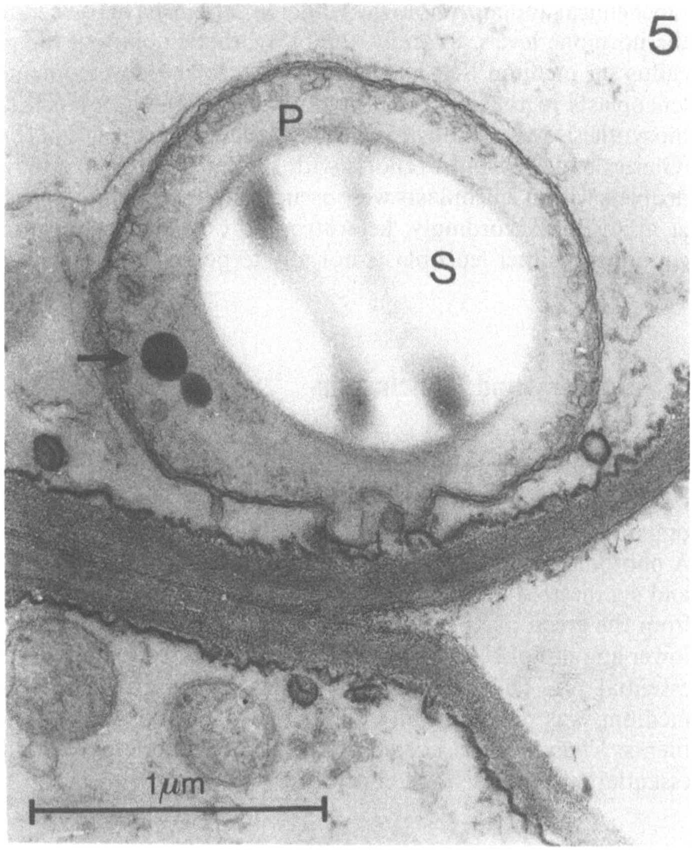


**Fig. 5**

which secondary metabolites were induced after greening of the cells (Charlwood et al. 1986). Common biosynthetic routes can be formulated for terpenoid flavor compounds and the phytyl side chains of chlorophylls. Several light-dependent enzymes of isoprenoid biosynthesis are localized in chloroplasts (Simpson and Herrera-Estrella 1990).

The results with the photomixotrophic cell lines on original SH medium and on media with different growth factor concentrations show that light is a crucial, but not the only, factor for the formation of volatile terpenes. Accumulation markedly depends on the levels of hormones present in the medium. The comparison of the productivity on media SH 2 and SH 3 demonstrates that not only the ratio of concentrations, but also the absolute amounts of the growth factors are important. Both media have the same hormone ratio, but only the elevated hormone concentrations resulted in terpenoid production.

Thus, in *C. album* cell culture, at least two conditions have to be met for secondary product accumulation to occur. In addition to illumination, further biochemical and morphological differentiations have to be induced by changing the hormone levels. Possibly, the biosynthetic ability of the photomixotrophic callus on medium SH 3 is associated with the development of characteristic leucoplasts in the cells. The involvement of these organelles in monoterpene biosynthesis was demonstrated using isolated leucoplasts of *Citrofortunella mitis* (Gleizes et al. 1983). In reports with similar electron micrographs, osmiophilic droplets within leucoplasts were assumed to represent terpenes (Bernard-Dagan et al. 1979). Accordingly, heterotrophic *C. album* cultures on SH 3 medium contained neither leucoplasts nor any terpenoid constituents.

## 3 Summary and Conclusions

The volatile oil from shoots of *Coleonėma album* was found to be composed of monoterpenes, sesquiterpenes, and phenylpropanoids. Heterotrophic and photomixotrophic callus cultures were established from the green parts of the plant. A photomixotrophic cell line contained a heterogeneous population of plastids and synthesized an essential oil which differed qualitatively and quantitatively from the green parts of the intact plant. Substitution of 2,4-D and pCPA by a lower amount of NAA and high kinetin concentrations favored the formation of essential oil. The heterotrophic callus, when grown on the same nutrient medium, was devoid of any differentiated plastids and of essential oil constituents. Thus, a close correlation of light-induced plastid differentiation and essential oil synthesis was concluded.

## References

Ambid C, Moisseeff M, Fallot J (1983) Interconversion of monoterpene aldehydes and alcohols by a cell suspension culture of Muscat grape. Physiol Vég 21:87–92

Aron D (1949) Copper enzymes in isolated chloroplasts; phenoloxidase in *Beta vulgaris*. Plant Physiol 24:1–15

Banthorpe DV, Njar VCO (1984) Light-dependent monoterpene synthesis in *Pinus radiata* cultures. Phytochemistry 23:295–299

Berger RG, Akkan Z, Drawert F (1990) The essential oil of *Coleonėma album* (Rutaceae) and of a photomixotrophic cell culture derived thereof. Z Naturforsch 45c:187–195

Bernard-Dagan C, Carde JP, Gleizes M (1979) Study of terpene compounds during the growth of maritime pine needles: comparison of biochemical and ultrastructural data. Can J Bot 57:255–263

Brophy J, Lassak E (1986) The volatile oil of *Coleonėma pulchellum*. Flavour Fragrance J 1:155–158

Brown JT, Charlwood BV (1986) The control of callus formation and differentiation in scented pelargoniums. J Plant Physiol 123:409–417

Charlwood BV, Hegarty PK, Charlwood KA (1986) The synthesis and biotransformation of monoterpenes by plant cell culture. In: Morris P, Scragg A, Stafford A, Fowler M (eds) Secondary metabolism in plant cell cultures. Cambridge University Press, Cambridge, pp 25–34

Drawert F, Heimann W, Emberger R, Tressl R (1969) Anreicherung, Trennung und Identifizierung von Apfelaromastoffen. Chromatographia 2:57–66

Dreyer DL, Pickering MV, Cohan P (1972) Distribution of limonoids in the Rutaceae-D. Phytochemistry 11:705–713

Gamborg OL, Miller RA, Ojima K (1968) Nutrient requirements of suspension cultures of soybean root cells. Exp Cell Res 50:151–158

Gleizes M, Pauly G, Carde JP, Marpeau A, Bernard-Dagan C (1983) Monoterpene hydrocarbon biosynthesis by isolated leucoplasts of *Citrofortunella mitis*. Planta 159:373–381

Gray AI (1981) New coumarins from *Coleonema album*. Phytochemistry 20:1711–1714

Guerra MDS (1984) Cytogenetics of Rutaceae 2. Nuclear DNA content. Caryologia 37:219–226

Knobloch K, Weigang H, Weis N, Schwarm H-M, Vigenschow H (1986) Action of terpenes on energy metabolism. In: Brunke EJ (ed) Progress in Essential Oil Research. de Gruyter, Berlin, pp 429–445

Koch-Heitzmann I, Schultze W (1988) Compilation of volatile compounds found in plant cell cultures. In: Schreier P (ed) Bioflavour '87. de Gruyter, Berlin, pp 365–398

Lappin GJ, Stride JD, Tampion J (1987) Biotransformation of monoterpenoids by suspension cultures of *Lavandula angustifolia*. Phytochemistry 26:995–997

Linsmaier EM, Skoog F (1965) Organic growth factor requirements of tobacco tissue cultures. Physiol Plant 18:100–117

Muller C, Creveling R, Jennings WG (1968) Some minor sesquiterpene hydrocarbons of black pepper. J Agric Food Chem 16:113–117

Nitz S, Berger RG, Leupold G, Drawert F (1984) Detektion geruchsaktiver Substanzen durch Kombination einer SCOT-Glaskapillarsäule mit einem variablen Ausgangsteiler. Chem Mikrobiol Technol Lebensm 8:121–124

Pech JC, Latché A, Austruy M, Fallot J (1975) In-vitro growth of tissues and cell suspension cultures from apple fruits. Bull Soc Bot Fr 122:183–194

Reichling J, Becker H, Martin R, Burkhardt G (1985) Comparative studies on the production and accumulation of essential oil in the whole plant and cell culture of *Pimpinella anisum*. Z Naturforsch 40c:165–168

Robins RJ (1987) Progress towards producing food flavour compounds by the biotechnological exploitation of plant cell cultures, In: Martens M, Dalen G, Russwurm H (eds) Flavour science and technology. Wiley, Chichester pp 217–230

Schenk RU, Hildebrandt AC (1972) Medium and techniques for induction and growth of monocotyledonous and dicotyledonous plant cell cultures. Can J Bot 50:199–209

Simpson J, Herrera-Estrella M (1990) Light-regulated gene expression. Crit Rev Plant Sci 9:95–109

Sugisawa H, Miwa K, Matsuo T, Tamura H (1988) Volatile compounds produced from the cultured cells of thyme (*Thymus vulgaris* L.). In: Schreier P (ed) Bioflavour '87. de Gruyter, Berlin, pp 327–340

Watts MJ, Galpin IJ, Collin HA (1984) The effect of growth regulators, light and temperature on flavour production in celery tissue cultures. New Phytol 98:583–591

Webb JK, Banthorpe DV, Watson DG (1984) Monoterpene synthesis in shoots regenerated from callus cultures. Phytochemistry 23:903–904

Williams I (1981) Studies in the genera of the Diomeae Rutaceae 9. A revision of the genus *Coleonéma*. J S Afr Bot 47:63–102

# XII *Dianthus* Species (Carnation): In Vitro Culture and the Biosynthesis of Dianthalexin and Other Secondary Metabolites

U. MATERN[1]

## 1 Introduction

### 1.1 Distribution and Importance of the Plant

The genus *Dianthus* comprises a relatively large group of some 300 species, which have attracted attention because of spectacular flower color combinations ranging from white to yellow, red, and deep purple (Helm 1975; Rechinger 1979). Particular morphological traits and pigmentation distinguish *Dianthus* from other genera within the family Caryophyllaceae, although the evolutionary phyllogenetic background and the subdivision of the genus have remained controversial (Rechinger 1979). The genus was thought to have originated in the Mediterranean hillsites, but is now believed to have inherited traits from various nontropical locations of Europe and Asia (Rechinger 1979). Some species spread along the southeast African continent into South Africa and even into the Far East, which is recognized in their taxonomic designation, *D. chinensis* being one example. *Dianthus* generally prefers moderately dry and warm conditions, high light intensities, and mineral-rich soils (Helm 1975: Rechinger 1979), but a few species such as *D. alpinus* and *D. glacialis* have managed to colonize rock soils in the Austrian and Italian Alps at 2000 to 2800 m altitude. Crosses between species may occur spontaneously, and numerous hybrids are known to exist in their natural habitat as well as under cultivation (Helm 1975; Rechinger 1979). The plants show perennial growth, but there is a trend towards annual cultivars in breeding for ornamental varieties.

### 1.2 Conventional Practices for Propagation, Potential and Demand on the World Market

Since the 16th century *D. caryophyllus*, *D. barbatus*, *D. plumarius*, *D. carthusianorum*, and *D. chinensis* have been chosen in particular as ornamental plants because of their appealing coloration and growth characteristics (Rechinger 1979). Nevertheless, the existence of a truly wild form of *D. caryo-*

[1] Plant Biochemistry, Biological Institute II, University of Freiburg, Schaenzlestrasse 1, 79104 Freiburg, Germany

Biotechnology in Agriculture and Forestry, Vol. 28
Medicinal and Aromatic Plants VII (ed. by Y.P.S. Bajaj)

*phyllus* has remained questionable, and the commercial hexaploid cultural variants subsumed nowadays under *D. caryophyllus* presumably trace back to the Mediterranean *D. sylvestris* and a broadleaved relative such as *D. rupicola*. The latter is also responsible for delayed and repeated blossoms. Similarly, the shorter growing *D. plumarius* used today in rock gardens most likely descended from hybrids of *D. plumarius* and *D. gratianopolitanus* (Rechinger 1979).

Carnation cuttings, mostly of *D. caryophyllus*, make up a considerable share of the world's cut flower market, which, in economic terms, has steadily increased over recent years (Sparnaaij 1987; Lu et al. 1991; Tzuri et al. 1991). Large-scale cultivation and breeding for worldwide sales are being pursued in countries such as Columbia, France, Holland, Israel, Italy, and Korea (Sparnaaij 1987). Therefore, the generation and use of cultivars with enhanced natural pest resistance is definitely needed. Field-grown *D. caryophyllus* blooms during the summer, the plant dries down in autumn to the root that stores "lactosin"-type galactanes as carbohydrate reserve rather than starch (Hegnauer 1964), and new shoots emerge in the following spring. Commercial growers, however, desire high prices during the wintertime and, therefore, explant side shoots which are harvested from the autumn roots. These shoots root within a few weeks in the greenhouse and grow into plants that yield flowers from December through March. Shoot initiation from cryopreserved shoot apices (Seibert 1976; Dereuddre et al. 1988) would provide for storage of germplasm, but has rarely been pursued in commercial nurseries.

Carnation breeding has focused on the ornamental characteristics, and major goals are the "filling" of the flower head by petaloid transformation of stamen and anthers but, in particular, modifications in pigmentation as well as an extended vase life of the cuttings. Chalcones (Harborne 1966), flavonoids (Cordell et al. 1977), and anthocyanins (Terahara et al. 1986) are the natural pigments in the petals, and it is noteworthy in this context that the Caryophyllaceae represents the only family in the Caryophyllales ( = Centrospermae) which lack the otherwise typical betaine pigments (Hegnauer 1964). Accordingly, the genetics of pigmentation of carnation was investigated (Geissman and Mehlquist 1947, Mehlquist and Geissman 1947; Geissman et al. 1956; Forkmann and Dangelmayr 1980). Natural mutation appears to occur frequently, and the selection of spontaneous carnation mutants as well as irradiation mutation (Silvy 1978) have been employed in addition to conventional breeding techniques in the search for new cultivars. The durability of carnation cuttings, on the other hand, depends immensely on the amount of ethylene released by the petals, which causes rapid senescence of the flowers. Indeed, the cut carnation has become a standard model system for the investigation of the role of plant hormones in flower senescence (Cook and van Staden 1988), and numerous reports have been published on the enzymes involved in ethylene biosynthesis (Hanley et al. 1989; Larsen and Woodson 1991; Henskens et al. 1992; Park et al. 1992), on ethylene-regulated flower genes (Raghothama et al. 1991), on the degradation of phospholipids (Engelmann-Sylvestre et al. 1989; Brown et al. 1991a; Rouet-Mayer et al. 1992) and on changes in ultrastructure (Smith et al. 1992) and selected enzyme activities (Droillard and Paulin 1990; De Vetten et al. 1991).

Although *Dianthus* flowers were occasionally used as a medical stimulant (Rechinger 1979), only a few components from this plant other than pigments have been studied in detail (Hegnauer 1964, 1989). Among these were benzoic and salicylic esters that give rise to the flower fragrance, triterpene saponins, pyran-type glycosides, seed oil, and protein contents, etc., none of which has aroused any commercial interest. A few years ago, however, the accumulation of unusual and novel anthranilates was reported from carnation plants infected by fungi. These compounds, which will be dealt with below, show great bioactive potential, and, as suggested previously (Rechinger 1979), the re-investigation of carnation for alkaloid contents appears necessary.

## 2 In Vitro Culture Studies

### 2.1 Review of Biotechnological Studies

The regeneration of carnation from stem and petal (Miller et al. 1991a; Nugent et al. 1991) or axillary bud explant tissues (Miller et al. 1991b) has been reported. Such techniques have become common practice with tissues from other plant sources and are a prerequisite for the generation of transgenic plants. Major emphasis in carnation studies was placed on cultivar and tissue differences, on the effects of cytokinin and growth medium compositions, and on the direct adventitious shoot formation (for review see Mii et al. 1990), thus avoiding the cumbersome screening of somaclonal variants that may arise on passage through callus tissue (Matern et al. 1978). In vitro regeneration of carnation plantlets (Kevers and Gaspar 1985; Leshem et al. 1988), however, revealed low survival upon transplanting due to the vitrification of leaves, which yields non-functional stomata, hypolignification, and heterotrophic growth characteristics (Ziv et al. 1987). The regeneration methodology may also distinguish the chimeric tissue layers, which are known to compose the flowers of some cultural varieties introduced by mutation breeding (Pereau-Leroy 1974), or lead to novel somatic hybrids (Nakano and Mii 1993). All these investigations clearly aimed at the transformation of carnation callus via the *Agrobacterium* Ti-vector and the generation of transgenic plants. Meanwhile, the feasibility of this approach has been proven experimentally by introduction of the GUS reporter gene (Lu et al. 1991). An obvious immediate goal of this research is the delay or suppression of ethylene release by carnation petals, which would confer longevity to carnation cuttings. The integration of useful genes by recombinant DNA technology is considered to complement and expand the scope of traditional carnation breeding. Along these lines, the transient expression of chimeric CAT and GUS constructs was accomplished also in carnation protoplasts (Ming et al. 1990), which comprises a step further towards the selection of promotors that eventually lead to the controlled expression of foreign genes in carnation transformants. Stable inheritage of introduced genes may be checked by quick

DNA fingerprint analysis that has been developed recently for carnation (Tzuri et al. 1991).

Once the carnation transformants have been established, the technique may find wider application, e.g., in the modulation of flower pigmentation. The enzymology and the control of pigmentation of carnation flowers is being unraveled with the in vitro measurements of chalcone synthase (Spribille and Forkmann 1982), dihydroflavonol reductase, and flavonol synthase activities (Stich et al. 1992a, b), and further insight into the regulation of biosynthesis will presumably be gained. Analogous to petunia (Meyer et al. 1987), carnation might eventually qualify for a molecular approach to achieve delphinidin-type coloration by introduction of dihydrokaempferol 3′,5′-hydroxylase, because it was shown that the endogeneous dihydroflavonol reductase accepts dihydromyricetin in vitro (Stich et al. 1992a). The generation of "designer" carnation flowers was envisaged as a priority goal (Lu et al. 1991).

## 2.2 Tissue and Cell Suspension Cultures

Heterotrophic callus cultures of carnation grow readily on standard culture media from the base of petals, and suspension cultures have been established in different laboratories (Buiatti et al. 1985, 1987; Gay 1985; Rebéillé 1988; Reinhard and Matern 1989; Avelange et al. 1991; Palet et al. 1991), but no information is available on their ploidy level. Such cultures have been used in particular for the investigation of phytoalexin synthesis (Gay 1985; Reinhard and Matern 1989), and there are correlations of phytoalexin pattern and disease resistance character when compared with the parent plants (Buiatti et al. 1985, 1987). Besides, photoautotrophic suspension cultures of carnation were also established, which were the first plant cells capable of growing in atmospheric air rather than in an atmosphere enriched with $CO_2$ (Rebéillé 1988). These cultures have been used mainly for the determination of their physiological performance, i.e., measurements of assimilatory versus photorespiratory rates including correlations with mitochondria numbers and the comparison of efficiency ratios with intact carnation leaves. Photoautotrophic cells served furthermore for mass spectrometric measurements of gas exchanges on dark-to-light transition (Avelange et al. 1991) to determine the degree of coupling of $CO_2$ efflux and $O_2$ uptake. A stimulation of respiration in the dark was observed by the addition of bicarbonate and was confirmed for heterotrophic as well as photomixotrophic carnation callus cultures (Palet et al. 1991).

## 2.3 Structure and Enzymology of Bioactive Compounds

A few reports attributed antiviral (Ostermann et al. 1987) and protein synthesis inhibitory activities (Stirpe et al. 1981) to carnation extracts. With the exception of the dianthins (Reisbig and Bruland 1983), which have been classified as RIPs (Stirpe et al. 1992), however, the active components have not been completely characterized and their physiological relevance remains obscure. New interest in the bioactive capabilities of carnation has arisen in recent years by numerous investigations focusing on fungal infections and on the degree of resistance of different new carnation cultivars to diseases. *Fusarium oxysporum* f.sp. *dianthi*, *Botrytis cinerea*, *Alternaria dianthi*, and*Rhizoctonia solani* are the major pathogens, causing vascular wilt (Baayen and van der Plas 1992), gray mold (Elad and Volpin 1991), blight (Yu et al. 1989), and stem rot (Trujillo et al. 1988), respectively. *Fusarium* appears to cause the most severe economic losses in carnation nurseries, and considerable research has been devoted to RFLP analysis of *Fusarium* pathogenicity (Manicomb et al. 1990), to the biological control of this pathogen (Freier et al. 1990; Postma and Rattink 1992), to the biochemistry of wilt disease resistance (Niemann 1992), and to the correlation of the resistance character of carnation cultivars with pathogen-induced phenolic constituents (Baayen and Niemann 1989; van Peer et al. 1991) as well as with the cell wall's hydroxyproline contents (Krátká 1989).

From 1983 to 1988, a number of phytoalexins were described from infected carnations (Hauteville et al. 1988; Ponchet et al. 1988a, b), all of which derive from anthranilic acid (Fig. 1) and can be classified as either 2-aryl-1,3-benzoxazin-4H-ones or N-aroylanthranilates. Trivial designations (dianthalexins and dianthramides) and a simple letter code (D for dianthramides, HD for hydroxydianthramides, MD for methoxydianthramide) specifying the N-aroyl moiety (A, 4-anisoyl; B, benzoyl; M, 4-methoxysalicyloyl; P, 4-coumaroyl; R, β-resorcyl, S, salicyloyl) (Fig. 1) was proposed for the individual compounds (Ponchet et al. 1988a; Niemann et al. 1991a), and this will be followed here for convenience. Pyrolysis mass spectrometry of infected carnation tissues revealed later (Niemann et al. 1991a, 1992) that the in vivo proportion of benzoxazinones had been underestimated due to their chemical lability upon extraction (Williams and Salvadori 1971; Errede et al. 1980) and that N-aroylanilines (HAnB, HAnS, and HanR, respectively, Fig. 1) are additionally present in the infected carnation tissue. The phytoalexin contents were directly correlated with the degree of resistance to *Fusarium* (Baayen and Niemann 1989) with HDB and MDS (Fig. 1) showing the greatest antimycotic potential (van Peer et al. 1991). Pyrolysis mass spectrometry detected another anthranilic derivative, dianthramine (Fig. 1), that, on the contrary, was correlated with susceptibility to *Fusarium* (Niemann et al. 1991b). The relevant data have been recently reviewed elsewhere (Niemann 1993).

Aside from the phytopathological context, the inducible N-aroylanthranilates and benzoxazinones hold promise for bioactivity towards other targets. For example, N-(3,4-dimethoxycinnamoyl)anthranilic acid, initially isolated from *Nandina domestica* (Berberidaceae), shows remarkable antihistaminic (Azuma et al. 1976; Koda et al. 1985), antiallergic (Kakegawa et al. 1985), and anti-inflammatory properties (Isaji et al. 1987), and is being used commercially in antiasthmatic therapy. Anti-inflammatory activity was furthermore reported for benzoxazinoids that had been isolated from *Coix lachryma-jobi* (Otsuka et al. 1988) with structures similar to the dianthalexins (Fig. 1). It is obvious therefore that new, adequate bioactivity testing of carnation metabolites is necessary and that this will conceivably disclose new applications. Recent in vitro studies on dianthramide biosynthesis in our laboratory have revealed the very potent

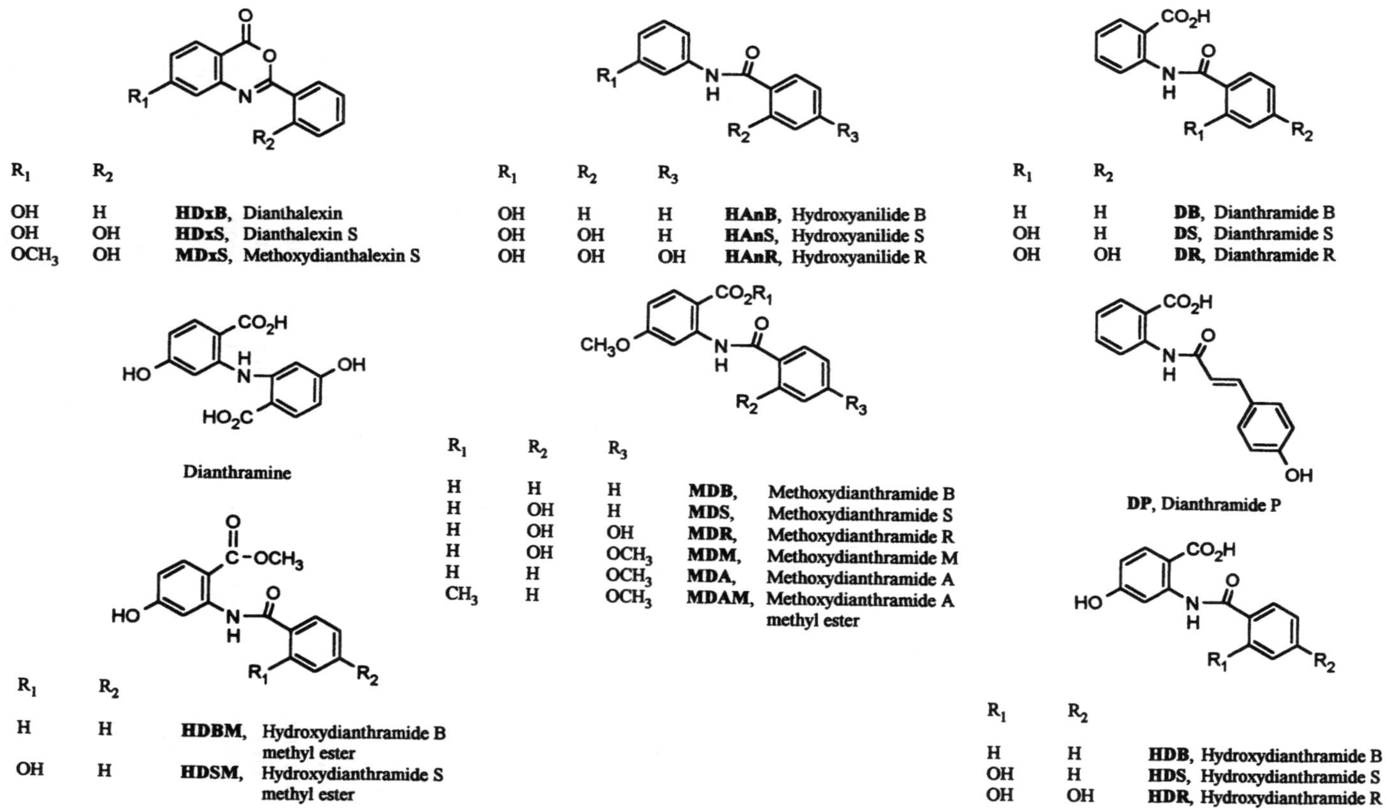


**Fig. 1.** Derivatives of anthranilic acid isolated from carnation plants infected with either *Phytophthora parasitica* or *Fusarium oxysporum* f.sp. *dianthi*. Some of these metabolites were also isolated from elicitor-induced, heterotrophic carnation cell cultures

inhibitory activity of DS on cytochrome-P-450-dependent monoxygenases (Reinhard and Matern 1991). Neither its immediate biosynthetic precursor (DB; Fig. 2) nor the hydroxylated product (HDS; Fig. 2) shares such activity, and the apparent dissimilarity of DS structure to those of commercial cytochrome P-450 inhibitors (Berg and Plempel 1988) make this carnation metabolite an interesting primary compound for pharmaceutical and agrochemical research. Initial greenhouse experiments on the protection of plants from fungal infection by spraying DS have indeed been successful. It is apparent from these very preliminary studies that the anthranilic metabolites deserve more attention than, for example, the analgesic and anti-inflammatory steroidal saponins described previously from carnations (Cordell et al. 1977).

The biosynthesis of hydroxydianthramides was investigated in cell suspension cultures initiated from the nodal or petal tissue of a commercial carnation cutting. These cultures were elicited with either a crude cell wall extract from *Phytophthora megasperma* f.sp. *glycinea* or commercial yeast extract for the induction of anthranilamide phytoalexins (Reinhard and Matern 1989). Yeast extract proved to be as effective in this regard as the *Phytophthora* elicitor, although much larger amounts were required (Reinhard 1992). Surprisingly, only the cultures derived from petals could be induced efficiently. Benzoyl-CoA: anthranilate N-benzoyltransferase, catalyzing the reaction at the branch from primary to the specific anthranilamide metabolism (Fig. 2), was shown to be induced about seven fold to one transient maximum about 6 h following the addition of elicitor (Reinhard and Matern 1989). The enzyme was purified to homogeneity, specific polyclonal antisera were raised in mouse and rabbit and roughly 20% of the amino acid sequence of the N-terminally blocked enzyme protein was determined from tryptic fragments (Reinhard 1992). The cloning and sequencing of cDNA is under way. The enzyme is unusually rich in proline residues, which might confer a nonglobular structure on the native enzyme and be a reason for its rapid proteolytic degradation in cellular extracts (Yaron and Naider 1993). In fact, gel permeation chromatography repeatedly indicated a lower molecular mass (44 kDa) than was determined by SDS-PAGE (54 kDa) (Reinhard 1992). Furthermore, the catalytic activity of the isolated enzyme in dilute solution was activated up to 25-fold by the addition of immunoglobulines, bovine serum albumin, or other proteins, a phenomenon that might point to conformational transitions. Immunoprecipitations revealed that the native, monomeric enzyme of 54 kDa was rapidly digested to 31 and 22 kDa fragments in crude cell extracts, which nevertheless retained their full enzymatic activity (Reinhard 1992). The transferase appears to be highly specific for anthranilate while rather low specificity was observed for the cosubstrate; cinnamoyl-CoA or 4-coumaroyl-CoA was even preferred to benzoyl-CoA. This low specificity opens the possibility of producing enzymatically greatly different aroylanthramides for bioactivity testing.

Anthranilate synthase (Fig. 2) was also purified from the elicited carnation cells, yielding two protein bands of 72 and 68 kDa, respectively, upon SDS-PAGE (Schmoeckel 1993), which complies with the native molecular mass determined at about 140 kDa (Schmoeckel 1993). Major problems in the purification were posed by the need to stabilize the enzyme activity in vitro and

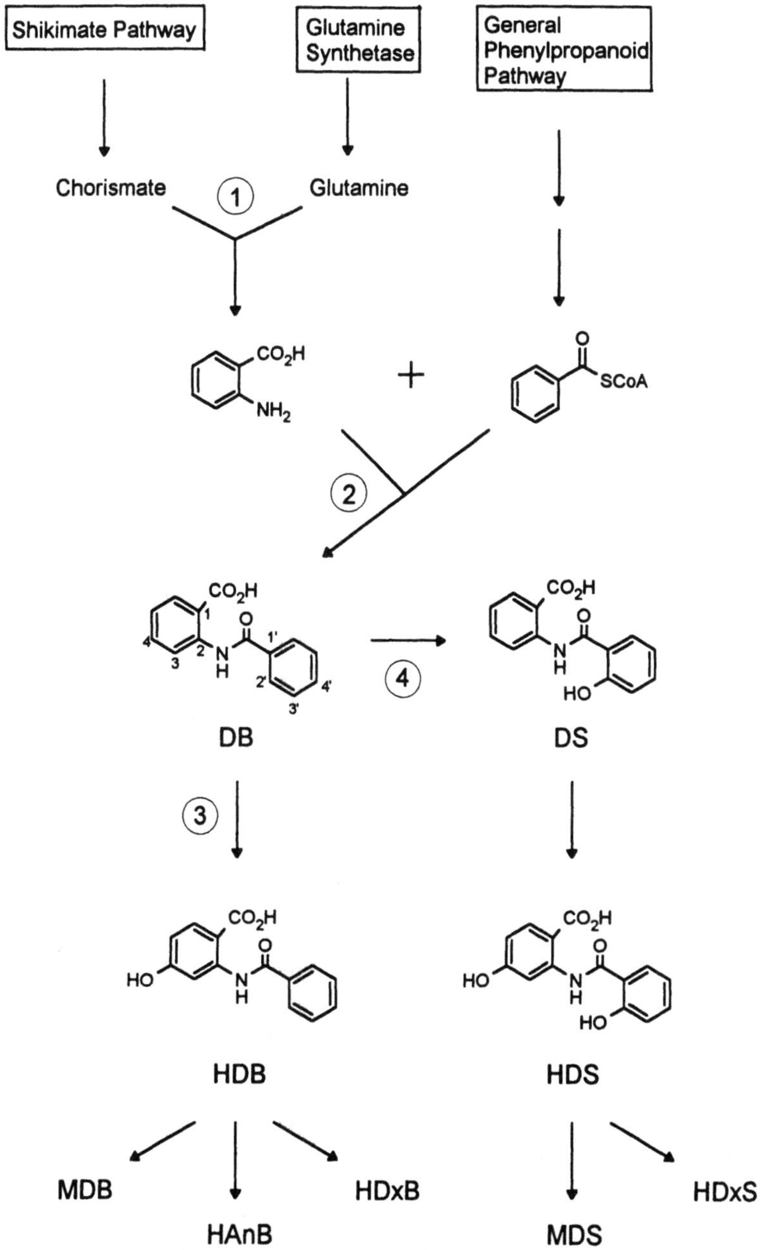


**Fig. 2.** Biosynthetic pattern of carnation phytoalexins. *Letter codes* for the individual compounds refer to Fig. 1 . The enzymes catalyzing the phytoalexin biosynthesis and studied in vitro are 1 anthranilate synthase, 2 benzoyl-CoA : anthranilate N-benzoyltransferase; 3 N-benzoylanthranilate 4-monooxygenase (4-hydroxylase); 4 N-benzoylanthranilate 2′-hydroxylase (2′-hydroxylase)

by the extremely low abundance of the enzyme protein in cellular extracts. The kinetics of induction upon elicitation of cells, showing a biphasic pattern with maxima at 6 and 10 h, as well as the fractionation pattern of enzyme activity upon purification, suggested the presence of multiple anthranilate synthases in the carnation extract, in contrast to *Arabidopsis* (Niyogi and Fink 1992) or *Catharanthus* (Poulson et al. 1993). Furthermore, the dimeric carnation enzyme did not accept ammonium as a nitrogen source instead of glutamine, which contrasts to the tetrameric enzyme from *Catharanthus* (Poulson et al. 1993). The heterodimeric composition of the carnation enzyme resembles that of the corresponding enzyme from *Neurospora crassa* (Hulett and DeMoss 1975) and other sources. The inducible anthranilate synthase from carnation appears to be an attractive research object, because dianthramide synthesis requires the synthesis of anthranilate rather than tryptophan (Fig. 2). This seems to exclude the multifunctional association of this enzyme with anthranilate-5-phosphoribosylpyrophosphate phosphoribosyl transferase or indole glycerol-3-phosphate synthetase, which is known to occur in microrganisms and would require sophisticated deletion analysis for distinction (Walker and Demoss 1990). For the same reason, the anthranilate synthase from carnation appears particularly suitable for mechanistic investigations that need to be done complementarily to those on 4-aminobenzoate synthase from *E. coli* (Walsh et al. 1987).

The phytoalexin strategy of carnation shows analogy to that of oat plants, which accumulate avenalumin phytoalexins derived from N-aroyl-5-hydroxy-anthranilate upon infection (Mayama et al. 1981). The avenalumins were the first nitrogen-containing phytoalexins described and were considered as metabolites unique to the infection sites. However, it has been discovered meanwhile that oat groats and hulls contain numerous closely related N-aroylanthranilates (Collins et al. 1991), suggesting that, aside from the defence, such compounds are physiologically relevant for flowering or fruiting. If this were also true in carnation, it might help to explain the differential inducibility of dianthramides in the different cell cultures.

Both the hydroxylations leading to 4-hydroxydianthramides (4-hydroxylase) and the salicyloyl-type dianthramides (2′-hydroxylase), respectively (Fig. 2), are catalyzed in vitro by the microsomal fraction from elicited carnation cells (Reinhard and Matern 1991). The strict dependence of catalytic activity on molecular oxygen and NADPH, as well as the synergism of NADPH with NADH and the apparent inhibition by cytochrome c, seemed to classify both enzymes as cytochrome-P-450-dependent monooxygenases (Reinhard and Matern 1991). However, inhibition experiments with carbon monoxide or commercial inhibitors for such monooxygenases later distinguished the cytochrome-P-450-dependent 4-hydroxylase from the 2′-hydroxylase, which does not depend on cytochrome P-450 (Reinhard and Matern 1991). This difference was further substantiated by incubation in imidazole buffer, which completely inhibited the 4-hydroxylase but not the 2′-hydroxylase activity. MOPS or MOPPSO buffers, on the other hand, fully inhibited the 2′-hydroxylase activity, whereas the 4-hydroxylase activity even increased considerably in comparison to standard incubations in MES buffer (Reinhard and Matern 1991; Reinhard 1992). The 2′-hydroxylase appears therefore to represent a novel type of plant hydroxylase, which may belong to the flavine-dependent enzymes known from microbial sources (Powlowski et al. 1987). The 2′-hydroxylation generates phytoalexins with higher antimycotic activity (Ponchet et al. 1988b; van Peer et al. 1991) and may be crucial for the resistance response of carnation. This activity might be attributed to the inhibition of cytochrome-P-450-dependent monooxygenases confirmed in vitro, which points to a mode of action analogous to that of commercial antisteroidal antimycotics (Berg and Plempel 1988).

## 3 Commercial Aspects and Future Prospects

The major importance of carnation will remain with the cut flower market. Consumer preferences regarding pigmentation and growth characteristics is changing so rapidly, however, that breeders are forced to reduce the time for release of new cultivars (Lu et al. 1991). The improvement of natural resistance, which would require time-consuming field testing, is neglected under these conditions, and, consequently, commercial cultivation of carnation in nurseries is accompanied by heavy preventative pesticide treatments. Molecular markers for horizontal resistance would greatly speed up the selection of valuable shoots for propagation and strengthen the natural resistance of commercial carnation cultivars. Phytoalexin accumulation is considered as a major factor in horizontal resistance and antisera specific for benzoyl-CoA:anthranilate N-benzoyltransferase and anthranilate synthase (Fig. 2) or the corresponding DNA probes would be sensitive tools to determine the resistance quality of plants from single leaves and at a very early stage of growth by Western or Northern blotting techniques. Provided that homologous enzymes catalyze phytoalexin biosynthesis in oat plants, the probes might also be employed for the selection of oat plants.

Elicited carnation cells may serve as a simple model system to study the regulation of anthranilate metabolism at the molecular level. Such studies would provide the basis for further investigations on acridone or pharmacologically important indole alkaloids. The enzymes from carnation (Schmoeckel 1993) and from *Catharanthus* (Poulson et al. 1993) are the first anthranilate synthases purified from plants. Although these studies are in their very beginning, an improved knowledge of anthranilate synthase characteristics may pave the way for the development of selective inhibitors of tryptophan biosynthesis. Synthetic inhibitors of amino acid biosynthesis account for a multi-million dollar share of the world's herbicide market and are being used, for example, for postharvest field cleaning. None so far have been developed for interference with the tryptophan pathway.

The successful infection of carnation plants with *Agrobacterium tumefaciens* and the ease with which carnation plants can be regenerated from callus tissue suggest that the stable transformation with useful genes will be accomplished soon. This reveals new ways to improve the ornamental, disease resistance, and metabolite qualities of carnation plants for commercial use. It is to be expected that antisense constructs will be introduced first for inhibition of ethylene biosynthesis and delayed senescence of carnation cuttings. The enhancement of resistance to fungal diseases might be accomplished by the introduction of foreign phytoalexin synthesis genes, as has been demonstrated recently with stilbene synthase in tobacco (Hain et al. 1993). The modification of carnation N-aroylanthranilates by introduction of, for example, the N-aroylanthranilate 5-hydroxylase from oat may yield new metabolites upon elicitation. This may further attract interest to use carnation as a source of valuable medicinal metabolites, thus giving the plant a new role in addition to its ornamental value.

Carnation phytoalexins are composed of the anthranilate and benzoyl moieties, both of which derive from the shikimate pathway. Whereas the biosynthesis of anthranilate branches off early and drains away chorismate as the substrate, the benzoyl moiety is formed later in the sequence from cinnamate. Provided that all reactions proceed in the same cellular compartment, the scenario requires that the flow of metabolites through the shikimate pathway and particularly the percentage of chorismate channeled into anthranilate is tightly controlled so that sufficient amounts of both anthranilate and benzoate are produced during the elicitation of cells. The investigation of branch-point enzymes may reveal further information on the modes of control of anthranilate-indole and benzoate metabolism, which might eventually contribute to an understanding of alkaloid biosynthesis regulation.

## References


Avelange M-H, Thiéry JM, Sarrey F, Gans P, Rébeillé F (1991) Mass-spectrometric determination of $O_2$ and $CO_2$ gas exchange in illuminated higher-plant cells. Planta 183:150–157

Azuma H, Banno K, Yoshimura T (1976) Pharmacological properties of N-(3′,4′-dimethoxycinnamoyl)anthranilic acid (N-5′), a new anti-atropic agent. BrJ Pharmacol 58:483–488

Baayen RP, Niemann GJ (1989) Correlations between accumulation of dianthramides, dianthalexin and unknown compounds, and partial resistance to *Fusarium oxysporum* f.sp. *dianthi* in 11 carnation cultivars. J Phytopathol 126:281–292

Baayen RP, van der Plas CH (1992) Localization ability, latent period and wilting rate in eleven carnation cultivars with partial resistance to fusarium wilt. Euphytica 59:165-174

Berg D, Plempel M (eds) (1988) Sterol biosynthesis inhibitors: pharmaceutical and agrochemical aspects. Ellis Horwood, Chichester

Brown JH, Chambers JA, Thompson JE (1991a) Acyl chain and head group regulation of phospholipid catabolism in senescing carnation flowers. Plant Physiol 95:909–916

Brown JH, Chambers JA, Thompson JE (1991b) Distinguishable patterns of phospholipid susceptibility to catabolism in senescing carnation petals. Phytochemistry 30:2537–2543

Buiatti M, Scala A, Bettini P, Nascari G, Morpurgo R, Bogani P, Pellegrini G, Gimelli F, Venturo R (1985) Correlations between in vivo resistance to *Fusarium* and in vitro response to fungal elicitors and toxic substances in carnation. Theor Appl Genet 70:42–47

Buiatti M, Marcheschi G, Venturo R, Bettini P, Bogani P, Morpurgo R, Nachmias B, Pellegrini MG (1987) In vitro response to *Fusarium* elicitor and toxic substances in crosses between resistant and susceptible carnation cultivars Plant Breed 98:346–348

Collins FW, McLachlan DC, Blackwell BA (1991) Oat phenolics: avenalumic acids, a new group of bound phenolic acids from oat groats and hulls. Cereal Chem 68:184–189

Cook EL, van Staden J (1988) The carnation as a model for hormonal studies in flower senescence. Plant Physiol Biochem 26:793–807

Cordell GA, Lyon RL, Fong HHS, Benoit PS, Farnsworth NR (1977) Biological and phytochemical investigations of *Dianthus barbatus cv.* China Doll (Caryophyllaceae). Lloydia 40:361–363

Dereuddre J, Fabre J, Bassaglia C (1988) Resistance to freezing in liquid nitrogen of carnation (*Dianthus caryophyllus* L. var Eolo) apical and axillary shoot tips excised from different aged in vitro plantlets. Plant Cell Rep 7:170–173

De Vetten NC, Huber DJ, Gross KC (1991) Endoglycanase-catalyzed degradation of hemicellulose during development of carnation (*Dianthus caryophyllus* L.) petals. Plant Physiol 95:853–860

Droillard M-J, Paulin A (1990) Isozymes of superoxide dismutase in mitochondria and peroxysomes isolated from petals of carnation (*Dianthus caryophyllus*) during senescence. Plant Physiol 94:1187–1192

Elad Y, Volpin H (1991) Heat treatment for the control of rose and carnation gray mold (*Botrytis cinerea*). Plant Pathol 40:278–286

Engelmann-Sylvestre I, Bureau J-M, Trémolières A, Paulin A (1989) Changes in membrane phospholipids and galactolipids during senescence of cut carnation. Connection with ethylene rise. Plant Physiol Biochem 27:931–937

Errede LA, McBrady JJ, Tiers GVD (1980) Acylanthranils. 10. Influence of hydrogen bonding on hydrolysis of acylanthranil in organic solvents. J Org Chem 45:3868–3875

Forkmann G, Dangelmayr B (1980) Genetic control of chalcone isomerase activity in flowers of *Dianthus caryophyllus*. Biochem Genet 18:519–527

Freier K, Krebs B, Junge H, Bochow H, Huber J, Hirte W (1990) Dose-effect-relationship and population dynamics for the antagonist *Bacillus subtilis* used for biological control of *Fusarium oxysporum* f.sp. *dianthi*. Zentralbl Mikrobiol 145:563–578

Gay L (1985) Phytoalexin formation in cell cultures of *Dianthus caryophyllus* treated by an extract from the culture medium of *Phytophthora parasitica*. Physiol Plant Pathol 26:143–150

Geissman TA, Mehlquist GAL (1947) Inheritance in the carnation, *Dianthus caryophyllus*. IV. The chemistry of flower color variation, I. Genetics 32:410–433

Geissman TA, Hinreiner EF, Jorgensen EC (1956) Inheritance in the carnation, *Dianthus caryophyllus*. V. The chemistry of flower color variation, II. Genetics 41:93–115

Hain R, Reif H-J, Krause E, Langebartels R, Kindl H, Vornam B, Wiese W, Schmelzer E, Schreier PH, Stöcker RH, Stenzel K (1993) Disease resistance results from foreign phytoalexin expression in a novel plant. Nature 361:153–156

Hanley KM, Meir S, Bramlage WJ (1989) Activity of ageing carnation flower parts and the effects of 1-(malonylamino)cyclopropane-1-carboxylic acid-induced ethylene. Plant Physiol 91:1126–1130

Harborne JB (1966) Comparative biochemistry of flavonoids–I. Distribution of chalcone and aurone pigments in plants. Phytochemistry 5:111–115

Hauteville M, Ponchet M. Ricci P, Favre-Bonvin J (1988) Novel synthesis of dianthalexin (phytoalexin) analogues preparation. Heterocyclic Chem 25:715–718

Hegnauer R (ed) (1964) Chemotaxonomie der Pflanzen, vol 3. Birkhäuser, Basel

Hegnauer R (ed) (1989) Chemotaxonomie der Pflanzen, vol 8. Birkhäuser, Basel

Helm J (1975) Unterfamilie Caryophylloideae (Silenoideae). In: Urania Pflanzenreich, Höhere Pflanzen I. Urania, Leipzig, pp 278–281

Henskens H, Somhorst D, Woltering E (1992) Molecular cloning and tissue expression of ACC synthase mRNAs in carnation flowers. Physiol Plant 85:A61 (Abstr 344)

Hulett FM, DeMoss JA (1975) Subunit structure of anthranilate synthetase from *Neurospora crassa*. J Biol Chem 250:6648–6652

Isaji M, Nakajoh M, Naito J (1987) Selective inhibition of collagen accumulation by N-(3,4-dimethoxycinnamoyl)anthranilic acid (N-5′) in granulation tissue. Biochem Pharmacol 36:469-474

Kakegawa H, Mitsuo N, Matsumoto H, Satoh T, Akagi M, Tasaka K (1985) Hyaluronidase-inhibitory and anti-allergic activities of the photo-irradiated products of Tranilast, Chem Pharm Bull 33:3738–3744

Kevers C, Gaspar Th (1985) Soluble, membrane and wall peroxidases, phenylalanine ammonia-lyase, and lignin changes in relation to vitrification of carnation tissues cultured in vitro. J Plant Physiol 118:41–48

Koda A, Kurashina Y, Nakazawa M (1985) The inhibition mechanism of histamine release by N-(3,4-dimethoxycinnamoyl)anthranilic acid. Int Arch Allergy Appl Immunol 77:244–245
Krátká J (1989) Changes of hydroxyproline content in the cell wall of carnation after inoculation with *Fusarium oxysporum* f.sp. dianthi. Zentralbl Mikrobiol 144:485–488
Larsen PB. Woodson WR (1991) Cloning and nucleotide sequence of a S-adenosyl-methionine synthetase cDNA from carnation. Plant Physiol 96:997–999
Leshem B, Werker E, Shalev DP (1988) The effect of cytokinins on vitrification in melon and carnation. Ann Bot 62:271–276
Lu C-Y, Nugent G, Wardley-Richardson T, Chandler SF, Young R, Dalling MJ (1991) *Agrobacterium*-mediated transformation of carnation (*Dianthus caryophyllus* L.). Bio/Technology 9: 864–868
Manicomb BQ, Bar-Joseph M, Kotze JM, Becker MM (1990) A restriction fragment length polymorphism probe relating vegetative compatibility groups and pathogenicity in *Fusarium oxysporum* f.sp. *dianthi.* Phytopathology 80:336–339
Matern U, Strobel G, Shepard J (1978) Reaction to phytotoxins in a potato population derived from mesophyll protoplasts. Proc Natl Acad Sci USA 75:4935–4939
Mayama S, Tani T, Ueno T, Hirabayashi K, Nakashima T, Fukami H, Mizuno Y, Irie H (1981) Isolation and structure elucidation of genuine oat phytoalexin, avenalumin I. Tetrahedron Lett 22:2103–2106
Mehlquist GAL, Geissman TA (1947) Inheritance in the carnation (*Dianthus caryophyllus*) III, Inheritance of flower color. Ann M Bot Gard 34:39–75
Meyer P, Heidmann I, Forkmann G, Saedler H (1987) A new petunia flower colour generated by transformation of a mutant with a maize gene. Nature 330:677–678
Mii M, Buiatti M, Gimeli F (1990) Carnation. In: Ammirato PV, Evans DA, Sharp WR, Bajaj YPS (eds) Handbook of plant cell culture, vol 5. Ornamental species. McGraw-Hill, New York, pp 284–318
Miller RM, Kaul V, Hutchinson JF, Maheswaran G, Richards D (1991a) Shoot regeneration from fragmented flower buds of carnation (*Dianthus caryophyllus*). Ann Bot 68:563–568
Miller RM, Kaul V, Hutchinson JF, Richards D (1991b): Adventitious shoot regeneration in carnation (*Dianthus caryophyllus*) from axillary bud explants. Ann Bot 67:35–42
Ming X-T, Jing-Jiu MI, Nai-sui P, Zhang-Liang C (1990) Transient expression of CAT and GUS genes in protoplasts isolated from rice and corn. Acta Bot Sin 32:443–447
Nakano M, Mii M (1993) Interspecific somatic hybridization in *Dianthus*: selection of hybrids by the use of iodoacetamide inactivation and regeneration ability. Plant Sci 88:203–208
Niemann GJ (1992) The mechanism of resistance of carnation to wilt diseases. Acta Hortic 307:29–36
Niemann GJ (1993) The anthranilamide phytoalexins of the Caryophyllaceae and related compounds – Phytochemistry 34:319–328
Niemann GJ, Liem J, Pureveen JBM, Boon JJ (1991a) The amide-type phytoalexin activity of carnation extracts is partly due to an artifact. Phytochemistry 30:3923–3927
Niemann GJ, van der Kerk A, Niessen WMA, Versluis K (1991b) Free and cell wall-bound phenolics and other constituents from healthy and fungus-infected carnation (*Dianthus caryophyllus* L.) stems. Physiol Mol Plant Pathol 38:417–432
Niemann GJ, Liem J, van der Kerk A, van Hoof A, Niessen WMA (1992) Phytoalexins, benzoxazinones, N-aroylanthranilates and N-aroylanilines from *Fusarium*-infected carnation stems. Phytochemistry 31:3761–3767
Niyogi KK, Fink GR (1992) Two anthranilate synthase genes in *Arabidopsis*: Defense-related regulation of the tryptophan pathway. Plant Cell 4:721–733
Nugent G, Wardley-Richardson T, Lu C-Y (1991) Plant regeneration from stem and petal of carnation (*Dianthus caryophyllus* L.). Plant Cell Rep 10:477–480
Ostermann W-D, Meyer U, Leiser R-M (1987) Induction of plant virus resistance: 2. Leaf extract from carnation plants (*Dianthus caryophyllus* L.) as inducers of resistance. Zentralbl Mikrobiol 142:229–238
Otsuka H, Hirai Y, Nagao T, Yamasaki K (1988) Anti-inflammatory activity of benzoxazinoids from roots of *Coix lachryma-jobi* var. Ma-Yuen. J Nat Prod 51:74–79
Palet A, Ribas-Carbó M, Argilés JM, Azcón-Bieto J (1991) Short-term effects of carbon dioxide on carnation callus cell respiration. Plant Physiol 96:467–472

Park KY, Drory A, Woodson WR (1992) Molecular cloning of an 1-aminocyclopropane-1-carboxylate synthase from senescing carnation flower petals. Plant Mol Biol 18:377–386
Pereau-Leroy P (1974) Genetic interaction between the tissues of carnation petals as periclinal chimeras. Radiat Bot 14:109–116
Ponchet M, Favre-Bonvin J, Hauteville M, Ricci P (1988a) Dianthramides (N-benzoyl and N-paracoumaroylanthranilic acid derivatives) from elicited tissues of *Dianthus caryophyllus*. Phytochemistry 27:725–730
Ponchet M, Ricci P, Hauteville M, Auge G (1988b) Activé antifongique in vitro de la dianthramide A. CR Acad Sci Paris Ser III 306:173–178
Postma J, Rattink H (1992) Biological control of fusarium wilt of carnation with a nonpathogenic isolate of *Fusarium oxysporum*. Can J Bot 70:1199–1205
Poulson C, Bongaerts RJM, Verpoorte R (1993) Purification and characterization of anthranilate synthase from *Catharanthus roseus*. Eur J Biochem 212:431–440
Powlowski JB, Dagley S, Massey V, Ballou DP (1987) Properties of anthranilate hydroxylase (deaminating), a flavoprotein from *Trichosporon cutaneum*. J Biol Chem 262:69–74
Raghothama KG, Lawton KA, Goldsbrough PB, Woodson WR (1991) Characterization of an ethylene-regulated flower senescence-related gene from carnation. Plant Mol Biol 17:61–71
Rebéillé F (1988) Photosynthesis and respiration in air-grown and $CO_2$-grown photoautotrophic cell suspension cultures of carnation. Plant Sci 54:11–21
Rechinger KH (ed) (1979) Gustav Hegi, Illustrierte Flora von Mitteleuropa, Vol III/2 Paul Parey, Berlin, pp 984–1037
Reinhard K (1992) Enzymologie der Phytoalexinsynthese in Zellkulturen der Nelke (*Dianthus caryophyllus* L.). PhD Thesis, University of Freiburg
Reinhard K, Matern U (1989) The biosynthesis of phytoalexins in *Dianthus caryophyllus* L. cell cultures: induction of benzoyl-CoA:anthranilate N-benzoyltransferase activity. Arch Biochem Biophys 275:295–301
Reinhard K, Matern U (1991) Different types of microsomal enzymes catalyze *ortho*- or *para*-hydroxylation in the biosynthesis of carnation phytoalexins. FEBS Lett 294:67–72
Reisbig RR, Bruland O (1983) Dianthin 30 and 32 from *Dianthus caryophyllus*: two inhibitors of plant protein synthesis and their tissue distribution. Arch Biochem Biophys 224:700–706
Rouet-Mayer M-A, Bureau J-M, Laurière C (1992) Identification and characterization of lipoxygenase isoforms in senescing carnation petals. Plant Physiol 98:971–978
Schmoeckel A (1993) Die Anthranilatsynthase aus Zellkulturen der Nelke (*Dianthus caryophyllus* L.). Diploma Thesis, University of Freiburg
Seibert M (1976) Shoot initiation from carnation shoot apices frozen to − 196°C. Science 191:1178-1179
Silvy A (1978) Mutation breeding in carnation. In: Quagliotti L, Baldi A (eds) Proc Eucarpia Meet on Carnation and Gerbera, Alassio. Institute of Plant Breeding and Seed Production, Turin, Italy, pp 91–102
Smith MT, Saks Y, van Staden J (1992) Ultrastructural changes in the petals of senescing flowers of *Dianthus caryophyllus* L. Ann Bot 69:277–285
Sparnaaij L (ed) (1987) Third Int Symp carnation culture. Acta Hortic 216, Int Soc Hortic Sci, Wageningen
Spribille R, Forkmann G (1982) Chalcone synthesis and hydroxylation of flavonoids in 3′-position with enzyme preparations from flowers of *Dianthus caryophyllus* L. (carnation). Planta 155:176–182
Stich K, Eidenberger T, Wurst F, Forkmann G (1992a) Enzymatic conversion of dihydroflavonols to flavan-3,4-diols using flower extracts of *Dianthus caryophyllus* L. (carnation). Planta 187:193–208
Stich K, Eidenberger T, Wurst F, Forkmann G (1992b) Flavonol synthase activity and the regulation of flavonol and anthocyanin biosynthesis during flower development in *Dianthus caryophyllus* L. (carnation). Z Naturforsch 47c:553–560
Stirpe F, Williams DG, Onyon LJ, Legg RF, Stevens WA (1981): Dianthins, ribosome-damaging proteins with anti-viral properties from *Dianthus caryophyllus* L. (carnation). Biochem J 195:399–405
Stirpe F, Barbieri L, Battelli MG, Soria M, Lappi DA (1992) Ribosome-inactivating proteins from plants: present status and future prospects. Bio/Technology 10:405–412

Terahara N, Yamaguchi M-A (1986) 1H NMR Spectral analysis of the malylated anthocyanins from *Dianthus*. Phytochemistry 25: 2906–2907

Terahara N, Yamaguchi M, Takeda K, Harborne JB, Self R (1986) Anthocyanins acylated with malic acid in *Dianthus caryophyllus* and *Dianthus deltoides*. Phytochemistry 25: 1715–1717

Trujillo EE, Shimabuku R, Cavin CA, Aragaki M (1988) *Rhizoctonia solani* anastomosis groupings in carnation fields and their pathogenicity to carnation. Plant Dis 72: 863–865

Tzuri G, Hillel J, Lavi U, Haberfeld A, Vainstein A (1991) DNA fingerprint analysis of ornamental plants. Plant Sci 79: 91–97

van Peer R, Niemann GJ, Schippers B (1991) Induced resistance and phytoalexin accumulation in biological control of fusarium wilt of carnation by *Pseudomonas* sp. strain WCS417r. Phytopathology 81: 728–734

Walker MS, DeMoss JA (1990) Deletion analysis of domain independence in the TRP1 gene product of *Neurospora crassa*. Mol Gen Genet 223: 49–57

Walsh CT, Erion MD, Walts AE, Delany III JJ, Berchtold GA (1987) Chorismate aminations: partial purification of *Escherichia coli* PABA synthase and mechanistic comparison with anthranilate synthase. Biochemistry 26: 4734–4745

Williams A, Salvadori G (1971) Studies on the hydrolysis of 3,1-benzoxazin-4-ones. J Chem Soc Phys Org: 1105–1110

Yaron A, Naider F (1993) Proline-dependent structural and biological properties of peptides and proteins. Crit Rev Biochem Mol Biol 28: 31–81

Yu SH, Kang HW, Lee HB, Kim HG (1989) Occurrence of *Alternaria dianthi* on carnation in Korea. Korean J Mycol 17: 229–232

Ziv M, Schwartz A, Fleminger D (1987) Malfunctioning stomata in vitreous leaves of carnation (*Dianthus caryophyllus*) plants propagated in vitro; implications for hardening. Plant Sci 52: 127–134

# XIII *Dunaliella Tertiolecta* (Green Microalga): Culture and Biotransformation of Terpenoids and Related Compounds

Y. NOMA[1] and Y. ASAKAWA[2]

## 1 Introduction

Many microorganisms have been used for manufacturing foods, drinks, medicinal drugs, etc. We are continuing study on the biotransformation of plant secondary metabolites and synthetic compounds using bacteria (Noma and Tatsumi 1973; Noma et al. 1974a, b, 1975; Noma 1976, 1977), yeast (Noma 1976), fungi (Noma and Nonomura 1974; Noma 1976; Nishimura et al. 1982; Noma et al. 1985, 1990, 1992b; Miyazawa et al. 1990a, b, 1991, 1992a, b, c; Asakawa et al. 1991), actinomycetes (Noma 1976, 1978a, b, 1979, 1980; Noma et al. 1982; Nishimura et al. 1983; Noma and Sakai 1984; Noma and Nishimura 1987), and *Euglena* (Noma et al. 1991b, c; Noma and Asakawa 1992) from the viewpoint of their application in drugs and perfumery sources as well as for the resolution of racemates.

We chose *Dunaliella tertiolecta* (LB 999), obtained from the Algal Collection of the University of Texas (Starr 1978), and studied the biotransformation of monoterpenoids and related compounds (Noma et al. 1991a, 1992a). *Dunaliella* was named in honor of Dunal (1837) by Teodoresco (1905), who first described this genus. Although *Dunaliella* is traditionally placed in the order Volvocales, family Polyblechariodaceae, Ettl (1983) has proposed the new order Dunaliellales, with the family Dunaliellaceae (Christensen 1962). It is an obligate halophilic unicellular, bi-flagellate (equal length), naked, green alga (Chlorophyta, Chlorophyceae) (McLachlan 1960), which requires a saline environment. Cell shape is very variable, being oval, spherical, cylindrical, ellipsoidal, egg-, pear-, or spindle-shaped with radial, bilateral, or dorsiventral symmetry, or asymmetrical, and cells may change shape with changing conditions, often becoming spherical under unfavorable conditions. Cell size may also vary with growth conditions and light intensity (Riisgard 1981). *Dunaliella* is motile with two flagella and contains a large cup-shaped chloroplast which occupies about half the intercellular space (Hoshaw and Maluf 1981). Since Teodoresco's initial description, many species from a very wide range of habitats have been described, and Massyuk (1973) recognized 29 species of the genus and a number of varieties and forms. *Dunaliella* spp. have a wide tolerance for salinity

[1] Faculty of Domestic Sciences, [2] Faculty of Pharmaceutical Sciences, Tokushima Bunri University, Yamashiro-cho, Tokushima 770, Japan

Biotechnology in Agriculture and Forestry, Vol. 28
Medicinal and Aromatic Plants VII (ed. by Y.P.S. Bajaj)

(0.4–35%), pH (1–11), and other environmental conditions, and are widely distributed in oceans, salt pond, brine lakes, salt marshes, and salt water ditches near the sea (Pierce 1914; Cifuentes 1992; Baas-Becking 1931; Lerche, 1936) and also in highly saline environments such as the Great Salt Lake in the state of Utah, USA (Post 1977), and The Dead Sea in Israel (Butcher 1959; Kaplan and Friedmann 1970). Based on the linearly increasing intracellular glycerol content in *Dunaliella* with increasing extracellular NaCl concentration, many investigators have believed that the ability of the *Dunaliella* cell to live in high salt concentrations is due to the ability of the cells to produce intracellular glycerol to maintain an osmotic balance with the extracellular salt concentration (Craigie and Mclachlan 1964; Wegmann 1971; Ben-Amotz and Avron 1973; Borowitzka and Brown 1974; Frank and Wegmann 1974).

## 2 Cultivation and Biotransformation

Aliquots (ca. 0.5 ml) of the cultured broth of *D. tertiolecta* were taken and added to 100 ml of the medium in an Erlenmeyer flask containing $MgCl.6H_2O$ (1.5 g), $MgSO_4.7H_2O$ (0.5 g), KCl (0.2 g), $CaCl_2$ (0.2 g), $KNO_3$ (1 g), $NaHCO_3$ (0.043 g), Tris (2.45 g), $K_2HPO_4$ (0.045 g), NaCl (58.5 g), Fe-EDTA (3.64 mg), EDTA.2Na (1.89 mg), $ZnSO_4.7H_2O$ (0.087 mg), $H_3BO_4$ (0.61 mg), $CoCl_2.H_2O$ (0.015 mg), $CuSO_4.5H_2O$ (0.06 mg), $MnCl_2$ (0.23 mg), and $(NH_4)_6Mo_7O_{24}.4H_2O$ (0.38 mg) in 1 l distilled water; pH of the medium was adjusted at 8 with HCl (Noro 1978) and cultivated both photoautotrophically (light illumination at ca. 2000–3000 lx) and statically at 15–20 °C for 7–10 days. On obtaining full growth of *Dunaliella*, we carried out the biotransformation of terpenoids and related compounds at ca. 50–200 ppm (average 100 ppm) for 1–7 days under the same conditions as described above. For time course changes, aliquots (6 ml) of the cultured broth were taken and centrifuged at 1000 rpm for 10 min, and the broth was extracted with 5 ml of $Et_2O$. Each extract was analyzed by GC and GC-MS. For acquisition of metabolites, large-scale culture was carried out repeatedly. After reaction, the organisms were removed by centrifugation at 1000 rpm for 10 min, and the broth was extracted with $Et_2O$. The total recovery ratios of metabolic products were more than 90%.

The metabolites were separated and purified by a combination of CC on silica gel and prep. GC. The products were identified by comparison of GC Rts, mass, IR, $^1H$-(400 MHz), and $^{13}C$-NMR (100 MHz) spectra with those of authentic specimens.

When substances toxic to the survival of *Dunaliella* are added to the cultured broth, the phytoflagellate sinks to the bottom and dies in due course of time. By measuring the absorption at 640 nm as turbidity, we studied the relationship

**Fig. 1.** Relationship between structure of terpenoids and survival of *Dunaliella tertiolecta* after 1 day, (Noma et al. 1992a)

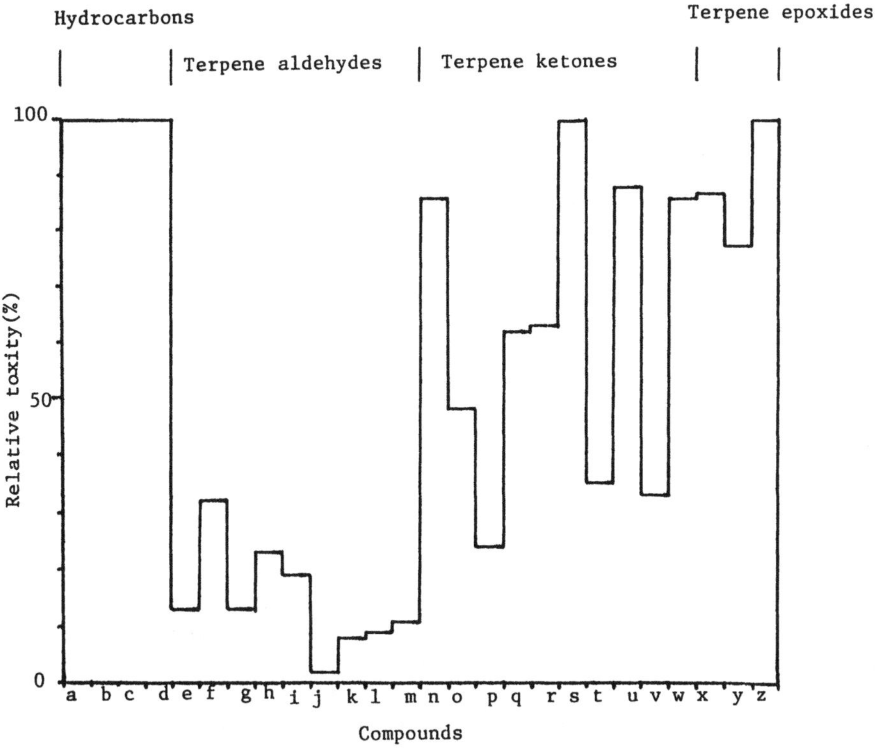


| | |
|---|---|
| Hydrocarbons | a *d*-Limonene |
| | b *p*-Menthane |
| | c α-Pinene |
| | d β-Pinene |
| Terpene aldehydes | e Myrtenal(**1**) |
| | f *l*-Perillaldehyde(**2**) |
| | g *l*-Phellandral(**3**) |
| | h Mixture of *trans*- and *cis*-1,2-dihydroperillaldehyde(**5a** and **5b**) |
| | i Cumin aldehyde(**6**) |
| | j Citral(**7**) |
| | k *d*-Citronellal(**8**) |
| | l *l*-Citronellal(**9**) |
| | m *dl*-Citronellal(**10**) |
| Terpene ketones | n Verbenone |
| | o *l*-Carvone(**65**) |
| | p *d*-Carvone(**66**) |
| | q *dl*-Piperitone |
| | r Isopiperitenone |
| | s *n*-Piperitenone |
| | t *dl*-Isomenthone |
| | u *l*-Menthone |
| | v Mixture of *d*-*n*- and isodihydrocarvones (**73** and **74**, 80:20) |
| | w Mixture of *l*-*n*- and isodihydrocarvones (**75** and **76**, 80:20) |
| Terpene epoxides | x 1,4-Cineole |
| | y 1,8-Cineole |
| | z Mixture of *trans*- and *cis*-Shisool-8,9-epoxides |

between 26 monoterpenes and the survival of *Dunaliella* (Fig. 1). Most of the terpene aldehydes and some of the saturated and α, β-unsaturated ketones strongly inhibited the survival of *Dunaliella* at ca. 200 ppm. Therefore, the biotransformation of monoterpenes was carried out at 100 ppm. The biotransformations of monoterpene aldehydes such as myrtena(**1**), *l*-(**2**) and *dl*-perillaldehydes(**3**), *l*-phellandral(**4**), the mixture of *trans*- and *cis*-1,2-dihydroperillaldehydes(**5a**, **b**), cuminaldehyde(**6**), citral(**7**), *d*-(**8**), *l*-(**9**), and *dl*-citronellals(**10**) are summarized in Table 1 with the substrate concentration when it was added continuously till the death of the phytoflagellate. All monoterpene aldehydes were reduced to the corresponding primary alcohol as shown in Fig. 2. However, although a small amount of *trans*-shisool(**53**) was formed from *l*-(**51**) and *dl*-perillyl alcohol, myrtenol(**50**), *l*-phellandrol(**52**), *trans*-(**53**) and *cis*-shisools(**54**) and cumin alcohol(**55**) were difficult or impossible to transform. On the other hand, when either geraniol(**56**) or nerol(**57**) was used as a substrate, isomerization between both compounds were observed in about equal ratio in the GC peak area. The terpene aldehydes strongly inhibited the survival of the phytoflagellate at ca. 100–400 ppm when they were added continuously to the medium every 100 ppm and every day (Table 1 and Fig. 3).

The biotransformation of aromatic aldehydes and related aldehydes by *Dunaliella* is summarized in Table 2, together with the substrate concentration for the death of the phytoflagellate. The transformation yield, except for salicyl aldehyde(**12**), *p*-hydroxy-(**14**) and *o*-nitrobenzaldehyde(**23**), vanillin(**36**), and 2-hydroxy-5-methoxybenzaldehyde(**40**), was ca. 100% after 1 or 2 days. Although most aromatic aldehydes and related aldehydes were easily reduced to the corresponding primary alcohols as major products, *Dunaliella* showed high

**Table 1.** The summary of biotransformation of monoterpene aldehydes and death concentration of substrate for *Dunaliella*, when it was added continuously in the medium (Noma et al. 1991a, 1992a)

| Substrates | Products | [c]D.C. (μl or mg/100 ml) |
|---|---|---|
| Myrtenal(**1**) | Myrtenol(**50**, M[a]) | 40 |
| *l*-Perillaldehyde(**2**) | *l*-Perilly alcohol(**51**, M) | 10 |
| | *trans*-Shisool(**53**, m[b]) | |
| *dl*-Perillaldehyde(**3**) | *dl*-Perillyl alcohol(M) | 20 |
| | *trans*-Shisool(**53**, m) | |
| *l*-Phellandral(**4**) | *l*-Phellandrol(**52**, M) | 20 |
| 1,2-Dihydroperillaldehyde | *trans*-Shisool(**53**, M) | 40 |
| (*trans*-(**5a**) and *cis*-(**5b**) = 87:17) | *cis*-Shisool(**54**, m) | |
| Cumin aldehyde(**6**) | Cumin alcohol(**55**, M) | 40 |
| Citral(**7**) | Geraniol(**56**, M) | 20 |
| (Geranial(**7a**):neral(**7b**) = 56:44) | Nerol(**57**, M) | |
| *d*-Citronellal(**8**) | *d*-Citronellol(**58**, M) | 20 |
| *l*-Citronellal(**9**) | *l*-Citronellol(**59**, M) | 20 |
| *dl*-Citronellal(**10**) | *dl*-Citronellol(M) | 30 |

[a] M: major product
[b] m: minor product
[c] D.C.: concentration of substrate when substrate was added to the medium continuously till the death of *Dunaliella*

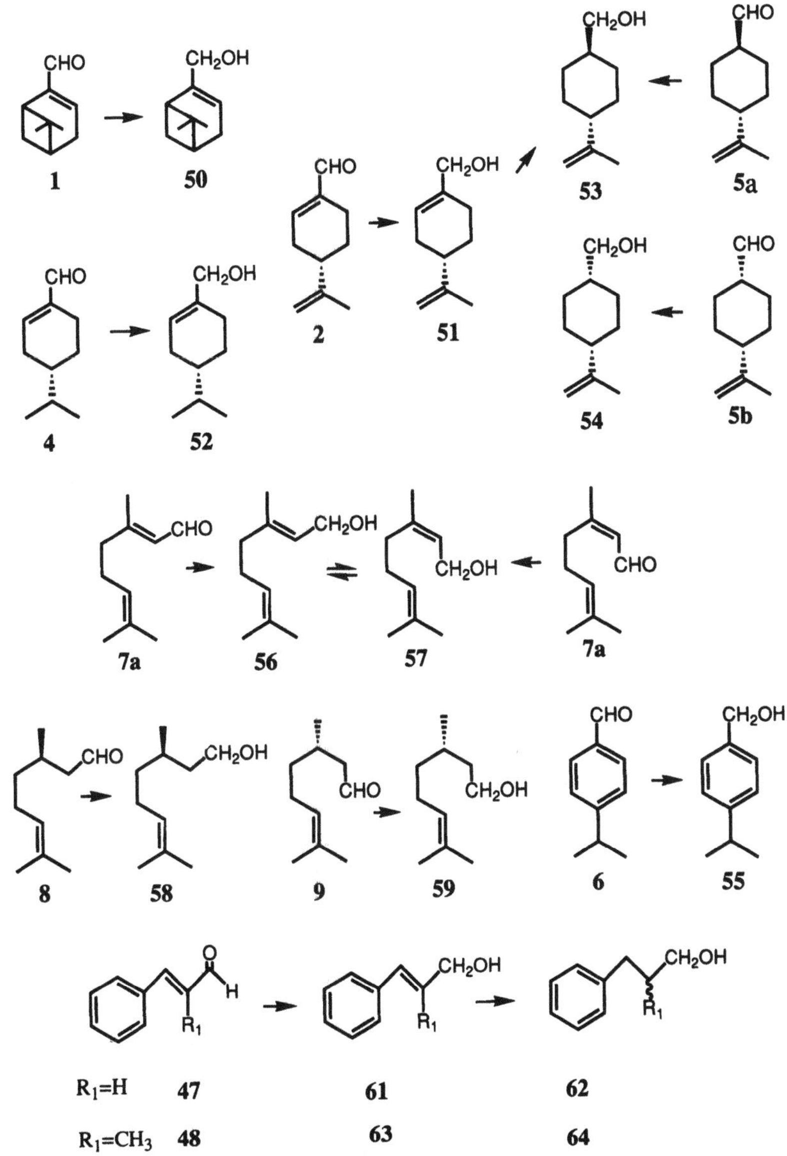


**Fig. 2.** The biotransformation of monoterpene aldehydes and related aldehydes by *Dunaliella tertiolecta*. (Noma et al. 1992a)

selectivity for reduction of mono- and disubstituted aromatic aldehydes such as hydroxybenzaldehydes(**12–14**), nitrobenzaldehydes(**23–25**), and disubstituted aromatic aldehydes(**34–40**). Compounds such as **12, 14, 23, 36,** and **40** possessing hydroxyl or nitro groups at the *ortho* or *para* position were difficult to transform or not transformed at all. On the other hand, compounds such as **37** and **38**

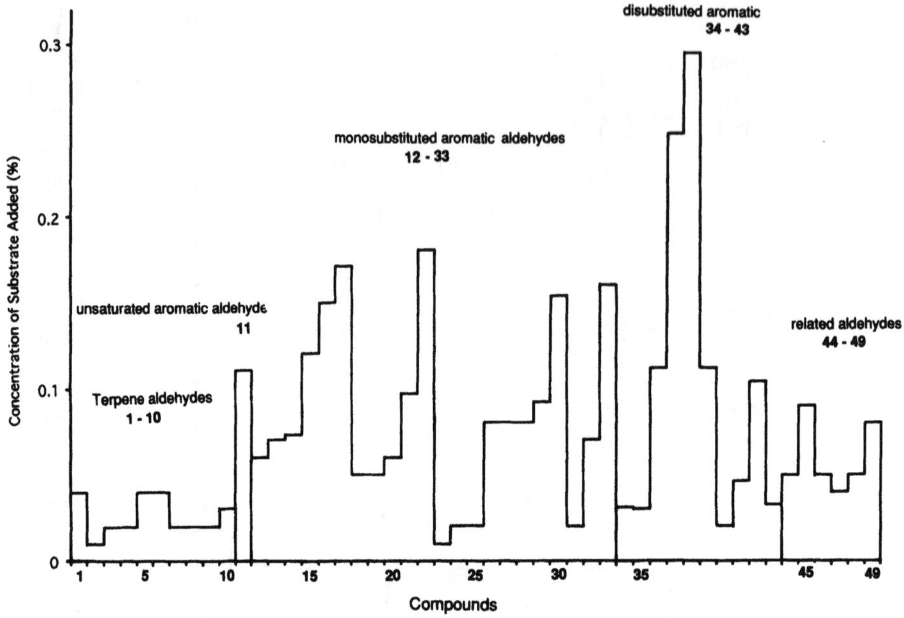


**Fig. 3.** The relationship of terpene aldehydes, aromatic aldehydes, and related aldehydes and the survival of *Dunaliella tertiolecta*, when the substrate was added continuously every day and every 100 ppm. Compounds are same as in Tables 1 and 2. (Noma et al. 1991a)

**Table 2.** The summary of biotransformation of aromatic aldehydes and related aldehydes and substrate concentration for the death of *Dunaliella*. (Noma et al. 1991a, 1992a)

| Substrates | Products [T.r.(%)] | [c]D.C. (μl or mg/100 ml) |
|---|---|---|
| Benzaldehyde(**11**) | Benzyl alcohol(M[a], 100) | 110 |
| Salicyl aldehyde(**12**, *o*-Hydroxybenzaldehyde) | Salicyl alcohol(m[b],4 and 42 after 1 and 4 days, respectively) | 50 |
| *m*-Hydroxybenzaldehyde(**13**) | *m*-Hydroxybenzyl alcohol (M, 15 and 88 after 1 and 4 days, respectively) | |
| *p*-Hydroxybenzaldehyde(**14**) | *p*-Hydroxybenzyl alcohol(M, 0 and 100 after 4 and 10 days, respectively) | |
| *o*-Anisaldehyde(**15**) | *o*-Anisalcohol(M, 100) | 120 |
| *m*-Anisaldehyde(**16**) | *m*-Anisalcohol(M, 100) | 150 |
| *p*-Anisaldehyde(**17**) | *p*-Anisalcohol(M, 100) | 170 |
| *o*-Chlorobenzaldehyde(**18**) | *o*-Chlorobenzyl alcohol(M, 100) | 50 |
| *m*-Chrolobenzalaldehyde(**19**) | *m*-Chlorobenzyl alcohol(M, 100) | 50 |
| *p*-Chlorobenzaldehyde(**20**) | *p*-Chlorobenzyl alcohol(M, 100) | 60 |
| 2-Cyanobenzaldehyde(**21**) | 2-Cyanobenzyl alcohol(m) Phthalide(**60**, M) | 95 |
| 3-Cyanobenzaldehyde(**22**) | 3-Cyanobenzyl alcohol(M, 100) | 180 |
| *o*-Nitrobenzaldehyde(**23**) | *o*-Nitrobenzyl alcohol(m, 4) | 10 |
| *m*-Nitrobenzaldehyde(**24**) | *m*-Nitrobenzyl alcohol(M, 100) | 20 |

**Table 2.** (*Contd.*)

| Substrates | Products [T.r.(%)] | [c] D.C. ($\mu$l or mg/100 ml) |
|---|---|---|
| *p*-Nitrobenzaldehyde(**25**) | *p*-Nitrobenzyl alcohol(M, 100) | 20 |
| *o*-Tolualdehyde(**26**) | *o*-Tolualcohol(M, 100) | 80 |
| *m*-Tolualdehyde(**27**) | *m*-Tolualcohol(M, 100) | 80 |
| *p*-Tolualdehyde(**28**) | *p*-Tolualcohol(M, 100) | 80 |
| *o*-Phthalaldehydic acid(**29**) | | 90 |
| Terephthalaldehydic acid(**30**) | | 180 |
| *o*-Phthalaldehyde(**31**) | | 20 |
| Isophthalaldehyde(**32**) | Isophthalaldehydic alcohol(M, 100 after 1 day)<br>Isophthalalcohol(M, 100 after 2 days) | 70 |
| Terephthaldehyde(**33**) | Terephthalaldehydic alcohol (M, 100 after 1 day)<br>Terephthalalcohol(M, 100 after 2 days) | 160 |
| *o*-Vanillin(**34**) | *o*-Vanillyl alcohol(M, 100 after 5 days) | 32 |
| 3-Nitrosalicyl aldehyde(**35**) | 3-Nitrosalicyl alcohol(M, 100) | 29 |
| Vanillin(**36**) | Vanillyl alcohol(M, 25 after 9 days) | 114 |
| Veratraldehyde(**37**) | Veratralcohol(M, 100) | 249 |
| Isovanillin(**38**) | Isovanillyl alcohol(M, 100) | 296 |
| Ethylvanillin(**39**) | Ethylvanillyl alcohol(M, 50 after 9 days) | 112 |
| 2-Hydroxy-5-methoxybenzaldehyde(**40**) | Not transformed | 20 |
| 2,4-Dihydroxybenzaldehyde(**41**) | | 47 |
| 3,4-Dihydroxybenzaldehyde(**42**) | | 105 |
| 2,5-Dihydroxybenzaldehyde(**43**) | | 33 |
| Phenylacetaldehyde(**44**) | β-Phenylethanol(M, 100) | 50 |
| 2-Phenylpropionaldehyde(**45**) | 2-Phenylpropanol(M, 100) | 90 |
| 3-Phenylpropionaldehyde(**46**) | 3-Phenylpropanol(**62**, M, 100) | 50 |
| Cinnamaldehyde(**47**) | Cinnamyl alcohol(**61**, M, 100)<br>3-Phenylpropanol(**62**, M, 100 after 9 days) | 40 |
| α-Methyl cinnamaldehyde(**48**) | 2-Methylcinnamyl alcohol(**63**, M, 100)<br>2-Methyl-3-phenylpropanol(**64**, M, 80 after 8 days | 50 |
| Nicotin aldehyde(**49**) | Nicotin alcohol(M, 100) | 80 |

T.r.: transformation ratio
[a] M: major product
[b] m: minor product
[c] D.C.: Concentration of substrate when substrate was added to the medium continuously till the death of *Dunaliella*.

having a methoxyl group at the *para*-position were good substrates for transformation. High selectivity in the reduction of the aldehyde group by *Dunaliella* is one of the characteristic features. 2-Cyanobenzaldehyde(**21**) was transformed to phthalide(**60**) as one of the products which might be formed via 2-cyanobenzyl alcohol and an imine as intermediates. However, **60** was not formed from *o*-phthalaldehyde(**31**) and *o*-phthalaldehydic acid(**29**). Iso- (**32**) and terephthalaldehydes(**33**) were reduced stepwise, and the corresponding diols were formed via the aldehydic alcohols. All related compounds(**44**–**48**) of the aromatic aldehydes and nicotin aldehyde(**49**) were firstly transformed to the corresponding primary alcohols, while cinnamaldehyde(**47**) and α-methyl cinnamaldehyde(**48**) were reduced to 3-phenylpropanol(**62**) and 2-methyl-3-phenylpropanol(**64**) via cinnamyl alcohol(**61**) and 2-methyl cinnamyl alcohol(**63**), respectively (Fig. 2).

The biotransformations of terpene ketones and related compounds are summarized in Table 3. *l*-Carvone(**65**) was transformed to *d*-neodihydrocar-

**Table 3.** The summary of biotransformation of monoterpene ketones and related compounds (Noma et al. 1991a)

| Substrates | Products [T.r. (%)] |
|---|---|
| *l*-Carvone(**65**) | *d*-Dihydrocarvone(**73**, M[a])<br>*d*-Neodihydrocarveol(**78**, M) |
| *d*-Carvone(**66**) | *l*-Isodihydrocarvone(**76**, M)<br>*l*-Dihydrocarvone(**75**, m[b])<br>*l*-Isodihydrocarveol(**79**, M)<br>*l*-Neoisodihydrocarveol(**80**, M) |
| *l*-Carvotanacetone(**67**) | *d*-Carvomenthone(**81**, M)<br>*d*-Neocarvomenthol(**82**, M) |
| *d*-Carvotanacetone(**68**) | *l*-Isocarvomenthone(**83**, M)<br>*l*-Neoisocarvomenthol(**84**, M) |
| *l*-Carvone-8,9-epoxide(**69**) | *d*-Dihydrocarvone-8,9-epoxides(**77**)<br>*d*-Neodihydrocarveol-8,9-epoxides(**85**) |
| *d*-Carvone-8,9-epoxide(**70**) | *l*-Isodihydrocarvone-8,9-epoxides(**86**) |
| 2-Methyl-2-cyclohexenone(**71**) | *l*-2-Methylcyclohexanone(**87**)<br>*d*-2-Methylcyclohexanol(**88**) |
| 2-Cyclohexenone(**72**) | Cyclohexanone(**89**)<br>Cyclohexanol(**90**) |
| *d*-Dihydrocarvone and *d*-isodihydrocarvone (**73** and**74**, 80:20) | *d*-Neodihydrocarveol(**78**)<br>*d*-Neoisodihydrocarveol |
| *l*-Dihydrocarvone and *l*-isodihydrocarvone (**75** and **76**, 80:20) | *l*-Neodihydrocarveol<br>*l*-Neoisodihydrocarveol |
| *d*-Dihydrocarvone-8,9-epoxides(**77**) | *d*-Neodihydrocarveol-8,9-epoxides(**85**) |
| *l*-Dihydrocarvone-8,9-epoxides | Neodihydrocarveol-8,9-epoxides |
| *dl*-2-Methylcyclohexanones | *cis*- and *trans*-2-Methylcyclohexanol(55:28) |
| *dl*-3-Methylcyclohexanones | *cis*- and *trans*-3-Methylcyclohexanol(73:16) |
| *dl*-4-Methylcyclohexanones | *trans*- and *cis*-4-Methylcyclohexanol(71:29) |
| Cyclohexanone | Cyclohexanol(M, 100%, 4 days) |
| Cyclobutanone | Cyclobutanol(M, 100%, 2 days) |
| Cyclopentanone | Cyclopentanol(M, 90%, 4 days) |
| Cycloheptanone | Cycloheptanol(M, 55%, 8 days) |
| Cyclooctanone | Cyclooctanol(m, 20%, 8 days) |
| Cyclodecanone | Cyclodecanol(m, 49%, 8 days) |
| Cyclododecanone | Cyclododecanol(M, 50%, 8 days) |

T.r.: transformation ratio.
[a] M: major product.
[b] m: minor product.

veol(**78**) via *d*-dihydrocarvone(**73**) stereospecifically, whereas *d*-carvone(**66**) was transformed mainly to *l*-iso- and *l*-neoisodihydrocarveols(**79** and **80**) via *l*-isodihydrocarvone(**76**) (Fig. 2). *l*-Carvotanacetone(**67**) was also hydrogenated to *d*-neocarvomenthol(**82**) via *d*-carvomenthone (**81**). On the other hand, *d*-isomer(**68**) was also transformed to *l*-neoisocarvomenthol(**84**) via *l*-isocarvomenthone(**83**). α,β-Unsaturated ketones such as *l*-(**69**) and *d*-carvone-8,9-epoxides(**70**), 2-methyl-2-cyclohexenone(**71**), and 2-cyclohexenone(**72**) were also transformed in the same manner. However, *l*-isodihydrocarvone-8,9-epoxides(**86**) formed from **70** were difficult to transform. Although compounds **65**–**72**, dihydrocarvones(**73**–**76**), 2-, 3-, and 4-methylcyclohexanones, and cyclohexanone were easily reduced, α,β-unsaturated ketones such as *dl*-piperitone, *dl*-isopiperitenone, *n*-piperitenone, verbenone, pulegone, and 3-methyl-2-cyclohexenone and saturated ketones such as *dl*-isomenthone, *l*-menthone, *d*-, *l*-, and *dl*-camphor were not transformed at all. The efficient formation of compounds **73**, **79**, *d*-dihydrocarvone-8,9-epoxides(**77**), **81**, **83**, **86**, and *l*-2-methylcyclohexanone(**87**) from the corresponding α,β-unsaturated ketones(**65**–**71**) suggested that the C=C double bond may be hydrogenated from behind (*Si*-face), if the compounds possessing 2-methyl-2-cyclohexenone skeleton are presented as shown in Fig. 4. The stereospecific hydrogenation

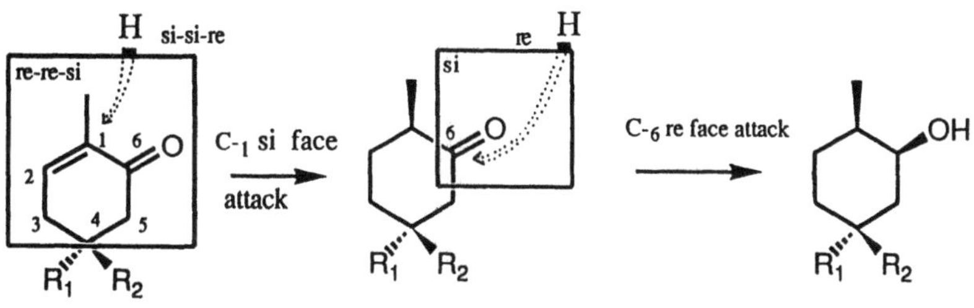


| Substrates | R1 | R2 | Products |
|---|---|---|---|
| *l*-Carvone(**65**) | Isopropenyl | H | *d*-Dihydrocarvone(**73**)<br>*d*-Neodihydrocarveol(**78**) |
| *d*-Carvone(**66**) | H | Isopropenyl | *l*-Isodihydrocarvone(**76**)<br>*l*-Neoisodihydrocarveol(**80**) |
| *l*-Carvotanacetone(**67**) | Isopropyl | H | *d*-Carvmenthone(**81**)<br>*d*-Neocarvomenthol(**82**) |
| *d*-Carvontanacetone(**68**) | H | Isopropyl | *l*-Isocarvomenthone(**83**)<br>*l*-Neoisocarvomenthol(**84**) |
| *l*-Carvone-8,9-epoxide (**69**) | Epoxyisopropenyl | H | *d*-Dihydrocarvone-8,9-epoxide(**77**)<br>*d*-Neodihydrocarveol-8,9-epoxide(**85**) |
| *d*-Carvone-8,9-epoxide(**70**) | H | Epoxyisopropenyl | *l*-Isodihydrocarvone-8,9-epoxide (**86**) |
| 2-Methyl-2-cyclohexenone(**71**) | H | H | *l*-2-Methylcyclohexanone(**87**)<br>*d*-2-Methylcyclohexanol(**88**) |

**Fig. 4.** The stereospecific hydrogenation of C=C double bond of α,β-unsaturated ketones and reduction of saturated ketones by *Dunaliella tertiolecta*. (Noma et al. 1991a)

occurs independently of the configuration and the kinds of the substituent at C-4 position, so that the methyl group at C-1 position is fixed mainly at the *R*-configuration. On the other hand, from the efficient formation of **78**, **80**, **82**, **84**, and *d*-2-methylcyclohexanol(**88**), the reduction of carbonyl group in **73**, **76**, **81**, **83**, and **87** also occurs stereospecifically from the *Re* face.

The biotransformation of cycloalkanones (C4-C8, C10, and C12) and cycloalkanols (C4-C8, C10, and C12) by *D. tertiolecta* was carried out, and the results

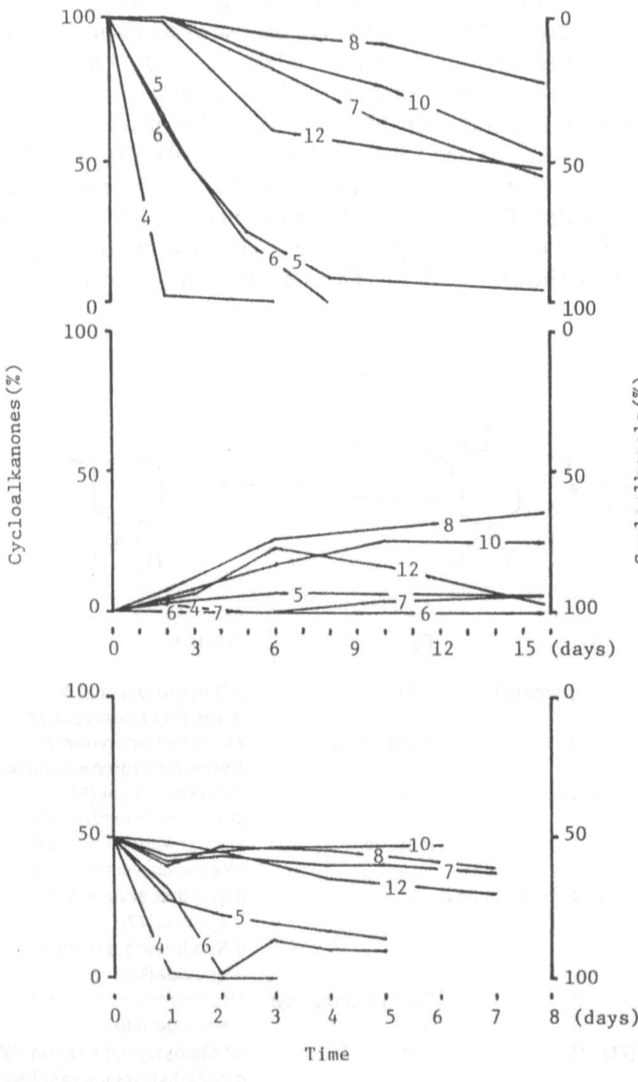


**Fig. 5.** Biotransformation of cycloalkanones and cycloalkanols by *Dunaliella tertiolecta.* (Noma et al. 1991a)

are shown in Fig. 5. *Dunaliella* reduced all cycloalkanones to the corresponding cycloalkanols. The order to be reduced was C4 > C6 > C5 > C12 > C7 > C10 > C8. On the other hand, in the case of cycloalkanols, although C6 and C4 were not transformed, a small amount of cycloalkanones having C8, C10, C12, C5, and C7 were also dehydrogenated to give the corresponding ketones. In the

**Table 4.** The summary of biotransformation of α- and β-ionones and related compounds and substrate concentration for the death of *Dunaliella*, when it was added continuously to the medium (Noma et al. unpubl.)

| Substrates | Products [T.r. (%)] | [c] D.C. (mg or μl/100 ml) |
|---|---|---|
| *dl*-α-Ionone(**86**) | α-Ionol(**108**, M[a]) | 4 |
| | 7,8-Dihydro-α-ionone(**109**, m[b]) | |
| | 7,8-Dihydro-α-ionone(**110**, m) | |
| β-Ionone(**87**) | β-Ionol(**111**, m) | 4 |
| | 7,8-Dihydro-β-ionone(**112**, M) | |
| | 7,8-Dihydro-β-ionone(**113**, M) | |
| Benzalacetone(**88**) | 7,8-Dihydrobenzalacetone(**114**, 90%, 3 days) | 30 |
| | 4-Phenyl-2-butanol(**115**, 10%, 3 days) | |
| Benzalacetophenone(**89**) | 7,8-Dihydrobenzalacetophenone(**116**, 73%, 3 days) | 10 |
| 3-Buten-2-one(**90**) | 2-Butanone(**94**, m) | 4 |
| | 2-Butanol(**117**, m) | |
| Cinnamaldehyde(**47**) | Cinnamyl alcohol(**61**, M, 100) | 10 |
| | 3-Phenylpropanol(**62**, M, 100) | |
| α-Methylcinnamaldehyde(**48**) | 2-Methylcinnamyl alcohol(**63**, M, 100) | 10 |
| | 2-Methyl-3-phenylpropanol(**64**, M, 100) | |
| Acetophenone(**91**) | 1-Phenylethanol(m, 13%, 23 days) | 110 |
| *p*-Methylacetophenone(**92**) | *p*-Methyl-1-phenylethanol(m, 9%, 23 days) | 70 |
| Benzoyl acetone(**93**) | 1-Benzoyl-2-propanol(m, 4%, 16 days) | 45 |
| 2-Butanone(**94**) | 2-Butanol(**117**, M) | 1830 |
| Methyl-*n*-propyl ketone(**95**) | 2-Pentanol(M, 54%, 4 days) | 1310 |
| Methyl-*n*-butyl ketone(**96**) | 2-Hexanol(M, 63%, 4 days) | 430 |
| Methyl-*n*-pentyl ketone(**97**) | 2-Heptanol(M, 100%, 4 days) | 230 |
| Methyl-*n*-hexyl ketone(**98**) | 2-Octanol(M, 100%, 4 days) | 70 |
| Methyl-*n*-heptyl ketone(**99**) | 2-Nonanol(M, 86%, 2 days) | 25 |
| Methyl-*n*-noryl ketone(**100**) | 2-Undecanol(11%, 2 days, 59% 5 μl) | 10 |
| Ethyl-*n*-butyl ketone(**101**) | 3-Heptanol(M, 57%, 5 days) | 330 |
| Di-*n*-propyl ketone(**102**) | 4-Heptanol(M, 66%, 5 days) | 330 |
| Geranyl acetone and neryl acetone (**103a** and **b**) | Not transformed | 4 |
| Acetyl acetone(**104**) | Not transformed | 100 |
| Acetonyl acetone(**105**) | Not transformed | 230 |
| 5-Methyl-3-heptanone(**106**) | Not transformed | 190 |
| 1-Phenyl-2-butanone(**107**) | Not transformed | 10 |

T.r.: transformation ratio
[a] M: major product
[b] m: minor product
[c] D.C.: Substrate concentration for the death of *Dunaliella*.

case of the biotransformations of an equal mixture of cycloalkanones and cycloalkanols, the same phenomenon as described above was observed.

The biotransformations of α- and β-ionones(**86** and **87**) as one of the norterpenes and related compounds by *Dunaliella* are summarized in Table 4. Compounds **86** and **87** showed high toxity even at a low concentration of ca.

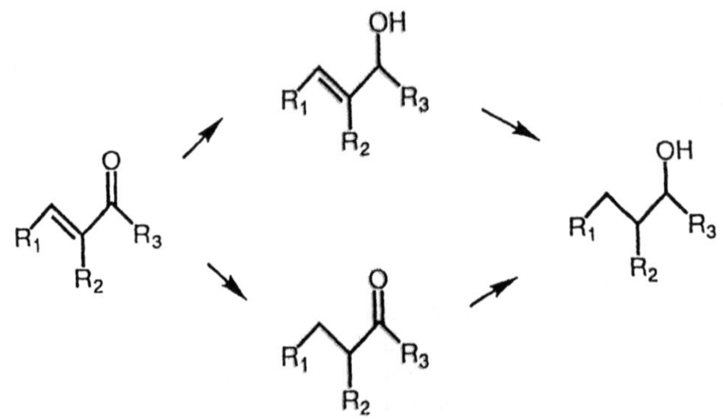


| Substrates | $R_1$ | $R_2$ | $R_3$ | Products |
|---|---|---|---|---|
| α-Ionone(86) | 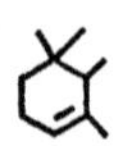 | H | Me | α-Ionol(108)<br>7,8–Dihydro–α-ionol(109)<br>7,8–Dihydro–α-ionone(110) |
| β-Ionone(87) |  | H | Me | β-Ionol(111)<br>7,8–Dihydro–β-ionol(112)<br>7,8–Dihydro–β-ionone(113) |
| Benzalalcetone(88) |  | H | Me | 7,8–Dihydrobenzalacetone(114)<br>4-Phenyl-2-butanol(115) |
| Benzalalcetophenone(89) |  | H |  | 7,8–Dihydrobenzalacetophenone (116) |
| Cinnamaldehyde(47) |  | H | H | Cinnamyl alcohol(61)<br>3-Phenylpropanol(62) |
| α-Methylcinnamaldehyde(48) |  | Me | H | 2-Methylcinnamyl alcohol(63)<br>2-Methyl-3-phenylpropanol(64) |
| 3-Buten-2-one(90) | H | H | Mc | 2-Butanone(94)<br>2-Butanol(117) |

**Fig. 6.** Metabolic pathways of α- and β-ionones and related compounds by *Dunaliella tertiolecta*. (Noma et al. 1991a)

40 ppm. However, compound **86** was biotransformed to α-ionol(**108**) as the main product and 7,8-dihydro-α-ionone(**109**) and 7,8-dihydro-α-ionol(**110**) as the minor products at a very low concentration below 40 ppm (Fig. 6). Compound **87** was biotransformed to 7,8-dihydro-β-ionone(**112**) and 7,8-dihydro-β-ionol(**113**) as the major products, with β-ionol(**111**) as the minor product at a low concentration. Benzalacetone(**88**) was transformed to 7,8-dihydrobenzalacetone(**114**) and 4-phenyl-2-butanol(**115**) as the major products. On the other hand, benzalacetophenone (chalcone, **89**) was transformed to 7,8-dihydrobenzalacetophenone (dihydrochalcone, known as one of sweeteners, **116**), which was not transformed at all. 3-Buten-2-one(**90**) also proved to be very toxic to *Dunaliella*. However, a small amount of 2-butanone(**94**) and 2-butanol(**117**) were formed at a low concentration of the addition of **90**. On the other hand, **94** was

**Table 5.** The summary of biotransformation of terpene alcohols and related compounds. (Noma et al. 1991a)

| Substrates | Products [T.r. (%)] |
|---|---|
| *l-cis*-Carveol(**118**) | *l*-Carvone(**65**, m[b])<br>*d*-Dihydrocarvone(**73**, m)<br>*d*-Neodihydrocarveol(**78**, m, 6%) |
| *d-cis*-Carveol(**119**) | Not transformed |
| *dl-cis*-Carveol(**118** and **119**) | *l*-Carvone(**65**, m)<br>*d*-Dihydrocarvone(**73**, m)<br>*d*-Neodihydrocarveol(**78**, m) |
| *l-trans*-Carveol(**120**) | Not transformed |
| *d-trans*-Carveol(**121**) | Not transformed |
| *dl-trans*-Carveol(**120** and **121**) | Not transformed |
| The mixture of *l-trans-* and *cis*-Carveol (**120** and **119**) | *l*-Carvone(**65**, m)<br>*d*-Dihydrocarvone(**73**, m)<br>*d*-Neodihydrocarveol(**78**, m) |
| The mixture of *d-trans-* and *cis*-Carveol (**121** and **119**) | Not transformed |
| *l*-Perillyl alcohol(**51**) | *trans*-Shisool(**53**, m) |
| *dl*-Perillyl alcohol | *trans*-Shisool(**53**, m) |
| Cinnamyl alcohol(**61**) | 3-Phenyl propanol(**62**) |
| 2-Methylcinnamyl alcohol(**63**) | 2-Methyl-3-phenylpropanol(**64**, M[a]) |
| Geraniol(**56**) | Nerol(**57**, M) |
| Nerol(**57**) | Geraniol(**56**, M) |
| *l*-Borneol | Not transformed |
| *d*-Borneol | Not transformed |
| *dl*-Borneol | Not transformed |
| *l*-Isoborneol | Not transformed |
| *d*-Isoborneol | Not transformed |
| *dl*-Isoborneol | Not transformed |
| Myrtenol(**50**) | Not transformed |
| *trans-* and *cis*-Shisool(**53** and **54**) | Not transformed |
| Cumin alcohol(**55**) | Not transformed |

T.r.: transformation ratio
[a] M: major product
[b] m: minor product

easily transformed to **117** at a high concentration. As shown in Fig. 6, in the biotransformation of compounds **47** and **48** (where $R_1$ = phenyl, $R_2$ = H, and $CH_3$, respectively, and $R_3$ = H), the reduction of the aldehyde group occurs predominantly, and then the resulting alcohols(**61**, **63**) were further hydrogenated to the corresponding saturated alcohols(**62**, **64**). In the case of **86** (where $R_1$ = 2,6,6-trimethylcyclohexene, $R_2$ = H, and $R_3$ = $CH_3$), the reduction of carbonyl group occurred predominantly to give **108** as the major product. In the case of compounds **87** (where $R_1$ = 2,6,6-trimethylcyclohexene, $R_2$ = H, and $R_3$ = $CH_3$), **88** (where $R_1$ = phenyl, $R_2$ = H, and $R_3$ = $CH_3$), **89** (where $R_1$ = phenyl, $R_2$ = H, and $R_3$ = phenyl) and **90** (where $R_1$ and $R_2$ = H and $R_3$ = $CH_3$), the hydrogenation of the C=C double bond in the side chain

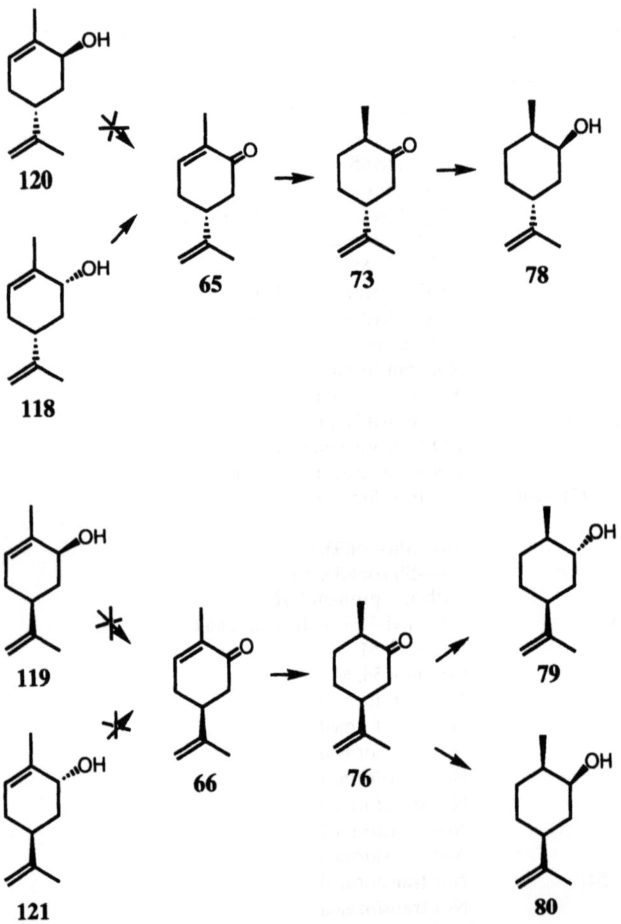


**Fig. 7.** Enantio- and diastereoselective dehydrogenation of *l-trans-* and *cis*-carveol and *d-trans-* and *cis*-carveol and metabolic pathways of *l-* and *d*-carvone by *Dunaliella tertiolecta*. Compounds as in Tables 3 and 5. (Noma et al. 1991a)

occurred predominantly rather than the reduction of the carbonyl group. On the other hand, in the biotransformation of aliphatic methyl ketones such as 2-butanone (**94**), methyl-*n*-propyl ketones(**95**), methyl-*n*-butyl ketone(**96**), methyl-*n*-pentyl ketone(**97**), methyl-*n*-hexyl ketone(**98**), methyl-*n*-heptyl ketone(**99**), and methyl-*n*-noryl ketone(**100**), ethyl-*n*-butyl ketone(**101**) and di-*n*-propyl ketone(**102**) were also transformed easily to the corresponding alcohols. On the other hand, although acetophenone(**91**), *p*-methylacetophenone(**92**), and benzoyl acetone(**93**) were biotransformed to some extent, the mixture of geranyl acetone(**103a**) and neryl acetone(**103b**), acetyl acetone(**104**), acetonyl acetone(**105**), 5-methyl-3-heptanone(**106**), and 1-phenyl-2-butanone(**107**) were difficult to transform or not transformed at all. *Dunaliella* showed high selectivity in the reduction of aliphatic and aromatic methyl ketones and related ketones. It is considered that the longer the side chain of aliphatic methyl ketones(**94**–**100**), the stronger is the toxity to *Dunaliella* (Table 4).

The biotransformation of terpene alcohols is summarized in Table 5. In the biotransformation of carveols, a small amount of *l*-*cis*-carveol(**118**) was dehydrogenated to *l*-carvone(**65**), *d*-dihydrocarvone(**73**), and *d*-neodihydrocarveol(**78**). Enantio- and diastereoselective dehydrogenation of **118** in the biotransformation of *dl*-*cis*-carveol and the mixture of *l*-*trans*-carveol(**120**) and **118** was observed to some extent as shown in Fig. 7. However, *l*-, *d*-, and *dl*-borneols and isoborneols were not transformed at all.

The present method has a short-term reaction, and the work-up is very simple. On the basis of the present results, we consider that *Dunaliella* is one of the good bioreactors to reduce aldehydes and ketones to the corresponding alcohols, and the hydrogenate C=C bond of α,β-unsaturated ketones and unsaturated alcohols to the corresponding saturated ketones and saturated alcohols, respectively.

## 3 Summary and Conclusions

The biotransformation of terpenoids and related compounds by *Dunaliella tertiolecta* was carried out. Terpene aldehydes and related aldehydes were generally reduced to the corresponding primary alcohols. The terpene aldehydes strongly inhibited the survival of the phytoflagellate at ca. 100–400 ppm. In biotransformation of aromatic aldehydes and related aldehydes, *Dunaliella* showed high selectivity for reduction of monosubstituted aromatic aldehydes such as hydroxy- and nitrobenzaldehydes and disubstituted aromatic aldehydes. On the other hand, compounds having a methoxyl group at the *para*-position were good substrates for transformation. α,β-Unsaturated ketones such as carvone and carvotanacetone were reduced to the corresponding saturated ketones and secondary alcohols stereospecifically. In the biotransformation of α- and β-ionones, benzalacetone, benzalacetophenone, and 3-buten-2-one, both/either hydrogenation of C=C double bond and/or reduction of C=O bond were observed at a very low concentration. Aliphatic methyl ketones were also transformed easily to the corresponding alcohols. *Dunaliella* showed high selectivity in the reduction of aliphatic and aromatic methyl ketones and related compounds. Enantio- and diastereoselective dehydrogenation of *l-cis*-carveol was observed to some extent in the biotransformation of *dl-cis*-carveol and the mixture of *l-trans*-carveol and *l-cis*-carveol.

*Acknowledgments.* The authors wish to express their gratitude to Nippon Terpene Chemical Co. Ltd. for supplying terpenes and Dr. Shuhei Fujii, Tezukayama College, for supplying *Dunaliella tertiolecta.* The authors would also like to thank Dr. Toshihiro Hashimoto, Tokushima Bunri University, for useful discussion, and Ms. Naoko Miki, Etsuko Akehi, and Emiko Manabe for their technical assistance.

## References

Asakawa Y, Takahashi H, Toyota M, Noma Y (1991) Biotransformation of monoterpenoids ( − )- and ( + )-menthols, terpinolene and carvotanacetone by *Asepergillus* species. Phytochemistry 30 (12): 3981–3987

Baas-Becking LGM (1931) Salt effect on swarmers of *Dunaliella vorodos* Teod. J Gen Physiol 14: 765–779

Ben-Amotz A, Avron M (1973) The role of glycerol in the osmotic regulation in the halophilic alga, *Dunaliella parva.* Plant Physiol 51: 875–878

Ben-Amotz A, Avron M (1981) Glycerol and β-carotene metabolism in the halotolerant alga *Dunaliella*: a model system for biosolar energy conversion. Trends Biochem Sci 6(11): 297–299

Borowitzka LJ, Brown AD (1974) The salt relations of marine and halophilic strains of the unicellular green alga, *Dunaliella.* The role of glycerol as a compatible solute. Arch Microbiol 96: 37–52

Butcher RW (1959) An introductory account of the smaller algae of British coastal water. I. Introduction and Chlorophyceae. Fish Invest Ser 4: 1–74

Christensen T (1962) Alger. In: Böcher TW, Lange M, Sőresen T (eds) Systematik Botanik, vol 2, 2nd edn. Munksgaard, Kopenhagen, pp 1–178

Cifuentes AS, Gonzalez M, Conejeros M, Dellarossa V, Parra O (1992) Growth and carotenogenesis in eight strains of *Dunaliella salina* Teodoresco from Chile. J Appl Phycol 4: 111–118

Craigie GS, McLachlan J (1964) Glycerol as a photosynthetic products of *Dunaliella tertiolecta.* Can J Bot 42: 777–778

Dunal MF (1837) Note sur les algues qui colorent en rouge certaines eaux des marais selants méditerranéens. C R Acad Sci, Paris 15: 585–587

Ettl H (1983) Taxonomische Bemerkungen zu der Phytomonadina. Nova Hedwigia 35: 731–736

Frank G, Wegmann K (1974) Physiology and biochemistry of glycerol biosynthesis in *Dunaliella.* Biol Zentral 93: 707–723

Hoshaw RW, Maluf LY (1981) Ultrastructure of the green flagellate *Dunaliella tertiolecta* (Chlorophyceae, Volvocales) with comparative notes on three other species. Phycologia 20: 199–206

Kaplan IR, Fredmann A (1970) Biological productivity in the Dead Sea. I. Microorganisms in the water column. Isr J Chem 8: 513–533

Lerche W (1936) Untersuchungen über Entwicklung und Fortpflanzung in der Gattung *Dunaliella.* Arch Protistenkd 88: 236–268

Massyuk NP (1973) Morphology, taxonomy, ecology and geographic distribution of the genus *Dunaliella* Teod. and prospects for its potential utilization. Naukova Dumka, Kiev

McLachlan J (1960) The culutre of *Dunaliella tertiolecta* Butcher – a euryhaline organism. Can J Microbiol 6: 367–379

Miyazawa M, Yamamoto K, Noma Y, Kameoka H (1990a) Bioconversion of ( + )-fenchone to ( + )-6-*endo-hydroxyfenchone* by *Aspergillus niger.* Chem Express 5(4): 237–240

Miyazawa M, Yamamoto K, Noma Y, Kameoka H (1990b) Bioconversion of ( + )-fenchone to 5-*endo-hydroxyfenchone* by *Aspergillus niger*. Chem Express 5(6): 407–410
Miyazawa M, Noma Y, Yamamoto K, Kameoka H (1991) Biotransformation of 1,4-cineole to 2-*endo*-hydroxy-1,4-cineole by *Aspergillus niger*. Chem Express 6(10): 771–774
Miyazawa M, Noma Y, Yamamoto K, Kameoka H (1992a) Biotransformation of 1,4-cineole to 3-*endo*-hydroxy-1,4-cineole by *Aspergillus niger*. Chem Express 7(4): 305–308
Miyazawa M, Noma Y, Yamamoto K, Kameoka H (1992b) Biohydroxylation of 1,4-cineole to 9-hydroxy-1,4-cineole by *Aspergillus niger*. Chem Express 7(4): 305–308
Miyazawa M, Noma Y, Yamamoto K, Kameoka H (1992c) Biohydroxylation of 1,4-cineole to 8-hydroxy-1,4-cineole by *Aspergillus niger*. Chem Express 7(9): 721–724
Nishimura H, Noma Y, Mizutani J (1982) *Eucalyptus* as biomass. Novel compounds from microbial conversion of 1,8-cineole. Agric Biol Chem 46(10) : 2601–2604
Nishimura H, Hiramoto S, Mizutani J, Noma Y, Furusaki A, Matsumoto T (1983) Structure and biological activity of bottrospicatol, a novel monoterpene produced by microbial transformation of ( – )-*cis*-carveol. Agric Biol Chem 47(11) : 2697–2699
Noma Y (1976) Microbiological conversion of carvone. Ann Res Stud Osaka Joshigakuen Junior College 20 : 33–47
Noma Y (1977) Conversion of the analogues of carvone and dihydrocarvone by *Pseudomonas ovalis*, strain 6–1. Nippon Nogeikagaku Kaishi 51(7) : 463–470
Noma Y (1978a) On the pattern of reaction mechanism of ( – )-carvone conversion by *Actinomycetes*. Ann Res Stud Osaka Joshigakuen Junior College 22 : 57–62
Noma Y (1978b) Conversion of ( – )-carvone by *Nocardia lurida* A-0141 and *Streptosporangium roseum* IFO 3776. Nippon Nogeikagaku Kaishi 53(1) : 35–39
Noma Y (1979) On the pattern of reaction mechanism of ( + )-carvone by *Actinomycetes*. Ann Res Stud Osaka Joshigakuen Junior College 23 : 27–31
Noma Y (1980) Conversion of ( – )-carvone by strains of *Streptomyces*, A-5-1, and *Nocardia*, 1-3-11. Agric Biol Chem 44(4) : 807–812
Noma Y, Asakawa Y (1992) Enantio- and diastereoselectivity in the biotransformation of carveols by *Euglena gracilis* Z. Phytochemistry 31(6): 2009–2011
Noma Y, Nishimura H (1987) Bottrospicatols, novel monoterpenes produced on conversion of ( – )- and ( + )-*cis*-carveol by *Streptomyces*. Agric Biol Chem 51(7): 1845–1849
Noma Y, Nonomura S (1974) Conversion of ( – )-carvone and ( + )-carvone by a strain of *Aspergillus niger*. Agric Biol Chem 38(4): 741–744
Noma Y, Sakai H (1984) The conversion of ( – )-perillyl alcohol, 1,8-cineole, ( + )-carvone and ( – )-carvone by rare *Actinomycetes*. Ann Res Stud Osaka Joshigakuen Junior College 28: 7–18
Noma Y, Tatsumi C (1973) Conversion of ( – )-carvone by *Pseudomonas ovalis*, strain 6–1. Nippon Nogeikagaku Kaishi 47(11): 705–711
Noma Y, Nonomura S, Ueda H, Tatsumi C (1974a) Conversion of ( + )-carvone by *Pseutomonas ovalis*, strain 6–1(1). Agric Biol Chem 38(4): 735–740
Noma Y, Nonomura S, Sakai H (1974b) Conversion of ( – )-carvotanacetone and ( + )-carvotanacetone by *Pseudonomas ovalis*, strain 6–1. Agric Biol Chem 38(9): 1637–1642
Noma Y, Nonomura S, Sakai H (1975) Epimerization of ( – )-isodihydrocarvone to ( – )-dihydrocarvone by *Pseudomonas fragi* IFO 3458. Agric Biol Chem 39(2)437–441
Noma Y, Nishimura H, Hiramoto S, Iwami M, Tatsumi C (1982) A new compound, (4R, 6R)-( + )-6,8-oxidomenth-1-en-9-ol produced by microbial conversion of ( – )-*cis*-carveol. Agric Biol Chem 46(11): 2871–2872
Noma Y, Toyata M, Asakawa Y (1985) Biotransformation of ( – )-carvone and ( + )-carvone by *Aspergillus niger*. 29th Symp Chemistry of terpenes, essential oils and aromatics of Japan. Symp Pap pp 235–237 (Tsu)
Noma Y, Takahashi H, Asakawa Y (1990) Formation of p-menthan-1,9-diol from *p*-menthane by *A. niger*. 34th Symp Chemistry of terpenes, essential oils and aromatics of Japan, Symp Pap, pp 253–255 (Takamatsu)
Noma Y, Miki N, Akehi E, Manabe E, Asakawa Y (1991a) Biotransformation of terpenoids and related compounds by *D. tertiolecta*. 35th Symp Chemistry of terpenes, essential oils and aromatics of Japan, Symp Pap, pp 112–114 (Nagoya)
Noma Y, Okajima Y, Takahashi H, Asakawa Y (1991b) Biotransformation of aromatic aldehydes and related compounds by *Euglena gracilis* Z. Phytochemistry 30(9): 2969–2972

Noma Y, Takahashi H, Asakawa Y (1991c) Biotransformation of terpene aldehydes by *Euglena gracilis* Z. Phytochemistry 30(4):1147–1151

Noma Y, Akehi E, Miki N, Asakawa Y (1992a) Biotransformation of terpene aldehydes, aromatic aldehydes and related compounds by *Dunaliella tertiolecta*. Phytochemistry 31(2):515–517

Noma Y, Yamasaki S, Asakawa Y (1992b) Biotransformation of limonene and related compounds by *Aspergillus cellulosae*. Phytochemistry 31(8):2725–2727

Noro T (1978) Effects of Mn on the growth of a marine alga *Dunaliella tertiolecta*. Jpn J Phycol 26:69–72

Pierce GJ (1914) The behavior of certain microorganisms in brine. In: The Salton Sea. Carnegie Inst Wash Publ 193:49–69

Post FJ (1977) The microbial ecology of the Great Salt Lake. Microb Ecol 3:143–165

Riisgard HU (1981) Cell-volume responses in the naked marine flagellate *Dunaliella marina* transferred from darkness to light of different intensities. Bot Mar 24:657–659

Satrr RC (1978) The culture collection of algae at the University of Texas at Austin. J Phycol 14 (Suppl):47–100

Teodoresco EC (1905) Organisation et développement du *Dunaliella*, nouveau genre de Volvacacée-Polyblepharidée. Beih Bot Centralbl 18:215–232

# XIV *Epimedium diphyllum* Morr. et Decne. (Baika-ikariso): In Vitro Cultures and the Production of Isoprenylflavonol Glycosides

H. YAMAMOTO and M. IINUMA[1]

## 1 Introduction

### 1.1 Distribution and Importance of the Plant

The genus *Epimedium* (family Berberidaceae), is distributed widely in eastern Asia, the Middle East, the Himalayas, and Europe (Stearn 1938). In Europe, these plants are mainly cultivated for horticulture, whereas in eastern Asia, some *Epimedium* species have been used as a crude drug. In China, the aerial parts of *E. brevicornum* Maxim, *E. saggitatum* (Sieb. et Zucc.) Maxim, *E. pubescens* Maxim, and *E. koreanum* Nakai are described in The Chinese Pharmacopeia as 'Yinyanghuo' and used as a tonic, and for curing rheumatism, paralysis, and involutional hypertension (Jiang Su Sin Medical College 1977a). In Japan, the aerial parts of *E. grandiflorum* Morr. var. *thunbergianum* (Miq.) Nakai, *E. sempervirens* Nakai, *E. koreanum* Nakai, and *E. diphyllum* Morr. et Decne. (Fig. 1) are used mainly as a tonic (Namba 1980). In China, the underground parts of *Epimedium* species are also used for treating asthmatic fits and menstrual irregularity (Jiang Su Sin Medical College, 1977b).

### 1.2 Chemical Constituents of *Epimedium*

The main constituents in *Epimedium* plants are flavonol glycosides, which have a γ,γ-dimethylallyl (isoprenyl) group at the C-8 position. More than 20 isoprenylflavonol glycosides are isolated from both the aerial and the underground parts of *Epimedium* species (Fig. 2). As a general characteristic of the glycosides, the aglycone moiety is 5,7,4′-trihydroxyflavonol (kaempferol) which has an isoprenyl group at its C-8 position (des-*O*-methylanhydroicaritin); a hydroxy group at the C-4′ position is sometimes methylated (anhydroicaritin). As regards the sugar moiety, a rhamnose is always linked at C-3 as an endo-sugar, and another sugar, such as rhamnose, xylose, or glucose, is attached to the rhamnose mainly at its C-2, or sometimes at its C-3 or C-4 position as the terminal sugar. A glucose is also linked at the C-7 position. Some glycosides, such as epimedins

[1]Department of Pharmacognosy, Gifu Pharmaceutical University, 6-1 Mitahora-higashi 5 chome, Gifu 502, Japan

Biotechnology in Agriculture and Forestry, Vol. 28
Medicinal and Aromatic Plants VII (ed. by Y.P.S. Bajaj)

**Fig. 1.** Flowers of *Epimedium diphyllum* (April 1991)

A-C, are easily hydrolyzed by heat, light, or oxygen, and yield their corresponding di- and monoglycosides, and sometimes aglycones (Mizuno et al. 1988a). This suggests that some of mono- and diglycosides are artifacts caused by the preservation and the extraction stages of original materials.

Benzylisoquinoline alkaloid magnoflorine (Tomita and Ishii 1957a, b), lignans (Tokuoka et al. 1975b; Miyase et al. 1987a), and phenolic glycosides (Miyase et al. 1987a, b, 1988) were also isolated.

It is noteworthy that des-*O*-methylanhydroicaritin glycosides were rich in the underground parts of *Epimedium* species (Takemoto et al. 1975; Tokuoka et al. 1975a, c, Fukai and Nomura 1988; Mizuno et al. 1988b, 1989a), whereas anhydroicaritin glycosides were abundant in the aerial parts (Tokuoka et al. 1975d; Mizuno et al. 1987, 1988c, 1989b, 1990; Oshima et al. 1987; Ito et al. 1988a, b), and sometimes in the underground parts of the plants (Fukai and Nomura 1988; Mizuno et al. 1988b).

Several pharmaceutical activities of crude extracts of *Epimedium* have been reviewed by Tang and Eisenbrand (1992), and it was shown that anhydroicaritin glycosides such as icariin and epimedin C have biological characteristics, such as, myocardial activities (Liu et al. 1982), an anticomplementary activity (Oshima et al. 1987) and a phagocytic activity (Iinuma et al. 1990).

## 2 In Vitro Culture Studies

There have been many reports on flavonoid production by plant tissue cultures (Hinderer and Seitz 1988), and the chemicals and metabolism of the compounds

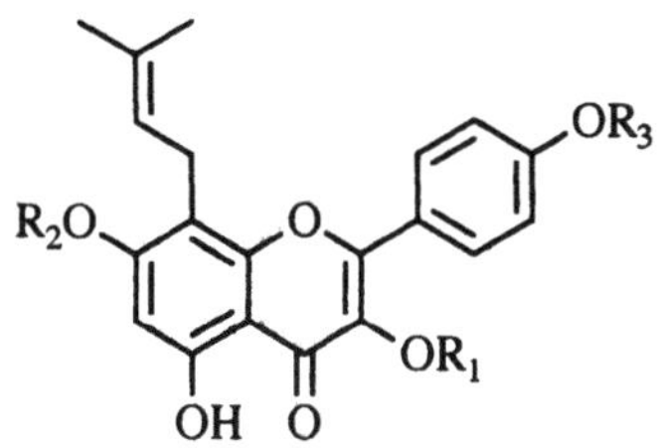


| | $R_1$ | $R_2$ | $R_3$ |
|---|---|---|---|
| epimedin A | rha$^2$-glc | glc | Me |
| epimedin B | rha$^2$-xyl | glc | Me |
| epimedin C | rha$^2$-rha | glc | Me |
| sempervirenoside A | rha$^3$-xyl | glc | Me |
| | $^4$Ac $^3$Ac | | |
| sempervirenoside B | rha$^3$-xyl | glc | Me |
| | $^4$Ac | | |
| baohuoside VI | rha$^4$-rha | glc | Me |
| icariin | rha | glc | Me |
| baohuoside VII | rha$^4$-rha | H | Me |
| sagittatoside A | rha$^2$-glc | H | Me |
| sagittatoside B | rha$^2$-xyl | H | Me |
| sagittatoside C | rha$^2$-glc | H | Me |
| | $^3$Ac | | |
| icarisid I | H | glc | Me |
| icarisid II | rha | H | Me |
| anhydroicaritin | H | H | Me |

| | $R_1$ | $R_2$ | $R_3$ |
|---|---|---|---|
| diphylloside C | rha$^2$-glc | glc$^2$-glc | H |
| ikarisoside C | rha$^2$-glc | glc | H |
| epimedoside E | rha$^2$-xyl | glc | H |
| diphylloside B | rha$^2$-rha | glc | H |
| baohuoside V | rha$^4$-rha | glc | H |
| epimedoside A | rha | glc | H |
| baohuoside IV | rha | rha | H |
| ikarisoside B | rha$^2$-glc | H | H |
| ikarisoside F | rha$^2$-xyl | H | H |
| baohuoside III | rha$^4$-rha | H | |
| ikarisoside A | rha | H | H |
| ikarisoside D | rha | | H |
| | $^4$Ac | H | H |
| des-*O*-methyl-anhydroicaritin | H | H | H |

**Fig. 2.** Isoprenylflavonol glycosides isolated from *Epimedium* species

have also been reviewed (Harborne 1988; Stafford 1990). However, there are only few reports concerning the production of prenylflavonoids (Hargreaves and Selby 1978; Ayabe et al. 1980; Zähringer et al. 1981; Kobayashi et al. 1985; Ikuta et al. 1986; Yamamoto et al. 1991), and the production of prenylflavonoid glycosides has never been reported. To investigate the production mechanism of isoprenylflavonol glycosides, we established the callus cultures of *Epimedium diphyllum* and reported the production of these compounds (Yamamoto et al. 1992). To our knowledge, this is the first study of in vitro cultures of *Epimedium* species.

## 2.1 Establishment of Cell Culture

For the induction of callus cultures, segments of surface-sterilized flower buds of *Epimedium diphyllum* Morr. et Decne. collected in March at the Botanical Garden of Gifu Pharmaceutical University were placed on LS medium

(Linsmaier and Skoog 1965) containing 10 μM 2,4-dichlorophenoxyacetic acid (2,4-D), solidified with 0.3% gellan gum, and cultured at 25° C in the dark. Yellow callus tissues established were then subcultured on the same medium under the same culture conditions, at intervals of 1 month for over 2 years.

Cell suspension cultures were started by transferring callus (3 g fr. wt.) to LS liquid medium (20 ml) containing 10 μM 2,4-D in 100-ml Erlenmeyer flasks. The flasks were agitated on a rotary shaker at a speed of 100 rpm at 25°C in the dark, and were subcultured (1g fresh cells) every 3 weeks for more than 1 year.

## 2.2 Identification of Chemical Constituents

Lyophilized callus tissues (30 g) were Soxhlet-extracted with MeOH for 36 h. The MeOH extract was evaporated to dryness, and the residue dissolved in $H_2O$ was washed with $CHCl_3$, and then extracted with *n*-BuOH. The *n*-BuOH fraction (4.4 g) was chromatographed on silica gel eluted with $CHCl_3$-MeOH stepwise gradients as a solvent system. A fraction containing isoprenylflavonol glycosides ($CHCl_3$:MeOH = 3:1) was rechromatographed on silica gel using $CHCl_3$:MeOH:$H_2O$ = 13:7:2 (lower phase) as a solvent. Epimedoside A (80 mg) was recrystallized from MeOH. Other flavonol glycosides were further purified by Sephadex LH-20 (MeOH as a solvent) and by reverse phase HPLC (MeCN-$H_2O$ solvent systems). Ikarisoside C (20 mg), epimedoside E (8 mg), diphylloside B (11 mg), icariin (1 mg), epimedin A (3 mg), epimedin B (3 mg), and epimedin C (1 mg) were also recrystallized from MeOH, respectively. Each compound was identified by direct comparison with an authentic sample (TLC, HPLC, MS,$^1$H-NMR, and $^{13}$C-NMR).

By the use of TLC (detected by Dragendorff reagent) and the ion pair reverse phase HPLC-photodiode array SPDM-6A system (Shimadzu, Japan), it was clarified that benzylisoquinoline alkaloid magnoflorine was also produced in the cultured cells (unpubl.). The conditions of HPLC were as follows; column: Cosmosil $_5C_{18}$ (4 × 250 mm, Nacalai Tesque, Japan), solvent: MeCN: MeOH:$H_2O$ = 32:8:60 (containing 0.05 M tartaric acid and 0.01 M sodium dodecyl sulfate), oven temperature: 40 °C, flow rate: 0.8 ml/min. HPLC-photodiode array analysis indicated that cells also produced some phenolic glycosides, but these were not identified.

The content of isoprenylflavonol glycosides was analyzed by reverse phage HPLC. Lyophilized callus (100 mg) was extracted with 5 ml of MeOH by ultrasonication (15 min). After centrifugation, cell debris were extracted twice with 2 ml of MeOH. The supernatants were combined and subjected to HPLC using Cosmosil $_5C_{18}$ column (4 × 250 mm) in an oven at 40°C, with a MeCN/$H_2O$ gradient solvent system (20 % MeCN to 27% MeCN in 10 min, to 50% MeCN in another 20 min, then to 70% MeCN in 4 min) at a flow rate of 0.9 ml/min, monitoring with UV absorption at 272 nm. The quantities were calculated from the peak areas recorded by Chromatopac C-R6A (Shimadzu, Japan) using *p*-tert-butylphenol as an internal standard.

Contents of each isoprenylflavonol glycoside in the callus tissues, in the underground parts, and in the aerial parts of *E. diphyllum* plants collected in

May at the same botanical garden, are shown in Table 1. The total content of isoprenylflavonol glycosides in the callus tissues was higher than that in the original plant, but the composition of each glycoside in the callus tissues was different from that of the original plant. In the intact plant, 3-*O*-biglycosyl-7-*O*-glucosides of anhydroicaritin and des-*O*-methylanhydroicaritin were the main components, whereas epimedoside A, 3-*O*-rhamnoside-7-*O*-glucoside of des-*O*-methylanhydroicaritin was the main compound in the callus tissues. Moreover,

**Table 1.** Contents of isoprenylflavonol glycosides in callus cultures, in the underground parts, and in the aerial parts of *Epimedium diphyllum*

| | Flavonol glycosides (mg/g dry wt.) | | |
|---|---|---|---|
| | Callus | Underground | Aerial |
| Epimedoside A | 36.84 | 3.82 | 0.67 |
| Ikarisoside C | 7.15 | 13.35 | 0.63 |
| Epimedoside E | 3.46 | 2.20 | 0.27 |
| Diphylloside B | 8.13 | 7.49 | 0.33 |
| Icariin | 0.60 | 0.00 | 1.73 |
| Epimedin A | 1.69 | 1.46 | 1.02 |
| Epimedin B | 1.96 | 17.27 | 2.73 |
| Epimedin C | 0.96 | 0.54 | 2.86 |
| Total contents | 60.79 | 46.13 | 10.24 |
| % of anhydroicaritin | 8.57 | 41.77 | 81.45 |

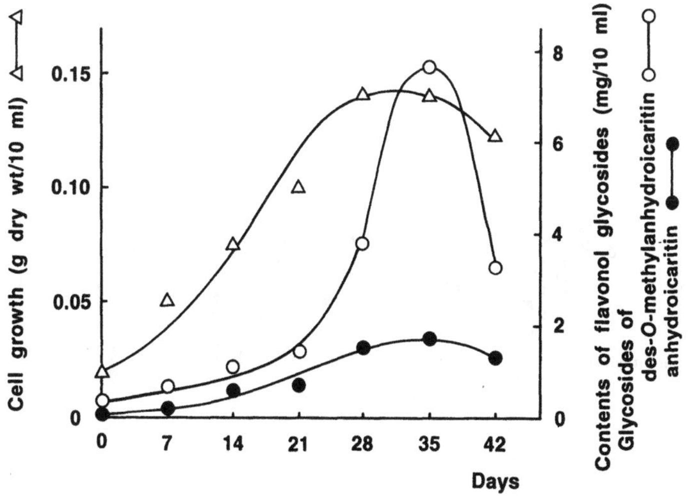


**Fig. 3.** Time course of cell growth and isoprenylflavonol glycoside production in callus cultures of *Epimedium diphyllum*; 0.65 g of cells were inoculated on 10 ml of LS gellan gum medium and cultured for 6 weeks in the dark. (Yamamoto et al. 1992)

the anhydroicaritin glycoside content in the original plant was higher than that in the callus tissues.

The time-course of cell growth and isoprenylflavonol glycoside production in static cultures is shown in Fig. 3. Glycoside production started at the early straight growth stage and reached its maximum at the early stationary growth stage, then decreased immediately. During the culture period, the ratio of des-*O*-methylanhydroicaritin glycosides and anhydroicaritin glycosides did not change much.

## 2.3 Effects of Phytohormones on Isoprenylflavonol Glycoside Production

In general, 2,4-D inhibits the production of many kinds of secondary metabolites (Sakuta and Komamine 1987). We investigated the effects of phytohormones in different concentrations on cell growth and isoprenylflavonol glycoside production. Effects of various combinations of 2,4-D and kinetin are shown in Fig. 4. Effects of combinations of 2,4-D and 6-benzylaminopurine gave results similar to those of 2,4-D and kinetin (data not shown). As shown in Fig. 4, kinetin was not needed for isoprenylflavonol glycoside production, and at higher concentration, the production was inhibited, whereas 2,4-D was needed for optimum isoprenylflavonol glycoside production, the best concentration being around 10 μM. Other auxins, such as 1-naphthaleneacetic acid and indole-3-acetic acid, were unsuitable for both cell growth and isoprenylflavonol glycoside production (data not shown).

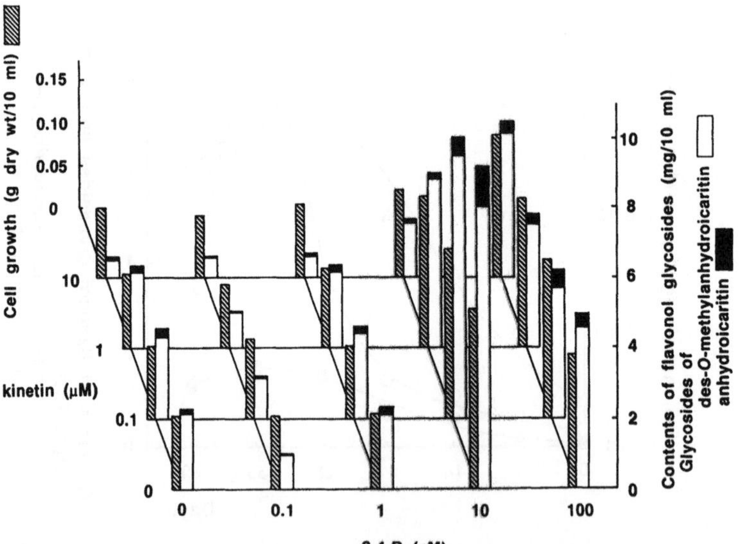


**Fig. 4.** Effects of 2,4-D and kinetin on cell growth and isoprenylflavonol glycoside production in callus cultures of *Epimedium diphyllum*. (Yamamoto et al. 1992)

## 2.4 Effects of Sugars on Isoprenylflavonol Glycoside Production

To investigate the effect of carbon sources on isoprenylflavonol glycoside production, 3% of five monosaccharides and three disaccharides were added to LS medium and cultured for 1 month (Fig. 5). Sucrose was the best for both cell growth and isoprenylflavonol glycoside production, followed by glucose. Galactose, rhamnose, and xylose, which were utilized as sugar moieties of isoprenylflavonol glycosides of *Epimedium* species, have inhibitory effects on both cell growth and isoprenylflavonol glycoside production, which indicated that *E. diphyllum* cells could not use these sugars directly. The effect of sucrose concentration was also examined. (Fig. 6). Optimum cell growth and highest isoprenylflavonol glycoside production were observed with 3% sucrose. Many workers have reported that the increase of sucrose concentration enhances both cell growth and the production of secondary metabolites (Sakuta and Komamine 1987), but in *E. diphyllum* callus, higher sucrose concentration inhibited both cell growth and isoprenylflavonol glycoside production. This result is probably caused by osmotic stress.

## 2.5 Effects of Nutrients on Isoprenylflavonol Glycoside Production

As shown in Table 2, LS and NN medium (Nitsch and Nitsch 1969) were the best for cell growth, whereas the highest isoprenylflavonol glycoside production was observed with NN medium. White's medium (White 1963), which had been

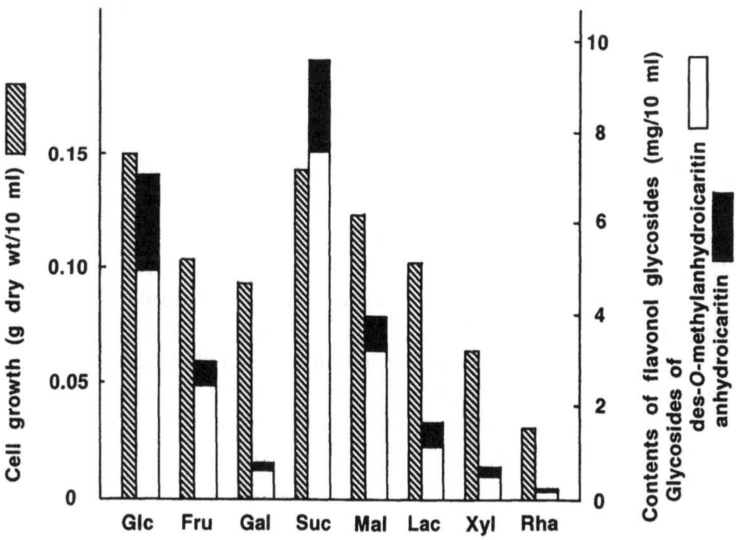


**Fig. 5.** Effects of various sugars on cell growth and isoprenylflavonol glycoside production in callus cultures of *Epimedium diphyllum*; 3% of each sugar was added to 10 ml of LS medium solidified with 0.3% gellan gum and cultured for 1 month in the dark. (Yamamoto et al. 1992). Glc Glucose; Fru fructose; Gal galactose; Suc sucrose; Mal maltose; Lac lactose; Xyl xylose; Rha L-rhamnose

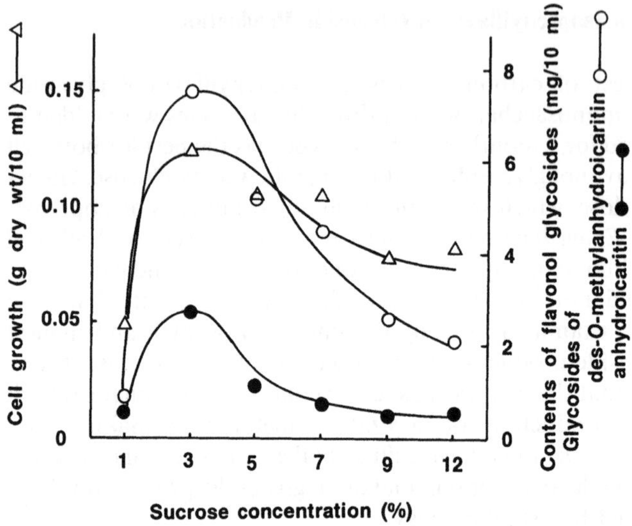


**Fig. 6.** Effects of sucrose concentrations on cell growth and isoprenylflavonol glycoside production in callus cultures of *Epimedium diphyllum.* (Yamamoto et al. 1992)

**Table 2.** Effects of media on cell growth and prenylflavonol glycoside production in callus cultures of *Epimedium diphyllum*

| Medium | Cell growth (g dry wt./10 ml) | Anhydroicaritin glycosides (mg/10 ml) | Des-*O*-methyl-anhydroicaritin glycosides (mg/10 ml) |
|---|---|---|---|
| LS | 0.145 | 0.23 | 8.52 |
| NN | 0.145 | 0.33 | 11.05 |
| White's | 0.099 | 0.13 | 6.44 |

established for root cultures and increased lavandulyl flavanone production in *Sophora flavescens* callus cultures (Yamamota et al. 1991), was not suitable for either cell growth or isoprenylflavonol glycoside production. NN medium contains almost half the amount of macroelements (nitrogen, phosphate, calcium, and magnesium) compared with LS medium. Therefore, some of these nutrients play an important role in isoprenylflavonol glycoside production.

To clarify the kind of macroelements which affect cell growth and isoprenylflavonol glycoside production, *E. diphyllum* cell suspension cultures were used for the experiment. Fresh cells (1 g) were inoculated in different media (20 ml) and cultured for 3 weeks in the dark.

### 2.5.1 *Effect of Phosphate*

Half the amount of phosphate (0.625 mM) compared with LS medium was suitable for cell growth and the production of isoprenylflavonol glycoside. Lower or higher phosphate levels inhibited both (Fig. 7).

## 2.5.2 *Effect of Nitrogen*

LS medium contains nitrogen as a mixture of $NO_3^-$ (39.4 mM) and $NH_4^+$ (20.6 mM). The effect of the total amount of nitrogen (the ratio of $NO_3^-/NH_4^+$ was fixed at 39.4/20.6) was examined. As shown in Fig. 8, both cell growth and isoprenylflavonol glycoside production were optimized at concentrations between 7.5 to 15 mM (1/8–1/4 compared with LS medium).

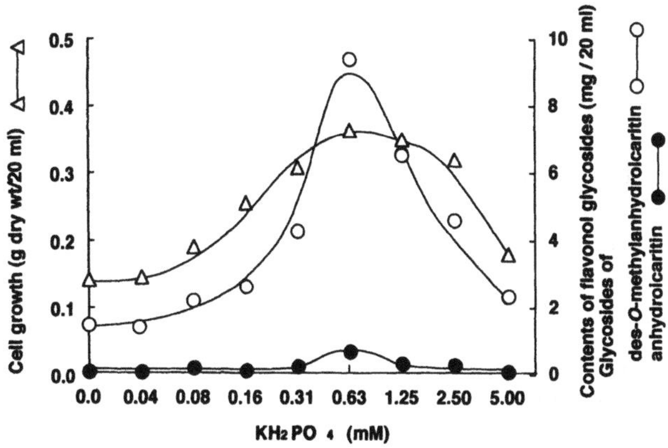


**Fig. 7.** Effects of phosphate concentrations on cell growth and isoprenylflavonol glycoside production in cell suspension cultures of *Epimedium diphyllum*. Concentrations of other nutrients were the same as in LS medium; 1 g cells were inoculated in 20 ml of each medium (three replicates), and cultured for 3 weeks in the dark at a speed of 100 rpm (Yamamoto et al. unpubl.)

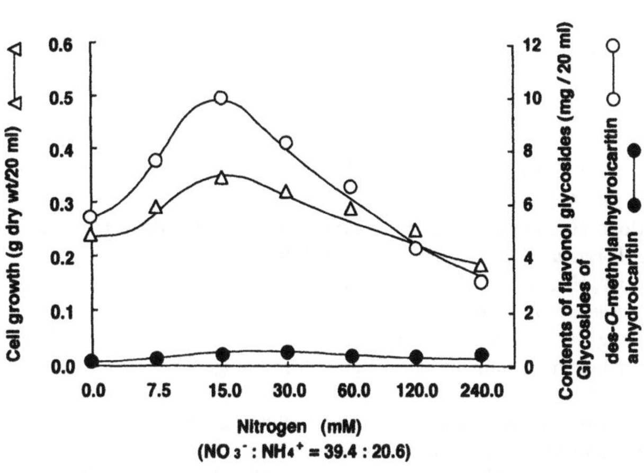


**Fig. 8.** Effects of total nitrogen concentrations on cell growth and isoprenylflavonol glycoside production in cell suspension cultures of *Epimedium diphyllum*. The ratio of $KNO_3^-/NH_4^+$ was fixed at 39.4/20.6. Concentrations of other nutrients were the same as in LS medium. (Yamamoto et al. unpubl.)

## 3 Conclusion

*Epimedium diphyllum* callus produced isoprenylflavonol glycosides in larger amounts than the intact plant. The cell cultures required 2,4-D for both cell growth and isoprenylflavonol glycoside production, and the results of nutrient effects indicated that cell growth and the production of isoprenylflavonol glycosides are closely correlated. It is noteworthy that in the original plant, anhydroicaritin glycosides are produced in a considerable amount, whereas des-*O*-methylanhydroicaritin glycosides were the main product in cell cultures, and anhydroicaritin glycosides were accumulated in a small amount. The ratio of des-*O*-methylanhydroicaritin/anhydroicaritin in cell cultures was not changed in the present study. This indicates that the methylation of 4′-OH is suppressed in cell cultures and might be controlled in a manner different from the isoprenylflavonol skeleton production. This may partly explain the difference of des-*O*-methylanhydroicaritin/anhydroicaritin patterns between the aerial and underground parts of the original plant. This *Epimedium diphyllum* cell culture system would be suitable for the study of the mechanism of the prenylation of flavonoids.

## References

Ayabe S, Kobayashi M, Hikichi M, Matsumoto K, Furuya T (1980) Flavonoids from cultured cells of *Glycyrrhiza echinata*. Phytochemistry 19:2179–2183

Fukai T, Nomura T (1988) Seven prenylated flavonol glycosides from two *Epimedium* species. Phytochemistry 27:259–266

Harborne JB (1988) The flavonoids, advances in research since 1980. Chapman and Hall, London

Hargreaves JA, Selby C (1978) Phytoalexin formation in cell suspensions of *Phaseolus vulgaris* in response to an extract of bean hypocotyls. Phytochemistry 17:1099–1102

Hinderer W, Seitz HU (1988) Flavonoids. In: Constabel F, Vasil IK (eds) Cell culture and somatic cell genetics of plants, vol V. Academic Press, San Diego, pp 23–48

Iinuma M, Tanaka T, Sakakibara N, Mizuno M, Matsuda H, Shiomoto H, Kubo M (1990) Phagocytic activity of leaves of *Epimedium* species on mouse reticuloendotherial system. Yakugaku Zasshi 110:179–185

Ikuta J, Fukai T, Nomura T, Ueda S (1986) Constituents of *Morus alba* L. cell cultures (1). Structures of four new natural Diels-Alder type adducts, kuwanons J, Q, R, and V. Chem Pharm Bull 34:2471–2478

Ito Y, Hirayama F, Suto K, Sagara K, Yoshida T (1988a) Three flavonol glycosides from *Epimedium koreanum*. Phytochemistry 27:911–913

Ito Y, Hirayama F, Suto K, Oshima T, Sagara K, Yoshida T (1988b) Three flavonol glycosides from *Epimedium koreanum*. J Chromatogr 456:392–397

Jiang Su Sin Medical College (ed) (1977a) Yinyanghuo. In: Encyclopedia of traditional Chinese medicine. Shanghai Science and Technology, Shanghai, pp 2250–2253

Jiang Su Sin Medical College (ed) (1977b) Yinyanghuogen. In: Encyclopedia of traditional Chinese medicine. Shanghai Science and Technology, Shanghai, pp 2253

Kobayashi M, Noguchi H, Sankawa U (1985) Formation of chalcones and isoflavones by callus cultures of *Glycyrrhiza uralensis* with different production patterns. Chem Pharm Bull 33:3811–3816

Linsmaier EM, Skoog F (1965) Organic growth factor requirements of tobacco tissue cultures. Physiol Plant 18:100–127

Liu CM, Yu QH, Zhang LM (1982) Effect of icariin on heart. Chin Trad Herb Drugs 13:414–416

Miyase T, Ueno A, Takizawa N, Kobayashi H, Karasawa H (1987a) Studies on the glycosides of *Epimedium grandiflorum* Morr. var. *thunbergianum* (Miq.) Nakai. I. Chem Pharm Bull 35:1109–1117
Miyase T, Ueno A, Takizawa N, Kobayashi H, Oguchi H (1987b) Studies on the glycosides of *Epimedium gandiflorum* Morr. var. *thunbergianum* (Miq.) Nakai. II. Chem Pharm Bull 35:3713–3719
Miyase T, Ueno A, Takizawa N, Kobayashi H, Oguchi H (1988) Studies on the glycosides of *Epimedium grandiflorum* Morr. var. *thunbergianum* (Miq.) Nakai. III. Chem Pharm Bull 36:2475–2484
Mizuno M, Hanioka S, Suzuki N, Iinuma M, Tanaka T, Liu XS, Min ZD (1987) Flavonol glycosides from *Epimedium sagittatum*. Phytochemistry 26:861–863
Mizuno M, Iinuma M, Tanaka T, Sakakibara N, Hanioka S, Liu XS (1988a) Flavonol glycosides in *Epimedium* species. Chem Pharm Bull 36:3487–3490
Mizuno M, Iinuma M, Tanaka T, Sakakibara N, Fujikawa T, Hanioka S, Ishida Y, Liu XS, Murata H (1988b) Flavonol glycosides in the roots of *Epimedium diphyllum*. Phytochemistry 27:3645–3647
Mizuno M, Sakakibara N, Hanioka S, Iinuma M, Tanaka T, Liu XS, Shi DW (1988c) Flavonol glycosides from *Epimedium sagittatum*. Phytochemistry 27:3641–3643
Mizuno M, Iinuma M, Tanaka T, Sakakibara N, Nishi M, Inada A, Nakanishi T (1989a) A flavonol glycoside from *Epimedium diphyllum*. Phytochemistry 28:2527–2529
Mizuno M, Iinuma M, Tanaka T, Sakakibara N, Nakanishi T, Inada A, Nishi M (1989b) A novel flavonol glycoside in the leaves of *Epimedium sempervirens*. Chem Pharm Bull 37:2241–2242
Mizuno M, Iinuma M, Tanaka T, Sakakibara N (1990) New flavonol glycoside from the leaves of *Epimedium sempervirens*. J Nat Prod 53:744–746
Namba T (1980) Epimedii herba, Epimedii folium. In: The crude drugs in Japan, China and neighbouring countries. Hoikusha, Osaka, pp 48–50
Nitsch JP, Nitsch C (1969) Haploid plants from pollen grains. Science 163:85–87
Oshima Y, Okamoto M, Hikino H (1987) Epimedins A, B and C, flavonoid glycosides of *Epimedium koreanum* herbs. Heterocycles 26:935–938
Sakuta M, Komamine A (1987) Cell growth and accumulation of secondary metabolites. In: Constabel F, Vasil IK (eds) Cell culture and somatic cell genetics of plants, vol IV. Academic Press, San Diego, pp 97–114
Stafford HA (1990) Flavonoid metabolism. CRC Press, Boca Raton
Stearn WT (1938) *Epimedium* and *Vancouveria*. J Linn Soc Bot 41:409–534
Takemoto T, Daigo K, Tokuoka Y (1975) Studies on the constituents of *Epimedium*. I. Flavonoids of *Epimedium grandiflorum* Morr. (1). Yakugaku Zasshi 95:312–320
Tang W, Eisenbrand G (1992) Chinese drugs of plant origin. Springer, Berlin Heidelberg New York, pp 491–498
Tokuoka Y, Daigo K, Takemoto T (1975a) Studies on the constituents of *Epimeidum*. II. Flavonoids of *Epimedium grandiflorum* Morr. (2). Yakugaku Zasshi 95:321–325
Tokuoka Y, Daigo K, Takemoto T (1975b) Studies on the constituents of *Epimedium*. III. Lignoids of *Epimedium grandiflorum* Morr. Yakugaku Zasshi 95:557–563
Tokuoka Y, Daigo K, Takemoto T (1975c) Studies on the constituents of *Epimedium*. IV. Flavonoids of *Epimedium grandiflorum* Morr. (3). Yakugaku Zasshi 95:698–705
Tokuoka Y, Daigo K, Takemoto T (1975d) Studies on the constituents of *Epimedium*. V. Flavonoids of *Epimedium grandiflorum* Morr. (4). Yakugaku Zasshi 95:825–829
Tomita M, Ishii H (1957a) Studies on the alkaloids of berberidaceous plants, XII. Alkaloids of *Epimedium rugosum*. Yakugaku Zasshi 77:114–116
Tomita M, Ishii H (1957b) Studies on the alkaloids of berberidaceous plants, XIII. Alkaloids of *Epimedium grandiflorum* Morren var. thunbergianum Nakai. Yakugaku Zasshi 77:212–213
White PR (1963) The cultivation of animal and plant cells, 2nd edn. Ronald Press, New York
Yamamoto H, Kawai S, Mayumi J, Tanaka T, Iinuma M, Mizuno M (1991) Prenylated flavonone production in callus cultures of *Sophora flavescens* var. *angustifolia*. Z Naturforsch 46c:172–176
Yamamoto H, Ieda K, Tsuchiya S, Kuang Y, Tanaka T, Iinuma M, Mizuno M (1992) Flavonol glycosides production in callus cultures of *Epimedium diphyllum*. Phytochemistry 31:837–840
Zähringer U, Schaller E, Grisebach H (1981) Induction of phytoalexin synthesis in soybean. Structrue and reactions of naturally occurring and enzymatically prepared prenylated pterocarpans from elicitor-treated cotyledons and cell cultures of soybean. Z Naturforsch 36c:234–241

# XV *Euphorbia characias* L.: In Vitro Culture and the Production of Sterols, Triterpenols, and Hydrocarbon-Like Compounds

M. FERNANDES-FERREIRA[1], J. M. NOVAIS[2], and M. S. S. PAIS[3]

## 1 General Account

Since antiquity, *Euphorbia* species have been used for multiple purposes. The leaves and branchlets of *Euphorbia lancifolia* Schlecht were used by Mayam Indians to produce a tea named *Ixbut* which is reported to act as a galactogogue, increasing the flow or volume of milk in postpartem women (Rosengarten 1978; Schultes 1987). Some species have been used for treatment of cancer, tumors, and warts for more than 2000 years (Kupchan et al. 1976). This is the case for *E. fischeriana* Steud., that was used in traditional Chinese medicine as an antitumor drug (Schroeder et al. 1980). Medicinal uses of *Euphorbia* species include treatment of skin diseases, warts, intestinal parasites, and gonorrhea. Table 1 summarizes the uses in folk medicine.

The latex of some plants of *Euphorbia* is toxic, causing poisoning in human beings and livestock, skin dermatitis, and inflammations of mucous membranes, conjunctivitis, tumor promotion, and cancer (Kingsbury 1964; Füstenberger and Hecker 1977, 1985; Seip and Hecker 1983; Adolf et al. 1984; Evans and Edwards 1987; Speroni et al. 1991).

### 1.1 Phytoconstituents Isolated from *Euphorbia* Species

Diterpenoid esters with diterpene skeletons of the tigliane (e.g., phorbol esters), daphnane (e.g., resiniferonol esters), and ingenane (e.g., ingenol esters) (Fig. 1) are responsible for most of the biological effects of the latex, including tumor promotion and cell proliferation (Seip and Hecker 1983; Aitken 1987; Evans and Edwards 1987; Hecker 1987). In contrast to the diterpenoid esters of the tigliane, ingenane, and daphnane type, the polyfunctional macrocyclic diterpenoids with skeletons of the lathyrane (e.g., lathyrol, ingol, and jolkinol) and jatrophane (e.g., jatrophone, kansuinines A and B, euphornin, euphoscopins A to D, characiol, and derivatives thereof) (Fig. 1) are inactive as irritants or as tumor promotors, as tested so far (Seip and Hecker 1983, 1984). Some derivatives of these

[1] Dep. de Biologia, Universidade do Minho, 4719 Braga Codex, Portugal
[2] Lab. de Eng. Bioquímica, Instituto Superior Técnico, Av. Rovisco Pais, 1000 Lisboa, Portugal
[3] Dep. de Biologia Vegetal, Fac. de Ciências de Lisboa, 1294 Lisboa Codex, Portugal

Biotechnology in Agriculture and Forestry, Vol. 28
Medicinal and Aromatic Plants VII (ed. by Y.P.S. Bajaj)

**Table 1.** Some species of *Euphorbia* used in folk medicine

| Species | Used as treatment of | Reference |
|---|---|---|
| *E. antiquorum* L. | Dyspepsia | Ayensu (1978) |
| *E. caudicifolia* Haines | Purgative, expectorant | Afza et al. (1988) |
| *E. fischeriana* Steud. | Antitumor | Schroeder et al. (1980) |
| *E. genistoides* Berg. | Diaphoretic | Watt and Breyer-Brandwijk (1962) |
| *E. helioscopia* L. | Bronchitis | Chen et al. (1979) |
| *E. hirta* L. | Antihistaminic | Duke and Ayensu (1985) |
| *E. hirta* L. | Heart diseases | Ayensu (1978) |
| *E. humifusa* Willd. | Jaundice | Ayensu (1978) |
| *E. lancifolia* Schelect | Galactogogue | Rosengarten (1978) |
| *E. maddeni* | Hypotensive | Sahai et al. (1981) |
| *E. millii* Desmoulins | Hepatitis | Duke and Ayensu (1985) |
| *E. nerifolia* L. | Antipyretic | Watt and Breyer-Brandwijk (1962) |
| *E. pilulifera* | Antispasmodic | Hellett and Parks (1953) |
| *E. ruspolii* Chiov. | Disinfectant collyrium | Passannanti et al. (1985) |
| *E. splendens* Bojer | Anticancer (antileukemic) | Lee et al. (1982) |
| *E. supina* Rafin. | Diarrhea, purulent swellings | Nanba and Mikage (1983) Okuda (1986) |
| *E. thymifolia* L. | Galactogogue | Ayensu (1978) |
| *E. tirucalli* L. | Neuralgia | Baslas (1982) |

macrocyclic diterpenes (e.g., jatrophone) exhibit antileukemic activity instead of acting as irritants (Kupchan et al. 1976).

According to Hecker (1987), the concentrations of diterpenoids in the latex are very small. Carbohydrates, lipids, amino acids, alkaloids, enzymes, and haemagglutinins have been isolated from the latex of *Euphorbia* species (Lynn and Clevette-Radford 1985, 1986a, b, 1987a, b). Uemura and Hirata (1971) isolated two new alkaloids, milliamines A and B from *E. millii.* Nonutilizable starch (Biesboer and Mahlberg 1978) and conjugated fatty acids (Warnaar 1981) were also isolated from laticifers of some *Euphorbia* species. According to Calvin (1984), about one third of the latex compounds from species of this genus are apolar, type hydrocarbons. The hydrocarbon fraction of the latex of some *Euphorbia* species includes polyisoprenes (Watt and Breyer-Brandwijk 1962). However, the major compounds of the hydrocarbon fraction of the latex of almost all the *Euphorbia* species studied up to now are triterpenoids (Fig. 2). Triterpenoids are also the major constituents of the hydrocarbon fraction obtained from the biomass of *Euphorbia* plants after extraction by hexane (Nemethy et al. 1981; Fernandes-Ferreira et al. 1991). Along with the latex triterpenols, this hydrocarbon fraction includes the constituents of epicuticular waxes (phytosterols, wax triterpenols, wax esters, alkanes, alkanols, fatty acids) (Hemmers and Gülz 1986; Gülz et al. 1987, 1988; Hemmers et al. 1988, 1989; Koops et al. 1991). The phytoconstituents isolated from *Euphorbia* species were reviewed by Singla and Pathak (1990).

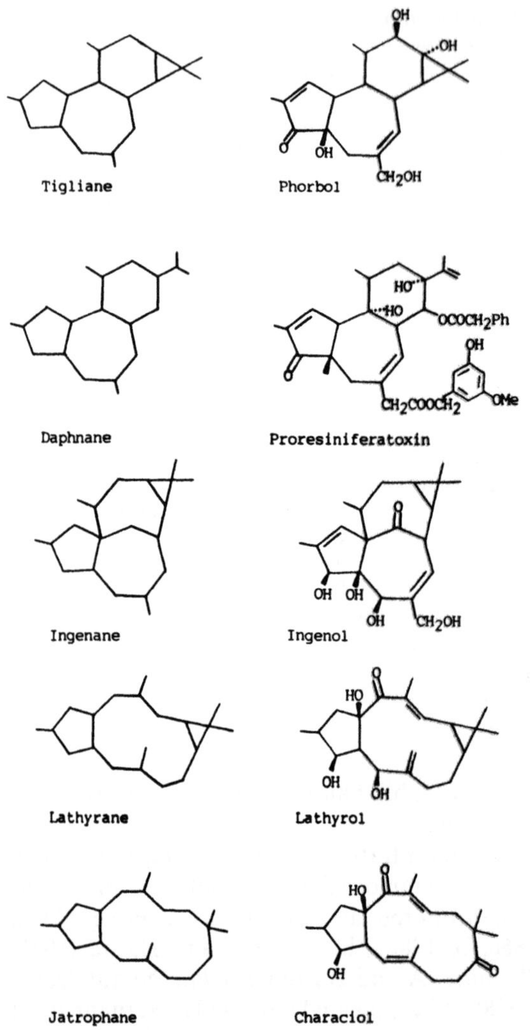


**Fig. 1.** Diterpene skeletons and chemical structures of five diterpenoids isolated from the latex of some *Euphorbia* species

## 1.2 Potential of *Euphorbia* species as Energy and Raw Chemicals Resources

After the oil embargo of the early 1970s, hundreds of species of *Euphorbia* and other genera were screened for their content in biocrude, with the assumption that one of the most promising options for producing liquid fuels from biomass is the direct extraction of the low molecular weight nonpolar constituents of plants (Buchanan et al. 1978a, b; Adams and McChesney 1983; Bhatia et al. 1984; Ng'eny-Mengech and Kiumba 1985; Adams et al. 1986; Sharma and Prasad 1986; Gnecco et al. 1988). The criteria for the selection of the best candidates for energy cultures were: high concentrations of hydrocarbon-like compounds, and high production of biomass in marginal lands (McLaughlin et al. 1983). Following these criteria, *Euphorbia lathyris* was selected for establishment and maintenance of experimental cultures in California (Calvin 1987) and

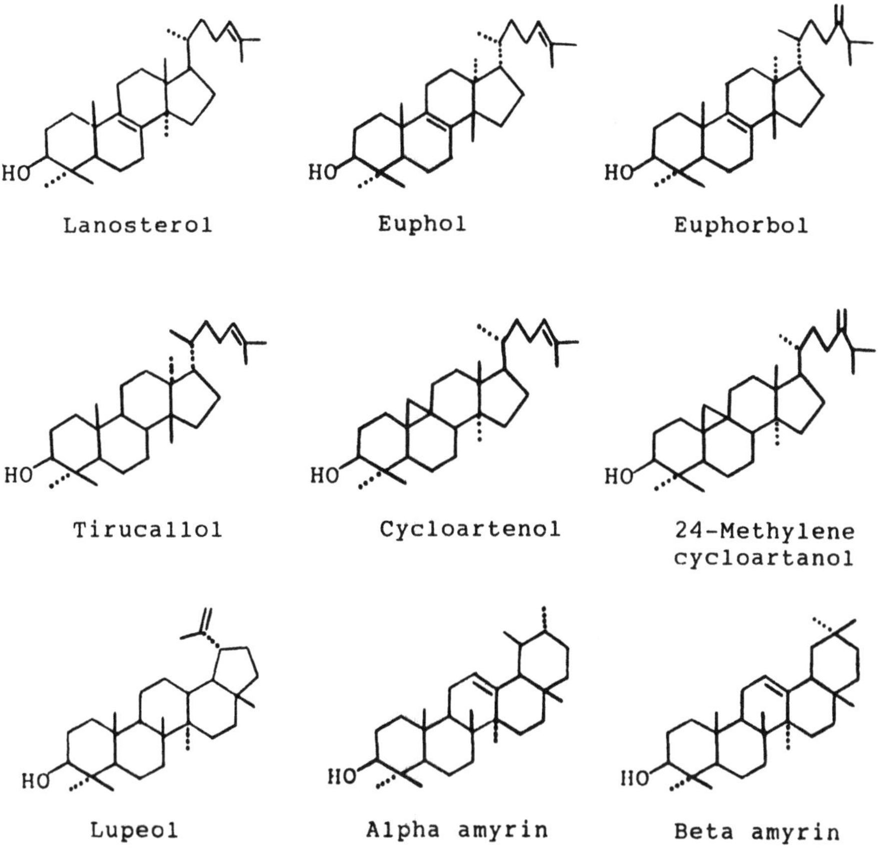


**Fig. 2.** Chemical structures of some of the most common triterpenoids isolated from the latex of *Euphorbia* species

Spain (Ayerbe et al. 1984a, b). *E. tirucalli* has been studied in experimental plantations, namely in Kenya (100 ha) (Declerck et al. 1985). These field cultures, established by traditional methods of sowing and/or transplantation, have allowed the study of the ability of these crops to produce fuel and chemicals under different conditions of irrigation and plant density (Ayerbe et al. 1984b; Ventas et al. 1985).

## 1.3 *Euphorbia characias* Latex in Mediterranean Rural Middle Ages Communities

*Euphorbia characias* is a native species of mediterranean regions, growing wild in Portugal in marginal lands, namely at the south of the Mondego River. During the Middle Ages the seeds and latex of wild plants of this species were used as powerful purgatives (Quer 1981). The latex was taken as pills or as lenticular tablets after mixing with the flour of some oleaginous seeds. The latex was also used to destroy warts and, mixed with olive-oil, as a hair-remover (Quer 1981).

The crushed and ground plants of *E. characias* were also used in lakes and rivers by fishermen, enabling the capture of the fish immobilized by the rapidly dissolved poison. As with other species of *Euphorbia, E. characias* is highly toxic and extremely harmful to the skin, mucous membranes, heart, liver, and stomach (Quer 1981).

Diterpenoid esters with diterpene skeletons of tigliane, ingenane, and daphnane, responsible for the irritant and tumor-promoting activities, have been isolated from the latex of this species (Seip and Hecker 1983). Nonirritant macrocyclic diterpenoids belonging to the lathyrane (Seip and Hecker 1983) and jatrophane (Seip and Hecker 1984) types were also identified. Lectins (Lynn and Clevette-Radford, 1986a), proteins, carbohydrates, and enzymes such as esterases (Lynn and Clevette-Radford 1987a), diamine oxidase (Floris et al. 1983), and peroxidase (Floris et al. 1984) were also isolated from polar fractions of the latex of *E. characias.*

### 1.4 *Euphorbia characias* as Source of Lipid Compounds and Biomass

Although in the Middle Ages *E. characias* was an important species used in folk medicine by rural communities of the mediterranean regions, the local populations do not use it nowadays as a natural resource. However, taking into account the growing importance of the *Euphorbia* species as alternative sources of raw chemicals and biofuel, the production of lipid compounds by wild and micropropagated plants, calli, and suspended cells of this species was studied.

The production of plant biomass can be obtained by in vivo or in vitro techniques. In vivo, by sowing or by plant propagation following the traditional methods of cuttings and transplantation, in vitro, by micropropagation and transfer of the micropropagated plantlets to the field or by establishment and permanent maintenance of cultures of calli or suspended cells.

The study of biomass growth during the developmental cycle of *E. characias* plants propagated by traditional methods was performed in large plots by Coppola and Brunori (1984). In our studies, wild field-growing plants were used either for determination of crude oil content and composition or as source of primary explants for initiation of cultures of micropropagated plants, calli, and suspended cells.

## 2 In Vitro Culture and Micropropagation of *Euphorbia* Species

In vitro culture and micropropagation studies on various species of *Euphorbia* have recently been reviewed (see Preece and Ripley 1992). *Euphorbia lathyris* has been vegetatively propagated by stem and leaf bud cuttings (Preece and Wollbrink 1983), and also micropropagated by somatic embryogenesis (Kumar and Joshi 1983) and by shoot tips (Ripley and Preece 1984). The micropropagation of other *Euphorbia* species, namely, *E. pulcherrima* (Langhe et al. 1974),

*E. hirta* (Baburaj et al. 1987), and *E. fulgens* (Zhang et al. 1987) has also been reported. However, as far as we know, no study on the production of hydrocarbon-like compounds by these micropropagated plants has been reported. On the contrary, studies on the production of this type of compounds by calli and/or suspended cells have been reported for *E. esula* and *E. cyparissias* (Lee and Starratt 1972); *E. pulcherrima* (Biesboer et al. 1982); *E. lathyris* (Adolf et al. 1984); *E. tirucalli* (Biesboer and Mahlberg 1979; Ohyama et al. 1984; Yamamoto et al. 1981); *E. millii* (Yamamoto et al. 1981). Anthocyanin production by suspended cells of *E. millii* was also reported (Yamamoto et al. 1981, 1989).

## 2.1 Micropropagation of *Euphorbia characias*

Axillary buds of wild-growing plants of *E. characias* were cultured on solidified MS medium (Murashige and Skoog 1962) supplemented with 0.5 mg/l gibberellic acid. Rooted plantlets obtained by immersion of shoots in a solution of 1 g/l indole-3-butyric acid were transferred to a mixture of two-thirds soil and one-third peat. After 45 to 60 days in the greenhouse, the plants were transferred to the field at intervals of 0.5 m corresponding to a density of 40 000 plants/ha. Ninety days after transfer to the field, ten of those plants were harvested, freeze dried, and the dry weight determined. One year later, another sample of ten plants was harvested, freeze dried, and weighed. Figure 3 shows the micropropagated plants 15 months after transfer to the field. Table 2 shows the mean biomass produced per micropropagated field-growing plant from July, 1989 to July, 1990.

The mean growth recorded for micropropagated plants was lower than that reported by Coppola and Brunori (1984) for plants of the same species maintained in large plots. However, these results show that the dried biomass produced by micropropagated plants of this species would be 10.3 tons/ha/year. This production is similar to those obtained for traditionally propagated cultures of *E. lathyris* (Sacks et al. 1981; Ayerbe et al. 1984a) and *E. tirucalli* (Declerck et al. 1985).

**Table 2.** Biomass produced per micropropagated plant of *E. characias*. The values are the average of the dry weight of ten plants ( ± SD,)

| | Root | Stem (g) | Leaves | Total |
|---|---|---|---|---|
| 3 months in the field (July 1989) | 0.9 ± 0.2 | 2.3 ± 1 | 3.5 ± 1 | 6.7 ± 2 |
| 15 months in the field (July 1990) | 43.2 ± 16 | 117.4 ± 41 | 104.2 ± 5 | 264.8 ± 57 |
| Net production (July/1989 to July/1990) | 42.3 ± 16 | 115.1 ± 41 | 100.7 ± 6 | 258.0 ± 57 |

**Fig. 3.** Micropropagated plants of *Euphorbia characias* 15 months after transfer from the greenhouse to the field

## 2.2 Callus Cultures and Cell Suspension

Young leaves of wild *E. characias* collected in January before flowering were used as primary explants for calli induction on MS medium solidified with 8 g/l agar and supplemented with 30 g/l sucrose, 1 g/l edamine, 5 mg/l ascorbic acid, 1 mg/l 2,4-dichlorophenoxyacetic acid (2,4-D) and 0.75 mg/l benzyl adenine (BA). The MS medium with this supplementation was also used for the maintenance of calli cultures. The same liquid medium was used for initiation and maintenance of cell suspension cultures. Calli were subcultured at intervals of 4 weeks. Subculture of suspended cells was carried out periodically every 10–12 days (Fernandes-Ferreira et al. 1989). The effects of the type and concentrations of phytohormones, as well as the effects of sucrose and edamine on the growth and production of hydrocarbon-like compounds by *E. characias* calli were studied (Fernandes–Ferreira et al. 1989, 1990b, 1992a, b). The effects of different basal media on biomass growth and production of hydrocarbon-like compounds by calli of this species have also been studied. Starting the subcultures with calli fragments of 35–38 mg dry wt., the specific growth rate of calli maintained on MS medium supplemented with 30 g/l sucrose, 1 g/l edamine, 5 mg/l ascorbic acid, 0.5 mg/l BA, and 0.75 mg/l 2,4-D was $\mu = 0.066\ day^{-1}$. Under these conditions the doubling time was td = 10.5 days, and the average of the dry weight reached 4 weeks after subculturing was 110 mg dry wt./callus. Reduction of 2,4-D concentration to 0.25 mg/l increased the specific growth rate to 0.085 $day^{-1}$ (td = 8.2 days). The dry weight reached at the end of 4 weeks of subculture under these conditions was 132 mg dry wt./callus. Figure 4 shows the growth profiles of calli maintained under these two conditions.

Calli are able to grow in a wide range of different basal media. Half-strength Murashige and Skoog (1962) medium (1/2 MS) was the most suitable for growth, followed by Murashige and Skoog (1962) and Gamborg et al. (1968) media. The less efficient for growth of calli of this species were the media of Campbell and Durzan (1975); Dutcher and Powell (1972); Eeuwens (1978) and Miller and Skoog (1953) (Table 3).

Starting with initial cell densities of about $8.5 \times 10^4$ cells/ml, the specific growth rate of suspended cells maintained in liquid MS medium supplemented with 30 g/l sucrose, 5 mg/l ascorbic acid, 0.75 mg/l BA and 1 mg/l 2,4-D was $\mu = 0.216\ day^{-1}$ (td = 3.2 days). The addition of 1 g/l edamine to the same medium increased the specific growth rate of suspended cells to $\mu = 0.256\ day^{-1}$ corresponding to a doubling time td = 2.7 days (Fernandes-Ferreira et al. 1989, 1992a).

Our studies on calli and suspended cells were carried out with heterotrophic cultures. However, even with the initial concentration of 30 g/l sucrose and in the absence of the exogenous supply of $CO_2$, the calli grown under a photoperiod of 16 hlight/8 hdark (2000 lx from daylight fluorescent tubes) synthesized

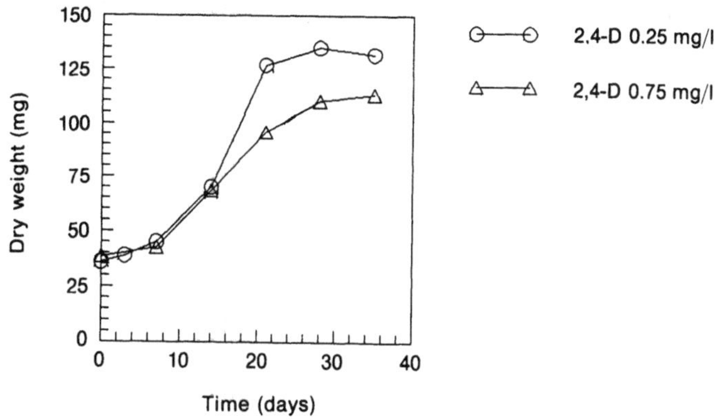


**Fig. 4.** Growth profiles of *Euphorbia characias* calli maintained on MS medium supplemented with 30 g/l sucrose, 1 g/l edamine, 5 mg/l ascorbic acid, 0.75 mg/l benzyl adenine, and 2,4-dichlorophenoxyacetic acid (2,4-D) at concentrations of 0.25 and 0.75 g/l

**Table 3.** Biomass produced by calli of *E. characias* maintained on different basal media. All media were supplemented with 30 g/l sucrose, 1 g/l edamine, 5 mg/l ascorbic acid, 0.5 mg/l BA, and 0.25 mg/l 2,4-D. Subcultures were initiated with calli fragments of 35 mg d.w. The values of biomass yield are the average of the dry weight of 14 calli cultured during 30 days on the different basal media

| Basal medium | Biomass yield (mg/callus) | Basal medium | Biomass yield (mg/callus) |
|---|---|---|---|
| Murashige and Skoog (1962) | 129 ± 4 (C) | Schenk and Hildebrandt (1972) | 100 ± 7 |
| 1/2 Murashige and Skoog[a] | 141 ± 2 | Eeuwens (1978) | 97 ± 6 |
| Heller (1953)[b] | 112 ± 2 | Campbell and Durzan (1975) | 88 ± 2 |
| Nitsch and Nitsch (1956) | 121 ± 4* | Linsmaier and Skoog (1965) | 102 ± 2 |
| White (1963) | 106 ± 1 | Dutcher and Powell (1972) | 97 ± 2 |
| Hildebrandt et al. (1946) | 109 ± 2 | Murashige and Tucker (1969) | 107 ± 3 |
| Gautheret (1942)[c] | 120 ± 5* | Miller and Skoog (1953)[g] | 95 ± 3 |
| White (1943)[d] | 103 ± 4 | Nitsch (1951) | 101 ± 2 |
| Wood and Braun (1961)[e] | 115 ± 4 | Nitsch (1972) | 103 ± 2 |
| Eriksson (1965)[f] | 116 ± 4 | Nitsch et al. (1968) | 103 ± 7 |
| Gamborg et al. (1968) | 128 ± 4[=] | | |

All data are significantly different from control (C) at $p < 0.001$ except those with superscription (*significantly different at level $p < 0.05$;[=] statistically not different than control at level $p < 0.1$).
[a] All nutrients of MS medium were reduced to half with the exception of sucrose, edamine, ascorbic acid, and phytohormones.
[b] KCl and $NaNO_3$ were substituted by NaCl and $KNO_3$.
[c] $Ti(SO_4)_3$, $BeSO_4$, and $H_2SO_4$ were removed.
[d] $Na_2MoO_4 \cdot 2H_2O$ at 0.01 mg/l was used instead of 0.0017 mg/l $H_2MoO_4$.
[e] 910 mg/l KCl and 1800 mg/l $NaNO_3$ were substituted by 910 mg/l NaCl and 1720 mg/l $KNO_3$.
[f] 1.5 mg/l $ZnSO_4 \cdot 7H_2O$ was used instead of 15 mg/l $ZnNa_2EDTA$.
[g] 10 mg/l Fe citrate was used instead of 2.5 mg/l $Fe(SO_4)_3$.

high amounts of chlorophylls in almost all the media used. In calli maintained on MS medium supplemented with 30 g/l sucrose, 1 g/l edamine, 5 mg/l ascorbic acid, 0.5 mg/l BA, and 0.25 mg/l 2,4-D, chlorophyll concentration recorded was 3.4 mg per gram of dry weight (Fernandes-Ferreira et al. 1989). This ability to synthesize chlorophylls led some authors to establish and maintain photoautotrophic suspension cultures of *E. characias* (Hardy et al. 1987a, b). Some physiological and biochemical studies on these photoautotrophic suspension cultures have been reported (Chagvardieff et al. 1988; Rebeille et al. 1988; Carriere et al. 1989, 1990).

## 3 Crude Oil and Hydrocarbon-Like Compounds Produced by Wild and Micropropagated Plants, Calli, and Suspended Cells of *E. characias*

### 3.1 Crude Oil Contents

All the extractions were performed according to the method of Buchanan et al. (1978b) modified in the partition step (Fernandes-Ferreira et al. 1990a, b, 1991).

Figure 5 shows the procedure followed for extraction and analysis of the crude oil. Calli and suspended cells were not post-extracted by cyclohexane because the preliminary assays showed that no compounds were extracted by this solvent.

Extractions performed over 1 year on the whole aerial part of *E. characias* wild plants revealed a seasonal variation of the content of the neutral fraction of the crude oil. The content of the other fractions was constant during that period (acid crude oil fraction, 0.5% dry wt.; polyphenol fraction, 2.4% dry wt.; hydrocarbon fraction, 0.4% dry wt.). The highest content of the neutral fraction of the crude oil was obtained in November (7.5% dry wt.) and the lowest in May (4.5% dry wt.) (Fernandes-Ferreira et al. 1990a, b). Similar results were obtained by Conti et al. (1982) and by Coppola and Brunori (1984) for plants of *E. characias* grown in Sardinia and Italy, respectively. The seasonal variation of the crude oil content recorded for plants of *E. characias* is similar to those reported for *E. lathyris* (Coppola and Brunori, 1984) and *E. esula* (Wiatr 1985). Such variation is inversely correlated with the vegetative growth (Coppola and

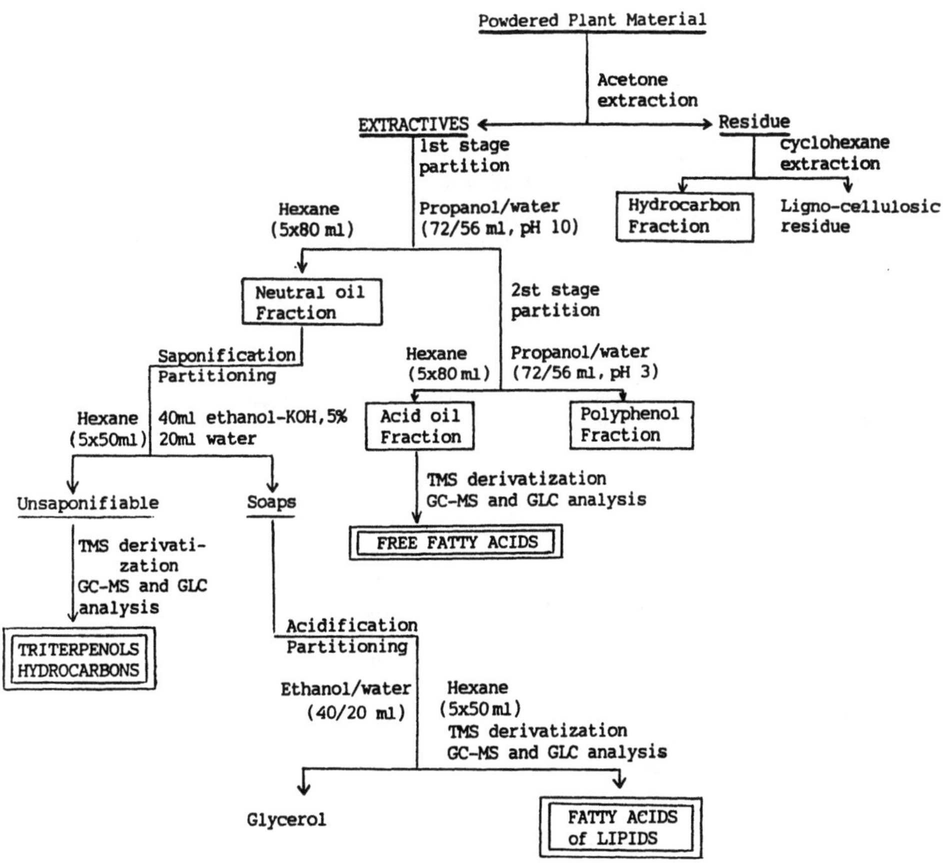


**Fig. 5.** Scheme of the partitioning and analysis of the extractives of the powdered biomass of *Euphorbia characias*

Brunori 1984). The yields recorded for wild plants of *E. characias* are similar to those recorded for other *Euphorbia* species considered as suitable candidates for energy crops (Buchanan et al. 1978a, b; Adams and McChesney 1983; McLaughlin et al. 1983; Wiatr 1985; Adams et al. 1986; Calvin 1987; Gnecco et al. 1988).

Two determinations performed on the micropropagated plants of *E. characias* with 1 year's interval showed that crude oil content of these plants was higher than the best obtained from the wild ones (Table 4). The neutral crude oil fraction contents recorded for calli and suspended cells of *E. characias* were lower than those recorded for wild and micropropagated plants. On the contrary, the acid fraction contents of calli and suspended cells were significantly higher than those obtained for wild and micropropagated plants (Table 4).

The concentration of the crude oil in leaves is considerably higher than that recorded in stems. The contents of the neutral fraction, prior and after purification, by activated charcoal, as well as the contents of the acid fraction of the crude oil obtained from leaves and stems of wild and micropropagated plants are shown in Table 5.

As discussed before (Fernandes-Ferreira et al. 1990b, 1991; Fernandes-Ferreira 1992), the reasons for the higher contents of the neutral fraction in leaves and stems of the micropropagated plants of *E. characias*, when compared with those recorded for the same organs of wild plants, are not yet fully known. The less lignification of the micropropropagated plants and/or the increased ability to produce crude oil compounds induced as a consequence of the in vitro manipulation, are two likely hypotheses for the explanation of those higher yields. The specific production of crude oil compounds, namely sterols and triterpenols, by calli of *E. characias* is strongly influenced by the type and concentration of phytohormones and other medium components (Fernandes-

**Table 4.** Crude oil contents of whole aerial part of wild and micropropagated plants, calli, and suspended cells of *Euphorbia characias*

| Harvest | Neutral fraction (% dry wt.) | Acid fraction (% dry wt.) |
|---|---|---|
| Wild plants | | |
| Nov. | 7.5 | 0.60 |
| Feb. | 5.1 | 0.48 |
| May | 4.5 | 0.44 |
| Aug. | 5.8 | 0.49 |
| Nov. | 7.6 | 0.58 |
| Micropropagated plants | | |
| Jul. (3 months in the field) | 8.3 | 0.21 |
| Jul. (15 months in the field) | 8.2 | 0.41 |
| Calli[a] | 1.7 | 1.06 |
| Suspended cells[a] | 1.9 | 1.32 |

[a] Calli and suspended cells at stationary phase grown in MS maintenance medium.

**Table 5.** Contents of the neutral and acid fractions of the crude oil in leaves and stems of wild and micropropagated *E. characias* plants

| | Neutral fraction | | Acid fraction |
|---|---|---|---|
| | Total | Purified (% dry wt.) | Total |
| Leaves of wild plants | 8.5 | 4.5 | 0.76 |
| Stems of wild plants | 4.0 | 2.9 | 0.37 |
| Leaves of micropropagated plants | 9.9[a] | 5.5[a] | 0.24[a] |
| | 12.2[b] | 6.4[b] | 0.48[b] |
| Stems of micropropagated plants | 5.9[a] | 4.4[a] | 0.17[a] |
| | 4.7[b] | 3.5[b] | 0.34[b] |

[a] 3 months in the field.
[b] 15 months in the field.

Ferreira 1992; Fernandes-Ferreira et al. 1992a, b). Some differences in micro climatic or soil factors would also be responsible, at least in part, for these differences on crude oil yields.

The low specific accumulation of the neutral fraction of the crude oil recorded for calli and suspended cells is certainly correlated with the lack of laticifer and cuticular differentiation, since laticifers and cuticules are the structures where most of the hydrocarbon-like compounds of this fraction are accumulated.

## 3.2 Main Constituents of the Crude Oil

Hydrocarbons, sterols and triterpenols, and esterified fatty acids were the main groups of compounds identified in the purified neutral fractions of the crude oil obtained from wild and micropropagated plants, calli, and suspended cells from *E. characias*. Free fatty acids were the main compounds identified in the acid fractions (Fernandes-Ferreira et al. 1991, 1992b). The concentrations of these groups of compounds in leaves and stems of wild and micropropagated plants, calli, and suspended cells are shown in Table 6.

Calli and suspended cells produce the same fatty acids as leaves and stems of wild and micropropagated plants of this species. The fatty acids identified by GC-MS and GLC were: lauric acid (12:0), myristic (14:0), pentadecanoic (15:0), palmitic (16:0), stearic (18:0), oleic (18:1), linoleic (18:2), linolenic (18:3), arachidic (20:0), behenic (22:0), and lignoceric (24:0) (Fernandes-Ferreira et al. 1991; Fernandes-Ferreira 1992). These fatty acids are major constituents either from the acid fractions or from the saponified fractions obtained from plants and from in vitro cultures of *E. characias*.

The hydrocarbons nonacosane, hentriacontane, tritriacontane, and squalene, and the triterpenoids lanosterol, lanosterol isomer, cycloartenol, 24-methyl enecycloartanol, lupen-3-one, and α-amyrin were identified by GC-MS and GLC as major constituents of the unsaponifiable fractions of the crude oil

**Table 6.** Concentrations of the main groups of compounds identified in the purified neutral fractions and acid fractions of the crude oil in leaves and stems of wild and micropropagated plants, calli, and suspended cells of *E. characias*

| | Sterols and triterpenols | Hydrocarbons | Esterified fatty acids | Free fatty acids |
|---|---|---|---|---|
| | | (mg/g dry wt.) | | |
| Leaves of wild plants | 14.01 | 2.54 | 2.06 | 0.92 |
| Stems of wild plants | 10.93 | 1.94 | 1.29 | 0.33 |
| Leaves of microprop. plants[a] | 17.52 | 2.99 | 3.35 | 0.15 |
| Stems of microprop. plants[a] | 22.14 | 2.54 | 1.73 | 0.41 |
| Calli[b] | 1.06 | 0.05 | 0.86 | 3.12 |
| Suspended cells[b] | 2.22 | 0.09 | 2.67 | 6.37 |

[a] 3 months in the field.
[b] Calli and suspended cells at stationary phase grown in MS maintenance medium.

obtained from leaves and stems of wild and micropropagated plants (Fernandes-Ferreira et al. 1991; Fernandes-Ferreira 1992; Fig. 6). The sterols campesterol, β-sitosterol, $\Delta^5$-avenasterol (Fig. 7), and the triterpenol α-amyrin are the main constituents of the unsaponifiable fraction obtained from calli and suspended cells of this species (Fernandes-Ferreira et al. 1990c, 1991; Fernandes-Ferreira 1992; Fig. 8). Squalene and trace amounts of hydrocarbons, namely nonacosane and hentriacontane, were also identified in the unsaponifiable fractions obtained from these in vitro cultures (Fernandes-Ferreira et al. 1991; Fernandes-Ferreira 1992).

Nonacosane, hentriacontane, tritriacontane, lupene-3-one, and α-amyrin were identified as components of the cuticular waxes of *E. characias* (Hemmers and Gülz 1986; Hemmers et al. 1988). The low concentrations of hydrocarbons in calli and suspended cells of *E. characias* may be correlated to the absence of cuticular structures in these in vitro cultures. The production of long straight chain saturated hydrocarbons by suspended cells of *E. characias* was also reported by other authors (Carriere et al. 1990).

Lanosterol, lanosterol isomer, cycloartenol, and 24-methylenecycloartanol are constituents of the latex of *E. characias* (Nielsen et al. 1977, 1979; Fernandes-Ferreira et al. 1990c; Fernandes–Ferreira 1992). The role of these compounds in the latex is still obscure. Apparently they are accumulated in the latex as end products; however, in nonlaticifer cells, these compounds, namely cycloartenol and 24-methylenecycloartanol, can be converted to phytosterols (Grunwald 1980).

The absence of lanosterol, lanosterol isomer, cycloartenol, and 24-methylenecycloartanol, in calli and suspended cells of *E. characias* may be due to the absence of laticifers in in vitro cultures. A ready conversion of the cycloartenol and 24-methylenecycloartanol to phytosterols, namely campesterol, β-sitosterol and $\Delta^5$-avenasterol, could explain the apparent absence of these compounds in the unsaponifiable fractions obtained from the calli and suspended cells.

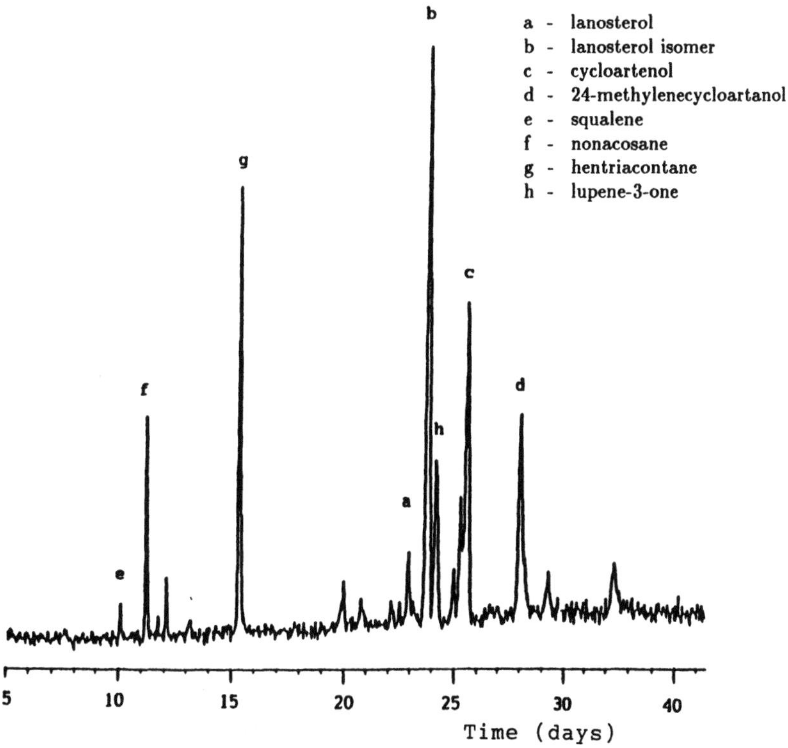


**Fig. 6.** Reconstructed total ion current trace obtained from the unsaponifiable fraction of the crude oil of *Euphorbia characias* plants using a GC-MS quadrupole instrument operated in the EI mode; ion source, interface, and injector temperatures at 250, 300, and 350 °C respectively; oven temperature 250–280 °C, rate 5 °C/min; electron energy, 70 eV

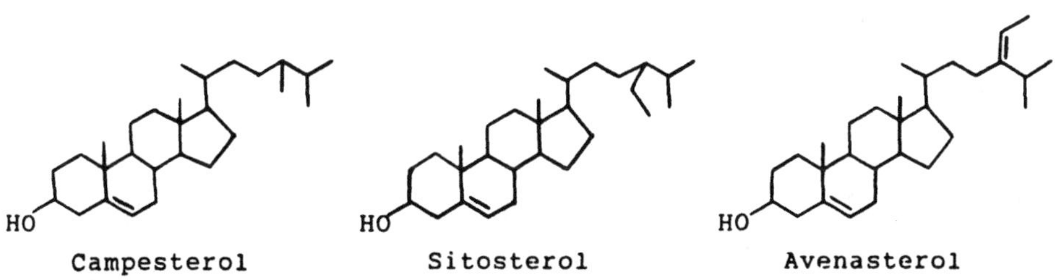


**Fig. 7.** Chemical structures of the sterols produced by calli and suspended cells of *Euphorbia characias*

## 3.3 Influence of Some Factors on the In Vitro Production of Sterols

The specific content of the total sterols in calli and suspended cells vary considerably during the growth cycle. At the exponential phase, the biosynthesis of these compounds increases markedly, reaching the maximum specific accumulation during the transition to the stationary phase (Fig. 9).

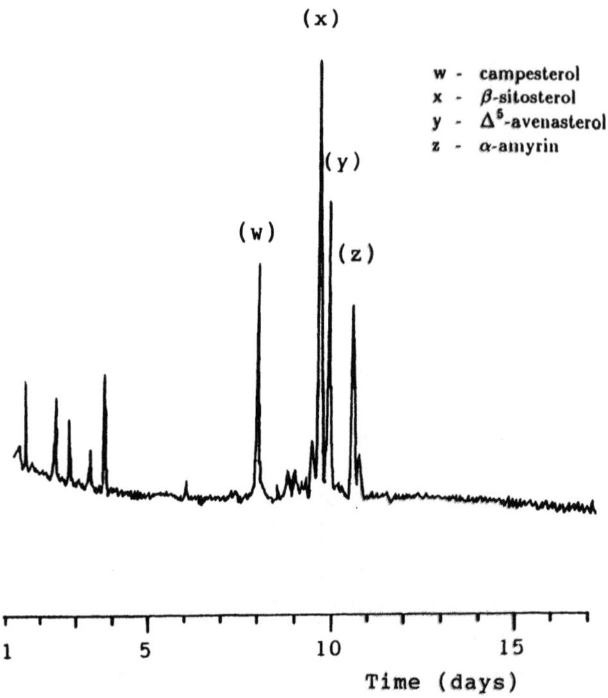


**Fig. 8.** Reconstructed total ion current trace obtained from the unsaponifiable fraction of the crude oil of *Euphorbia characias* calli and suspended cells using a GC-MS quadrupole instrument operated in the EI mode; ion source, interface, and injector temperatures at 250, 300, and 350 °C respectively; oven temperature 250–280 °C, rate 5 °C min; electron energy, 70 eV

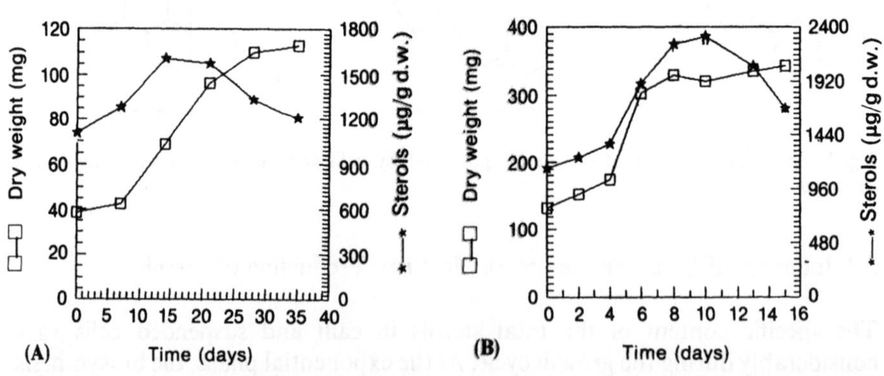


**Fig. 9A, B.** Variation of the specific content of the total sterols in calli (**A**) and in suspended cells (**B**) of *Euphorbia characias* during the growth cycles

**Table 7.** Sterols produced by calli of *E. characias* maintained on different basal media. Subcultures were initiated with calli fragments of 35 mg dry wt. The values of sterols contents are the average of six determinations

| Basal medium | Sterols (μg/g dry wt.) | Basal medium | Sterols (μg/g dry wt.) |
|---|---|---|---|
| Murashige and Skoog (1962) | 948 | Schenk and Hildbrandt (1972) | 615 |
| 1/2 Murashige and Skoog[a] | 519 | Eeuwens (1978) | 860 |
| Heller (1953)[b] | 1417 | Campbell and Durzan (1975) | 519 |
| Nitsch and Nitsch (1956) | 1265 | Linsmaier and Skoog (1965) | 517 |
| White (1963) | 1047 | Dutcher and Powell (1972) | 1161 |
| Hildebrandt et al. (1946) | 1256 | Murashige and Tucker (1969) | 1110 |
| Gautheret (1942)[c] | 1381 | Miller and Skoog (1953)[g] | 874 |
| White (1943)[d] | 1097 | Nitsch (1951) | 672 |
| Wood and Braun (1961)[e] | 1286 | Nitsch (1972) | 1105 |
| Eriksson (1965)[f] | 1339 | Nitsch et al. (1968) | 808 |
| Gamborg et al. (1968) | 726 | | |

[a–g] See Table 3 for media supplements and modifications.

About 50% of the basal media tested are more efficient for sterol production by calli of *E. characias* than that of Murashige and Skoog (1962), used as maintenance medium (Table 7). 2,4-D and IBA, at concentrations up to $5 \times 10^{-6}$ M, are more efficient for sterol production than IAA and NAA used at the same concentrations. Contrarily to what happens with kinetin and zeatin, BA at $5 \times 10^{-5}$ M strongly inhibits sterols biosynthesis by calli, and the increase of the sucrose concentration decreases the specific content of sterols in the same cultures (Fernandes–Ferreira 1992; Fernandes–Ferreira et al. 1992a, b).

### 3.4 Importance of the Sterols Produced by Calli and Suspended Cells

At least two functions have been attributed to campesterol, β-sitosterol, $\Delta^5$-avenasterol, and other 4,4-demethyl sterols: (1) as components of membranes, where by interaction with the phospholipids they contribute to membrane stabilization and control of permeability; and (2) as precursors in the formation of other steroids (Heftmann 1975; Grunwald 1980). Some hormonal-like effects have been attributed to some sterols, namely to β-sitosterol. According to Kopcewicz (1969), β-sitosterol stimulates the growth of dwarf plants of *Pisum sativum*. This sterol seems also to stimulate the induction of the floral buds of *Chrysanthemum* sp. (Biswas et al. 1967). According to Herbert (1981), β-sitosterol has estrogenic effects and, in high concentrations, can interfere with the breeding of herbivores. It is likely that sterols by themselves do not act as hormones but they certainly are degraded to compounds with hormonal activity in animals (Grunwald 1980; Herbert 1981). The production of steroids, such as diosgenin, by biotransformation of β-sitosterol, has been reported by some authors (Stohs and Rosenberg 1975).

The practical importance of β-sitosterol and campesterol for the steroid industry was negligible until the mid-1970s. Diosgenin, from *Dioscorea* plants, and stigmasterol were, thus far, used as primary raw materials for the production of pharmacologically active steroids (e.g., contraceptives, corticosteroids) (Martin 1977). However, as a result of the growing demand for these therapeutic steroids, these sources of raw material became insufficient (Sedlaczek 1988). Over the last 15 years, microbiological methods of selective side-chain cleavage of phytosterols, such as β-sitosterol, campesterol, and related $\Delta^5$-sterols, were developed, and these materials gained technical significance as suitable precursors of therapeutic steroids (Martin 1977, 1984; Schoemer and Martin 1980; Maxon 1985; Sedlaczek 1988).

As a result of the growing demand for phytosterols and of the new applications that have been found for the sterol-producing plants, such as soybean, the prices of these compounds have been steadily rising (Malik 1982; Sedlaczek 1988). On the other hand, the production of diosgenin from *Dioscorea* plants is greatly dependent on the climatic influences. The production of sterols by plant tissue and cell cultures makes the supply of plant sterols independent of the climate. According to Sedlaczek (1988), this technology is potentially a new way to fulfill the future requirements for raw materials used in the steroid industry.

## 4 Summary and Conclusions

The total crude oil content of the aerial part of wild plants of *E. characias* varies from 5 to 8% of the dry weight, depending on the season. The major compounds identified in this fraction were the 4,4-dimethyl sterols lanosterol, lanosterol isomer, cycloartenol, and 24-methylene cycloartanol, the triterpenone lupene-3-one, the hydrocarbons nonacosane, hentriacontane, and tritriacontane, and the fatty acids 12:0, 14:0, 15:0, 16:0, 18:0, 18:1, 18:2, 18:3, 20:0, 22:0, and 24:0. The same compounds are also produced by micropropagated plants. Micropropagated plants gave crude oil yields higher than those obtained from wild plants. The latex of this species contains compounds with high biological activity, some of which, namely diterpenoids, may, in the future, become important for the chemical and pharmaceutic industry. Calli and suspended cells do not accumulate the 4,4-dimethyl sterols found in the latex of the parent plants but they produce and accumulate high amounts of 4-demethyl sterols (β-sitosterol, campesterol, and $\Delta^5$-avenasterol). The specific content of these compounds varies considerably during the growth cycle of calli and suspended cells. The maximum specific accumulation of these compounds occurs at the end of the exponential phase.

Plants of *E. characias* are potential sources of compounds for the petrochemical industry, with a value similar to plants of other species of this genus, namely *E. lathyris*. The utilization of heterotrophic cultures of calli and suspended cells for energy purposes seems to be a remote hypothesis, since the production of lipidic compounds depends on the exogenous supply of another energy-resource (e.g., sucrose) as well as on other expensive resources. However, contrarily to the triterpenols produced by in vivo plants of *E. characias*, the sterols produced by calli and suspended cells of this species can be used by the pharmaceutical industry as raw chemicals for the production of therapeutic steroids.

## References


Adams RP, McChesney JD (1983) Phytochemicals for liquid fuels and petrochemical substitutions: extraction procedures and screening results. Econ Bot 37(2) : 207–215

Adams RP, Balandrin MF, Brown KJ, Stone GA, Gruel SM (1986) Extraction of liquid fuels and chemicals from terrestrial higher plants. Part I. Yields from a survey of 614 Western United States plant taxa. Biomass 9: 255–292

Adolf W, Hecker E, Becker H (1984) Macrocyclic lathyrane type diterpene esters (jolkinols) from callus cultures and roots of *Euphorbia lathyris*. Planta Med 47: 259–261

Afza N, Khan AQ, Malik A, Badar Y (1988) Triterpenoids from the latex of *Euphorbia caudicifolia*. Planta Med 53: 482

Aitken A (1987) The activation of protein kinase C by daphnane, ingenane and tigliane diterpenoid esters. In: Jury SL, Reynolds T, Cutler DF, Evans EJ (eds) The Euphorbiales – chemistry, taxonomy and economic botany. Academic Press, London, pp 247–263

Ayensu ES (ed) (1978) Medicinal plants of West Africa. Reference, Algonac, Michigan

Ayerbe L, Tenorio JL, Ventas P, Funes E, Mellado L (1984a) *Euphorbia lathyris* as an energy crop – Part 1. Vegetative matter and seed productivity. Biomass 4: 283–293

Ayerbe L, Funes E, Tenorio JL, Ventas P, Mellado L (1984b) *Euphorbia lathyris* as an energy crop – Part. II. Hydrocarbon and sugar productivity. Biomass, 5: 37–42

Baburaj S, Dhamotharan R, Santhaguru K (1987) Regeneration in leaf callus cultures of *Euphorbia hirta* Linn. Curr Sci 56: 194

Baslas RK (1982) Phytochemical studies of the plants of the genera *Euphorbia*. Part II. Herba Hung 21: 115–126

Bhatia VK, Srivastava GS, Garg VK, Gupta YK, Rawat SS (1984) Petro-crops for fuel. Biomass 4: 151–154

Biesboer DD, Mahlberg PG (1978) Accumulation of non-utilizable starch in laticifers of *Euphorbia heterophylla* and *E. myrsinites*. Planta 143: 5–10

Biesboer DD, Mahlberg PG (1979) The effect of medium modification and selected precursors on sterol production by short-term callus cultures of *Euphorbia tirucalli*. J Nat Prod 42: 648–657

Biesboer DD, D'Amour P, Wilson SR, Mahlberg P (1982) Sterols and triterpenols in latex and cultured tissues of *Euphorbia pulcherrima*. Phytochemistry 21: 1115–1118

Biswas PK, Paul KB, Henderson JHM (1967) Effects of steroids on *Chrysanthemum* in relation to growth and flowering. Nature (Lond) 213: 917–918

Buchanan RA, Cull IM, Otey FH, Russell CR (1978a) Hydrocarbon- and rubber-producing crops. Evaluation of U. S. plant species. Econ Bot 32: 131–145

Buchanan RA, Cull IM, Otey FH, Russell CR (1978b) Hydrocarbon- and rubber-producing crops. Evaluation of 100 U. S. plant species. Econ Bot 32: 146–153

Calvin M (1984) Oil from plants. In: Sharp WR, Evans DA, Ammirato PV, Yamada Y (eds) Handbook of plant cell culture vol 2. Crop species Macmillan, London, pp 1–18

Calvin M (1987) Fuel oils from euphorbs and other plants. In: Jury SL, Reynolds T, Cutler DF, Evans FJ (eds) The Euphorbiales. Chemistry, taxonomy and economic botany. Academic Press, London, pp 97–110

Campbell RA, Durzan DJ (1975) Induction of multiple buds and needles in tissue cultures of *Picea glauca*. Can J Bot 53: 1652–1656

Carriere F, Gil G, Tapie P, Chagvardieff P (1989) Biotransformation of geraniol by photo-autotrophic, photomixotrophic and heterotrophic plant cell suspensions. Phytochemistry 28:1087–1090

Carriere F, Chagvardieff P, Gil G, Pean M, Sigoillot JC, Tapie P (1990) Paraffinic hydrocarbons in heterotrophic, photomixotrophic and photo-autotrophic cell suspensions of *Euphorbia characias* L. Plant Sci 71:93–98

Chagvardieff P, Pean M, Carrier P, Dimon B (1988) Oxygen exchange during growth of photo-mixotrophic cell suspensions of *Euphorbia characias* L. Plant Cell Tissue Organ Cult 12:243–251

Chen Y, Tang Z-J, Jiang F-X, Zhang X-X, Lao AI-NA (1979) Studies on the active principles of Ze–Qi (*Euphorbia helioscopia* L.), a drug used for chronic bronchitis. Yao Hsueh Hsueh Pao 14 (2):91–95 (via Chem Abstr 92:72680)

Conti L, Botteghi C, Alamanni U, Luca GD (1982) Oils from *Euphorbia lathyris* and *E. characias*. Chim Ind 64:705–707

Coppola F, Brunori A (1984) The effect of the herbicide Gramixel on the heptane fraction of *Euphorbia* biomass. Biomass 4:59–68

Declerck M, Smets PH, Smets J, Roman J (1985) Euphorbia project: renewable energy production through the cultivation and processing of semi arid land biomass in Kenya. In: Palz W, Coombs J, Hall DO (eds) Energy from Biomass. 3rd E C Conf Commission of the European Communities. Elsevier Amsterdam, pp 310–314

Duke JA, Ayensu ES (eds) (1985) Medicinal plants of China, 1. Reference, Algonac, Michigan

Dutcher RD, Powell LE (1972) Culture of apple shoots from buds in vitro. J Am Soc Hortic Sci 97:511–515

Eriksson KB (1965) Crossing experiments with races of *Ditylenchus dipsaci* on callus tissue cultures. Nematologica 11:244–248

Eeuwens CJ (1978) Effects of nutrients and hormones on growth and development of tissue explants from coconut (*Cocos nucifera*) and date (*Phoenix dactylifera*) palms cultured in vitro. Physiol Plant 42:173–177

Evans FJ, Edwards MC (1987) Activity correlations in the phorbol ester series. In: Jury SL, Reynolds T, Cutler DF, Evans EJ (eds) The Euphorbiales–chemistry, taxonomy and economic botany. Academic Press, London pp 231–246

Fernandes-Ferreira M (1992) Constituintes lipídicos produzidos por plantas espontâneas e micro-propagadas, calli e células em suspensão de *Euphorbia characias* L. subsp. *characias*. PhD Thesis, Universidade do Minho, Portugal

Fernandes-Ferreira M, Novais JM, Pais MSS (1989) Calli and suspension cultures for biomass production of *Euphorbia characias* L. subsp. *characias*. Biotechnol Lett 11:259–264

Fernandes-Ferreira M, Pais MSS, Novais JM (1990a) Production of lipid compounds by wild and micropropagated in nature growing plants of *Euphorbia characias*. In: Grassi G, Gosse G, Dos Santos G (eds) Biomass for energy and industry. Elsevier, London, pp 517–521

Fernandes-Ferreira M, Pais MSS, Novais JM (1990b) Energetic potentialities of wild and micropro-pagated plants of *Euphorbia characias* L vol 3. In: Sayigh AAM (ed) Energy and environment into the 1990s. Pergamon Press, Oxford, pp 2022–2027

Fernandes-Ferreira M, Novais JM, Pais MSS (1990c) Free triterpenols and sterols produced by in vitro cultures and laticifer cells from *Euphorbia characias*. Phytochemistry 29:1855–1860

Fernandes-Ferreira M, Pais MSS, Novais JM (1991) Oils from wild, micropropagated plants, calli, and suspended cells of *Euphorbia characias* L. Appl Biochem Biotechnol 28/29:169–181

Fernandes-Ferreira M, Novais JM, Pais MSS (1992a) Hormonal control of triterpenols synthesis in *Euphorbia characias* calli. Bioresource Technol 39:31–37

Fernandes-Ferreira M, Pais MSS, Novais JM (1992b) The effects of medium composition on biomass, sterols and triterpenols production by in-vitro cultures of *Euphorbia characias*. Bio-resource Technol 42:67–73

Floris G, Giartosio A, Rinaldi A (1983) Essential sulfhydryl groups in diamine oxidase from *Euphorbia characias* latex. Arch Biochem Biophys 220:623–627

Floris G, Medda R, Rinaldi A (1984) Peroxidase from *Euphorbia characias* latex:purification and properties. Phytochemistry 23:953–956

Füstenberger G, Hecker E (1977) New highly irritant euphorbia factors from latex of *Euphorbia tirucally* L. Experientia 33(8):986–988

Füstenberger G, Hecker E (1985) On the active principles of the spurge family (Euphorbiaceae) XI. (1) The skin irritant and tumour promoting diterpene esters of *Euphorbia tirucalli* L. originating from South Africa. Z Naturforsch 40c: 631–646

Gamborg OL, Miller RA, Ojma K (1968) Nutrient requirements of suspension cultures of soybean root cells. Exp Cell Res 50: 151–158

Gautheret RJ (ed) (1942) Manuel technique de culture de tissues végétaux. Masson Cie, Paris

Gnecco S, Bartulín J, Marticorena C, Ramirez A (1988) Chilean Euphorbiaceae species of fuels and raw chemicals. Biomass 15: 165–173

Grunwald C (1980) Steroids. In:Bell EA, Chartwood BV (eds) Secondary plant products. Encyclopedia of plant physiology. New series vol 8. Springer, Berlin Heidelberg New York, pp 221–256

Gülz P-G, Hemmers H, Bodden J, Marner F-J (1987) Epicuticular leaf wax of *Euphorbia dendroides* L., Euphorbiaceae. Z Naturforsch 42c: 191–196

Gülz P-G, Bodden J, Müller E, Marner F-J (1988) Epicuticular wax of *Euphorbia aphylla* Brouss. ex. Willd., Euphorbiaceae. Z Naturforsch 43c: 19–23

Hardy T, Chaumont D, Brunel Gudin C (1987a) Photoautotrophic suspension cultures I-obtention of photoautotrophic cultures from *Euphorbia characias* L. J Plant Physiol 128: 11–19

Hardy T, Chaumont D, Wessinger ME, Bournat P (1987b) Photoautotrophic suspension cultures I – transition from photoheterotrophic to photo-autotrophic growth. J Plant Physiol 130: 351–361

Hecker E (1987) Tumour promoters of the irritant diterpene ester type as risk factors of cancer in man. In: Jury SL, Reynolds T, Cutler DF, Evans EJ (eds) The Euphorbiales – chemistry, taxonomy and economic botany. Academic Press, London, pp 197–219

Heftmann E (1975) Functions of steroids in plants. Phytochemistry 14:891–901

Heller R (1953) Recherches sur la nutrition minérale des tissus végétaux cultivés in vitro. Ann Sci Nat 14: 1–223

Hellet FP, Parks LM (1953) The antispasmodic principle of *Euphorbia pilulifera.* J Am Pharm Assoc 42: 607–609

Hemmers H, Gülz P-G (1986) Epicuticular waxes from leaves of five *Euphorbia* species. Phytochemistry 25: 2103–2107

Hemmers H, Gülz P-G, Marner F-J (1988) Triterpenoids in epicuticular waxes of three European *Euphorbia* species. Z Naturforsch 43c: 799–805

Hemmers H, Gülz P-G, Marner F-J (1989) Tetra- and pentacyclic triterpenoids from epicuticular wax of *Euphorbia cyparissias* L., Euphorbiaceae. Z Naturforsch 44c: 563–567

Herbert RB (1981) The acetate-mevalonate pathway. In: Herbert RB (ed) The biosynthesis of secondary metabolites. Chapman and Hall, London, pp 112–156

Hildebrandt AC, Riker AJ, Duggar BM (1946) The influence of the composition of the medium on growth in vitro of excised tobacco and sunflower tissue cultures. Am J Bot 33: 591–597

Kingsbury JM (ed) (1964) Poisonous plants of the United States and Canada. Prentice Hall, Trenton, NJ

Koops AJ, Baas WJ, Groeneveld HW (1991) The composition of phytosterols, latex triterpenols and wax triterpenoids in the seedling of *Euphorbia lathyris* L. Plant Sci 74: 185–191

Kopcewicz J (1969) Effect of estrone on the content of endogenous gibberellins in the dwarf pea. Naturwissenschaften 56: 334–338

Kumar A, Joshi B (1983) In vitro growth and differentiation of *Euphorbia lathyris:a* hydrocarbon yielding plant. In:Strub A, Chartier P, Schleser G (eds) Energy from biomass. 2nd E C Conf, Commission of the European Communities. Elsevier, London, pp 261–264

Kupchan SM, Uchida I, Branfman AR, Dayley RG Jr Fey BY (1976) Antileukemic principles isolated from Euphorbiaceae plants. Science 191: 571–572

Langhe ED, Debergh P, Rijk RV (1974) In vitro culture as a method for vegetative propagation of *Euphorbia pulcherrima.* Z Pflanzenphysiol 71: 271–274

Lee K-H, Hayashi N, Okano M, Hall IH, Wu R-Y, Mcphail A (1982): Lasiodiplodin, a potent antileukemic macrolide from *Euphorbia splendens.* Phytochemistry 21: 1119–1121

Lee TT, Starratt AN (1972) Growth substance requirements and major lipid constituents of tissue cultures of *Euphorbia esula* and *E. cyparissias.* Can J Bot 50:723–726

Linsmaier EM, Skoog F (1965) Organic growth factor requirements of tobacco tissue culture. Physiol Plant 18: 100–127

Lynn KR, Clevette-Radford NA (1985) Three serine proteases from the latex of *Euphorbia cyparissias.* Phytochemistry 24:925–928

Lynn KR, Clevette-Radford NA (1986a) Lectins from latices of *Euphorbia* and *Elaeophorbia* species. Phytochemistry 25:1553–1557

Lynn KR, Clevette-Radford NA (1986b) Isolation and characterization of proteases from *Euphorbia lactea* and *Euphorbia lactea cristata.* Phytochemistry 25:807–810

Lynn KR, Clevette-Radford NA (1987a) Biochemical properties of latices from the Euphorbiaceae. Phytochemistry 26:939–944

Lynn KR, Clevette-Radford NA (1987b) Acid phosphatases from latices of Euphorbiaceae. Phytochemistry 26:655–657

Malik VS (1982) Aspects of biotechnology in steroid biotransformation. Z Allg Mikrobiol 22:261–266

Martin CKA (1977) Microbial cleavage of sterol side chains. Adv Appl Microbiol 22:29–58

Martin CKA (1984) Sterols. In:Rehm HJ, Reed G (eds) Biotechnology. A comprehensive treatise in 8 volumes. V Ch, Weinheim, pp 79–95

Maxon WD (1985) Steroid bioconversions:One industrial perspective. In:Tsao GT, Flickinger MC, Finn RK (eds) Annual reports on fermentation processes, vol 8. Academic Press, London, pp 171–185

McLaughlin SP, Kingsolver BE, Hoffmann JJ (1983) Biocrude production in arid lands. Econ Bot 37(2) :150–158

Miller CO, Skoog F (1953) Chemical control of bud formation in tobacco stem segments. Am J Bot 40:768–773

Murashige T, Skoog F (1962) A revised medium for rapid growth and bioassays with tobacco tissue cultures. Physiol Plant 15:473–497

Murashige T, Tucker DPH (1969) Growth factor requirements of citrus tissue culture. Proc 1st Int Citrus Symp 3:1155–1160

Nanba T, Mikage M (1983) Poisonous plants Hoikusha, Tokyo, pp 74

Nemethy EK, Otvos JW, Calvin M (1981) Hydrocarbons from *Euphorbia lathyris.* Pure Appl Chem 53:1101–1108

Ng'Eny-Mengech A, Kiumba SN (1985) Renewable hydrocarbons and industrial chemicals from Kenyan plants. In:Palz W, Coombs J, Hall DO (eds) Energy from biomass. Commission of the European Communities. 3rd E C Con. Elsevier, London, pp 749–755

Nielsen PE, Nishimura H, Otvos JW, Calvin M (1977) Plant crops as a source of fuel and hydrocarbon-like materials. Science 198:942–944

Nielsen PE, Nishimura H, Liang Y, Calvin M (1979) Steroids from *Euphorbia* and other latex-bearing plants. Phytochemistry 18:103–104

Nitsch JP (1951) Growth and development in vitro of excised ovaries. Am J Bot 38:566–570

Nitsch JP (1972) Haploid plants from pollen. Z Pflanzenzuecht 64:3–8

Nitsch JP, Nitsch C (1956) Auxin-dependent growth of excised *Helianthus tuberosus* tissues. Am J Bot 43:839–851

Nitsch JP, Nitsch C, Hamon S (1968) Réalisation experimentale de l'androgeneѐse chez divers *Nicotiana.* C R Seances Soc Biol Paris 162:369–375

Ohyama K, Uchida Y, Misawa N, Konamo T, Fujita M, Ueno T (1984) Oil body formation in *Euphorbia tirucalli* L. cell suspension cultures. Plant Cell Rep 3:21–22.

Okuda T (1986) Encyclopedia of natural medicine. Hirokawa, Tokyo p 312

Passannanti S, Paternostro M, Piozzi F, Faarax MX (1985) Terpenoids from the latex of *Euphorbia ruspolii.* Fitoterapia 56:47–50

Preece JE, Ripley KP (1992) In vitro culture and micropropagation of *Euphorbia* species. In:Bajaj YPS (ed) Biotechnology in agriculture and forestry, vol 20. High-Tech and micropropagation IV. Springer, Berlin Neidelberg New York, pp 91–112

Preece JE, Wollbrink EB (1983) Vegetative propagation of *Euphorbia lathyris* by stem and leaf bud cuttings. HortScience 18(2):193–194

Quer F (ed) (1981) Plantas medicinales – el dioscórides renovado. Editorial Labor, S A Madrid

Rebeille F, Gans P, Chagvardieff P, Pean M, Tapie P, Thibault P (1988) Mass spectrometric determination of the inorganic carbon species assimilated by photoautotrophic cells of *Euphorbia characias* L. J Biol Chem 263:12373–12377

Ripley KP, Preece JE (1984) In vitro culture of *Euphorbia lathyris* shoot tips. HortScience 19(3): 575
Rosengarten F Jr (1978) A neglected Mayan galactogogue-*ixbut* (*Euphorbia lancifolia*). Bot Mus Leafl, Harv Univ, 26 (9–10) : 277–309
Sacks RM, Low CB, McDonald JD, Awad AR, Sully MJ (1981) *Euphorbia lathyris*: a potential source of petroleum-like products. Calif Agric 35 : 7–8, 29–32
Sahai R, Dube MP, Rastogi RP (1981) Chemical and pharmacological study of *Euphorbia maddeni*. Indian J Pharm Sci 43 : 216–219
Schenk RU, Hildebrandt AC (1972) Medium and techniques for induction and growth of monocotyledonous and dicotyledonous plant cell cultures. Can J Bot 50 : 199–203
Schoemer U, Martin CKA (1980) Microbial tyransformation of sterols. Biotechnol Bioeng XXII (Suppl 1) : 11–25
Schroeder G, Rohmer M, Beck JP Anton R (1980) 7-Oxo-, 7 α-hydroxy- and 7 β-hydroxysterols from *Euphorbia fischeriana*. Phytochemistry 19 : 2213–2215
Schultes RE (1987) Members of Euphorbiaceae in primitive and advanced societies. In:Jury SL, Reynolds T, Culter DF, Evans FJ (eds) The Euphorbiales – chemistry, taxonomy and economic botany. Academic Press, London, pp 79–95
Sedlaczek L (1988) Biotransformation of steroids. CRC Crit Rev Biotechnol 7 : 187–236
Seip EH, Hecker E (1983) Lathyrane type diterpenoid esters from *Euphorbia characias*. Phytochemistry 22 : 1791–1795
Seip EH, Hecker E (1984) Derivatives of characiol, macrocyclic diterpene esters of the jatrophane type from *Euphorbia characias*. Phytochemistry 23 : 1689–1694
Sharma DK, Prasad R (1986) Biocrude and solid fuel from laticiferous plants. Biomass 11 : 75–79
Singla AK, Pathak K (1990) Phytoconstituents of *Euphorbia* species. Fitoterapia 61 : 483–516
Speroni E, Coletti B, Minghetti A, Perellino NC, Guicciardi A, Vincieri FF (1991) Activity on the CNS of crude extracts and of some diterpenoids isolated from *Euphorbia calyptrata* suspended cultures. Planta Med 57 : 531–535
Stohs SJ, Rosenberg H (1975) Steroids and steroid metabolism in plant tissue cultures. Lloydia 38 : 181–194
Uemura D, Hirata Y (1971) The isolation and structures of two new alkaloids, milliamines A and B, obtained from *Euphorbia millii*. Tetrahedron Lett 18 : 3673–3676
Ventas P, Tenorio JL, Funes E, Ayerbe L (1985) Controlled environment growth of *Euphorbia lathyris* in relation with temperature and water stress. In:Palz W, Coombs J, Hall DO (eds) 3rd E C Conf Energy from Biomass. Elsevier, London, pp 422–426
Warnaar F (1981) Conjugated fatty acids from latex of *Euphorbia lathyris*. Phytochemistry 20 : 89–91
Watt JM Breyer-Brandwijk MG (eds) (1962) The medicinal and poisonous plants of Southern and Eastern Africa, 2nd edn. Livingstone, Edinburgh
White PR (ed) (1943) A handbook of plant tissue culture. Cattell, Lancaster
White PR (ed) (1963) The cultivation of animal and plant cells, 2nd edn. Ronald Press, New York
Wiatr SM (1985) The oil potential of leafy spurge (*Euphorbia esula* L.). Biomass 8 : 27–40
Wood HN, Braun AC (1961) Studies on the regulation of certain essential biosynthetic systems in normal and crown-gall tumor cells. Proc Natl Acad Sci USA 67 : 1283–1287
Yamamoto Y, Mizuguchi R, Yamada Y (1981) Chemical constituents of cultured cells of *Euphorbia tirucalli* and *E. millii*. Plant Cell Rep 1 : 29–30
Yamamoto Y, Kinoshita Y, Watanabe S, Yamada Y (1989) Anthocyanin production in suspension cultures of high-producing cells of *Euphorbia millii*. Agric Biol Chem 53 (2) : 417–423
Zhang B, Stolz LP, Snyder JC (1987) In vitro propagation of *Euphorbia fulgens*. HortScience 22(3) : 486–488

# XVI *Forsythia* Species: In Vitro Culture and the Production of Lignans and Other Secondary Metabolites

P. M. Dewick[1]

## 1 Introduction

### 1.1 Distribution, Importance and Propagation

The genus *Forsythia* (Oleaceae) contains seven species, six of which (*F. giraldiana* Lingelsh., *F. japonica* Makino, *F. koreana* Nakai, *F. ovata* Nakai, *F. suspensa* Vahl, and *F. viridissima* Lindl.) originated from E. Asia (China, Korea, Japan), whilst *F. europaea* Deg. & Bald. is a native of SE Europe (Albania, Yugoslavia) (Bean 1973). They are popular deciduous ornamental shrubs bearing conspicuous bright yellow flowers which emerge very early in the growing season before the leaves appear (Fig. 1). Many cultivars and hybrids have been introduced, and hybrids of *F. suspensa* × *F. viridissima* have yielded an important group of garden shrubs listed as *F.* × *intermedia* Zabel.

*Forsythia* plants are propagated easily by cuttings made of young, half-ripened, or ripened shoots. Layering is also readily achieved, and indeed, *F. suspensa* and its varieties often root themselves where drooping branches touch the soil.

### 1.2 Secondary Metabolites

A significant number of secondary metabolites have been isolated and characterized from *Forsythia* tissues as detailed in Table 1. Pre-eminent amongst these are the lignans and phenylethanol glycosides. The lignans, which are found both as free phenols and glucosides, fall into two groups, dibenzylbutyrolactones such as matairesinol (**1**) and arctigenin (**3**), and furofurans such as pinoresinol (**9**) and phillygenin (**7**). All tissues examined in all species contain one or more of these structures, but the profile of compounds present appears chemotaxonomically significant and characteristic of the species (Kitagawa et al. 1984a, 1988). Indeed, the hybrid *F.* × *intermedia* shows an additive pattern of lignans arising from the two parents *F. suspensa* and *F. viridissima* (Nishibe et al. 1988). Interestingly, *F.* × *intermedia* is the only member of the genus shown to contain furofuran

[1] Department of Pharmaceutical Sciences, University of Nottingham, Nottingham NG7 2RD, United Kingdom

Biotechnology in Agriculture and Forestry, Vol. 28
Medicinal and Aromatic Plants VII (ed. by Y.P.S. Bajaj)

**Fig. 1.** *Forsythia × intermedia*

lignans restricted to the *epi* series, i.e., epipinoresinol (**5**) and phillygenin (**7**), rather than a mixture of *epi* furofurans and symmetrical furofurans such as pinoresinol (**9**). Despite the structural and stereochemical differences, the *Forsythia* lignans share a common biosynthetic pathway (see Sect. 3). The monolignol coniferin (**13**) may function as a storage form of the lignan precursor coniferyl alcohol.

Many of the phenylethanol glycosides are esterified with caffeic acid, invariably on the 4-position of the glucose. Positions 3 and/or 6 on the glucose may be glycosylated further, usually with rhamnose, though the 2-hydroxyl is unsubstituted in all examples reported so far. A further variation is the presence of a hydroxyl on the β-carbon of the phenylethanol sidechain in some compounds, e.g., forsythoside D (**14**), suspensaside (**17**), and β-hydroxyacteoside (**19**). Trivial nomenclature has unfortunately become somewhat confused due to the various authors using different names for the same compounds. The hybrid nature of *F.* × *intermedia* is again reflected in the presence of both forsythiaside (**16**) and acteoside (**18**), whereas *F. suspensa* contains only the former, and *F. viridissima* only the latter (Nishibe et al. 1988).

Undoubtedly related to the phenylethanol glycosides are a group of cyclohexylethyl derivatives isolated from the fruits of *F. suspensa.* Rengyol (**13**) and related compounds could well be produced in nature by modification of the aromatic ring of a phenylethanol derivative, and the presence of salidroside (**32**)

**Table 1.** Secondary metabolites in *Forsythia* tissues

| Compound | Species | Reference |
|---|---|---|
| *Lignans* | | |
| Matairesinol (**1**) | *F.* × *intermedia* lf, st | Nishibe et al. (1988); Rahman et al. (1990c) |
| | *F. koreana* fr, lf | Chiba et al. (1979); Kitagawa et al. (1984a) |
| | *F. ovata* lf | Kitagawa et al. (1988) |
| | *F. viridissima* fr, lf | Chiba et al. (1978); Kitagawa et al. (1984a) |
| Matairesinol 4′-*O*-glucoside (matairesinoside) (**2**) | *F.* × *intermedia* lf, bk, st | Thieme and Winkler (1968b)[a]; Nishibe et al. (1988); Rahman et al. (1990c) |
| | *F. koreana* fr, lf | Chiba et al. (1979); Kitagawa et al. (1984a) |
| | *F. ovata* lf | Kitagawa et al. (1988) |
| | *F. viridissima* fr, lf | Chiba et al. (1978); Kitagawa et al. (1984a) |
| Arctigenin (**3**) | *F.* × *intermedia* lf, st | Nishibe et al. (1988); Rahman et al. (1990c) |
| | *F. koreana* fr, lf | Chiba et al. (1979); Kitagawa et al. (1984a) |
| | *F. ovata* lf | Kitagawa et al. (1988) |
| | *F. viridissima* fr, lf | Chiba et al. (1978); Kitagawa et al. (1984a) |
| Arctiin (**4**) | *F.* × *intermedia* lf, bk, st | Thieme and Winkler (1968a)[a]; Nishibe et al. (1988); Rahman et al. (1990c) |
| | *F. koreana* fr, lf | Nishibe et al. (1977b); Kitagawa et al. (1984a) |
| | *F. ovata* lf | Kitagawa et al. (1988) |
| | *F. viridissima* fr, lf | Nishibe et al. (1977b); Kitagawa et al. (1984a) |
| Epipinoresinol (**5**) | *F.* × *intermedia* lf, st | Nishibe et al. (1988); Rahman et al. (1990c) |
| Epipinoresinol 4′-*O*-glucoside (**6**) | *F.* × *intermedia* lf, bk, st | Thieme and Winkler (1969a)[a]; Nishibe et al. (1988); Rahman et al. (1990c) |
| Phillygenin (**7**) | *F. europaea* lf | Kitagawa et al. (1988) |
| | *F. giraldiana* lf | Kitagawa et al. (1988) |
| | *F.* × *intermedia* lf, st | Nishibe et al. (1988); Rahman et al. (1990c) |
| | *F. japonica* lf | Kitagawa et al. (1988) |
| | *F. koreana* fr, lf | Chiba et al. (1979); Kitagawa et al. (1984a) |
| | *F. suspensa* fr, lf | Nishibe et al. (1977a); Kitagawa et al. (1984a) |
| Phillyrin (**8**) | *F. europaea* lf | Kitagawa et al. (1988) |
| | *F. giraldiana* lf | Kitagawa et al. (1988) |
| | *F.* × *intermedia* lf, bk, st | Thieme and Winkler (1968a)[a]; Nishibe et al. (1988); Rahman et al. (1990c) |
| | *F. japonica* lf | Kitagawa et al. (1988) |
| | *F. koreana* fr,lf | Nishibe et al. (1977b); Kitagawa et al. (1984a) |

**Table 1.** (*Contd.*)

| Compound | Species | Reference |
|---|---|---|
| | *F. suspensa* fr, lf | Nishibe et al. (1977a); Kitagawa et al. (1984a) |
| Pinoresinol **(9)** | *F. europaea* lf | Kitagawa et al. (1988) |
| | *F. giraldiana* lf | Kitagawa et al. (1988) |
| | *F. japonica* lf | Kitagawa et al. (1988) |
| | *F. koreana* fr, lf | Chiba et al. (1979); Kitagawa et al. (1984a) |
| | *F. suspensa* fr, lf | Nishibe et al. (1977a); Kitagawa et al. (1984a) |
| Pinoresinol *O*-glucoside **(10)** | *F. europaea* lf | Kitagawa et al. (1988) |
| | *F. giraldiana* lf | Kitagawa et al. (1988) |
| | *F. japonica* lf | Kitagawa et al. (1988) |
| | *F. koreana* fr, lf | Chiba et al. (1979); Kitagawa et al. (1984a) |
| | *F. suspensa* fr, lf | Chiba et al. (1978); Kitagawa et al. (1984a) |
| Pinoresinol monomethyl-ether **(11)** | *F. giraldiana* lf | Kitagawa et al. (1988) |
| | *F. japonica* lf | Kitagawa et al. (1988) |
| Pinoresinol monomethylether *O*-glucoside **(12)** | *F. giraldiana* lf | Kitagawa et al. (1988) |
| | *F. japonica* lf | Kitagawa et al. (1988) |
| Coniferin **(13)** | *F. × intermedia* lf, bk | Thieme and Winkler (1969b)[a] |
| *Phenylethanol glycosides and related compounds* | | |
| Forsythoside D **(14)** | *F. suspensa* fr | Endo and Hikino (1982) |
| Forsythoside E **(15)** | *F. suspensa* fr | Endo and Hikino (1984) |
| Forsythiaside (forsythoside A) **(16)** | *F. europaea* lf | Kitagawa et al. (1988) |
| | *F. × intermedia* lf, fl | Nishibe et al. (1988); Chapple and Ellis (1984) |
| | *F. koreana* fr, lf | Kitagawa et al. (1987); Kitagawa et al. (1984a) |
| | *F. suspensa* fr, lf | Endo et al. (1981); Nishibe et al. (1982a); Kitagawa et al. (1984a) |
| Suspensaside (forsythoside C) **(17)** | *F. koreana* fr | Kitagawa et al. (1987) |
| | *F. suspensa* fr | Nishibe et al. (1982b); Endo and Hikino (1982) |
| Acteoside (verbascoside) **(18)** | *F. giraldiana* lf | Kitagawa et al. (1988) |
| | *F. × intermedia* lf | Nishibe et al. (1988) |
| | *F. japonica* lf | Kitagawa et al. (1988) |
| | *F. koreana* fr, lf | Kitagawa et al. (1984a, 1987) |
| | *F. ovata* lf | Kitagawa et al. (1988) |
| | *F. viridissima* fr, lf | Kitagawa et al. (1984a, b) |
| β-Hydroxyacteoside **(19)** | *F. koreana* fr | Kitagawa et al. (1987) |
| | *F. viridissima* fr | Kitagawa et al. (1984b) |
| Forsythoside F (arenarioside) **(20)** | *F. × intermedia* rt | Andary et al. (1992) |
| | *F. viridissima* st | Endo and Takahashi (1990) |
| Forsythoside B **(21)** | *F. viridissima* st[b] | Endo et al. (1982) |
| | *F.* spp. lf | Andary et al. (1992) |
| Forsythoside G **(22)** | *F. viridissima* st | Endo and Takahashi (1990) |
| Rengyol **(23)** | *F. suspensa* fr | Endo and Hikino (1984) |
| Isorengyol **(25)** | *F. suspensa* fr | Endo et al. (1987a) |
| Rengyoxide **(26)** | *F. suspensa* fr | Endo and Hikino (1984) |

**Table 1.** (*Contd.*)

| Compound | Species | Reference |
|---|---|---|
| Rengyolone **(27)** | *F. suspensa* fr | Endo and Hikino (1984) |
| Suspenol **(28)** | *F. suspensa* fr | Endo et al. (1987b) |
| Rengyoside A **(24)** | *F. suspensa* fr | Seya et al. (1989) |
| Rengyoside B **(29)** | *F. suspensa* fr | Seya et al. (1989) |
| Rengyoside C **(30)** | *F. suspensa* fr | Seya et al. (1989) |
| Cornoside **(31)** | *F. suspensa* fr | Endo and Hikino (1984) |
| Salidroside **(32)** | *F. suspensa* fr | Seya et al. (1989) |
| *Triterpenoids* | | |
| Betulinic acid | *F. koreana* fr | Chiba et al. (1979) |
| | *F. suspensa* fr | Nishibe et al. (1977a) |
| | *F. viridissima* fr | Chiba et al. (1978) |
| Oleanolic acid | *F. koreana* fr | Chiba et al. (1979) |
| | *F. suspensa* fr | Nishibe et al. (1977a) |
| | *F. viridissima* fr | Chiba et al. (1978) |
| Ursolic acid | *F. koreana* fr | Chang and Kim (1973); Chiba et al. (1979) |
| | *F. suspensa* fr | Chiba et al. (1978) |
| | *F. viridissima* fr | Chiba et al. (1978) |
| β-Amyrin 3-acetate | *F. suspensa* fr | He and Wang (1983) |
| *Monoterpenoids* | | |
| Linalool | *F.* spp. lf oil | Marion et al. (1991) |
| β-Terpineol | *F.* spp. lf oil | Marion et al. (1991) |
| Geraniol | *F.* spp. lf oil | Marion et al. (1991) |
| Geranial | *F.* spp. lf oil | Marion et al. (1991) |
| β-Pinene | *F. suspensa* fr | Okugawa et al. (1986) |
| Terpinen-4-ol | *F. suspensa* fr | Okugawa et al. (1986) |
| p-Cymene | *F. suspensa* fr | Okugawa et al. (1986) |
| *Iridoids* | | |
| Forsythide | *F. japonica* fr, lf | Inouye and Nishioka (1973) |
| | *F. ovata* fr, lf | Inouye and Nishioka (1973) |
| | *F. viridissima* fr, lf | Inouye and Nishioka (1973) |
| Forsythide methyl ester | *F. viridissima* lf | Inouye and Nishioka (1973) |
| *Flavonoids* | | |
| Rutin | *F. europaea* lf | Harborne and Green (1980); Kitagawa et al. (1988) |
| | *F. giraldiana* lf | Harborne and Green (1980); Kitagawa et al. (1988) |
| | *F.* × *intermedia* lf | Harborne and Green (1980); Nishibe et al. (1988) |
| | *F. japonica* lf | Harborne and Green (1980); Kitagawa et al. (1988) |
| | *F. koreana* fr, lf | Chiba et al. (1979); Kitagawa et al. (1984a) |
| | *F. ovata* lf | Harborne and Green (1980); Kitagawa et al. (1988) |
| | *F. suspensa* fr, lf | Nishibe et al. (1977a); Harborne and Green (1980); Kitagawa et al. (1984a) |
| | *F. viridissima* fr, lf | Chiba et al. (1978); Harborne and Green (1980); Kitagawa et al. (1984a) |

**Table 1.** (*Contd.*)

| Compound | Species | Reference |
|---|---|---|
| Isoquercetrin | *F. europaea* lf | Harborne and Green (1980) |
| (quercetin 3-*O*-glucoside) | *F. japonica* lf | Harborne and Green (1980) |
| | *F. suspensa* lf | Harborne and Green (1980) |
| | *F. viridissima* lf | Matsuo et al. (1972); Harborne and Green (1980) |
| Astragalin | *F. japonica* lf | Harborne and Green (1980) |
| (kaempferol 3-*O*-glucoside) | *F. suspensa* lf | Harborne and Green (1980) |
| | *F. viridissima* lf | Matsuo et al. (1972); Harborne and Green (1980) |
| *Others* | | |
| 3-Ethyl-7-hydroxyphthalide | *F. japonica* st oil | Kameoka et al. (1975) |

bk = bark; fl = flower; fr = fruit; lf = leaf; st = stem
[a] Species not assigned; lignan profile indicates it must be *F.* × *intermedia.*
[b] Species originally assigned as *F. koreana* (Endo and Takahashi 1990).

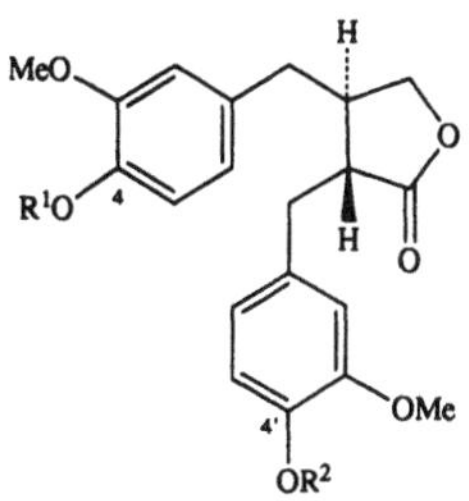


| | R¹ | R² |
|---|---|---|
| Matairesinol (1) | H | H |
| Matairesinol 4'-O-glucoside (2) | H | Glc |
| Arctigenin (3) | Me | H |
| Arctiin (4) | Me | Glc |

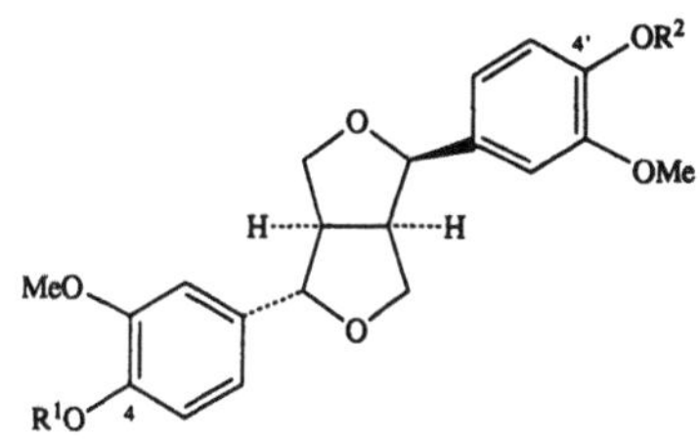


| | R¹ | R² |
|---|---|---|
| Epipinoresinol (5) | H | H |
| Epipinoresinol 4'-O-glucoside (6) | H | Glc |
| Phillygenin (7) | H | Me |
| Phillyrin (8) | Glc | Me |

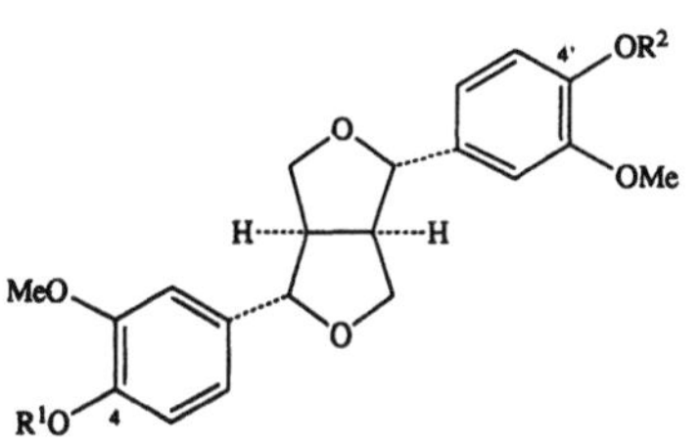


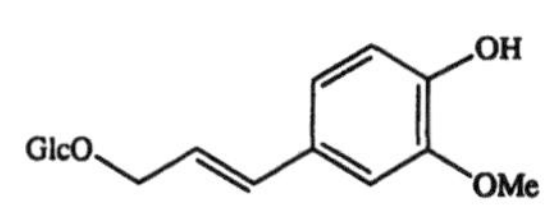


Coniferin (13)

| | R¹ | R² |
|---|---|---|
| Pinoresinol (9) | H | H |
| Pinoresinol O-glucoside (10) | H | Glc |
| Pinoresinol monomethyl ether (11) | H | Me |
| Pinoresinol monomethyl ether O-glucoside (12) | Glc | Me |

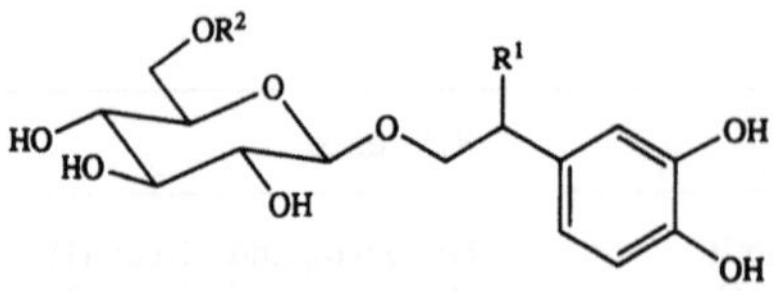

| | $R^1$ | $R^2$ |
|---|---|---|
| Forsythoside D (14) | OH | Rha |
| Forsythoside E (15) | H | Rha |

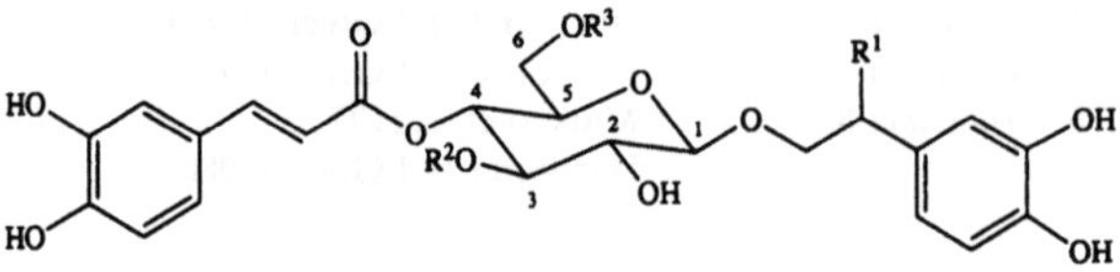

| | $R^1$ | $R^2$ | $R^3$ |
|---|---|---|---|
| Forsythiaside (= forsythoside A) (16) | H | H | Rha |
| Suspensaside (= forsythoside C) (17) | OH | H | Rha |
| Acteoside (= verbascoside) (18) | H | Rha | H |
| ß-Hydroxyacteoside (19) | OH | Rha | H |
| Forsythoside F (= arenarioside) (20) | H | Rha | Xyl |
| orsythoside B (21) | H | Rha | Api |
| Forsythoside G (22) | H | Rha | 2-O-MethylApi |

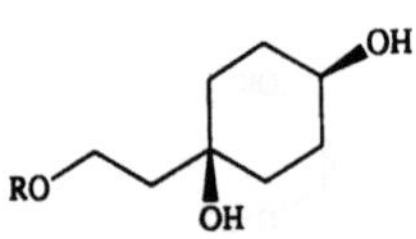

Rengyol (23) R=H
Rengyoside A (24) R=Glc

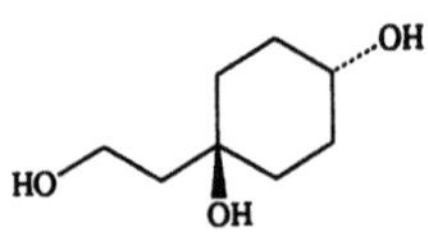

Isorengyol (25)

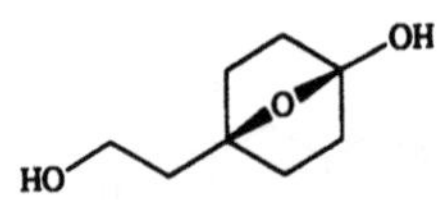

Rengyoxide (26)

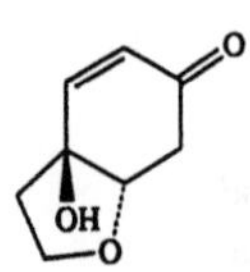

Rengyolone (27)

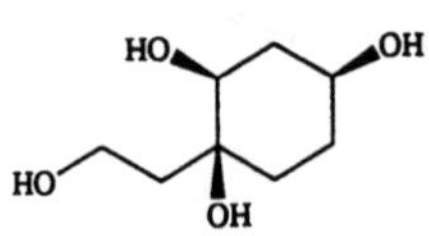

Suspenol (28)

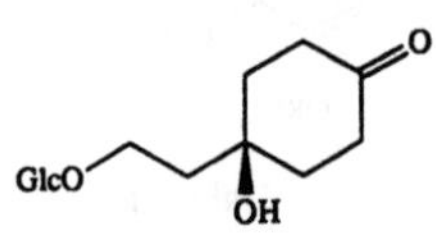

Rengyoside B (29)

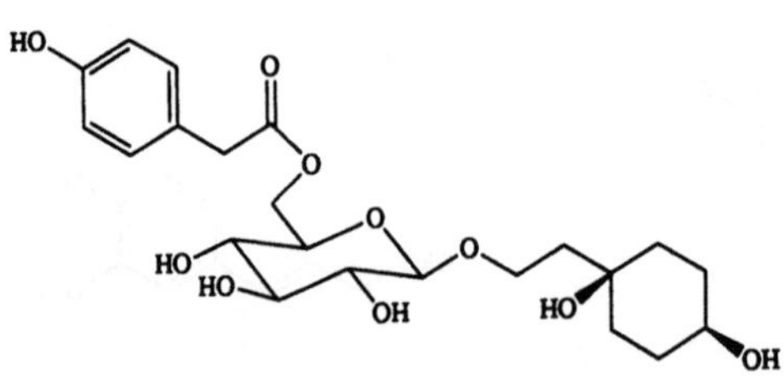

Rengyoside C (30)

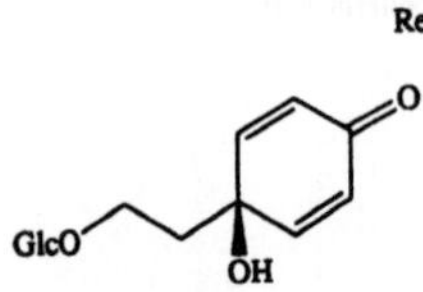

Cornoside (31)

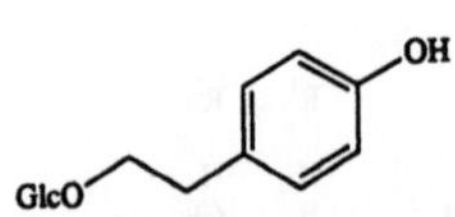

Salidroside (32)

offers substantial support for such reductive transformations (Endo et al. 1989; Seya et al. 1989). However, this group of compounds features the *para* hydroxyl of salidroside, whereas the forsythiaside group are all derivatives of 3,4-dihydroxyphenylethanol.

Other secondary metabolites characterized include triterpenoids, monoterpenoids, iridoids and flavonoids (Table 1). The flavonol glycoside rutin has been found in all *Forsythia* species examined.

### 1.3 Biological Activities and Medicinal Uses

Dried *Forsythia* fruits have long been used for the treatment of allergic and inflammatory diseases, as a diuretic and as an antidote in Japan, China, and Korea. The crude Oriental drug Forsythiae Fructus is listed in the Japanese Pharmacopoeia and is derived from *F. suspensa*, *F. viridissima*, or *F. koreana*, whilst fruits of *F. suspensa* feature in the Chinese Pharmacopoeia (Hikino 1989).

Aqueous extracts of *Forsythia* fruits exhibit antibacterial properties which have been traced mainly to the caffeoyl phenylethanol glycosides forsythiaside (**16**), suspensaside (**17**), acteoside (**18**), and β-hydroxyacteoside (**19**) (Endo et al. 1981, 1982; Endo and Hikino 1982; Nishibe et al. 1982b; Kitagawa et al. 1987).

Suspensaside and forsythiaside have also been shown to possess high inhibitory activity towards cyclic AMP phosphodiesterase, though acteoside and β-hydroxyacteoside were relatively inactive (Nishibe et al. 1982a, b; Kitagawa et al. 1984b). The lignans pinoresinol (**9**), pinoresinol *O*-glucoside (**10**), matairesinol (**1**), and arctigenin (**3**) were also active in this assay (Nikaido et al. 1981). Since many drugs active in the central nervous system are good inhibitors of cyclic AMP phosphodiesterase, the *Forsythia* metabolites active against this enzyme could well contribute to the crude drug's therapeutic effects, e.g., antihypertensive activity (Nishibe 1989).

Forsythiaside, suspensaside, acteoside, and β-hydroxyacteoside were found to affect arachidonate metabolism in leukocytes, selectively inhibiting formation of 5-lipoxygenase products 5-hydroxy-6,8,11,14-eicosatetraenoic acid and leukotriene $B_4$ (Kimura et al. 1987). Leukotrienes are involved in immunoregulation and in a variety of diseases including asthma, inflammation, and allergies, so these caffeoyl phenylethanol glycosides also have potential in the treatment of asthma and inflammatory diseases.

## 2 In Vitro Culture Studies

### 2.1 Previous Tissue Culture Studies on *Forsythia* Species

Prior to our studies described in Sections 2.2–2.4 (Rahman et al. 1990b), there was relatively little information available on secondary metabolites in tissue culture of *Forsythia*. The ready formation of callus on explants taken from young stem and buds of *F. suspensa* had been reported (Bader et al. 1982, 1983, 1984; Bader and Cachita-Cosma 1985), and high regeneration capacity allowing in vitro production of plantlets was noted (Bader and Cachita-Cosma 1985). Induction of plantlets from cotyledons, shoot tips, and hypocotyls of *F. suspensa* was also described by Qin et al. (1985). Meristem culture of *F.* × *intermedia* to eradicate viral diseases had been successful some years earlier (Duron 1977). The only information on the production of secondary metabolites in *Forsythia* tissue cultures was a brief report on the HPLC detection of acteoside (verbascoside) (**18**) in a *F.* × *intermedia* culture (Chapple and Ellis 1984), though no details pertaining to the culture were given. In a Japanese patent (Yamauchi et al. 1987), plant cells derived from *F. koreana* and *F. suspensa* cultures had been used to provide a deodorizing extract effective in the same way as extracts from leaves, presumably a result of secondary metabolite production.

## 2.2 Establishment of Callus and Cell Suspension Cultures of *Forsythia* × *intermedia*

Explants from leaf, stem, and shoot tip taken from young shoots of *F.* × *intermedia* and surface sterilized with sodium hypochlorite readily produced callus growth on media based on MS salts (Murashige and Skoog 1962). Growth hormones were essential. MS salts supplemented with 2,4-D at 1 mg/l (M1)[2] supported callus growth without regeneration, whilst media containing NAA (2 mg/l) and BAP (0.5 mg/l) initiated callus growth but also regeneration of plantlets from all types of explant. In media where BAP was replaced by kinetin (0.2 mg/l), proliferation of callus with regeneration of roots was observed. The best callus response was obtained on M8 medium[2] in which the concentration of MS salts was reduced to one third, where callus was initiated from explants within 14 days of culture. Although this medium contains kinetin, maintenance and subculture (every 4 weeks) of callus on M8 led to a reduction in, and ultimately complete loss of, root generation. M8 was also superior to White's (White 1954) and B5 (Gamborg et al. 1968) media for callus maintenance, while Heller's medium (Heller 1953) did not support callus growth. Callus growth was normally satisfactory both in the dark, or under constant illumination, though callus consistency and lignan production varied according to these conditions (see Table 2).

Cell suspension cultures were initiated from calli, transferring them to equivalent M1 or M8 liquid media and culturing under constant illumination. Cell cultures in M1 liquid medium were dark brown and contained large clumps of cells, whereas M8 medium gave green or white cells of a more dispersed nature. The cells were capable of lignan synthesis, but the amount of lignans produced was usually less than the amount produced in the parent callus (Table 3). Cell suspension cultures were generally used only after a stabilization period

[2] *Culture media* M1: MS salts, 2,4-D (1 mg/l), sucrose (30 g/l); pH 5.8 M8: one-third MS salts, casein hydrolysate (0.5 g/l), NAA (2 mg/l), kinetin (0.2 mg/l), thiamine (9 mg/l), sucrose (30 g/l); pH 5.8.

**Table 2.** Lignan production in callus cultures derived from *Forsythia* x *intermedia* leaf

| Culture medium | Conditions[a] | Yield (mg/g dry cell wt.)[b] Mataresinol (1) | Arctigenin (3) | Epipinoresinol (5) | Phillygenin (7) | Callus consistency |
|---|---|---|---|---|---|---|
| M1 | 3 | – | – | 6.2 | tr | Brown, friable |
| M8 | 1 | 11.0 | tr | – | – | White, friable |
| | 2 | 6.5 | – | – | – | White, friable |
| | 3 | 7.8 | – | – | – | Green or white, friable |
| B5 | 1 | 5.8 | – | – | – | White, friable |
| | 2 | 11.2 | tr | – | tr | White, friable |
| | 3 | 6.3 | – | tr | tr | Green, hard |
| White's | 1 | 1.5 | – | tr | 3.5 | White, hard |
| | 2 | 32.7 | tr | tr | tr | White, hard |
| | 3 | 43 | 7.2 | – | tr | Green, hard |

tr = trace.

[a] 1: 5-week-old callus grown in dark; 2: 12-week-old callus grown in dark; 3: 12-week-old callus grown in light.

[b] Lignan content measured in terms of aglycone after enzymic hydrolysis.

**Table 3.** Comparison of lignan production in callus and cell suspension cultures of *Forsythia* x *intermedia*[a]

| Culture | Medium | Lignans present (mg/g dry cell wt.) |
|---|---|---|
| Callus | M1 | Epipinoresinol 4'-*O*-glucoside (7.3)<br>Phillyrin (1.8)<br>Epipinoresinol (1.3) |
| Cell suspension | | Epipinoresinol 4'-*O*-glucoside (1.4) |
| Callus | M1 | Epipinoresinol 4'-*O*-glucoside (2) |
| Cell suspension | | None detected |
| Callus | M8 | Matairesinol 4'-*O*-glucoside (18) |
| Cell suspension | | Matairesinol 4'-*O*-glucoside (10) |

[a] Cell suspensions analyzed 14 days after initiation from callus tissue.

of three generations (subculture every 3 weeks). There was a gradual decline in production after four subcultures which made it difficult to keep a high yielding cell suspension line growing over long periods. To overcome this, the period between subcultures was extended to 3 months during which high-yielding suspensions could be maintained. *Forsythia* cells did not seem to be adversely affected by long-term maintenance in M8 liquid medium, based on their reasonable viability ( > 50% after 3 months), and healthy appearance (consistency and color).

## 2.3 Production of Lignans in Callus and Cell Suspension Cultures of *Forsythia* × *intermedia*

Analysis of leaf and stem tissues of *F.* × *intermedia* demonstrated that these were a rich source of dibenzylbutyrolactone and furofuran lignans (Rahman et al. 1990c). These were identified as matairesinol (**1**), arctigenin (**3**), epipinoresinol (**5**), and phillygenin (**7**), together with their *O*-glucosides matairesinol 4′-*O*-glucoside (2), arctiin (arctigenin 4′-*O*-glucoside) (**4**), epipinoresinol 4′-*O*-glucoside (**6**), and phillyrin (phillygenin 4-*O*-glucoside) (**8**). The lignan contents varied with season and also with tissue type, so that the furofuran derivatives epipinoresinol/epipinoresinol glucoside predominated in mature leaf, whilst the dibenzylbutyrolactones matairesinol/matairesinol glucoside represented the major lignan material in mature stems. In young shoot tips, however, arctigenin/arctiin were the main lignan constituents isolated from both leaves and stems.

For lignan analyses of callus or cell suspension cultures, the cell tissue was freeze dried, powdered, and extracted by stirring with methanol at 70 °C for 1 h. This extract was then analyzed by HPLC [Spherisorb S5 ODS2; methanol-water (9:11); UV detection 280 nm] to give satisfactory resolution and quantification of all major lignans and glycosides (Rahman et al. 1990c). Alternatively, the crude methanolic extract was evaporated to dryness, then treated with β-glucosidase in sodium acetate buffer (pH 5.0) at 37 °C for 24 h. Lignan aglycones, including those liberated by the enzymic hydrolysis, were extracted with dichloromethane, and HPLC analysis then quantified total lignans in terms of the aglycones (Rahman et al. 1990b). Individual lignans were identified by comparison with material isolated from intact plant tissues, purified via TLC, HPLC, and recrystallization, and authenticated by $^{1}H$ and $^{13}C$ NMR spectroscopy (Rahman et al. 1990c).

In preliminary experiments, callus was initiated from leaf explants and maintained on M1 medium. Analysis of this callus showed that epipinoresinol 4′-*O*-glucoside (**6**) was the predominant lignan produced. The other *Forsythia* lignans were absent, or could be detected in minor quantities only. Callus maintained on other media, namely M8, B5, and White's, yielded matairesinol 4′-*O*-glucoside (**2**) as the major product. Table 2 gives lignan analyses (in terms of total aglycones quantified after β-glucosidase treatment) for calli cultured on the various media, and maintained under different growth conditions. Thus, matairesinol (as the glucoside) was the sole product from calli maintained on M8 medium, except under conditions "1" when a trace amount of arctigenin was also detected. Calli grown on B5 medium produced trace amounts of other lignans as well as matairesinol, while calli cultured on White's medium produced reasonable quantities of arctigenin and phillygenin in addition to matairesinol and traces of epipinoresinol. In contrast, epipinoresinol was produced in significant quantities only when using M1 medium, and it then formed the vast bulk of the lignans, with only traces of phillygenin being detected. Lignan yields were highest from callus maintained on White's medium, especially when cultures were grown in the light. However, the slow growth rate on White's medium, coupled with the observed synthesis of predominantly a single lignan in other media, made M8 and M1 the media of choice for further studies.

Calli initiated and grown in different media produced lignan patterns quite different from those of the original explants (Rahman et al. 1990c). Accordingly, this change was monitored during subculture. Leaf and stem explants from young *Forsythia* shoots were divided into two parts, one for analysis and the other for culture. Callus was initiated and maintained on M8 medium either under constant illumination, or in the dark. Calli were subcultured onto fresh medium every 28 days and maintained under the same conditions, with samples removed for analysis at every subculture. Lignans were analyzed in terms of aglycones quantified after β-glucosidase hydrolysis of the crude extract.

The results (Fig. 2) showed pronounced but surprisingly similar changes in lignan patterns in each case. Leaf explant material contained high levels of arctigenin, medium amounts of epipinoresinol and phillygenin, but very low levels of matairesinol. On culture, there was a complete loss of epipinoresinol, a decrease to insignificant levels for phillygenin, and arctigenin content also fell to minimum values. In contrast, matairesinol production increased rapidly over three to four subcultures to a maximum level, and subsequently declined. Explants from young stem contained much more matairesinol than leaf explants, but arctigenin again predominated, and epipinoresinol was the minor lignan constituent. Calli from stem showed essentially the same pattern of change in lignans as did the leaf calli. Thus, there was a complete loss of epipinoresinol, a decrease to insignificant levels for phillygenin and arctigenin, but a large increase in matairesinol production, declining after three to four subcultures. Light-grown and dark-grown cultures were not markedly different, though stem cultures grown in the dark contained no detectable arctigenin, phillygenin, or epipinoresinol.

Although matairesinol was present in different amounts in the dried leaf and stem material (Rahman et al. 1990c), and in low concentration in the young growth of *Forsythia* plants used for explants, production of this lignan in culture was independent of the origin of the explant and reached even higher levels (ca. 18 mg/g dry wt.) than the maximum recorded for dry stem material (ca. 15 mg/g dry wt.) (Rahman et al. 1990c). The lignan profile in the callus appeared to be dependent only on the culture medium employed. Despite the subsequent decrease in matairesinol levels, none of the other lignans increased again, and matairesinol was presumably catabolized or directed towards synthesis of lignin.

Analysis of cells harvested from cell suspension cultures maintained on M1 or M8 media showed similar results (Table 3). Cultures grown in M1 liquid medium gave epipinoresinol 4′-*O*-glucoside in yields of up to 3 mg/g dry cell wt., depending on the cell line used. Cells cultured in M8 liquid medium were better lignan producers, giving mainly matairesinol 4′-*O*-glucoside, with typical yields of 10 mg/g dry cell wt. In each case, the amount of lignans produced was considerably less than that found in the parent callus (Table 3).

The lignan production rate in a matairesinol 4′-*O*-glucoside-producing cell line was related to cell growth rate, as shown in Fig. 3. Analysis for lignans was carried out on both cells and culture filtrates, but under normal conditions, lignans were not excreted into the medium until the culture was in the late stationary phase, when trace amounts could be detected. The cell suspensions gave a sigmoid curve with a lag phase of 5 to 6 days, and an exponential phase of 9 to 15 days (Fig. 3). The matairesinol content, measured as total aglycone after

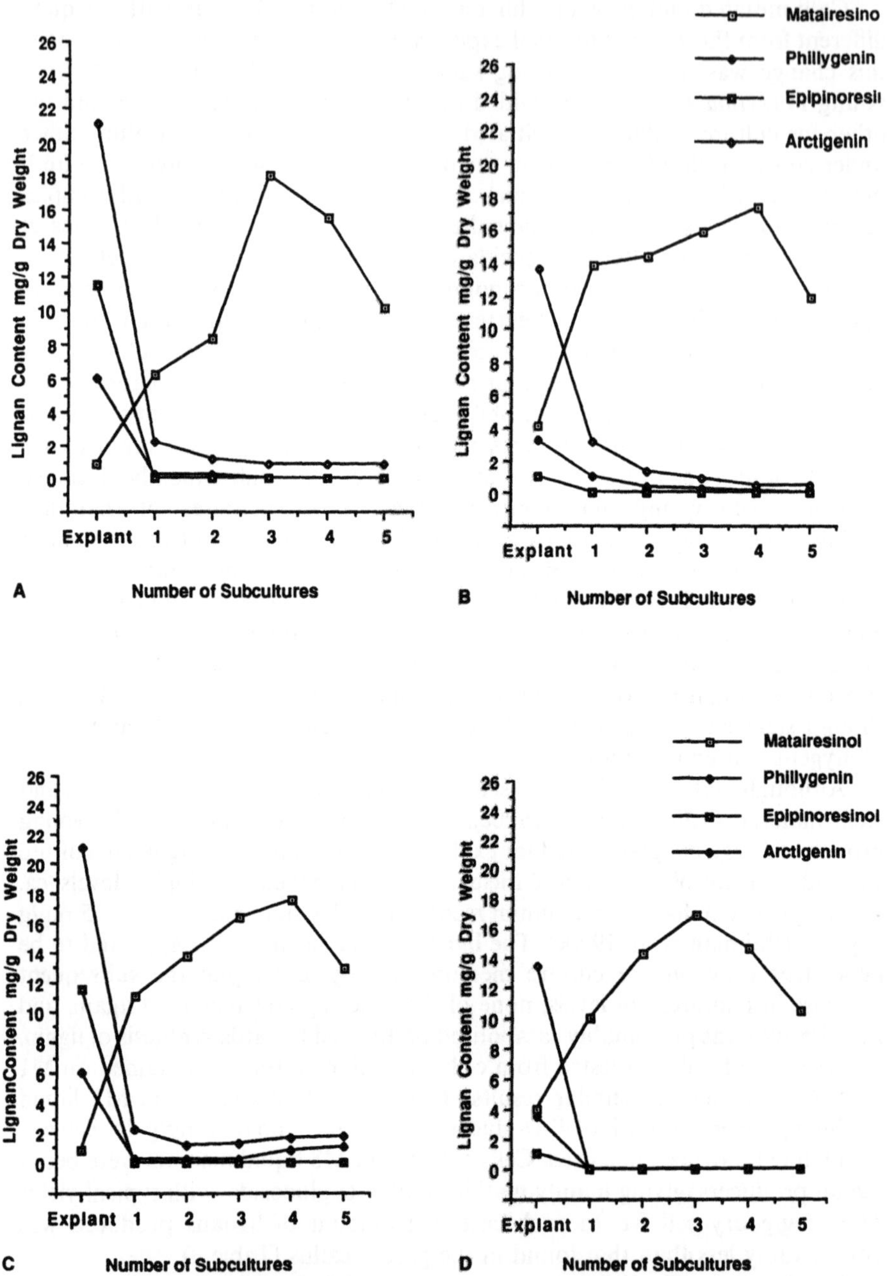


**Fig. 2A–D.** Effects of subculturing on lignan production in *Forsythia* × *intermedia* callus cultures (M8 medium). **A** Light-grown callus derived from leaf explants. **B** Light-grown callus derived from stem explants. **C** Dark-grown callus derived from leaf explants. **D** Dark-grown callus derived from stem explants. (Rahman et al. 1990b)

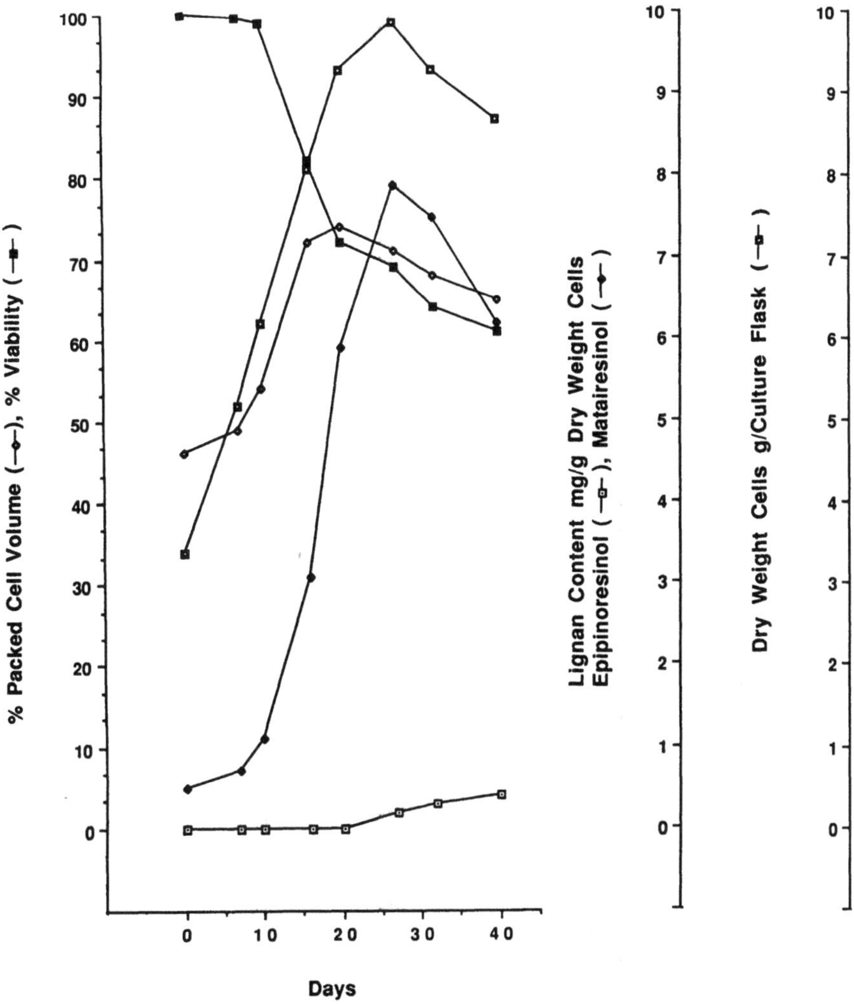


**Fig. 3.** Growth pattern and lignan production in *Forsythia* × *intermedia* cell suspension culture (M8 medium). (Rahman et al. 1990b)

hydrolysis, increased in parallel with the increase in biomass during the exponential phase. The percentage of viable cells decreased with time. Production of matairesinol continued to increase, reaching a maximum at the stationary phase, day 27, after which levels started to decrease, though minor quantities of epipinoresinol were then observed. No released matairesinol could be detected in the medium except for trace amounts at day 40.

Regular subculturing (every 3 weeks) also affected lignan production. The first generation of the cell suspension culture showed low yields of matairesinol (ca. 4 mg/g dry wt. cells), although in subsequent generations levels increased to a maximum of about 8 mg/g. There was then a gradual decline in production after four subcultures. Levels never reached those of the parent callus (ca.13 mg/g). Accordingly, cell suspension cultures were generally used only after a stabilization period of culture (three generations).

In these experiments, only lignan production was studied. No particular effort was made to analyze the tissues for other groups of secondary metabolites, e.g., phenylethanol glycoside derivatives. A brief report on the identification of acteoside (verbascoside) (**18**) in cell cultures of *F.* × *intermedia* has been published (Chapple and Ellis 1984).

## 2.4 Applications and Exploitation of *Forsythia* Tissue Cultures

The ability of *F.* × *intermedia* cell suspensions to synthesise high levels of matairesinol 4′-*O*-glucoside has been exploited to produce $^{14}C$-labeled matairesinol by feeding [U-$^{14}C$] phenylalanine to cell suspension cultures (Rahman et al. 1990a). The labeled precursor was added to a 16-day-old culture, corresponding to the beginning of the stationary phase when cell growth rate was declining and matairesinol production was increasing, and allowed to metabolize for 3 days. Matairesinol was isolated from the extract after β-glucosidase treatment and was obtained in good yield (9.9 mg from 65 ml culture) with high percentage incorporation (4.7%) of label. This material has been used in feeding experiments to investigate the biosynthesis of other lignans in *F.* × *intermedia* (Rahman et al. 1990a) and of tumor-inhibitory aryltetralin lactone lignans such as podophyllotoxin in *Podophyllum hexandrum*, *P. peltatum*, and *Diphylleia cymosa* (Broomhead et al. 1991).

Biotransformation studies using a low-producing cell suspension culture line of *F.* × *intermedia* in M8 medium have shown these were capable of rapidly oxidizing the aryltetralin lactone lignan podophyllotoxin (**34**) into podophyllotoxone (**35**) (Fig. 4) (Broomhead and Dewick 1991). Smaller amounts of the stereoisomer picropodophyllone (**36**) were also produced, probably due to pH changes in the culture during incubation. Transformation was optimal using cells in exponential phase and a 24-h feeding period, and the ketone products were obtained in up to 40% overall yield. Similar products were obtained when acetylpodophyllotoxin or podophyllotoxin 4-*O*-glucoside were fed to the cultures, demonstrating the presence of additional esterase and β-glucosidase activities. Desoxypodophyllotoxin (**33**) was also converted into podophyllotoxone, presumably via benzylic hydroxylation prior to the oxidation. Traces of the proposed hydroxy intermediate were detectable. Lignans containing phenolic groups when added to the cultures rapidly disappeared without yielding any biotransformation products, and were presumably catabolized (Rahman et al. 1990b; Broomhead and Dewick 1991). The cultures did not glucosylate podophyllotoxin or any other substrate.

The conversion of podophyllotoxin into podophyllotoxone by the cultures is fairly superficial and has no particular application, since the same transformation can be achieved chemically in much higher yields. However, the conversion of desoxypodophyllotoxin into podophyllotoxone, presumably via benzylic

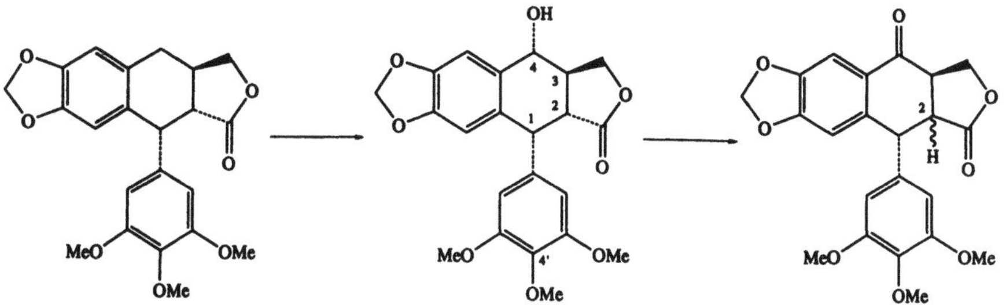


Desoxypodophyllotoxin (33) Podophyllotoxin (34) Podophyllotoxone (35) 2β-H Picropodophyllone (36) 2α-OH

**Fig. 4.** Biotransformation of *Podophyllum* lignans in *Forsythia* × *intermedia* cell suspension cultures

hydroxylation and then oxidation, is of greater significance. Although there appears to be no comparable hydroxylation/oxidation sequence involved in the biosynthesis of the *Forsythia* lignans (Rahman et al. 1990a), this conversion mimics the normal biosynthetic sequence in *Podophyllum*, where desoxypodophyllotoxin → podophyllotoxin → podophyllotoxone, and a similar sequence exists for the 4′-demethyl compounds (Jackson and Dewick 1984; Dewick 1989). We have yet to ascertain whether the proposed hydroxy intermediate in the *F. intermedia* transformation is podophyllotoxin or the 4-epimer epipodophyllotoxin, but in either case, the benzylic hydroxylation of desoxypodophyllotoxin may have useful applications. Desoxypodophyllotoxin is much more frequently encountered in nature than podophyllotoxin, and therefore could be employed as an alternative raw material for synthesis of the topoisomerase-inhibiting anticancer drugs etoposide and teniposide if a high yielding benzylic hydroxylation reaction were available. The observed biotransformation suggests that *F.* × *intermedia* contains a hydroxylase enzyme effective on desoxypodophyllotoxin.

Conventional vegetative propagation of *Forsythia* cultivars is relatively easy and reliable, but micropropagation undoubtedly offers considerable commercial potential. We have not studied regeneration of plantlets from callus cultures in any detail, but frequently noted shoot and root growth during the early media screens. Optimization of the conditions for regeneration should prove straightforward and allow rapid micropropagation (see also Sect. 2.1).

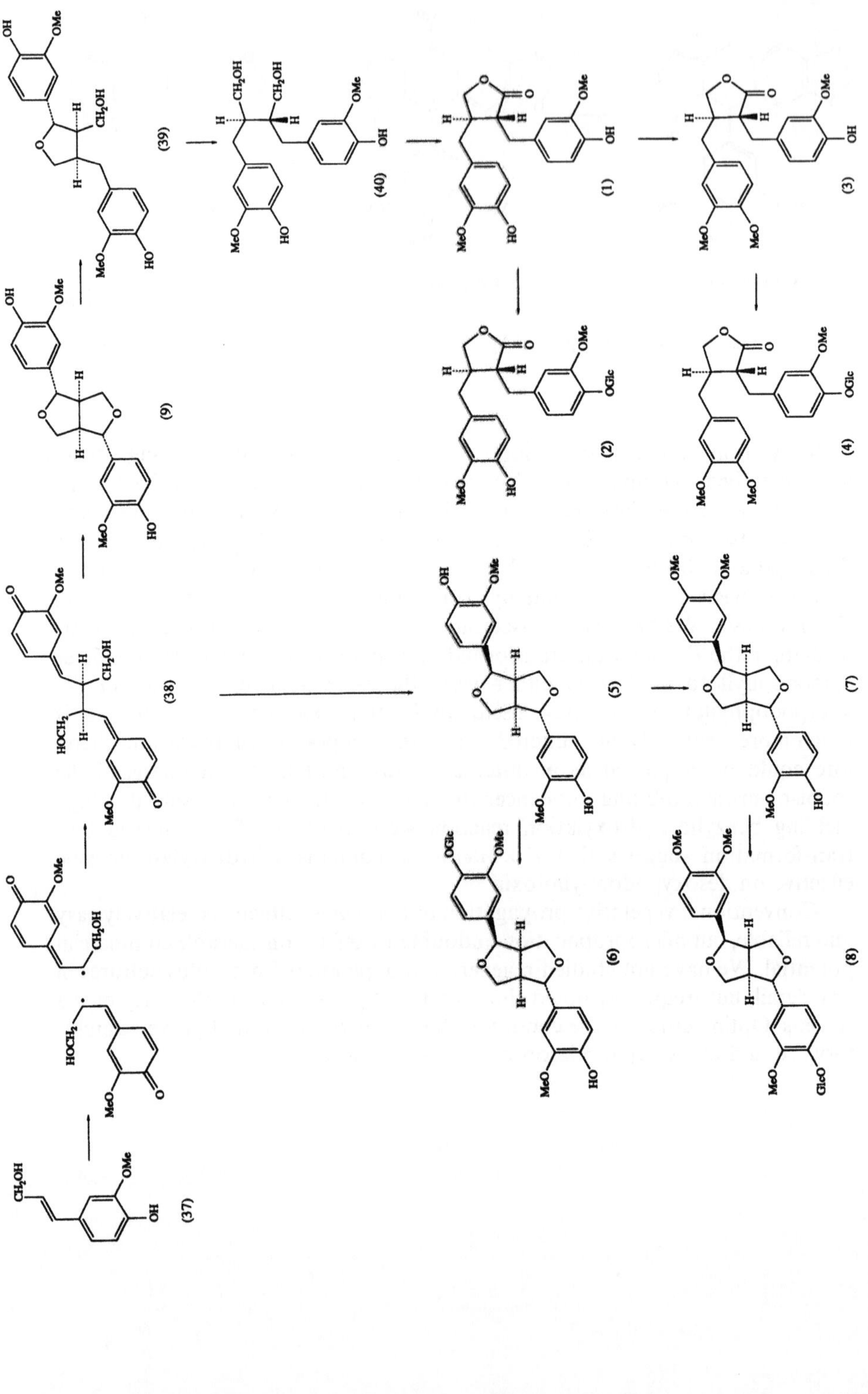

**Fig. 5.** Proposed biosynthetic pathway to lignans in *Forsythia* × *intermedia*

## 3 Conclusions and Prospects

Based on a structural analysis of the lignans found in *Forsythia* × *intermedia* and subsequent metabolic studies (Rahman et al. 1990a; Umezawa et al. 1991; Davin et al. 1992; Katayama et al. 1992, 1993), a biosynthetic sequence (Fig. 5) may be proposed. The lignans are envisaged as arising from a common quinone methide intermediate (**38**), derived by phenolic oxidative coupling of two coniferyl alcohol (**37**) units. The furofuran lignans epipinoresinol (**5**) or pinoresinol (**9**) are then produced from (**38**) by appropriate stereospecific nucleophilic additions. Pinoresinol does not accumulate to detectable levels in *F.* × *intermedia*, but appears to be used for the biosynthesis of dibenzylbutyrolactone lignans via the intermediates lariciresinol (**39**) and secoisolariciresinol (**40**), formed by successive reduction steps. Secoisolariciresinol is then transformed into matairesinol (**1**) by oxidation and lactone ring formation. In the latter stages of the pathways, epipinoresinol and matairesinol are methylated and/or glucosylated in an order yet to be ascertained fully.

The overall effect of culturing *F.* × *intermedia* cells appears to be channeling of biosynthesis to the production of epipinoresinol or matairesinol glucosides as the predominant lignans. This effect was noted in both callus and suspension cultures, and remarkably, was dependent on the culture medium employed. Thus, in the postulated pathway, one of the routes from the quinonemethide (**38**) is blocked, or highly inhibited in each case, resulting in formation of either dibenzylbutyrolactones or *epi* furofurans as the predominant lignans. The biosynthetic pathways to phillygenin or arctigenin, involving methylation reactions, are also blocked. However, glucosylation is not impaired.

These observations point to considerable scope for controlling secondary metabolite production in *Forsythia* cell cultures, with the prospect that single specific compounds may be obtained according to the conditions employed.

*Acknowlegments.* The contribution made by coworkers and former research students Drs. David Jackson, John Lucas, Maiada Rahman and Jane Feely (née Broomhead) is gratefully acknowledged.

## References

Andary C, Tahrouch S, Marion C, Wylde R, Heitz A (1992) Caffeic glycoside esters from *Jasminum nudiflorum* and some related species. Phytochemistry 31:885–886

Bader SM, Cachita-Cosma D (1985) In vitro growth of explants taken from buds of woody plants. Stud Cercet Biol Ser Biol Veg 37:34–40 (Biol Abstr 1986 81:23479)

Bader SM, Cachita-Cosma D, Osvath T, Deliu C (1982) Modification of some morphophysiological variables in the cauline callus of *Forsythia suspensa* grown in vitro in a varied hormonal balance medium. Rev Roum Biol Ser Biol Veg 27: 23–28 (Chem Abstr 1982 97:212867)

Bader SM, Cachita-Cosma D, Osvath T, Deliu C (1983) Test of the capacity for vegetative multiplication using techniques of micropropagation in vitro for some woody ornamentals. Stud Cercet Biol Ser Biol Veg 35:126–130 (Biol Abstr 1984 78:35447)

Bader SM, Cachita-Cosma D, Craciun C (1984) Cellular ultrastructure in calluses of *Forsythia* suspensa obtained from callus explants. Stud Cercet Biol Ser Biol Veg 36:53–56 (Biol Abstr 1986 81:13781)

Bean WJ (1973) Trees and shrubs hardy in the British Isles, vol 2. 8th edn. John Murray, London pp 197–205
Broomhead AJ, Dewick PM (1991) Biotransformation of *Podophyllum* lignans in cell suspension cultures of *Forsythia intermedia.* Phytochemistry 30:1511–1517
Broomhead AJ, Rahman MMA, Dewick PM, Jackson DE, Lucas JA (1991) Matairesinol as precursor of *Podophyllum* lignans. Phytochemistry 30:1489–1492
Chang SH, Kim JS (1973) Components of fruits of *Forsythia koreana* Nakai. III Occurrence of ursolic acid in the fruits of *Forsythia koreana.* Daehan Hwahak Hwoejee 17:444–449 (Chem Abstr 1974 80:80083)
Chapple CCS, Ellis BE (1984) Preparative and analytical liquid chromatography of complex caffeoyl esters. J Chromatogr 285:171–177
Chiba M, Hisada S, Nishibe S (1978) Studies on the Chinese crude drug Forsythiae Fructus. III On the constituents of fruits of *Forsythia viridissima* and *F. suspensa.* Shoyakugaku Zasshi 32:194–197 (Chem Abstr 1979 90:109843)
Chiba M, Tsukamoto H, Hisada S, Nishibe S (1979) Studies on the Chinese crude drug Forsythiae Fructus. IV On the constituents of fruits of *Forsythia koreana* and Forsythiae Fructus from Korea on the market. Shoyakugaku Zasshi 33:150–154 (Chem Abstr 1980 92:185787)
Davin LB, Bedgar DL, Katayama T, Lewis NG (1992) On the stereoselective synthesis of (+)-pinoresinol in *Forsythia suspensa* from its achiral precursor, coniferyl alcohol. Phytochemistry 31:3869–3874
Dewick PM (1989) Biosynthesis of lignans. In: Attar-Ur-Rahman (ed) Studies in natural products chemistry, vol 5. Elsevier, Amsterdam. pp 459–503
Duron M (1977) Utilisation de la culture in vitro pour l'amélioration de l'état sanitaire de cultivars de *Forsythia* (Vahl.). C R Acad Sci Paris 284:183–185
Endo K, Hikino H (1982) Validity of oriental medicine. Part 44. Structures of forsythoside C and D, antibacterial principles of *Forsythia suspensa* fruits. Heterocycles 19:2033–2036
Endo K, Hikino H (1984) Structures of rengyol, rengyoxide and rengyolone, new cyclohexylethane derivatives from *Forsythia suspensa* fruits. Can J Chem 62:2011–2014
Endo K, Takahashi K (1990) Constitution of forsythosides F and G, new phenol glycosides of *Forsythia viridissima* stems. Heterocycles 30:291–294
Endo K, Takahashi K, Abe T, Hikino H (1981) Structure of forsythoside A, an antibacterial principle of *Forsythia suspensa* leaves. Heterocycles 16:1311–1314
Endo K, Takahashi K, Abe T, Hikino H (1982) Structure of forsythoside B, an antibacterial principle of *Forsythia koreana* stems. Heterocycles 19:261–264
Endo K, Seya K, Hikino H (1987a) Stereostructure of rengyol and isorengyol, phenylethanoids of *Forsythia suspensa.* Tetrahedron 43:2681–2688
Endo K, Seya K, Hikino H (1987b) Structure and enantioselective synthesis of suspenol, a new polyol of *Forsythia suspensa.* Tennen Yuki Kagobutsu Toronkai Koen Yoshihu 29:660–667 (Chem Abstr 1988 109:110788)
Endo K, Seya K, Hikino H (1989) Biogenesis-like transformation of salidroside to rengyol and its related cyclohexylethanoids of *Forsythia suspensa.* Tetrahedron 45:3673–3682
Gamborg OL, Miller RA, Ojima K (1968) Nutrient requirements of suspension cultures of soybean root cells. Exp Cell Res 50:151–158
Harborne JB, Green PS (1980) A chemotaxonomic survey of flavonoids in leaves of the Oleaceae. Bot J Linn Soc 81:155–167
He M, Wang J (1983) Studies on chemical constituents in the seeds of *Forsythia suspensa.* I Isolation and identification of lipid soluble constituents. Zhongyao Tongbao 8:34 (Chem Abstr 1983 99:67502)
Heller R (1953) Recherches sur la nutrition minerale des tissus vegetaux cultives in vitro. Ann Sci Nat Bot Biol Veg 14:1–223
Hikino H (1989) Oriental drugs. In: Evans WC (ed) Trease and Evans' pharmacognosy, 13th edn. Bailliere Tindall, London, pp 726–738
Inouye H, Nishioka T (1973) Über die Monoterpenglucoside und verwandte Naturstoffe. XX. Über die Struktur des Forsythids, eines neuen Iridoidglucosides aus *Forsythia viridissima.* Chem Pharm Bull 21:497– 502
Jackson DE, Dewick PM (1984) Biosynthesis of *Podophyllum* lignans. II Interconversions of aryltetralin lignans in *Podophyllum hexandrum.* Phytochemistry 43:1037–1042

Kameoka H, Miyazawa M, Haze K (1975) 3-Ethyl-7-hydroxyphthalide from *Forsythia japonica*. Phytochemistry 14:1676–1677

Katayama T, Davin BD, Lewis NG (1992) An extraordinary accumulation of ( − )-pinoresinol in cell-free extracts of *Forsythia intermedia*: evidence for enantiospecific reduction of ( + )-pinoresinol. Phytochemistry 31:3875–3881

Katayama T, Davin LB, Chu A, Lewis NG (1993) Novel benzylic ether reductions in lignan biogenesis in *Forsythia intermedia*. Phytochemistry 33:581–591

Kimura Y, Okuda H, Nishibe S, Arichi S (1987) Effects of caffeoylglycosides on arachidonate metabolism in leukocytes. Planta Med 53:148–153

Kitagawa S, Hisada S, Nishibe S (1984a) Phenolic compounds from *Forsythia* leaves. Phytochemistry 23:1635–1636

Kitagawa S, Tsukamoto H, Hisada S, Nishibe S (1984b) Studies on the Chinese crude drug Forsythiae Fructus. VII A new caffeoyl glycoside from *Forsythia viridissima*. Chem Pharm Bull 32:1209–1213

Kitagawa S, Nishibe S, Baba H (1987) Studies on the Chinese crude drug Forsythiae Fructus. VIII. On the isolation of phenylpropanoïd glycosides from fruits of *Forsythia koreana* and their antibacterial activity. Yakugaku Zasshi 107:274–278 (Chem Abstr 1987 107:46141)

Kitagawa S, Nishibe S, Benecke R, Thieme H (1988) Phenolic compounds from *Forsythia* leaves II. Chem Pharm Bull 36:3667–3670

Marion C, Pelissier Y, Cadic A, Andary C, Bessiere JM (1991) Composition of essential oils from the genus *Forsythia* Vahl (Oleaceae). Plant Med Phytother 25:39–45 (Chem Abstr 1991 115:239289)

Matsuo K, Tokoroyama T, Kubota T (1972) Bitter constituents of *Forsythia viridissima*. Phytochemistry 11:1522–1523

Murashige T, Skoog F (1962) A revised medium for rapid growth and bioassays with tobacco tissue culture. Physiol Plant 15:473–497

Nikaido T, Ohmoto T, Kinoshita T, Sankawa U, Nishibe S, Hisada S (1981) Inhibition of cyclic AMP phosphodiesterase by lignans. Chem Pharm Bull 29:3586–3592

Nishibe S (1989) Structure elucidation and biological activities of phenylpropanoids, coumarins and lignans from medicinal plants In: Attar-Ur-Rahman (ed) Studies in natural products chemistry, vol 5. Elsevier, Amsterdam, pp 505–548

Nishibe S, Chiba M, Hisada S (1977a) Studies on the Chinese crude drug Forsythiae Fructus. I On the constituents of Forsythiae Fructus on the market. Yakugaku Zasshi 97:1134–1137 (Chem Abstr 1977 87:197360)

Nishibe S, Chiba M, Hisada S (1977b) Studies on the Chinese crude drug Forsythiae Fructus. II. Comparative examination on lignan glucosides of Forsythia fruits of the original plants listed in the Japanese Pharmacopoeia Ed IX. Shoyakugaku Zasshi 31:131–135 (Chem Abstr 1978 88:148945)

Nishibe S, Okabe K, Tsukamoto H, Sakushima A, Hisada S (1982a) The structure of forsythiaside isolated from *Forsythia suspensa*. Chem Pharm Bull 30:1048–1050

Nishibe S, Okabe K, Tsukamoto H, Sakushima A, Hisada S, Baba H, Akisada T (1982b) Studies on the Chinese crude drug Forsythiae Fructus. VI. The structure and antibacterial activity of suspensaside isolated from *Forsythia suspensa*. Chem Pharm Bull 30:4548–4553

Nishibe S, Sakushima A, Kitagawa S, Klimek B, Benecke R, Thieme H (1988) Phenolic compounds from *Forsythia* leaves. III On the comparison of constituents between hybrid and parents. Shoyakugaku Zasshi 42:324–328 (Chem Abstr 1989 111:36661)

Okugawa H, Moriyasu M, Saiki K, Kato A, Matsumoto K, Fujioka A, Hashimoto Y (1986) Evaluation of crude drugs by a combination of enfleurage and chromatography. II On flavour components in fruits of *Forsythia suspensa*. Shoyakugaku Zasshi 40:340–344; (Biol Abstr 1987 83:47303)

Ozawa S, Davin LB, Lewis NG (1993) Formation of (−)-arctigenin in *Forsythia intermedia*. Phytochemistry 32:643–652

Qin J, Wang L, Chen S (1985) Induction of plantlets in vitro of *Forsythia suspensa* Vahl. Plant Physiol Commun 1:36–37

Rahman MMA, Dewick PM, Jackson DE, Lucas JA (1990a) Biosynthesis of lignans in *Forsythia intermedia*. Phytochemistry 29:1841–1846

Rahman MMA, Dewick PM, Jackson DE, Lucas JA (1990b) Production of lignans in *Forsythia intermedia* cell cultures. Phytochemistry 29:1861–1866

Rahman MMA, Dewick PM, Jackson DE, Lucas JA (1990c) Lignans of *Forsythia intermedia*. Phytochemistry 29: 1971–1980
Seya K, Endo K, Hikino H (1989) Structures of rengyosides A, B and C, three glucosides of *Forsythia suspensa* fruits. Phytochemistry 28: 1495–1498
Thieme H, Winkler HJ (1968a) Occurrence of lignan glucosides in *Forsythia* species. I Isolation of phillyrin and arctiin. Pharmazie 23: 402–403
Thieme H, Winkler HJ (1968b) Occurrence of lignan glucosides in the genus *Forsythia*. II Isolation of (–)-matairesinol-4′-β-D-glucopyranoside. Pharmazie 23: 519
Thieme H, Winkler HJ (1969a) Occurrence of lignan glucosides in the genus *Forsythia*. III Isolation of ( + )-epipinoresinol-β-D-glucopyranoside. Pharmazie 24: 117–118
Thieme H, Winkler HJ (1969b) Occurrence of lignan glucosides in the genus *Forsythia*. IV Isolation of coniferin. Pharmazie 24: 292–293
Umezawa T, Davin LB, Lewis NG (1991) Formation of lignans ( – )-secoisolariciresinol and ( – )-matairesinol with *Forsythia intermedia* cell-free extracts. J Biol Chem 266: 10210–10217
White PR (1954) The cultivation of animal and plant cells. Ronald Press, New York
Yamauchi T, Date H, Saihara Y (1987) Japanese patent. Jpn Kokai Tokkyo Koho JP 62,122,666 (87,122,666) (Chem Abstr 1987 107: 120401)

# XVII *Gomphrena* Species (Globe Amaranth): In Vitro Culture and Production of Secondary Metabolites

C. C. J. Vieira[1], H. Mercier[2], E. P. Chu[1], and R. C. L. Figueiredo-Ribeiro[1]

## 1 General Account

### 1.1 Distribution and Importance

The genus *Gomphrena* (family Amaranthaceae) comprises approximately 120 species found in the Americas, Australia, and Indo-Malaysia; the most studied species is *G. globosa* (globe amaranth). Forty-six species occur in Brazil, in savanna vegetation (*cerrado*), napeadic grassland (*campo limpo*), high altitude grassland (*campo rupestre*), and *caatinga*; only a few species are found in forest. Most species are shrubs or herbs, usually with thickened underground systems of woody (xylopodia) or succulent nature (tuberous roots) (Siqueira 1991). There are numerous weedy species and some cultivated as garden ornamentals. The large purple inflorescences of *Gomphrena macrocephala* St.-Hil., in contrast to the hairy green leaves, are easily noted in the cerrado vegetation (Fig. 1A). *G. globosa* L., a tropical annual with white, red, or purple heads (Fig. 2), is grown as an "everlasting" plant (Heywood 1985). This species is also widely used as a test plant for virus propagation and detection, as it grows easily and rapidly and is susceptible to a wide variety of plant viruses, including TMV (tobacco mosaic virus), PVX (potato virus X) and TBRV (tomato black ring virus), (Smith 1974; Matthews 1981; Bos 1983). *G. globosa* is also a natural virus host (Kitajima and Costa 1966; Brunt et al. 1990) and is known to form spontaneous lesions under high light intensities (Francki 1967).

A number of Brazilian *Gomphrena* species are employed in the treatment of bronchial affections, diarrhea, and fever, and as an analgesic, tonic, or carminative. *G. pohlii* Moq., *G. arborescens* L.f., *G. macrocephala*, St.-Hil., *G. vaga* Mart., *G. mollis* Mart., *G. leucochephala* Mart., and *G. globosa* L. are among the most important medicinal plants of this genus (Siqueira 1985, 1987, 1991). Ancient manuscripts attributed widespread curative properties to *Gomphrena* roots, which were claimed to possess tonic properties, to be beneficial to the digestive and respiratory tracts, and to act as a stimulant (Caminhoá 1877). The medicinal properties of *G. macrocephala* have been known since the beginning of the 19th

[1] Laboratory of Plant Physiology and Biochemistry, Institute of Botany, C.P. 4005, 01061-970 São Paulo, SP, Brazil
[2] Laboratory of Physiology and Tissue Culture of Plants, Department of Botany, University of São Paulo, C.P. 11461, 05422-970 São Paulo, SP, Brazil

Biotechnology in Agriculture and Forestry, Vol. 28
Medicinal and Aromatic Plants VII (ed. by Y.P.S. Bajaj)

A B C

**Fig. 1A–C.** *Gomphrena macrocephala* St.-Hil. **A** Flowering plants in the *cerrado*. (Courtesy of Dr. A. Klein, UNESP, Assis, Brazil). **B** Tuberous root **C** Vegetative propagation by tuberous root cuttings. (Courtesy of Dr. T.T. Graziano, 1987, UNESP, Jaboticabal, Brazil)

**Fig. 2.** *Gomphrena globosa* L. flowering plant. (Courtesy of Dr. M.M. Barradas, Instituto Biológico, SP, Brazil)

century, when the species was first named "officinalis" in reference to its medicinal use (von Martius 1826). This perennial herb is commonly known in Brazil as paratudo ("for everything"), paratudo-do-campo, paratudo-de-ervas, and erva-do-Padre-Salerno. Leaf infusions are used for dysmenorrhea, and infusions of the fleshy tuberous root are employed as a tonic, a stimulant, and an antifebrile (Corrêa 1931; Hoehne 1939; Siqueira 1987). In previous papers (Figueiredo-Ribeiro et al. 1986, 1992; Vieira 1991; Mercier et al. 1992; Young et al. 1992), this species has been referred to as *Gomphrena officinalis* Mart. as recommended by Siqueira (1985). Recently, Siqueira (1991), in a revision of the genus *Gomphrena*, has reconsidered this opinion and recognized the specimens studied as *G. macrocephala* St.-Hil.

## 1.2 Methods of Propagation

Species of the genus *Gomphrena* growing naturally in *cerrado* or grasslands may reproduce by seed; natural dispersion is favored by fire, since dehiscence of fruits occurs just after burning. In the case of *G. macrocephala* fire appears to be beneficial; it promotes dispersion of the anemochoric seeds at a time when the soil surface is free of dry branches which might otherwise interfere with the dispersive action of the wind (Coutinho 1977).

*Gomphrena macrocephala* has a perennial underground tuberous structure (Fig. 1B) and displays a typical pattern of seasonal phenology, in common with other plants of the *cerrado*. The aerial parts of these plants wither away during autumn (March–May) and practically disappear in winter (June–July); new shoots are then formed, re-establishing the annual growth cycle in the following season, spring (Vieira and Figueiredo-Ribeiro 1993). Other perennial species from the *cerrado* and grasslands, including *G. graminea* Moq., *G. arborescens* L.f., and *G. prostrata* Mart., exhibit gemmiparous subterranean roots or xylopodia with a number of dormant buds in the upper region at ground level (Siqueira 1991; Figueiredo 1993). According to Coutinho (1990), the xylopodia are organs for mineral nutrient storage rather than water storage structures. On the basis of experimental data, Handro (1966) had considered *G. prostrata* as a calcium accumulator. Large amounts of zinc salt have been extracted from the tuberous root of *G. macrocephala* (Vieira 1991); this would seem to support the mineral storage theory. Fructose is the main nonstructural carbohydrate component in the tuberous root of *G. macrocephala* and represents approximately 50% of the dry weight (Figueiredo-Ribeiro et al. 1986; Vieira and Figueiredo-Ribeiro 1993). These underground reserve organs may also contribute to vegetative reproduction through their capacity to sprout vigorously after burning (Coutinho 1990) or as root cuttings.

Vegetative propagation of *G. macrocephala* by means of tuberous root cuttings (Fig. 1C) has been carried out as a routine in our laboratory. Even so, a complete study of the plant's propagation is needed, with a particular view to large clonal populations of selected plants. Propagation by tissue culture has already been reported (Mercier et al. 1992).

## 2 Secondary Metabolites

Investigations have been conducted into the chemical nature of betalains in the genus *Gomphrena*, i.e., the red to violet betacyanins and the yellow betaxanthins (Hegnauer 1964; Gibbs 1974). There have been comparatively few studies, however, of their overall chemical composition (Table 1, Fig. 3). Triterpenoids, saponins, ecdysteroids (phytoecdysones), and flavonoids are also well represented in *Gomphrena* species and in other related genera of the Amaranthaceae, including *Pfaffia*, which will be mentioned occasionally in the context of this topic (Brito 1986).

Ecdysterone [I] is the most widely occurring phytoecdysone. Plants possessing these compounds are mostly perennial woody rather than annual and herbaceous (Sláma 1979). The ecdysteroid concentration in plants is generally much higher than that in insects, from which these compounds were first isolated, and may reach levels of 0.1% dry wt. (Grundwald 1980). In some cases, concentrations as low as 0.0001% or as high as 2% have been detected, and the contents undergo considerable changes in accordance with seasonal variations (Sláma 1979). Flowering plants of *G. macrocephala* display high levels of ecdysterone (0.4% dry wt.) in the tuberous root. Lower amounts (0.025% dry wt.) have been isolated from whole plants of *G. celosioides* (Banerjii et al. 1971),

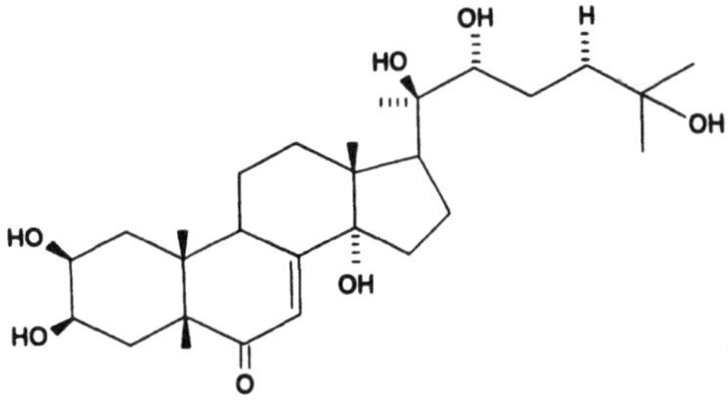


I. Ecdysterone

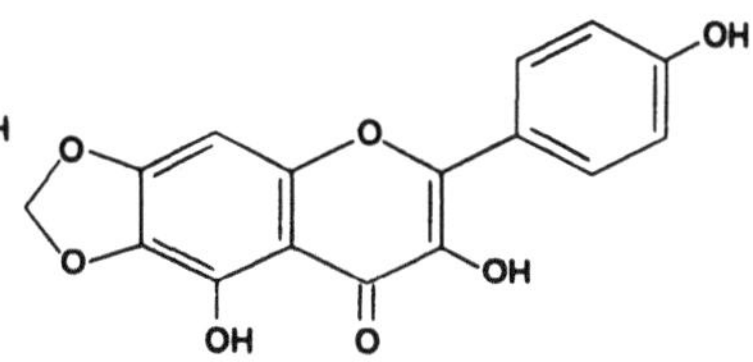


IV. Gomphrenol

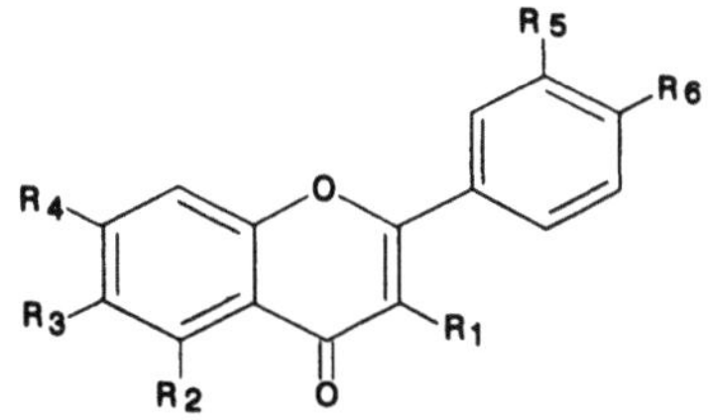


| | $R_1$ | $R_2$ | $R_3$ | $R_4$ | $R_5$ | $R_6$ |
|---|---|---|---|---|---|---|
| 1. | OMe | OMe | $-OCH_2O-$ | $-OCH_2O-$ | H | H |
| 2. | OMe | OH | OMe | OH | H | H |
| 3. | H | OH | OMe | OH | H | H |
| 4. | OMe | OMe | OMe | OMe | H | H |
| 5. | H | OMe | OMe | OH | H | H |
| 6. | OMe | OH | $-OCH_2O-$ | $-OCH_2O-$ | H | H |
| 7. | 0-Gal-0-Rham | OH | H | OH | OMe | OH |

V. Flavones and glycosylated flavonoid

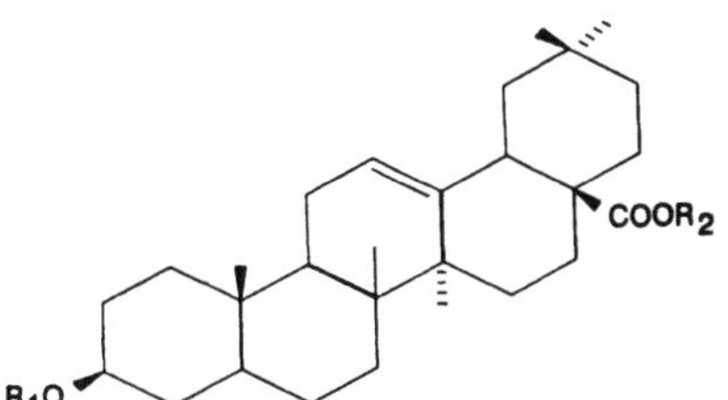


$R_1$ = β-D-glucuronopyranosyl
$R_2$ = β-D-glucopyranosyl

II. Chikusetsusaponin

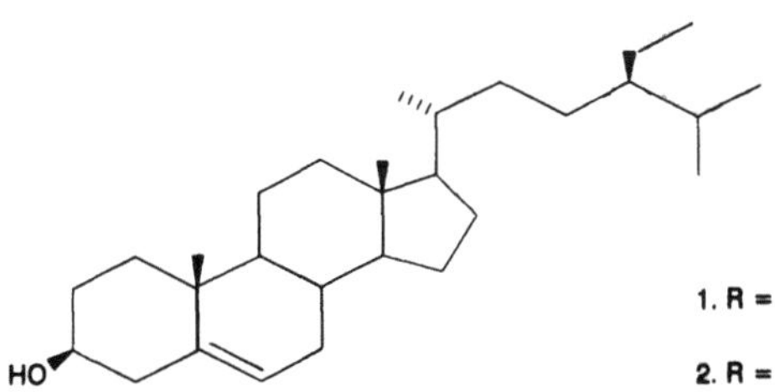


III. Sitosterol

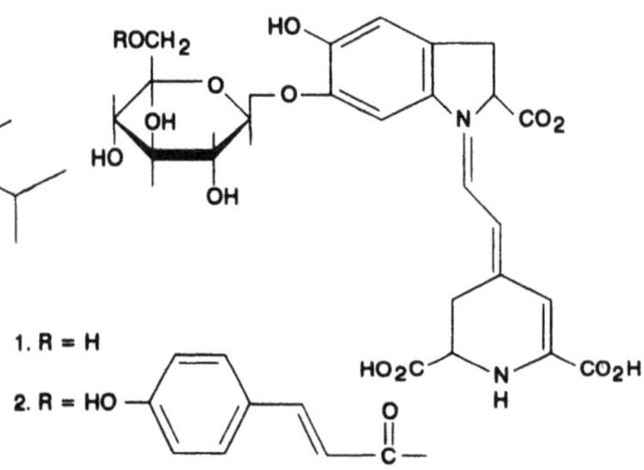


VI. Gomphrenin

**Fig. 3.** Structures of representative secondary compounds of *Gomphrena* species

0.16–0.31% dry wt. from the roots of *Pfaffia iresinoides* (Nishimoto et al. 1987, 1988), and 0.002% dry wt. from *P. pulverulenta* (Shiobara et al. 1992). Phytoecdysones act as potent chemo-sterilizers (Robbins et al. 1968) and as deterrents and antifeeding agents against insects (Grundwald 1980).

There are few reports of the presence of saponins in species of *Gomphrena*, despite the widespread distribution of these compounds in the Amaranthaceae (Hegnauer 1964). Botha and Gerritsma–Vander Vijver (1986) detected saponins in the roots of *G. celosioides*. A triterpenoidal saponin (chikusetsusaponin IVa [II]), composed of oleanolic acid, glucose, and glucuronic acid, has been

**Table 1.** Summary of reports on secondary metabolites in *Gomphrena* species

| Plant species | Isolated compound[a] | Reference |
|---|---|---|
| *G. boliviana* | –Terpenoids flavonoids protoalkaloids | Buschi and Pomilio (1982a) |
| | –Glycoside | Pomilio et al. (1992) |
| *G. celosioides* | –Ecdysterone | Banerjii et al. (1971) |
| | –Saponins | Botha and Gerritsma-Vander Vijver (1986) |
| *G. decumbens* | –Betalains | Veeranjaneyulu and Das (1984) |
| *G. dispersa*, *G. officinalis*, *G. sonorae* | –Sterols (campesterol, stigmasterol, sitosterol) | Patterson et al. (1991) |
| *G. globosa* | –Betacianins (gomphrenin) | Minale et al. (1967); Heuer et al. (1992) |
| | –Amarantin isoamarantin celosian | Minale et al. (1967) |
| | –Flavonol (gomphrenol) | Bouillant et al. (1978) |
| *G. haenkeana*, *G. meyeniana*, *G. perennis* | –Terpenoids flavonoids protoalkaloids | Buschi and Pomilio (1983) |
| *G. macrocephala* (ex. *Gomphrena officinalis*) | –Betalains | Vieira (1991) |
| | –Ecdysterone saponins (chikusetsu saponin) | Young et al. (1992) |
| *G. martiana* | –Glycosides sterols triterpenoids ecdysteroids | Buschi and Pomilio (1982b) |
| | –Flavonoids | Buschi et al. (1979, 1980) |
| | –Betacyanins | Seeligmann and Roncaglia (1973) |

[a] Some structures are shown in Fig. 3.

isolated from the tuberous root of *G. macrocephala* (Young et al. 1992). Considerable seasonal variation has been observed in the crude saponin mixture; plants in the dormant period accumulate higher levels (20% dry wt.) than those in a vegetative or reproductive phase (4% dry wt.). Chikusetsusaponin has also been isolated from the roots of *Panax japonicus*, Araliaceae (Lin et al. 1976), and *Pfaffia iresinoides* (Nishimoto et al. 1987). Saponins, partly because they are readily soluble in water, are particularly effective against snails, which are potential vectors of parasitic diseases. Saponins based on oleanolic acid are the most lethal to snails when doses varying between 1.5 and 32 ppm, according to the nature of the sugar side chain, are absorbed (Harborne 1988). Triterpenoidal saponins also exhibit an exciting potential for control over levels of serum cholesterol for therapeutic purposes (Applebaum and Birk 1979).

Sterols, such as campesterol, stigmasterol, and sitosterol, were detected in ten species (representing four genera) of the subfamily Gomphrenoideae (Patterson et al. 1991). *Gomphrena sonarae*, *G. dispersa*, and *G. officinalis* contained over 60% of sitosterol [III]. According to the authors, sterol composition appears to be a useful marker in the determination of taxonomic affinities in the order Caryophyllales.

Of the phenolic compounds present in *Gomphrena* species, the flavonoids are the best studied. Bouillant et al. (1978) first reported the isolation of a flavonol (gomphrenol [IV]), from the leaves of *G. globosa*; also a number of common phenolics, present in considerable quantities. Buschi and collaborators, in the late 1970s and early 1980s, reported the occurrence of flavonoids in numerous *Gomphrena* species (*G. martiana*, *G. boliviana*, *G. haenkeana*, *G. perennis*, *G. meyeniana* and *G. pulchella*), all native to Argentina (Pomilio et al. 1992). *Gomphrena martiana* and *G. boliviana* produce rare 5,6,7-trisubstituted flavones [V-1 to 6] and a glycosylated flavonoid, isorhamnetin 3-*O*-β-robinobioside [V-7], highly active against an acid-fast bacteria, *Mycobacterium phlei* (Pomilio et al. 1992). These species are used in an "over-the-counter" medicine for treatment of gastrointestinal disorders and of a number of bacterial diseases (Hieronymus 1929 in Pomilio et al. 1992). The medicinal effects of the roots and leaves of these plants against bacterial diseases, and perhaps also their antispasmodic and diuretic activity, are probably due to the presence of such compounds (Pomilio et al. 1992). These flavones, according to Pomilio et al. (1992), may also possess additional physiological and ecological significance, since they are considered to confer probable resistance to plant diseases. Petroleum ether extracts of *G. boliviana* and *G. meyeniana* were effective against *Mycobacterium phlei* (Tomes et al. 1986). However, no effect against Gram-negative bacteria has been detected in petroleum ether and ethanolic extracts of *G. perennis* and *G. haenkeana*, or in petroleum ether extracts of *G. boliviana*, *G. meyeniana*, and *G. pulchella* (Tomes et al. 1986).

The violet inflorescence of *G. globosa* contains, in addition to small amounts of amarantin, isoamarantin, and celosianin, seven other violet pigments (the gomphrenin group), which all belong to the betacyanins (Minale et al. 1967; Heuer et al. 1992). Gomphrenin-I [VI-1] and -II [VI-2] have been shown to be 6-O-β-D-glucopyranosides of betanidin and isobetanidin, respectively, the remaining five gomphrenin being acyl derivatives of the first two. For *G. decumbens*, Veeranjaneyulu and Das (1984) reported the presence of betalains in the leaves of plants growing under natural conditions; varying degrees of purple pigmentation were displayed, particularly in the mature and fully expanded leaves. According to the authors, the purple pigmentation in the leaves may be an adaptation on the part of the plant for survival under conditions of severe water stress and high irradiation; the intensity of such pigmentation was higher in $C_4$ than in $C_3$ plants, and in those plants growing on soil with low moisture levels. The purple pigmentation was not observed in plants growing on moist soils. Plants of *G. macrocephala* growing under natural conditions also show various spots of purple pigmentation in leaves and stems, particularly in the late reproductive and early dormancy periods (Vieira 1991).

## 3 In Vitro Culture Studies

Two species, *G. globosa* (Quintero and Valadez 1985) and *G. macrocephala* (Mercier et al. 1992), have successfully been induced to form callus and, in the case of the latter, to regenerate shoots and roots. In both studies seeds were used to initiate the in vitro cultures.

### 3.1 Establishment of Cultures

Seeds of *G. globosa* were sterilized with a 10% solution of calcium hypochlorite for 10 min, followed by several rinses with sterilized distilled water. The sterilized seeds were germinated on the basic culture medium of Murashige and Skoog (1962) containing 3–7% sucrose and 0.7% agar; hormonal regulators (NAA, BA, and 2,4-D) were added in various concentrations and combinations.

Ethanol (70%) and commercial chlorine bleach (20%) containing a few drops of Tween 20, were used for 1 and 20 min, respectively, for the sterilization of seeds of *G. macrocephala*. Seeds were washed several times in sterilized distilled water, then transferred to gelled MS medium (0.8% agar) with 3% sucrose, 100 mg/l myoinositol and 0.1 mg/l thiamine-HCL. The cultures were maintained at 20 °C, under 12-h light (57 $\mu mol/m^2/s$). Explants were taken from internodal and nodal stem sections and from leaves of 50-day-old aseptically grown seedlings.

### 3.2 Callus Cultures

The hypocotyl region of in vitro – germinated seedlings of *G. globosa* produced a considerable callus tissue. Callus induction was obtained when the cultures were kept under white (1200 lx), blue (80 lx), and red (75 lx) light, at different sucrose concentrations (3, 5, and 7%). The medium supplemented with 3% sucrose, $10^{-6}$ M 2,4-D, and $10^{-8}$ M BA formed friable calli with the best growth rate. Callus differentiation into shoots and roots was observed only on media containing 5–7% sucrose, $10^{-5}$ M NAA, and $10^{-8}$ or $10^{-9}$ M BA.

Cytokinin was essential for callus induction in *G. macrocephala*, and requirements were greater than those for auxin, which did not necessarily seem to be important. Callus was produced in leaf and internodal stem explants cultured on medium with BA alone (1, 5, and 10 mg/l) or combined with NAA (0.1, 0.5, and 1 mg/l) but not in media with NAA alone (Fig. 4A). After 10 to 15 days of culture, the callus formed was friable and exhibited a red-violet color, which was shown to be due to betalains. Callus induction and growth appeared to be better

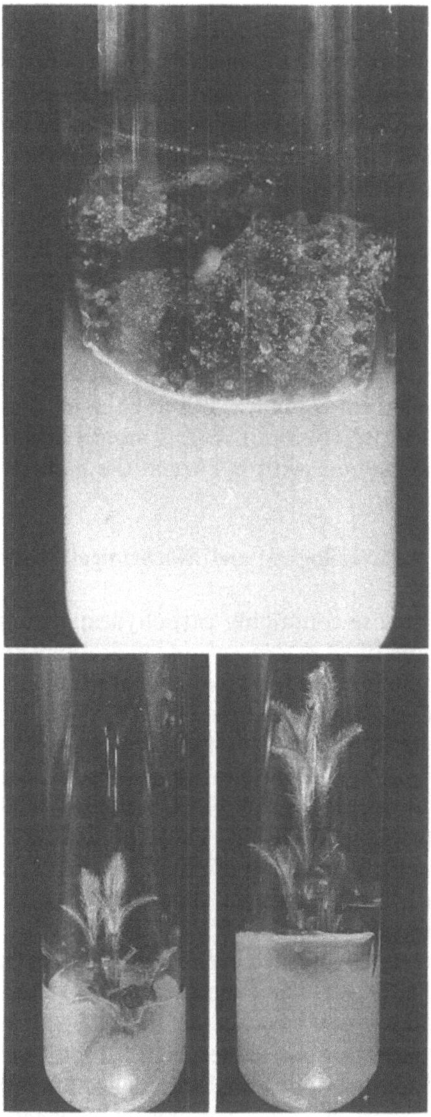

A B C

**Fig. 4A–C.** In vitro culture of *Gomphrena macrocephala* St.-Hil. (Photo M.E. Simões, 1991, Institute of Botany, São Paulo, Brazil). **A** Callus formation from stem explants in MS medium with 0.1 mg/l NAA and 5 mg/l BA after 30 days of culture **B** Adventitious shoot formation from callus on MS medium with 0.1 mg/l NAA and 10 mg/l BA after 60 days of culture **C** Axillary shoots developed in MS medium with 5 mg/l BA after 30 days of culture

under 12-h light than in conditions of darkness, and it was possible to start subcultures every 30 days. The callus, after 50 to 60 days of culture, regenerated green adventitious buds on the surface in the medium with BA alone (5 and 10 mg/l) or in combination with NAA (0.1 mg/l); these developed into shoots (Fig. 4B). Shoot-bud formation was not observed when only NAA or NAA combined with BA at concentrations higher than 0.1 mg/l were used. Approximately 10% of the regenerated shoots showed abnormalities in leaf morphology.

### 3.3 Culture of Axillary Buds and Root Formation

In vitro culture of nodal segments of *G. macrocephala* obtained from 50-day-old aseptically grown seedlings, developed vigorous and actively growing lateral shoots. In every case more than three shoots developed per nodal segment after 1 month of culture in the medium either with BA (0.1, 1, 5 and 10 mg/l) alone or in combination with NAA (0.1, 0.5 and 1 mg/l) (Fig. 4C). The best results of multiplication were achieved by supplementing the medium with low concentrations of NAA (0.1 mg/l) and with BA at concentrations of 5 to 10 mg/l. All regenerated shoots were morphologically normal.

Root formation of *G. macrocephala* took place satisfactorily in a variety of concentrations of NAA or IBA. No response was obtained in the absence of growth regulators or in the presence of IAA. A massive production of long pubescent roots was observed in all shoots, after 10 days of culture in medium with IBA (10 mg/l). After 1 month of culture, root development was observed in the medium with NAA and the highest rate of rooting was 58% at 5 mg/l.

### 3.4 Physiological and Biochemical Studies

Fructose-containing carbohydrates, similar to those accumulated in tuberous roots of *G. macrocephala* (Vieira and Figueirido–Ribeiro 1993), were found in calli obtained from explants of leaf and stem nodes, regardless of the cytokinin/auxin balance and the explant origin. However, the amounts detected were five fold lower than in tuberous roots (Vieira, Braga and Figueiredo–Ribeiro, unpubl.). The amount of fructose-containing carbohydrates in node calli rose with increasing BA/NAA balances, while those in leaf calli tended to show an inverse relationship. These preliminary results suggest that the carbohydrate pool is a reflection of physiological processes that are taking place in the callus rather than a direct effect of hormonal treatment. This work is currently in progress.

Betalains have been identified in callus cultures of *G. macrocephala* (Mercier et al. 1992). The concentration of these compounds varied with the hormonal balance and reached its highest value with NAA 0.1 mg/l and BA 5 mg/l. Concentrations higher than 0.5 mg/l of NAA in combination with BA (0.1 to 10 mg/l) reduced the betalain accumulation. Saponins and ecdysones were also detected in callus cultures of *G. macrocephala*. The large amounts of these compounds in leaves and underground reserve organs and the ability of producing these by tissue culture techniques are the subject of various studies.

In *G. globosa*, the optimum conditions for callus growth was established to verify the effect of light wavelength, hormonal balance, and carbon source concentration upon the production of betalains (Quintero and Valadez 1985). The highest betalain content was obtained, after 25 days of culture under blue light, in MS medium with $10^{-6}$ M 2,4-D and $10^{-8}$ M BA and with 3% sucrose. However, poor pigmentation was observed when NAA ($10^{-5}$ M) and BA ($10^{-8}$ M or $10^{-9}$ M) were combined or when the cultures were kept under red and white light. The activity of phenylalanine ammonia-lyase (PAL) was also detected in callus cultures of *G. globosa* (Castaneda and Quintero 1991). Maximum activity (three fold) was found in callus cultured for 8 days under blue light.

### 3.5 Protocol for Micropropagation of *G. macrocephala*

1. Immersion of seeds in 70% ethanol for 1 min.
2. Immersion in 20% commercial bleach (2% sodium hypochlorite) for 20 min.
3. Three rinses in sterile distilled water.
4. Aseptic germination on MS medium solidified with 0.8% agar.
5. Excision of explants from 50-day-old seedlings.
6. Inoculation of aseptically isolated stem nodes on solid MS medium containing 0.1 mg/l NAA and 5–10 mg/l BA.
7. Incubation in 12-h photoperiod (57 $\mu mol/m^2/s$) at 20°C for 1 month.
8. Rooting of aseptically isolated axillary shoots on solid MS medium containing 10 mg/l IBA.
9. Pot plants in *cerrado* soil and maintain under conditions of high humidity in greenhouse for 2–3 weeks.

## 4 Summary and Conclusions

The botanical description, distribution, biochemical composition, and cultivation practices of *Gomphrena* species have been discussed especially in terms of their potential economic significance. The optimum conditions for in vitro clonal propagation and callus cultures of *Gomphrena macrocephala* were also described. The need to improve research is emphasized in view of both plant survival and the utilization of many products of secondary metabolism, i.e., steroids, saponins, phytoecdysteroids, flavonoids, and betalains. The ecophysiological significance of these compounds in *Gomphrena* species, the economic importance of saponins, steroids, and ecdysteroids, and the wide use of many plants in popular medicine, are subjects of interest. A complete phytochemical screening of the genus *Gomphrena* will certainly be of importance, particularly in view of the gradual eradication of natural populations of many species in Brazil for folk medicinal use.

*Acknowledgments.* Thanks are due to the Conselho Nacional de Desenvolvimento Científico e Tecnológico (CNPq) for financial support and to Dr. Alasdair G. Burman for revision of the English text. Thanks are also due to FAPESP (Fundação de Amparo à Pesquisa do Estado de São Paulo), Proc. 91/3588-8.

## References

Applebaum SW, Birk Y (1979) Saponins. In: Rosenthal GA, Janzen DH (eds) Herbivores – their interaction with secondary plant metabolites. Academic Press, New York pp 539–566

Banerjii A, Chintalwar GJ, Joshi NK, Chadha MS (1971) Isolation of ecdysterone from Indian plants. Phytochemistry 10: 2225–2226

Bos L (1983) Introduction to plant virology. Centre for Agricultural Publishing and Documentation, Wageningen

Botha S Gerritsma-Vander Vijver LM (1986) Pharmacochemical study of *Gomphrena celosioides* (Amaranthaceae). SA Tydskr Naturwetenskap Tegnol 5: 40–45

Bouillant ML, Redolfi P, Cantisani A, Chopin J (1978) Gomphrenol, a new methylenedioxyflavonol from the leaves of *Gomphrena globosa* (Amaranthaceae). Phytochemistry 17: 2138–2140

Brito NRS (1986) Perfil químico de famílias de angiospermas. PhD Thesis, University of São Paulo, São Paulo

Brunt A, Crabtree K, Gibbs A (1990) Virus of tropical plants. Redwood Press, Melksham

Buschi CA, Pomilio AB (1982a) Terpenoids, flavonoids and protoalkaloids from *Gomphrena boliviana*. An Asoc Quím Argent 70: 855–861

Buschi CA, Pomilio AB (1982b) Isorhamnetin 3-O-robinobioside from *Gomphrena martiana*. J Nat Prod 45: 557–559

Buschi CA, Pomilio AB (1983) Comparative studies on *Gomphrena* species. An Asoc Quím Argent 71: 319–325

Buschi CA, Pomilio AB, Gros EG (1979) A new flavone from *Gomphrena martiana*. Phytochemistry 18: 1249–1250

Buschi CA, Pomilio AB, Gros EG (1980) New methylated flavones from *Gomphrena martiana*. Phytochemistry 19: 903–904

Caminhoá JM (1877) Elementos de botânica geral e médica, vol 2. Tipografia Nacional, Rio de Janeiro

Castañeda JL, Quintero A (1991) Regulation of phenylalanine ammonia-lyase (PAL) (E.C. 4.3.1.5.) activity by light in *Gomphrena globosa* tissue culture. Phyton 52: 151–154

Corrêa MP (1931) Dicionário das plantas úteis do Brasil e das exóticas cultivadas, vol 2. Ministério da Agricultura, Rio de Janeiro

Coutinho LM (1977) Aspectos ecológicos do fogo no cerrado. II. As queimadas e a dispersão de sementes em algumas espécies anemocóricas do estrato herbáceo subarbustivo. Bol Bot Univ São Paulo 5: 57–64

Coutinho LM (1990) Fire in the ecology of the Brazilian Cerrado. In: Goldammer JG (ed) Fire in the tropical biota. Ecological studies, vol 84. Springer, Berlin Heidelberg New York, pp 82–105

Figueiredo PS (1993) Efeito de fatores ambientais na dormência do xilopódio de *Gomphrena prostrata* Mart. PhD Thesis, University of Campinas, Campinas

Figueiredo-Ribeiro RCL, Dietrich SMC, Chu EP, Machado de Carvalho MA, Vieira CCJ, Graziano TT (1986) Reserve carbohydrates in underground organs of native brazilian plants. Rev Bras Bot 9: 159–166

Figueiredo-Ribeiro RCL, Dietrich SMC, Carvalho MAM, Vieira CCJ, Isejima EM, Dias-Tagliacozzo GM, Tertuliano MF (1992) As múltiplas utilidades dos frutanos. Ciênc Hoje 14: 16–18

Francki RIB (1967) Effect of high light intensities on spontaneous and virus-induced local lesions in *Gomphrena globosa*. Phytopathology 57: 329–333

Gibbs RD (1974) Chemotaxonomy of flowering plants, vol 2. McGill-Queen's University Press, Montreal

Grundwald C (1980) Steroids. In: Pirson A, Zimmermann MH (eds) Encyclopedia of plant physiology, vol 8, Springer Berlin Heidelberg New York, pp 221–256

Handro W (1966) Escleromorfismo foliar e nutrição mineral em *Gomphrena prostrata* Mart. An Acad Bras Ciênc 38: 225–242

Harborne JB (1988) Introduction to ecological biochemistry, 3rd edn. Academic Press, London

Hegnauer R (1964) Chemotaxonomie der Pflanzen, Band 3. Dicotyledoneae: Acanthaceae-Cyrillaceae. Birkhäuser, Basel

Heuer S, Wray V, Metzger JW, Strack D (1992) Betacyanins from flowers of *Gomphrena globosa*. Phytochemistry 31: 1801–1807

Heywood VH (1985) Flowering plants of the world. Croom Helm, London

Hoehne FC (1939) Plantas e substâncias vegetais tóxicas e medicinais. Departamento de Botânica do Estado de São Paulo, São Paulo

Kitajima EW, Costa AS (1966) Morphology and developmental stages of *Gomphrena* virus. Virology 29:523–539
Lin TD, Kondo N, Shoji J (1976) Studies on the constituents of *Panacis japonici* rhizoma. V. The structures of chikusetsusaponin I, Ia, Ib, IVa and glycoside P1. Chem Pharm Bull 24:253–261
Lorenti AS, Viale AA, Buschi CA, Gonzalez MD, Schteingart CD, Iribarren AM, Pomilio AB (1981) Antimicrobial activity of some Argentine higher plants. Fitoterapia 52:81–85
Matthews REF (1981) Plant virology, 2nd edn Academic Press, New York
Mercier H, Vieira CCJ, Figueiredo-Ribeiro RCL (1992) Tissue culture and plant propagation of *Gomphrena officinalis* a Brazilian medicinal plant. Plant Cell Tissue Organ Cult 28:249–254
Minale L, Piattelli M, De Stefano S (1967) Pigments of Centrospermae – VII. Betacyanins from *Gomphrena globosa* L. Phytochemistry 6:703–709
Murashige T, Skoog F (1962) A revised medium for rapid growth and bioassays with tobacco tissue culture. Physiol Plant 15:473–497
Nakai S, Takagi N, Miichi H, Hayashi S, Nishimoto N, Takemoto T, Kizu H (1984) Pfaffosides, nortriterpenoid saponins, from *Pfaffia paniculata*. Phytochemistry 23:1703–1705
Nishimoto N, Nakai S, Takagi N, Hayashi S, Takemoto T, Odashima S, Kizu H, Wada Y (1984) Pfaffosides and nortriterpenoid saponins from *Pfaffia paniculata*. Phytochemistry 23:139–142
Nishimoto N, Shiobara Y, Fujino M, Inoue SS Takemoto T, Oliveira F, Akizue G, Akizue MK, Tanaka O, Kasai R, Matsuura H (1987) Ecdysteroids from *Pfaffia iresinoides* and reassignment of some 13 C NMR chemical shifts. Phytochemistry 26:2505–2507
Nishimoto N, Shiobara Y, Inoue SS, Fujino M, Takemoto T, Yeoh CL, Oliveira F, Akizue G, Akizue MK, Hashimoto G (1988) Three ecdysteroid glycosides from *Pfaffia iresinoides*. Phytochemistry 27:1665–1668
Patterson GW, Xu S, Salt TA (1991) Sterols of caryophyllales with emphasis on Amaranthaceae. Phytochemistry 30:523–526
Pomilio AB, Buschi CA, Tomes CN, Viale AA (1992) Antimicrobial constituents of *Gomphrena martiana* and *Gomphrena boliviana*. J Ethnopharmacol 36:155–161
Quintero A, Valadez JM (1985) Effects of different growth factors on *Gomphrena globosa* callus tissue. J Nat Prod 48:976–980
Robbins WE, Kaplanis JN, Thompson MJ, Shortino TJ, Cohen CF, Joyner SC (1968) Ecdysones and analogs: effects on development and reproduction of insects. Science 161:1158–1160
Seeligmann P, Roncaglia R (1973) Distribución de betacianinas en especies argentinas de Centrospermales. I. Lilloa 33:299–303
Shiobara Y, Inoue SS, Nishiguchi Y, Kato K, Takemoto T, Nishimoto N, Oliveira F, Akisue G, Akisue MK, Hashimoto G (1992) Pfaffane-type nortriterpenoids from *Pfaffia pulverulenta*. Phytochemistry 31:1737–1740
Siqueira JC (1985) Contribuiçãoao conhecimento taxonômico das espécies do gênero *Gomphrena* L. (Amaranthaceae) que ocorrem nas regiões sudeste e centro-oeste do Brasil. Pesqui Bot 37: 1–111
Siqueira JC (1987) Importância alimentícia e medicinal das Amarantáceas do Brasil. Acta Biol Leopold 9:99–110
Siqueira JC (1991) O gênero *Gomphrena* L. (Amaranthaceae) no Brasil. PhD Thesis, University of Campinas, Campinas
Sláma K (1979) Insect hormones and antihormones in plants. In: Rosenthal A, Janzen DH (eds) Herbivores – their interaction with secondary plant metabolites. Academic Press, New York pp 683–700
Smith KM (1974) Plant viruses, 5th edn. Chapman and Hall, London
Takemoto T, Nishimoto N, Nakai S, Takagi N, Hayashi S, Odashima O, Wada Y (1983) Pfaffic acid, a novel nortriterpene from *Pfaffia paniculata* Kuntze. Tetrahedron Lett 24:1057–1060
Tomes CN, Viale AA, Buschi CA, Rofi RD, Schteingart CD, Inigo RPA, Zallocchi FM, Pomilio AB (1986) Antimicrobial screening of some Argentine higher plants, Part II. Fitoterapia 57: 46–50
Veeranjaneyulu K, Das VSR (1984) Purple pigmentation in leaves of some tropical weed species. Biol Plant 26:215–220
Vieira CCJ (1991) Flutuações sazonais e caracterizacão parcial dos carboidratos solúveis do órgão subterrâneo de *Gomphrena officinalis* Mart. (Amaranthaceae). MS Thesis, University of São Paulo, São Paulo

Vieira CCJ, Figueiredo-Ribeiro RCL (1993) Fructose-containing carbohydrates in the tuberous root of *Gomphrena macrocephala* St.-Hil. (Amaranthaceae) at different phenological phases. Plant Cell Environ 16:919–928

von Martius CFP (1826) Nova genera et species plantarum, vol 2. Monchii, Caroli Wolf

Young MCM, Vieira CCJ, Chu EP, Haraguchi M, Figueiredo-Ribeiro RCL (1992) Ecdysterone and saponins from tuberous roots of *Gomphrena officinalis* Mart. (Amaranthaceae). Rev Latinoam Quím 22/4 & 23/1:41–44

# XVIII *Larix leptolepis* (Japanese Larch): In Vitro Culture and the Production of Secondary Metabolites

K. Nabeta[1]

## 1 Introduction

The natural range of Japanese larch, *Larix leptolepis*, is restricted to highlands of 1000–2800 m in elevation in central Honshu, Japan. Its morphology can be seen from Fig. 1. Japanese larch, which reaches 30 m in height in favorable climate, is a deciduous conifer with a conical crown. This tree is planted on a large scale (40% of the total cultivated forest) in Hokkaido, the northern part of Japan. This conifer was originally planted at the end of the 19th century in northern Europe, and has now become one of the most important conifers (Niemann 1980). In northern America, it occurs in scattered plantation. Young trees of this species grow more rapidly than European larch. The species seems less subjected to canker (*Dasycypha willkommi*) and shoot blight caused by the fungus *Guignardia larcina* than European and American larches, but is susceptible to gray mold (*Botrytis cinera*) and honey fungus (*Armillaria mellea*) (Niemann 1980).

Japanese larch has many valuable uses; it is well fitted for the production of such items as timber, sleepers, props, poles, and piles. The large Japanese larch reserves, however, have remained out of reach of the pulping industry, mainly by reason of the high cost of cultivation, and partly by reason of the exceptionally high content of water-soluble materials such as arabinogalactan (3–8% of oven-dried weight) (Aspinall et al. 1968), and phenolic substances such as tannins (Brewerton 1956), flavonoids (Niemann 1973a, 1974, 1976; Nomura and Muto 1976), lignans (Sakakibara et al. 1987), and derivatives of hydroxybenzoic- and cinnamic acids (Niemann 1971, 1973b, 1976). The mean value, 14% of hot water-soluble extractives (Okuyama et al. 1968), which is at least double that of pine and spruce, would inevitably mean low pulping yield. The brightness, one of the most important concepts of optical pulp properties, of Japanese larch samples remains at a rather lower level than those of the corresponding pine kraft pulps due to the large content of phenolic substances.

This chapter deals with lignan (Nabeta et al. 1991) and lower terpene production (Nabeta et al. 1990a) by calli of *L. leptolepis*. In addition, biosynthesis of lower terpenes from specifically deuterated mevalonates (MVAs) and formation of cyclic sesquiterpenes from farnesyl pyrophosphate (FPP) with a cell-free preparation of the calli (Nabeta et al. 1990b) are discussed.

[1] Department of Bioresource Chemistry, Obihiro University of Agriculture and Veterinary Medicine, Inada-cho, Obihiro 080, Japan

Biotechnology in Agriculture and Forestry, Vol. 28
Medicinal and Aromatic Plants VII (ed. by Y.P.S. Bajaj)

**Fig. 1.** Mature trees of *Larix leptolepis*

## 2 In Vitro Culture Studies

### 2.1 Review of Tissue Culture Studies on *Larix* species

Extensive work has been done on various aspects of the in vitro culture and micropropagation of larch (see Chalupa 1991). Plantlet formation in tissue culture was described for hybrid larch (*Larix gmelinii* × *L. leptolepis* Gord) $F_1$ by Kuromatsu and Takao (1987). Adventitious buds were induced on cotyledon-epicotyl explants on Schenk and Hildebrandt medium (1972), and plantlets regenerated from adventitious buds on woody plant medium (Lloyd and McCown 1980) without hormones. Diploid and haploid embryogenesis was inducted in *L. leptolepis* and *L. decidua* and their reciprocal hybrids by Aderkas et al. (1990). Diploid embryogenic tissue was initiated in immature zygotic embryos and haploid embryogenesis was induced in isolated megagametophytes. There were few morphological differences between the diploid and haploid embryogenic tissues. Recently, entire plants have been regenerated from isolated protoplasts (Klimaszewska 1989), and genetic transformation studies conducted on *Larix* (see Shin et al. 1994). However, no investigation on the production of secondary metabolites in the cultured cells was conducted before our work.

### 2.2 Establishment of Tissue Cultures

The first leaves of the seedlings, (length 10–20 mm), were sterilized by soaking successively in a 0.1% benzalkonium chloride solution for 5 min, in a 70% EtOH for 5 s, and in a 2% sodium hypochlorite solution for 20 min. After being washed with sterile $H_2O$, the leaves were cut into 2–3-mm segments and placed on MS medium (Murashige and Skoog 1962) with 3% (w/v) sucrose, 1% (w/v) agar, various combinations of 1-naphthaleneacetic acid (NAA)-kinetin, indole-3-acetic acid (IAA)-kinetin and 2,4-dichlorophenoxyacetic acid (2,4-D)-kinetin at 1,5, and 10 mg/l concentrations and a pH of 5.8 before autoclaving to induce callus. The cultures were maintained at $25 \pm 1$ °C under continuous light at 5000 lx. After 35 days, vigorous green-callus formation was observed when the cultures were grown on the medium supplemented with 5 mg/l 2,4-D and 5 mg/l kinetin. The efficiency of callus induction on this medium was 65% (the efficiency of callus induction from the mature leaves was much lower: 10%). Callus was also formed on the medium supplemented with any combinations of NAA and kinetin, but the calli grown on the medium with IAA and kinetin became brownish and died. Undifferentiated green callus from the cut end surface was transferred to fresh medium supplemented with 5 mg/l 2,4-D and 5 mg/l kinetin and successively subcultured for over 5 years at intervals of 30–40 days.

Callus subcultured 16 times on fresh medium was then transferred into the liquid MS medium containing 5 mg/l 2,4-D and 5 mg/l kinetin. The cultures were kept on a rotatory shaker at 100 rpm under continuous light at 5000 lx at $25 \pm 1$ °C. The growth index (fr. wt. basis) of the suspension cells in 50 ml MS medium for 21 days was 4.1. However, the suspension cultures often became dark brown and died.

The calli maintained a stable capacity for producing chlorophyll over the long period of successive culturing. The content of chlorophyll a (78 μg/g fr. wt.) and chlorophyll b (29 μg) in the calli at the 29th generation was determined using the method reported by Sunderland (1966), whereas those in the suspension cells at the 2nd generation were 107 μg/g and 26 μg/g, respectively.

## 3 Procedures for Biosynthesis of Lignans and Lower Terpenes

### 3.1 Precursors

Deuterated MVAs (4,4-$^2H_2$-, 5,5-$^2H_2$-, and 6,6,6-$^2H_3$-) were synthesized by the methods reported by Cane and Levin (1975) and Suga and Shishibori (1975). Non-labeled and 1,1-$^2H_2$-FPP were prepared from geranylacetone, nerylacetone, and commercially available farnesol (2*Z*, 6*Z*-: 2*Z*, 6*E*- and 2*E*, 6*Z*-: 2*E*, 6*E* = 0.45 : 1.15 : 1) by the methods reported by Cornforth et al. (1962) and Dixit et al. (1981), and purified by the method reported by Davisson et al. (1985). 2*Z*, 6*Z*-, 2*E*, 6*Z*-, 2*Z*, 6*E*-, 2*E*, 6*E*-$^2H_2$-farnesol were separated before preparing FPP.

### 3.2 Synthesis of Pinoresinol by the Cell-Free Extracts

Fresh calli were ground in 0.05 M potassium phosphate buffer (pH 8.0) at 0 °C (Nabeta et al. 1991). The suspension was centrifuged at 2000*g* for 10 min. To the supernatant, guaiacol and $H_2O_2$ were added. Decrease in absorption at 470 nm was observed and used to calculate peroxidase activity (Chance and Maehly 1955). The cell-free extract from the calli was incubated with coniferyl alcohol and $H_2O_2$ in 0.1 M potassium phosphate buffer (pH 7.4). The reaction mixture was extracted with n-BuOH. Pinoresinol was isolated and its enantiometric composition was analyzed by high pressure liquid chromatography (HPLC) using a chiral column.

### 3.3 Feeding Experiment with MVA

Cultures were incubated with 24.8 mg of potassium MVA (non-labeled and deuterated) dissolved in 10 ml of MS agar medium (Nabeta et al. 1990a). The calli were grown at 25 ± 1 °C for 25 days under continuous light at 5000 lx. The volatiles were separated by a small-scale simultaneous-distillation-extraction apparatus (Nabeta 1985), and analyzed by gas chromatography (GLC) and gas chromatograph-mass spectrometry (GC/MS).

### 3.4 Synthesis of Sesquiterpenes by the Cell-Free Extracts

The calli were ground and stirred into a suspension containing polyvinylpyrrolidone [1% (w/w) of the calli] in potassium phosphate buffer (pH 8.0) (Nabeta et al. 1990b). The suspension was centrifuged (2000*g*, 10 min). The supernatant solution was incubated with non-labeled and deuterated FPPs (a mixture of four isomers and each of four isomers, 0.5 $\mu$M) at 25 ± 1 °C for 1, 3, and 24 h. The products were extracted with $Et_2O$ and then analyzed by GLC and GC/MS.

## 4 Analytical Methods

### 4.1 Analysis of Lignans from Calli

#### *4.1.1 Isolation of Lignans*

The calli were extracted with EtOH. The EtOH extracts were evaporated to dryness and redissolved in $H_2O$. The aqueous solution was successively extracted with hexane and n-BuOH. The n-BuOH solution was extracted with aqueous NaOH, and the aqueous NaOH solution was acidified with HCl and partitioned with n-BuOH. The BuOH extract (acidic fraction) was chromatographed with a silica gel column using benzene, benzene-ethyl acetate, and ethyl acetate-EtOH. Pinoresinol (**8** in Fig. 2) and dihydrobenzofuran derivative (**10**) were eluted in the benzene-ethyl acetate (2:3) and the ethyl acetate-EtOH (49:1) fractions, respectively. Lignans were finally purified by HPLC using a ODS column with MeOH-$H_2O$.

#### *4.1.2 Quantitative Analysis of Pinoresinol*

The acidic fraction was obtained as mentioned in Section 4.1.1. Pinoresinol in the acidic fraction was analyzed by HPLC on a ODS column (25 cm × 4.6 mm i.d.) using a solvent system of MeOH-$H_2O$ (2:3) at 2 ml/min (detection: UV at 280 nm). The quantities were determined on the basis of relative peak areas of an unknown amount of pinoresinol in the acidic fraction to the known amount.

#### *4.1.3 Separation of Optical Isomers of Pinoresinol*

( + ) and ( − ) isomers of pinoresinols isolated from the wood and the calli, and prepared from coniferyl alcohol by the cell-free preparation and horseradish peroxidase (EC 1.11.1.7) were separated by HPLC using a chiral column (Waters Opti-Pak XC, 30 cm × 3.9 mm i.d.) and a UV detector at 280 nm. The mobile phase was hexane-EtOH (1:1) at a flow of 0.5 ml/min. The ratios of the ( − )-isomer to the ( + )-isomer were determined on the basis of their relative peak areas with the consideration of relative peak area of a 1:1 mixture of the ( − ) and ( + )-isomers prepared from horseradish peroxidase.

### 4.2 Analysis of Volatiles

The volatiles obtained by simultaneous-distillation extraction and EtOH extraction were analyzed by GLC and GC/MS. GLC and GC/MS separation was carried out using a FS/WCOT capillary column (50 m × 0.25 mm, i.d.) coated with Silicone OV-101. The column temperature was kept at 60 °C for 5 min and then elevated to 220 °C at 2 °C/min. GLC was monitored by a single FID. The ionizing potential in GC/MS analysis was 70 eV.

## 5 Lignans in Callus Culture of *L. leptolepis*

### 5.1 Lignans from Callus

A variety of β–β-linked lignans have been isolated from wood and inner bark of *L. leptolepis* (Sakakibara et al. 1987). These include lariciresinol (**1**, type I in Fig. 2) and its derivatives (**2**–**4**), isolarisiresinol (**5**, type II) and its derivative (**6**), secoisolariciresinol (**7**, type III), and pinoresinol (**8**, type IV). Seven lignans (**9**–**15**) of dihydrobenzofuran type (V) and four (**16**–**20**) of arylglycerol-β-aryl ethers

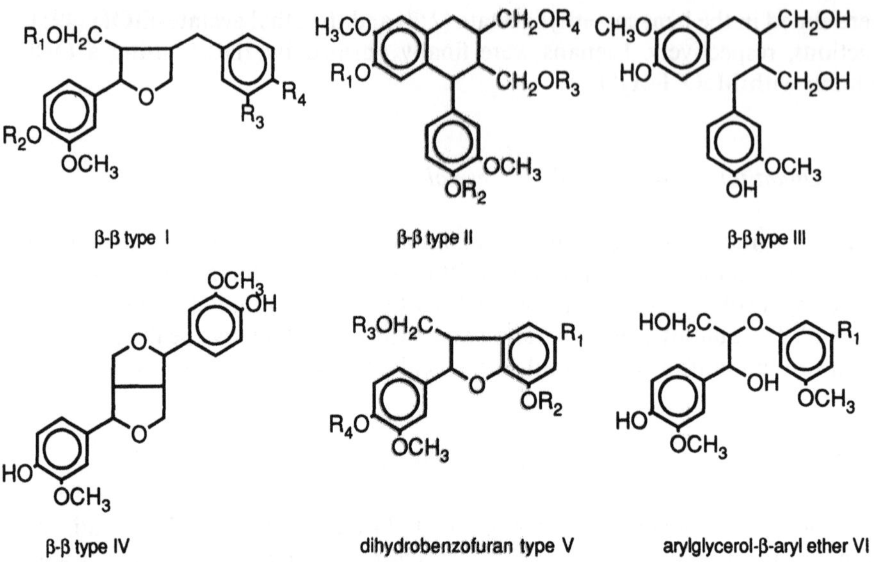


| Types | No. | $R_1$ | $R_2$ | $R_3$ | $R_4$ | Name of compound |
|---|---|---|---|---|---|---|
| I | 1 | H | H | $OCH_3$ | OH | Lariciresinol |
| | 2 | Ac | H | $OCH_3$ | OH | |
| | 3 | *p*-coumaryl | H | $OCH_3$ | OH | |
| | 4 | H | H | H | OH | |
| II | 5 | H | H | H | H | Isolariciresinol |
| | 6 | H | H | H | *p*-coumaryl | |
| | | | | or | | |
| | | H | H | *p*-coumaryl | H | |
| III | 7 | - | - | - | - | Secoisolariciresinol |
| IV | 8 | - | - | - | - | Pinoresinol |
| V | 9 | $CH_2CH_2CH_2OH$ | H | H | H | |
| | 10 | $CH_2CH_2CH_2OH$ | $CH_3$ | H | H | |
| | 11 | $CH_2CH_2CH_2OH$ | H | H | β-D-Glu | |
| | 12 | $CH_2CH_2CH_2OH$ | $CH_3$ | H | β-D-Glu | |
| | 13 | $CH_2CH_2CH2OH$ | H | β-L-Rham | H | |
| | 14 | CH=CHCHO | H | H | H | |
| | 15 | CH=CHCHO | H | H | $CH_3$ | |
| VI | 16 | $CH_2CH_2CH_2OH$ | - | - | - | |
| | 17 | $CHOHCHOHCH_2OH$ | - | - | - | |
| | 18 | $CH=CHCH_2OH$ | - | - | - | |
| | 19 | CH=CHCHO | - | - | - | |
| | 20 | CHO | - | - | - | |

**Fig. 2.** Lignans found in *Larix leptolepis*. (Sakakibara et al. 1987)

(type VI) were also obtained. Four trimeric phenylpropane derivatives (leptolepisol A to D) and one tetrameric derivatives (leptolepisol E) have been isolated. The above 25 compounds, except for compound **16** and leptopepisol D, were optically active.

We have isolated pinoresinol (0.5 mg) and 2,3-dihydro-2-(4-hydroxy-3-hydroxymethyl-5-(ω-hydroxypropyl)-7-methoxybenzofuran (**10**, 0.9 mg) from an ethanolic extract of the calli (the 20th to 28th generation, 35.56 g dry wt.) of *L. leptolepis* by fractionation in three steps: liquid-liquid partition, silica gel column chromatography, and HPLC separation with the ODS column (Nabeta et al. 1991). Lariciresinol, isolarisiresinol, and secoisolariciresinol were not detected in the ethanolic extract by TLC and HPLC analyses. Compounds **8** and **10** were identified by $^1$H NMR (500 MHz, see Fig. 3) and mass spectra, direct comparison with authentic samples (pinoresinol) and spectral data described for compound **10** (Takehara and Sasaya 1979).

The chemically related compound of pinoresinol, ( + )-(1*R*, 2*S*, 5*R*, 6*S*)-2,6-di-(4-hydroxyphenyl)-3,7-dioxabicyclo[3,3,0]octane (**21**, Fig. 4) was synthesized

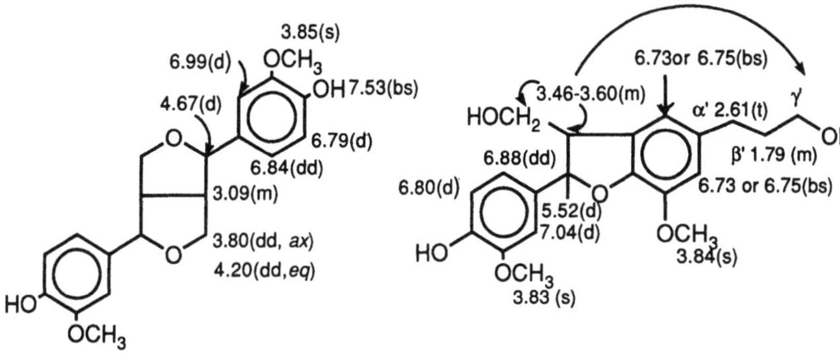


**8** **10**

**Fig. 3.** $^1$H-NMR data of compounds **8** and **10**

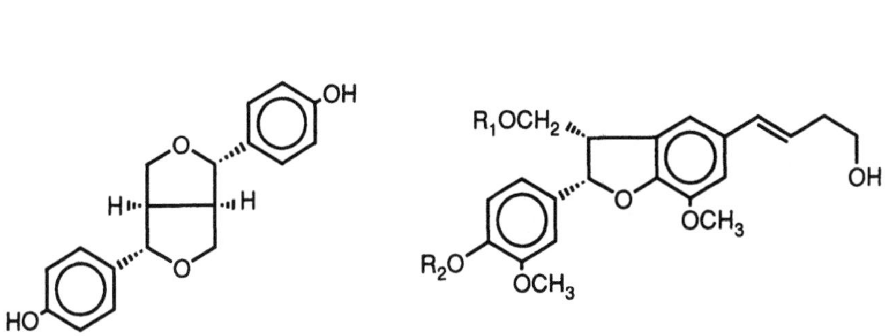


**21**

**22**: $R_1$= Glu, $R_2$=H

**23**: $R_1$=H, $R_2$=Glu

**Fig. 4.** Lignans isolated from plant cell culture

in cell suspension of *Vigna angularis* (Kobayashi and Ohta 1983). Dihydrobenzofuran derivatives (type IV), dehydrodiconyferyl alcohol 4-O-β-D glucoside (**22**) and the γ-O-β-D-glucoside (**23**) have been identified in extracts from suspension cultures of *Plarigorhegma dubium* (Arens et al. 1985).

## 5.2 Growth, Pinoresinol Content, and Peroxidase Activity in Calli

The growth and production characteristics of pinoresinol and sesquiterpenes in callus culture are given in Fig. 5. The time course of pinoresinol content was followed together with dry weight of the callus mass and peroxidase activity. Pinoresinol content was determined at 10-day intervals by two replicated cultures and HPLC analysis. It increased gradually in the early growth stage, increased sharply at the rapid growth stage (during days 20 to 60), and dropped rapidly at the stationary phase (after 60 days). The calli regularly accumulated more than 150 $\mu$g/g dry wt. pinoresinol at maximal level (day 60).

Peroxidase activity was determined by the guaiacol test (Chance and Maehly 1955) at days 10, 30, and 70, that is, the early, rapid, and stationary growth stages, respectively. Peroxidase activity was maximal in the cell-free preparation from 30-day-old culture ($210 \times 10^{-6}$ $\mu$mol $s^{-1} g^{-1}$ fr. wt.). Activity was 20 times higher than in the early and stationary growth stages. Pinoresinol was formed by the cell-free extracts from 50-day-old culture with 10.8 mg coniferyl alcohol and 30 mM $H_2O_2$ at 25 °C for 5 h. The purified pinoresinol (100 $\mu$g) was obtained by silica gel column chromatography and HPLC separation. These results suggested that pinoresinol was biosynthesized de novo at the rapid growth stage by peroxidase.

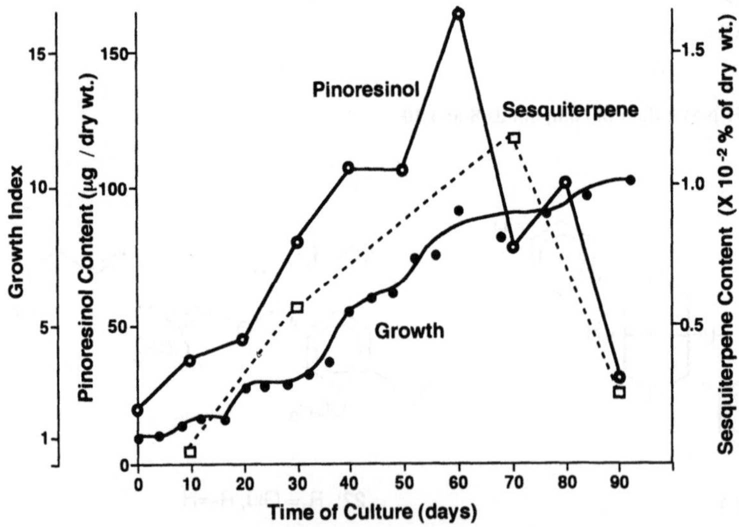


**Fig. 5.** Time course of pinoresinol content (○—○), sesquiterpene content (--□--□--), and growth (●—●) of *L. leptolepis* callus. (Nabeta et al. 1991)

## 5.3 Enantiometric Compositions of Pinoresinol

Separation of optical isomers of pinoresinol was carried out using a chiral column (Waters Opti-Pak XC). The mobile phase was n-hexane-ethanol (1 : 1) at a flow of 0.5 ml/min. Figure 6A to C shows the UV chromatograms of pinoresinols obtained from wood A and callus C, and by the preparation with horseradish peroxidase (EC 1.11.1.7,B). Table 1 summarizes the enantiometric compositions of these pinoresinols together with that prepared from cell-free extracts from the callus. These results conclusively indicate that the calli accumulate predominantly ( + )-pinoresinol, but the ratio of ( − )-isomer/( + )-isomer was much less than that from wood. This might support the assumption that peroxidase is the enzyme responsible for formation of pinoresinol in the callus. Interestingly, however, a slight predominance of formation of the ( − )-isomer was observed in incubation with the cell-free extracts from the callus. The reason for the selective accumulation of ( + )-pinoresinol in wood cannot be explained simply by the action of peroxidase. This might be rationalized by the explanation that (1) ( + )-pinoresinol is formed by peroxidase under the enantioselective condition present in wood and inner bark, (2) an enzyme(s) besides

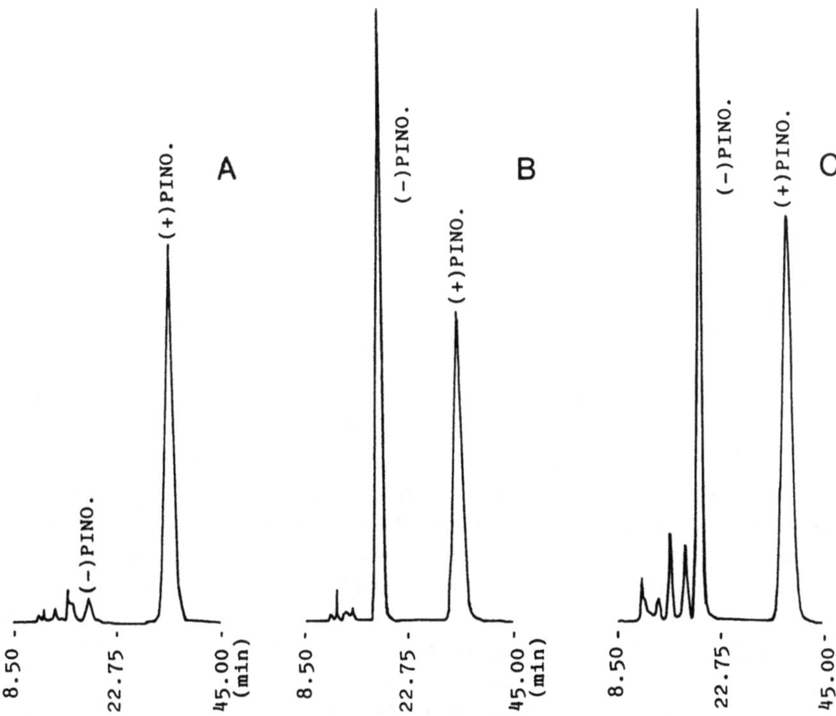


**Fig. 6A–C.** Optical resolution of pinoresinol by HPLC. Samples: Pinoresinols: **A** isolated from wood, **B** prepared from coniferyl alcohol with horseradish peroxidase, and **C** isolated from callus of *L. leptolepis*. Column: Waters OptiPak XC, 30 cm × 3.9 mm i.d., Solvent: hexane: ethanol 1:1, Detection: UV at 280 nm

**Table 1.** Enatiometric composition of pinoresinols. (Nabeta et al. 1991)

| *L. leptolepis* material | (−)-Isomer/(+)-isomer | |
|---|---|---|
| | Relative peak area | Relative content |
| Callus | 0.773 | 0.74 |
| Cell-free extract | 1.184 | 1.13 |
| Sapwood | 0.037 | 0.04 |
| Horseradish peroxidase | 1.050 | 1 |

peroxidase couples two coniferyl alcohol to afford optically pure (+)-pinoresinol, or (3) (−)-isomer is selectively depleted in wood and inner bark. Recently, the enantioselective formation of lignans by cell-free extracts has been reported (Umezawa et al. 1990).

# 6 Lower Terpenes in Callus Culture of *L. leptolepis*

## 6.1 Volatiles in Seedling Leaves, Callus, and Suspension Cells

Total ion-monitored (TIM) chromatogram of volatiles from the seedling leaves which were used as explant is reproduced in Fig. 7. The volatiles consisted mainly of monoterpenes (α-pinene, camphene, β-pinene, ocimene, α- and β-terpinene, terpinen-4-ol, α-terpineol, and bornyl acetate) and sesquiterpenes (β-caryophyllene, α-humulene, aromandendrene, δ-cadinene, and an unidentified oxygenated sesquiterpene). The quantity of terpenes in seedling leaves was almost the same as that in the mature leaves, but total quantity of volatiles [0.31% fresh wt. in the mature leaves (Nabeta et al. 1985), 0.029% in the seedling leaves] and the relative proportion of monoterpenes were much less than those in the mature leaves.

Figure 8 shows the TIM GC profile of volatiles accumulated in the calli of the 28th generation. The chromatogram obviously demonstrated the distinct differences between the compositions of volatiles from the calli and the seedling leaves. The total quantity (0.0031%) in the calli and the relative proportion of terpenes (below 15%) were much lower than in the seedling leaves. Monoterpenes were not detected in the calli, with the exception of *p*-cymene (**24** in Fig. 9). Although in the chromatogram, α-cedrene (**25**), β-cadinene (**26**), and 6*E*-β-farnesene (**27**) were observed, deviations in the qualitative patterns of the predominant sesquiterpenes were observed at the different generations. The optical isomerism of α-cedrene has been determined as an (+)-isomer, which was detected for the first time in nature by GLC analysis using a chiral capillary column (Nabeta et al. 1993).

Although in the suspension cells, phytol and methyl ionone (tentatively identified) were detected as terpenoids, no mono- and sesquiterpene were observed.

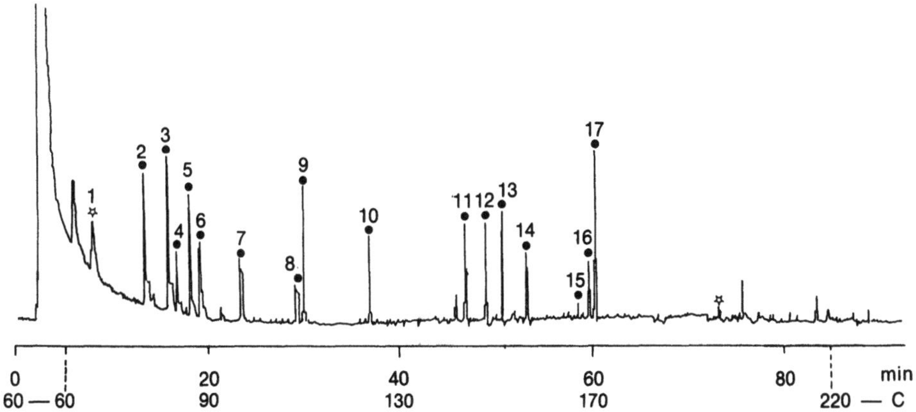


**Fig. 7.** TIM GC profile of volatiles from seedling leaves of *L. leptolepis* (Nabeta et al. 1990a). Column: an FS/WCOT capillary column (50 m) coated with Silicone OV-101. *Peak 1* Hexenal; *2* α-pinene; *3* camphene; *4* β-pinene; *5* ocimene; *6* β-terpinene; *7* α-terpinene; *8* terpinen-4-ol; *9* α-terpineol; *10* bornyl acetate; *11* β-caryophyllene; *12* α-humulene; *13* aromandendrene; *14* δ-cadinene; *15* to *17* oxygenated sesquiterpenes. ● and ☆ represent terpenoids and long-chain compounds, respectively

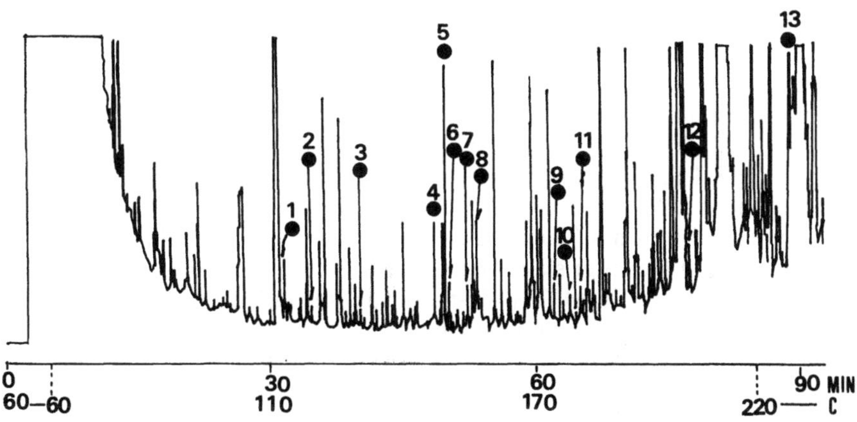


**Fig. 8.** TIM GC profile of volatiles from callus of *L. leptolepis* (Nabeta et al. 1990a). *Column* Given in Fig. 7. *Peak 1* p-cymene; *2* and *3* unknown monoterpenes; *4* acoradiene; *5* α-cedrene; *6* β-cadinene; *7* 6*E*-β-farnesene; *8* β-ionone; *9* to *13* oxygenated sesquiterpenes. ●, Terpenoids

## 6.2 De Novo Syntheses of Sesquiterpenes from Deuterated Mevalonate

Calli were cultured on potassium deuterated and non-labeled MVAs (ca. 25 mg/10 ml culture medium). After the 25-day culture period, volatiles were separated and subjected to GC/MS analysis (Nabeta et al. 1990a). Despite the higher concentration of dosages applied to the culture medium, growth index (4.45) changed little, but an increase in the relative proportion of the long-chain

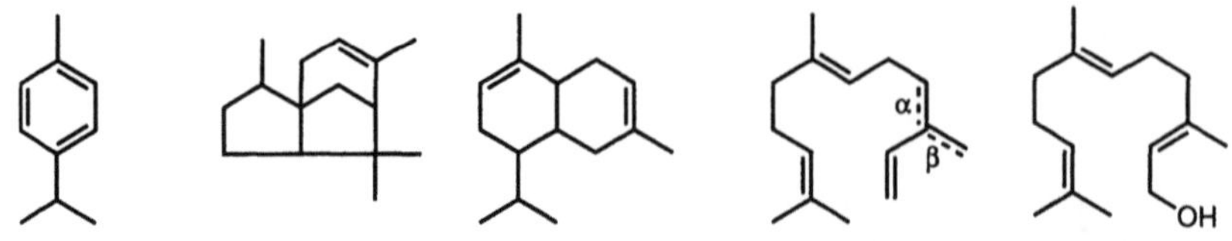

**24**: *p*-cymene **25**: α-cedrene **26**: β-cadinene **27**: β-farnesene **33**: α-farnesene **28**: farnesol

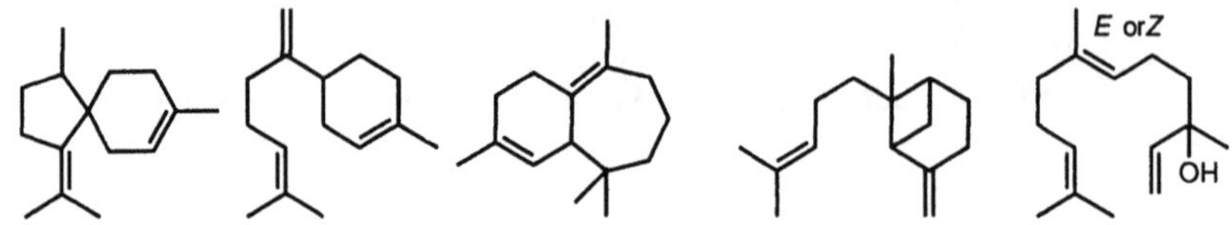


**29**: acoradiene **30**: β-bisabolene **31**:β-himachalene **32**:α-bergamotene **34**:*E*-nerolidol **35**: *Z*-nerolidol

**Fig. 9.** Lower terpenes found in *L. leptolepis* callus and those prepared from FPP with the cell-free extracts of callus

compounds such as fatty acid, their methyl and ethyl esters, aldehydes, and hydrocarbons was observed. However, volatile terpenoids **24** to **27** and farnesol (**28**) and some oxygenated sesquiterpenes were detected.

Mass spectra of *p*-cymene, α-cedrene, 6*E*-β-fanesene, and farnesol incorporating 6,6,6-$^2H_3$-MVA are shown in Fig. 10A, B, C and D, respectively. Mass spectra indeed indicate that almost 100% of these terpenes originated from the exogenous MVA. In contrast, deuterium atoms were not incorporated into the oxygenated terpenes. Thus, the calli might have synthesized the oxygenated sesquiterpene in the latter stage of growth. The mass spectrum of the *p*-cymene incorporating 6,6,6-$^2H_3$ MVA (6-D-*p*-cymene) shows [$M^+$] at *m*/*z* 140, indicating two of three methyl groups originating in the methyl groups of MVA. The 6-D-α-cedrene has [$M^+$] at *m*/*z* 213, which confirms the presence of three deuterated methyl group. Molecular ion peak of 6D-6*E*-β-farnesene shifted 8 mass units, clearly demonstrating the loss of the deuterium atoms for formation of a C = C double bond between the C-3 and C-15 positions.

Figure 11 indicates the generally accepted biosynthetic pathway of α-cedrene. The indicated 1,2-hydride shift in the bisabolane cation was confirmed by the comparison of the predominant fragment ions from the biosynthetically deuterated cedrene derived from deuterated and non-labeled MVA (Fig. 12). A base ion B was formed by the breaking of the C-2 and C-11, (our numbering), C-6/C-10, and C-7/C-8 bonds, with the elimination of two hydrogen atoms at the C-1 and C-5 positions. The ion B at *m*/*z* 121 in α-cedrene (4-D-α-cedrene), incorporating 4,4-$^2H_2$-MVA (*m*/*z* 119 in non-labeled compound), shifted by two mass units, whereas the predominant ion C at *m*/*z* 94 in 4-D-α-cedrene shifted by only one mass unit. The ion C was derived from the cyclohexene ring by breaking the C-2/C-11, C-6/C-10, and C-6/C-7 without elimination of any

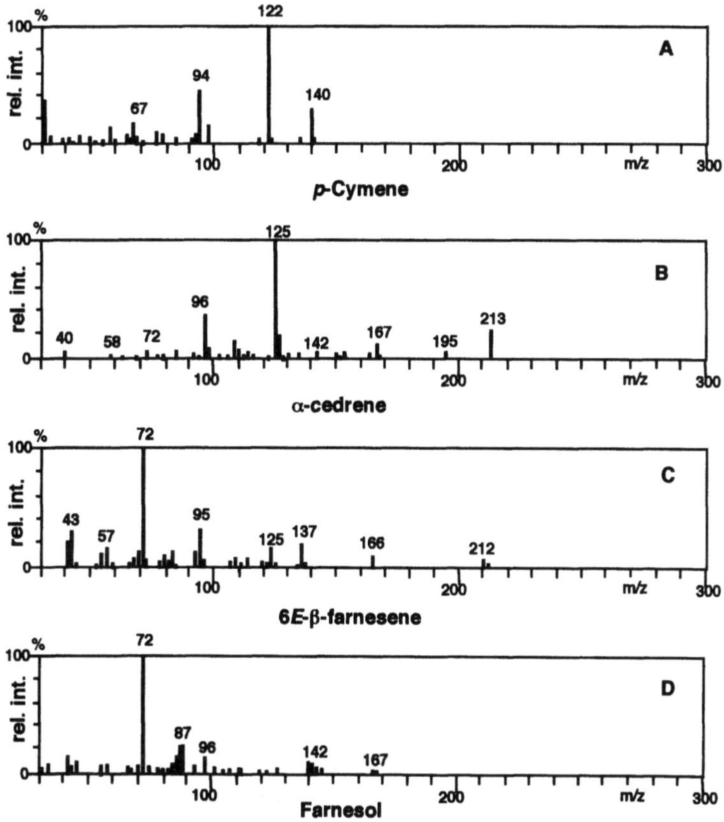


**Fig. 10A–D.** MS spectra of terpenes incorporating 6,6,6-$^2H_3$-mevalonate in *L. leptolepis* callus (Nabeta et al. 1990a). **A** *p*-Cymene. **B** α-Cedrene. **C** 6*E*-β-Farnesene. **D** Farnesol

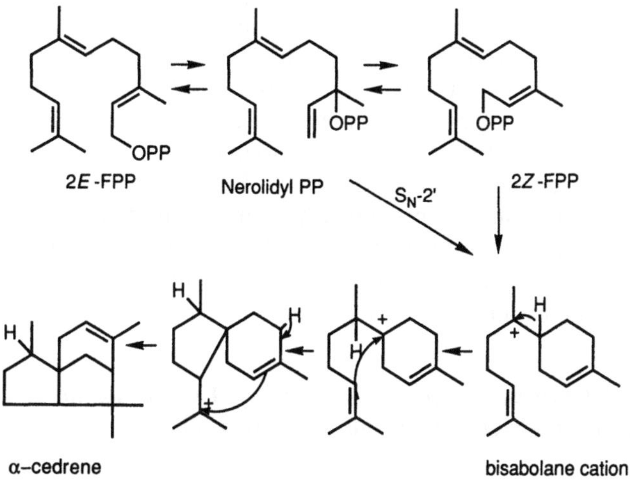


**Fig. 11.** Biosynthetic pathway of α-cedrene

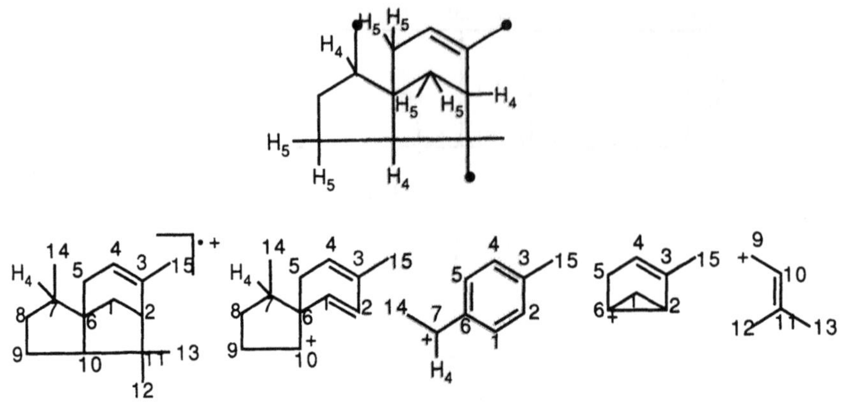

M⁺ A B C D

| Mevalonates | $M^+$ | A | B | C | D |
|---|---|---|---|---|---|
| 6,6,6-$^2H_3$- | 213 | 167 | 125 | 96 | 72 |
| 5,5-$^2H_2$- | 210 | 166 | 121 | 97 | 71 |
| 4,4-$^2H_2$- | 207 | 164 | 121 | 94 | 70 |
| non-labeled | 204 | 161 | 119 | 93 | 69 |

(Column headings M⁺ to D are grouped under *m/z*.)

**Fig. 12.** Fragment ions from the deuterated α-cedrenes (Nabeta et al. 1990a). $H_4$ and $H_5$ represent the positions of hydrogen atoms in mevalonate. ● represents methyl group in mevalonate

hydrogen atom. The comparison between ions B and C indicates conclusively that the hydrogen atom at the C-7 position originated from H-4 in MVA. Thus the 1,2-hydride shift in the cyclization of bisabolane cation to form α-cedrene is clearly demonstrated. Other predominant fragment ions from deuterated α-cedrenes support the labeling patterns, as indicated in Fig. 12.

Figure 13 lists the predominant fragment ions from the deuterated 6*E*-β-farnesene. A base ion at *m/z* 95 in the 6D-6*E*-β-farnesene is two mass units higher than that in the nonlabeled compound. These results support the assumption that formation of the C=C bond between the C-3 and C-15 position involves the elimination of the hydrogen atom at the C-15 position. The location of other deuterium atoms in deuterated β-farnesene was determined by the fragment ions, as indicated in Fig. 13.

The technique presented here, in which the location of deuterium atoms in terpenes deuterated at extremely high level was determined by GC/MS analysis, could be useful in cases where it is difficult to isolate each component from a complex mixture of terpenes.

## 6.3 Cyclization Products of Farnesyl Pyrophosphate with Cell-Free Preparation from the Calli

Incubation of the cell-free extracts was carried out with FPP at 25 °C for 1,3 and 24 h (Nabeta et al. 1990b). The enzymes in the cell-free extract catalyzed the

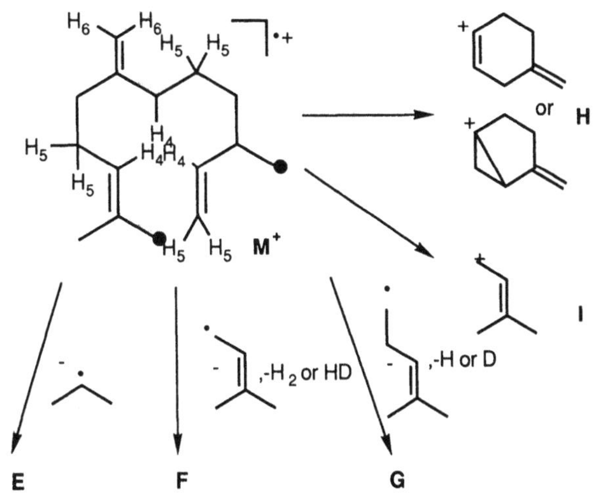


| | *m/z* | | | | | |
|---|---|---|---|---|---|---|
| Mevalonates | $M^+$ | E | F | G | H | I |
| 6,6,6-$^2H_3$- | 212 | 166 | 137 | 125 | 95 | 72 |
| 5,5-$^2H_2$- | 210 | 167 | 137&136 | 124&123 | 97 | 71 |
| 4,4-$^2H_2$- | 207 | 164 | 134 | 122 | 94 | 70 |
| non-labeled | 204 | 161 | 133 | 120 | 93 | 69 |

**Fig. 13.** Fragment ions from the deuterated 6*E*-β-farnesene. (Nabeta et al. 1991). *H4*, *H-5*, and ●: see legend to Fig. 12

cyclization of FFP (deuterated and non-labeled) to form a variety of sesquiterpenes including acoradiene (**29**), α-cedrene (**25**), β-bisabolene (**30**), β-himachalene (**31**), and α-bergamotene (**32**). The cell-free system also afforded acyclic compounds such as α- (**33**) and β-farnesenes and farnesols. The enzymes were assayed in phosphate buffer (pH 8.0, 0.05 M) containing polyvinylpyrrolidone (PVP, 10 mg/g callus). Table 2 summarizes the amounts of sesquiterpenes generated from 2*Z*, 6*E*- and 2*E*, 6*E*-FPP in the cell-free extracts. Dication $Mg^{2+}$ slightly improves the synthesis of sesquiterpene hydrocarbons, but depresses the formation of farnesol (Nabeta et al. unpubl. data). Interestingly, the cell-free extracts afforded the cyclic sesquiterpenes from 2*Z*, 6*E*- and 2*Z*, 6*Z*-FPP, but not from 2*E*-isomers. From 2*Z*, 6*E*- and 2*Z*, 6*Z*-FPP, acoradiene, α-cedrene, α-bergamotene, and β-bisabolene were produced together with nerolydol (*Z*-and *E*-**34** and **35**) and α, β-farnesenes and farnesols. In contrast, from 2*E*-isomers (6*Z* and 6*E*-), *Z* and *E*-nerolydols and farnesol were formed, but no cyclic terpenes. These results are seemingly opposed to the generally accepted cyclization mechanism, that is, $S_N$-2′ reaction (see Fig. 11) of nerolidyl pyrophosphate (nerolidyl PP in Fig. 11), which is preferentially converted from 2*E*-FPP, for affording the six-membered ring (Cane 1980). Thus the isomerization of 2*E*-FPP to the tertial pyrophosphate might be hindered by the presence of competing phosphatase. Alternatively, bisabolane cation might be formed directly from the

**Table 2.** Sesquiterpenes generated from four geometrical isomers of [1,1-$^2H_2$] - farnesyl pyrophosphates by the cell-free extracts of *Larix leptolepis* callus, their GLC retention times and incorporation

| | Rt (min) (% incorporation) | | | |
|---|---|---|---|---|
| Compounds | 2*Z*, 6*E*-FPP | 2*E*, 6*E*-FPP | Authentic | $M^+$ |
| Acoradiene | 59.55 (0.27%) | | | 206 |
| α-Cedrene | 60.22 (0.17%) | | 60.20 | 206 |
| *Z*-β-Farnesene | | | 60.53 | 206 |
| *E*-β-Farnesene | 62.20 (0.087%) | 62.15 (0.40%) | 62.13 | 206 |
| α-Bergamotene | 62.90 (1.02%) | | | 206 |
| β-Himachalene | 64.01 (0.087%) | | | 205 |
| β-Bisabolene | 65.81 (0.44%) | | 65.81 | 206 |
| *Z*-Nerolidol | | | 66.95 | 206 ($M^+$-$H_2O$) |
| *E*-Nerolidol | 68.64 (0.89%) | 68.62 (0.86%) | 68.81 | 206 ($M^+$-$H_2O$) |
| 2*Z*, 6*Z*-Farnesol | | | 75.02 | 224 |
| 2*E*, 6*Z*-Farnesol | | | 76.39 | 224 |
| 2*Z*, 6*E*-Farnesol | 76.74 (30.59%) | | 76.58 | 224 |
| 2*E*, 6*E*-Farnesol | | 77.90 (25.27%) | 77.74 | 224 |
| Total sesquiterpene hydrocarbons | (2.23%) | (0.49%) | | |

All figures in parentheses are means of three independent determinations

2*Z*-isomer via an electrophilic addition of the initially generated farnesyl cation to the C-6/C-7 double bond.

# 7 Summary and Conclusion

The in vitro production of lower terpenes and lignans in the calli and suspension cells of *Larix leptolepis* was studied. Callus culture was initiated from seedling leaves in an MS medium supplemented with sucrose (30 g/l), 1% agar, and various combination of NAA, 2,4-D, and IAA-kinetin at pH 5.8. Optimal combination of growth regulators for callus induction was 5 mg/l of 2,4-D and 5 mg/l of kinetin, and the callus stocks were subcultured in this medium. The composition of nutrient media used in the suspension culture was the same as for the callus culture.

Pinoresinol and 2,3-dihydro-2-dihydro-(4-hydroxy-3-methoxyphenyl)-3-hydroxymethyl-5-(ω-hydroxypropyl)-7-methoxybenzofuran were isolated from the calli. Pinoresinol was predominantly produced at the rapid growth stage. At this stage, the peroxidase activity was maximal. The ratio of the ( − )-isomer to the ( + )-isomer of pinoresinol from the calli was 0.77, whereas that of pinoresinol from wood is 0.04. Thus peroxidase may be an enzyme responsible for the formation of pinoresinol in the calli, although other enzyme(s) besides peroxidase may afford the ( + )-isomer in wood.

Sesquiterpenes and monoterpenes were also accumulated in the calli at an extremely low level. Feeding experiments of specifically deuterated MVAs to the calli demonstrated that the valuable information on biosyntheses of sesquiterpenes was obtained using plant cell culture technique. A cell-free extract from the calli could convert FPP to the cyclic sesquiterpenes. Suspension cells did not accumulate detectable amounts of lower terpenes.

# 8 Protocol

1. Explant; the first leaves of germinating seedlings grown on vermiculite under continuous light at 5000 lx at 25 ± 1 °C for 35 days.
2. Callus initiation and subculture; MS medium containing 30 g/l sucrose, 1% agar, 5 mg/l 2,4-D, and 5 mg/l kinetin. Cultures are maintained under continuous light at 5000 lx at 25 ± 1 °C. Culture periods: 30–40 days.
3. Suspension culture; same conditions as for callus culture, rotary shaker (100 rpm), culture period: 21 days.
4. Lignan and lower terpenoid production in the calli; same medium and conditions as for callus culture.

*Acknowledgments.* The author wishes to acknowledge the information provided by Dr. H. Okuyama (Obihiro University of Agriculture and Veterinary medicine). This work was supported in part by a Grant-in-aid (No. 03660123) from the Ministry of Education of Japan.

# References

Aderkas PV, Klimaszewska K, Bonga JM (1990) Diploid and haploid embryogenesis in *Larix leptolepis*, *L. decidua*, and their reciprocal hybrids. Can J For Res 20:9–14

Arens H, Fischer H, Leyck S, Romer A, Ulbrich B (1985) Antiflammatory compounds from *Plargiorhegma dubium* cell culture. Planta Med: 52–56

Aspinall GO, Fairweather M, Wood TM (1968) Arabinogalactan A from Japanese larch (*Larix leptolepis*). J Chem Soc (C): 2174–2179

Brewerton HV (1956) Extractives of *Larix decidua* and *Larix leprolepis*. N Z J Sci Technol 378:626–632

Cane DE (1980) The stereochemistry of allylic pyrophosphate metabolism. Tetrahedron 36:1109–1199

Cane DE, Levin RH (1975) Application of carbon-13 magnetic resonance to isoprenoid biosynthesis II, ovalicin and the use of doubly labeled mevalonate. J Am Chem Soc 98:1183–1188

Chalupa V (1991) Larch (*Larix decidua* Mill.). In: Bajaj YPS (ed) Biotechnology in agriculture and forestry, vol 16. Trees III. Springer, Berlin Heidelberg New York, pp 446–470

Chance B, Maehly AC (1955) Assays of catalases and peroxidases. Methods Enzymol 2:764–775

Cornforth JW, Cornforth RH, Hyhage R, Goodman DWS (1962) Biosynthesis of cholesterol XVI. Chemical synthesis of trans, trans-farnesyl-1-$D_2$-2-$C^{14}$ and 1-$H^3_2$-2-$C^{14}$ pyrophosphates and their utilization in squalene biosynthesis. J Biol Chem 237:56–61

Davisson VJ, Woodside AB, Dale CD (1985) Synthesis of allylic and homoallylic pyrophosphates. Methods Enzymol 110:130–144

Dixit VM, Laskovics FM, Noall WI, Poulter CD (1981) Tris (tetra-n-butylammonium) hydrogen pyrophosphate. A new reagent for the preparation of allylic pyrophosphate ester. J Org Chem 46:1967–1969

Klimaszewska KK (1989) Recovery of somatic embryos and plantlets from protoplast cultures of *Larix × eurolepsis*. Plant Cell Rep 8:440–444
Kobayashi M, Ohta Y (1983) Induction of stress metabolite formation in suspension cultures of *Vigna angularis*. Phytochemistry 22:1257–1261
Kuromatsu M, Takao S (1987) Plantlet formation from cotyledon-epicotyl explant of hybrid larch (*Larix gmelinii × L. leptolepis*) $F_1$ by tissue culture. J Jpn For Soc 69:355–358
Lloyd GB, McCown BH (1980) Commercially feasible micropropagation of mountain laurel, *Kalmia latifolia*, by use of shoot-tip culture. Proc Int Plant Prop Soc 30:421–427
Murashige T, Skoog F (1962) A revised medium for rapid growth and bioassay with tabacco tissue culture. Physiol Plant 15:473–497
Nabeta K (1985) Cultured cells producing volatiles and separation of volatiles. In: Yamada Y, Okada Y (eds) Plant Biotechnology. Tokyokagakudoujin, Tokyo, pp 89–96 (in Japanese)
Nabeta K, Toda N, Miyake M (1985) Analysis of volatiles from leaves and twigs of wood. Proc Hokkaido Branch, Jpn Wood Res Soc (Sapporo, Japan) 17:53–56
Nabeta K, Ara A, Aoki Y, Miyake M (1990a) Biosynthesis of monoterpenes and sesquiterpenes in *Larix leptolepis* callus from deuterated mevalonates. J Nat Prod 53:1241–1248
Nabeta K, Aoki H, Ara Y, Kurano H, Okuyama H, Miyake M (1990b) Biosynthesis of volatile terpenes in *Larix leptolepis callus*. In: Charalambous G (ed) Flavors and Off-flavors'89. Elsevier, Amsterdam, pp 641–657
Nabeta K, Nakahara K, Yonekubo J, Okuyama H, Miyake M (1991) Lignan biosynthesis in *Larix leptolepis* callus. Phytochemistry 30:3591–3593
Nabeta K, Katayama K, Aoki T, Honjoh K, Okuyama H (1993) Occurrence of (+)-α-cedrene in *Larix leptolepis* callus. Biosci Biotech Biochem 57:1022–1023
Niemann GJ (1971) Phenolic constituents from *Larix* needles III: Benzoic and cinnamic acids of *L. leptolepis*. Acta Bot Neerl 20:397–400
Niemann GJ (1973a) Phenolics from *Larix* needles. VII. Flavonoids from needles of *Larix leptolepis*. Phytochemistry 12:2056
Niemann GJ (1973b) Phenolic glucosides from needles of *Larix leptolepis*. Phytochemistry 12:723–724
Niemann GJ (1974) Phenolics of *Larix* needles. VIII. Flavonoids of *L. leptolepis*. Planta Med 26:101–103
Niemann GJ (1976) Phenolics from *Larix* needles. XII. Seasonal variation of main flavonoids in leaves of *L. leptolepis*. Acta Bot Neerl 25:345–359
Niemann GJ (1980) Phenolics from *Larix* needles. XVI. Inter- and intra-clonal variation in *Larix leptolepis*. Can J Bot 58:2313–2317
Nomura K, Muto K (1976) Studies on the constituents of the annual shoots of *Larix leptolepis* Gordon (1) Kaempferol-3-O-glucoside in the leaves during the growing season. J Jpn For Soc 58:86–91
Okuyama H, Ooshima M, Miyake M, Washimizu M (1968) Distribution of the extractives in a trunk of *Larix leptolepis* planted in Hokkaido. Res Bull Obihiro Univ 5:676–684
Sakakibara A, Sasaya T, Miki K, Takahashi H (1987) Lignans and Brauns' lignins from softwoods. Holzforschung 41:1–11
Schenk RU, Hildebrandt AC (1972) Medium and technique for induction and growth of monocotyledonous and dicotyledonous plant cell cultures. Can J Bot 50:199–204
Shin D, Podila GK, Karnosky DF (1994) Genetic transformation in *Larix* spp. (Larches). In: Bajaj YPS (ed) Biotechnology in agriculture and forestry, vol 29. Plant protoplasts and genetic engineering V. Springer, Berlin Heidelberg New York, pp 317–322
Suga T, Shishibiri T (1975) Stereospecificity of biosynthesis of squalene and β-amyrin in *Pinus sativum*. Phytochemistry 14:2411–2417
Sunderland N (1966) Pigment plant tissue (II). Growth, development and decline on chlorophyllous tissues. Ann Bot 31:573–591
Takehara T, Sasaya Y (1979) Lignans from the sapwood of *Larix leptolepis* Gord, Mokuzai Gakkaishi 25:516–517
Umezawa T, Davin LB, Lewis NG (1990) Formation of the lignan, (−) secoisolariciresinol by cell free extracts of *Forsythia intermedia*, Biochem Biophys Res Commun 171:1008–1014

# XIX *Lobelia inflata* L. (Lobelia): In Vitro Culture and the Production of Lobeline and Other Related Secondary Metabolites

E. SZÖKE[1]

## 1 General Account

### 1.1 Botany

The habitat of *Lobelia inflata* L. (family Lobeliaceae) is North America, where it grows wild in forests, pastures, and at roadsides, though in many countries it is also cultivated (Wagner 1980, Krajewska 1986) because of its active ingredients. It is a 30–60-cm-high annual plant (Fig. 1), flowering from June to August. Its flowers are blue, about 7 mm long and form a racemose inflorescence. The drug is supplied by the flowering shoot.

The plant was given its Latin name after Mathias von L'Obel, Flemish physician and botanist (1538–1616), who was the first to describe it (Heeger 1956). The term "inflata" refers to the bladder-like inflated capsule (Luckner et al. 1968). The drug is known under several designations: e.g., in America the leaves of the plant are used in pipe by the name of "Indian tobacco" (they recall the smell of tobacco), or are put into circulation as "emetic herba" owing to their nauseating effect (List and Hörhammer 1976).

### 1.2 Alkaloids of *Lobelia inflata*

The lobelia alkaloids were first detected by Procter in 1850 (in Berger, 1954). At the beginning of the 20th century, Wieland and Scheuing isolated the first alkaloids from *Lobelia inflata* (Berger 1954).

The most important alkaloid of the family Lobeliaceae is lobeline, which is an $\alpha$, $\alpha_1$-disubstituted piperidine. Further 20 alkaloids of *Lobelia inflata* have so far been isolated, with their structures described (Hegnauer 1966; Teusher 1979); (Fig. 2).

The typical alkaloids of the family Lobeliaceae have a piperidine skeleton. There are alkaloid bases disubstituted in the $\alpha$, $\alpha_1$-position (i.e., in $C_2$ and $C_6$) and those monosubstituted in the $\alpha$-position ($C_2$) (Schöpf et al. 1957).

[1] Institute of Pharmacognosy, Semmelweis University of Medicine, 1085 Budapest, Hungary

Biotechnology in Agriculture and Forestry, Vol. 28
Medicinal and Aromatic Plants VII (ed. by Y.P.S. Bajaj)

**Fig. 1.** ***Lobelia inflata*** **L. intact plant and in vitro regenerated plants after transferring to the soil**

*Group I. alkaloids, substituted in positions C-2 and C-6*

1. Lobelidione and norlobelidione analogues.
2. Lobelionol and norlobelionol analogues.
3. Lobelidiol and norlobelidiol analogues.

*Group II. alkaloids, substituted in position C-2*

1. Lobelol and norlobelol analogues.
2. Lobelone and norlobelone analogues.

These alkaloids are optically active compounds.
The following alkaloids were detected in the plant (Hegnauer 1966)

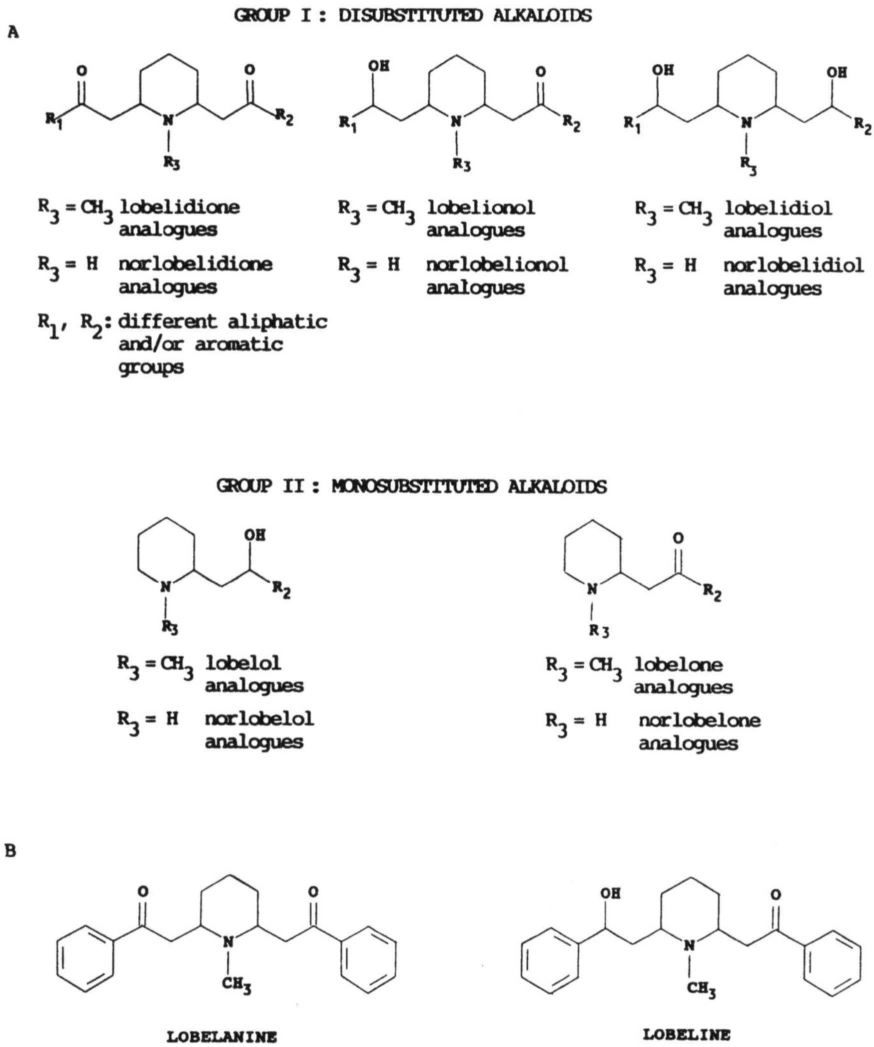


**Fig. 2A, B.** Alkaloids characteristic of the genus *Lobelia* (**A**), and their two characteristic representatives (**B**)

*Group I. disubstituted alkaloids*

*Lobelidione and norlobelidione analogues*

*cis*-8,10-diphenyl-lobelidione (lobelanine)

8-ethyl-10-phenyl-norlobelidione

*cis*-8,10-diphenyl-norlobelidione (norlobelanine = isolobelanine)

*Lobelionol and norlobelionol analogues*

( − )-*cis*-8,10-diphenyl-lobelionol (L-lobeline)

( ± )-*cis*-8,10-diphenyl-lobelionol (lobelidine)

*trans*-8-ethyl-10-phenyl-dehydrolobelionol (lobinine)

*cis*-8-ethyl-10-phenyl-dehydrolobelionol (isolobinine)
( − )-8-ethyl-10-phenyl-norlobelionol

*Lobelidiol and norlobelidiol analogues*
( ± )-8-methyl-10-ethyl-lobelidiol
8,10-diethyl-lobelidiol
( + )-8-methyl-10-phenyl-lobelidiol
( − )-*cis*-8-ethyl-10-phenyl-lobelidiol I [( − )-lelobanidine I]
( ± )-*cis*-8-ethyl-10-phenyl-lobelidiol I [( ± )-lelobanidine I]
( − )-*cis*-8-ethyl-10-phenyl-lobelidiol II (L-lelobanidine II)
*cis*-8,10-diphenyl-lobelidiol (lobelandine)
*trans*-8-ethyl-10-phenyl-dehydrolobelidiol (lobinandine)
*cis*-8-ethyl-10-phenyl-dehydrolobelidiol (isolobinandine)
*cis*-8-ethyl-10-phenyl-norlobelidiol I (D-norlelobanidine)
( ± )-8-ethyl-10-phenyl-norlobelidiol [( ± )-norlelobanidine]
*cis*-8,10-diphenyl-norlobelidiol (norlobelanidine)

*Group II. monosubstituted alkaloids*
*Lobelol and norlobelol analogues*
( − )-8-phenyl-lobelol I (allosedamine)
( + )-8-ethyl-norlobelol I
( + )-8-phenyl-norlobelol I

Another two alkaloids from *Lobelia inflata* having unknown structures have been published in the literature:
$C_{18}H_{27-29}O_3N$ (Schöpf et al. 1957)
$C_{19}H_{26}O_3N_2$ (Wieland et al. 1939)

Further alkaloids, which could be counted among the group of lobelia alkaloids: l-lelobanidine, isolated from *Lobelia nicotianaefolia* Heyne (Boit 1961); Lobinaline ($C_{27}H_{34}N_2$, where the two phenyl groups are condensation products) from *Lobelia cardinalis* L. (Robinson et al. 1964); and the alkaloids isolated from *Lobelia syphylitica* L. (Hegnauer 1966);
*cis*-8,10-diethyl-lobelionol,
*cis*-8,10-diethyl-norlobelionol,
*cis*-8,10-diethyl-norlobelidione,
( − )-*trans*-8,10-diethyl-dehydrolobelidiol,
( − )-*trans*-8-methyl-10-ethyl-dehydrolobelidiol, lofonine.

An alkaloid was isolated with a molecular formula $C_{14}H_{15}ON$ (Steinegger and Ochsner 1956b) from *Lobelia salicifolia*, and a mixture of alkaloids called lurenine (Steinegger and Grütter 1950) from *Lobelia urens* L.

## 1.3 Biosynthesis of Lobeline

Lobeline and the other *Lobelia* alkaloids with piperidine skeleton are placed among the alkaloids of lysine origin, as the part that contains the nitrogen heteroatom comes from lysine. In the synthesis of these alkaloids, phenylalanine

**Fig. 3.** Biosynthesis of *Lobelia* alkaloids. (O'Donovan and Forde 1971, 1975; Koukol and Conn 1961; Vagujfalvi 1993)

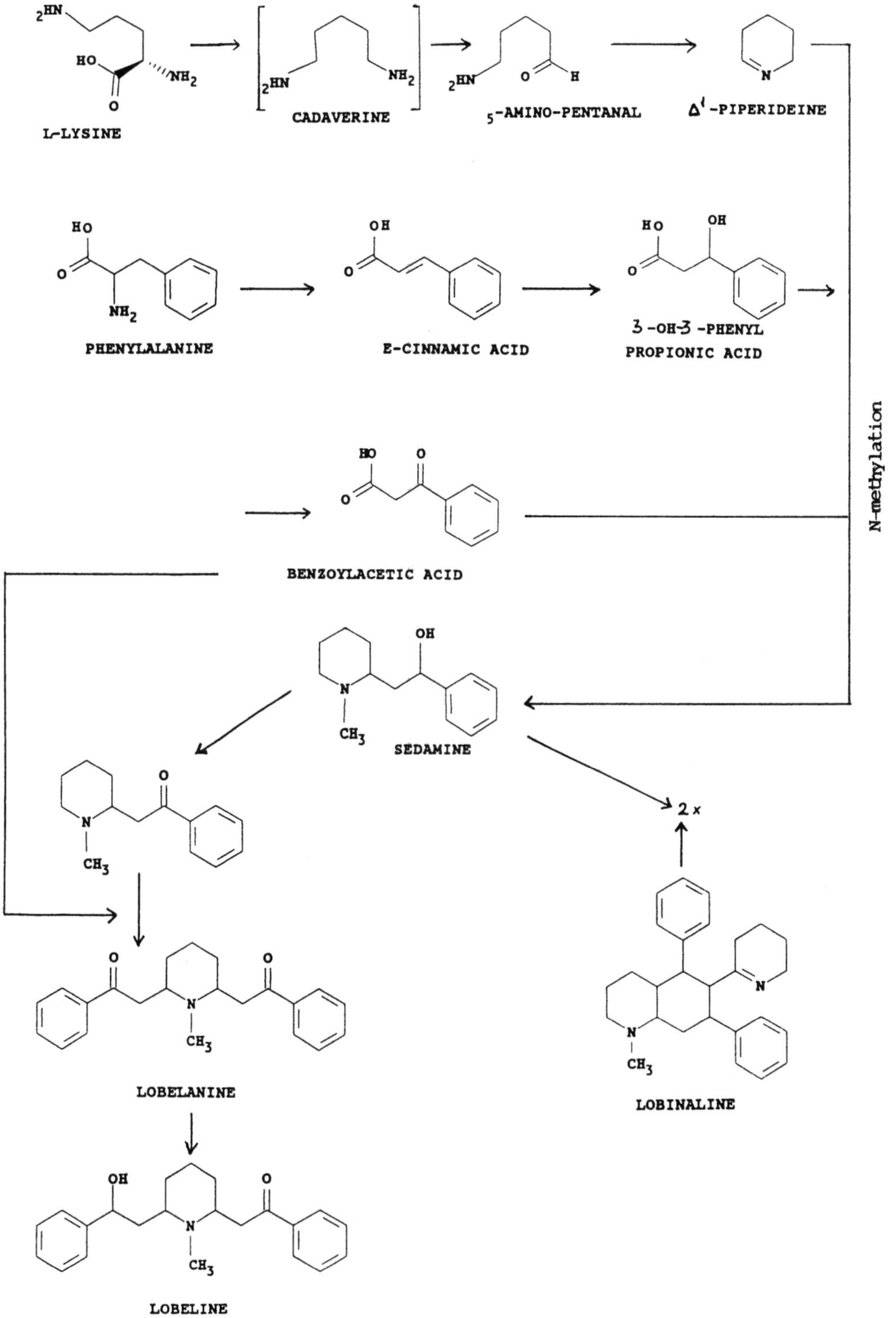
L-LYSINE
CADAVERINE
5-AMINO-PENTANAL
Δ¹-PIPERIDEINE
PHENYLALANINE
E-CINNAMIC ACID
3-OH-3-PHENYL PROPIONIC ACID
BENZOYLACETIC ACID
N-methylation
SEDAMINE
2 x
LOBELANINE
LOBINALINE
LOBELINE

as a precursor amino acid is also present. The individual stages of the synthesis have been dealt with by several authors.

In general, as a first step of the biosynthesis of alkaloids of lysine origin, cadaverine (pentane-1,5-diamine) is produced from the lysine (Mothes and Schütte 1963). However, according to O'Donovan et al. (1975), this process does not take place in *L. inflata.* The lysine is supposed to form directly into 5-aminopentanal, then from this $\Delta^1$ piperidein develops. With this the nitrogen-containing heterocycle has taken shape. The following authors succeeded in explaining the stages of transformation of phenylalanine, the other precursor amino acid. Wieland et al. (1939) then O'Donovan et al. (1975) studied the steps by which phenylalanine was transformed into 3-hydroxy-3-phenyl-propionic acid, among which there was a deamination reaction. As a result of this, phenylalanine first transforms into *trans*-cinnamic acid, and 3-hydroxy-3-phenyl-propionic acid is formed only subsequently. The deamination reaction is catalyzed by the phenylalanine-ammonialyase (Koukol and Conn 1961). From 3-hydroxy-3-phenyl-propionic acid benzoyl acetic acid is formed through an oxidation step. Keogh and O'Donovan (1970) proved with radioactive phenylalanine that the $C_2$–$C_6$ part of the two lateral phenyl rings of lobeline originated from this amino acid.

The $\Delta^1$ piperidein containing the nitrogen-heteroatom forms, and the benzoyl acetic acid becomes condensed and sedamine develops. After the oxidation of sedamine, with the participation of a further benzoyl acetic acid, lobelanine is produced and transforms into lobeline (O'Donovan and Forde 1971). With the condensation of two sedamines, lobinaline is formed, which occurs in *L. cardinalis* (Fig. 3). Among the enzymes that take part in the biosynthesis, the aminotransferases are important (Smogrovicova et al. 1972).

The concentration of the precursor amino acids in various parts of the plant was examined by Smogrovicova et al. (1966). The lysine and phenylalanine content of the plant changes with the stages of ontogenesis. At the time of flowering the amino acid content is low, while the lobeline level is high in the plant (Phe: 0.015%, Lys: 0.005%). After the flowering period, the lobeline content decreases and the amount of precursor amino acids increases in the plant (Phe: 0.09%, Lys: 0.02%). In the root, the correlation between amino acid and lobeline content is different. At the time of flowering the lobeline content of the root is high, but the concentration of amino acid can also be considered high (Phe: 0.07%, Lys: 0.01%); in autumn, on the other hand, the lobeline, phenylalanine, and lysine levels decrease equally.

## 1.4 Qualitative Analysis Methods for Alkaloids

### *1.4.1 Qualitative Identification of Lobeline*

- Chemical identification methods by lobeline (DAB-7, 1968; FP IV, 1970; ÖAB, 1981).
- Identification on the basis of UV spectrum ($\lambda$ max: 249 nm) and IR spectrum ($1/\lambda$ max: 1687 $cm^{-1}$, 700 $cm^{-1}$) (List and Hörhammer 1976).

### 1.4.2 Chromatographic Methods

*Thin Layer Chromatography.* According to the method of Luckner et al. (1968), after extraction with sulfuric acid the extract is alkalized and shaken with ether. The etheric phase is concentrated, then dissolved in methanol, and chromatographed. The developing mixture is chloroform-methanol (44:6), the developing reagent potassium tetraiodo-bismuthate; the alkaloids react with an orange color. The standard Rf value of lobeline is 0.4–0.45.

Many other methods can be found in the literature for the separation of lobelia alkaloids by thin layer chromatography (Waldi 1961; El-Masry et al. 1980).

The chromatograms were developed with Dragendorff reagent after Munier and Macheboeuf (1949). Having tested the developing systems described in the literature, we found the n-butanol-acetic-$H_2O$ mixture of 4:1:1 ratio used by the Pharmacopée Française (Ph.f 1972) to be the best for separating *Lobelia* alkaloids. However, for densitometric measuring used for the quantitative determination of lobeline following thin layer chromatography, the separation of the individual components was not satisfactory even with this system, so we found it necessary to elaborate a new developing system.

*Thin Layer Chromatographic Separation of Lobelia Alkaloids by the Application of the Prisma model.* The selective analytical thin layer chromatographic determination of the alkaloids of Lobelia tissue cultures was carried out by the so-called Prisma model (Nyiredy et al. 1985). By reducing the solvent concentration of the eluent to p = 1.21 the alkaloids could be properly separated.

To analyze the alkaloid content of *Lobelia* tissue culture and intact plants, respectively, a developer of the following composition is recommended:

| | |
|---|---|
| tetrahydro-furan | 1.1 ml |
| acetone | 1.1 ml |
| diethyl-amine | 2.8 ml |
| hexane | 15.0 ml |

Advantages of the System: In comparison to the eluents in the literature, it gives a better separation and is therefore, also suitable for densitometric determination (at 249 nm).

*High Pressure Liquid Chromatography.* Barker et al. (1979) used high pressure liquid chromatography to identify lobelia alkaloids. They chromatographed lobeline on $C_{18}$ column with 0.025 M $Na_2H_2PO_4$ and methanol-water (2:3) eluent.

Wheals (1980) chromatographed lobeline on silica gel modified with n-propyl-sulfonic acid, and employed methanol-2 M ammonium-hydroxide-1 M ammonium-nitrate (27:2:1) + sodium sulfite mobile phase.

Daldrup et al. (1981) elaborated a quick method for the identification of compounds, which used thin layer chromatography, gas chromatography, and high pressure liquid chromatography. In the case of high pressure liquid chromatography, lobeline was applied on RP 18, 10 μm, 250 × 4-mm column and treated with acetonitril, acetonitril-water, in acetonitril buffer (ortho-phosphoric acid + potassium-dihydrogen-phosphate) with mobile phase.

## 1.5 Quantitative Determination of Alkaloids

The methods for the quantitative determination of *Lobelia* alkaloids are divided into two groups: determination of total alkaloid content and determination of lobeline content.

1. Determination of total alkaloid content
   a) Titrimetric methods (B.Ph.Codex 1973; Ph.fr. 1972)
   b) Spectrophotometric methods (Luckner et al. 1968; El-Sayed et al. 1978; Mahmoud and El-Masry 1980 (Fig. 4); Krajewska 1986)

The quantity of total alkaloids was given in terms of dry tissue weight %. On taking up the concentration series for the determination of the total alkaloid content the methanolic solution of lobeline hydrochloride (Roth) was used. The reproducibility of the determination can be represented with a CV % = 3.81 value (n = 6).

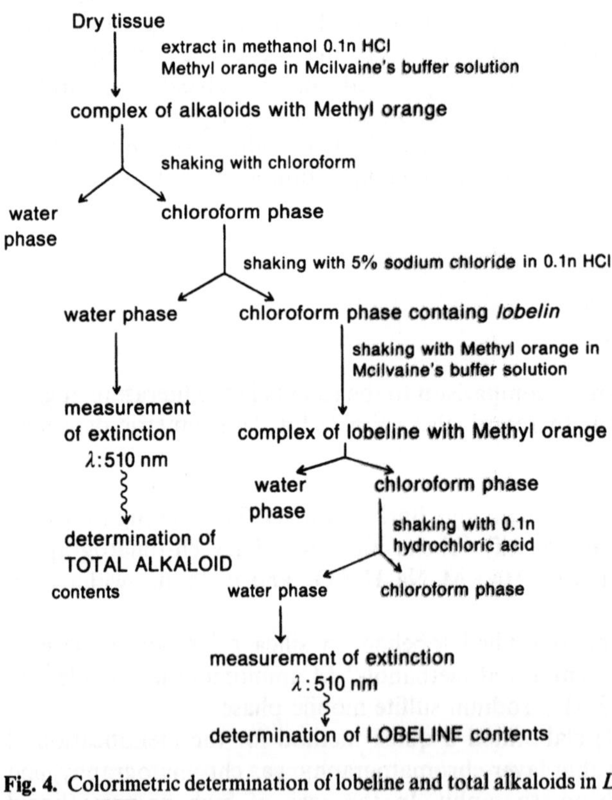


**Fig. 4.** Colorimetric determination of lobeline and total alkaloids in *Lobelia inflata*. (Mahmoud and El-Masry 1980)

2. *Determination of lobeline content*
   a) Titrimetric method (formation of acetophenone) after Uffelie (List and Hörhammer 1976).
   b) Spectroscopic methods (following chromatography): Krochmal et al. (1972); Steinegger and Ochsner (1956a); Kaczmarek (1960).
   c) Densitometrically: Krajewska et al. (1986).

## 1.6 Alkaloid Content in *Lobelia*

The total alkaloid content in the shoot of *L. inflata* varies greatly, depending on the population (Hlavackova 1972). According to Gedeon and Gedeon (1954) it is 0.59%, while Kaczmarek (1961) found it to be 0.48% (Table 1).

According to Smogrovicova et al. (1966) the lobeline content ranges between 0.01 and 0.035% depending on the ontogenetic stage (it is highest at the time of flowering); according to Wysokinska (1977) it is 0.06%.

In other parts of the plant the total alkaloid content is distributed as follows: 0.9–1.1% in the flower, 0.43% in the leaf, 0.38% in the stalk, 0.56% in the root (Gedeon and Gedeon 1954). In the root the lobeline content ranges from 0.02% to 0.07%, depending on the phase of ontogenesis (Smogrovicova et al. 1966).

Besides the alkaloids, the plant contains the following compounds: glycoside lobelacrine, wax, lobelic acid, chelidonic acid (List and Hörhammer 1976), l-phenoxy-propionic acid, d-phenoxy-propionic acid, resin, phytosterine, inflantine, and aromatic compounds (Berger 1954). The seed contains 30% fat (List and Hörhammer 1976).

## 1.7 Therapeutic Use of Lobeliae Herba

In Lobeliae herba, the most active pharmacological compound is lobeline, which has a stimulative effect on the respiratory center. Owing to this property, it is applied in cases of asthma, collapse, gas and narcotic poisoning. Lobeline

**Table 1.** Alkaloid contents in *Lobelia inflata*

| Plant part | Alkaloid content (%) | Lobeline content (%) | Reference |
|---|---|---|---|
| Shoot | 0.59 | | Gedeon and Gedeon (1954) |
| | 0.48 | | Kaczmarek (1961) |
| | 0.395 | | Krajewska (1986) |
| | | 0.01–0.035 | Smogrovicova et al. (1966) |
| | | 0.06 | Wysokinska (1977) |
| | | 0.76 | Krochmal et al. (1972) |
| Root | 0.56 | | Gedeon and Gedeon (1954) |
| | 0.308 | | Krajewska (1986) |
| | | 0.02–0.07 | Smogrovicova et al. (1966) |

also causes nausea. In large doses its effect resembles that of nicotine, so that it is therefore also used in preparations against smoking. It increases the effect of nicotine, thereby producing nausea (Borkowski 1973). Lobeline stimulates the secretion of catecholamine, and in this way increases the blood pressure. Lobeline is mostly applied parenterally [Lobelin hydrochloricum injection (FP IV, 1970)]. Isolobinin, when administered orally, has an antiasthmatic effect, the Lobeliae herba is therefore also used for asthma. Larger doses of isolobinin also cause nausea (Borkowski 1973). Lobelin hydrochloricum, Lobeliae herba, or Tinctura Lobeliae are official preparations in many dispensaries.

## 2 Tissue Culture Studies

### 2.1 Callus Tissues

Wysokinska (1977) was the first to obtain tissue culture from growing tip of shoot. Pieces of tissue sterilized with mercuric chloride were placed on solid culture medium (Schenk and Hildebrandt 1972) containing 5 mg/l 2,4-D and 0.25 mg/l kinetin. The callus produced was then transferred onto a culture medium containing less kinetin (0.1 mg/l) and 2,4-D (0.5 mg/l). The author examined the influence on growth and alkaloid production of the callus by the following regulators: 2,4-D ($10^{-5}$, $10^{-6}$ M), IAA ($10^{-5}$, $10^{6}$ M), NAA ($10^{-6}$, $10^{-7}$ M), kinetin ($10^{-5}$, $10^{-6}$ M). The alkaloids were extracted by the Kaczmarek (1961) method, chromatographed (Kaczmarek 1961; Krochmal et al. 1972), and determined for quantity (Wiewiorówski and Skolik 1959).

The tissues grew best on the culture medium containing 2,4-D ($10^{-6}$ M) (Table 2). The culture medium containing NAA ($10^{-7}$ M) induced rhizogenesis, while 2,4-D did not. The highest alkaloid content was attained under the

**Table 2.** Growth and lobeline content of *Lobelia inflata* tissue cultures. (Wysokinska 1977)

| Treatment | | Growth value | Root formation | Lobeline content, ug/g (dry wt.) |
|---|---|---|---|---|
| 2,4-D | $10^{-5}$ | 9.8 | – | 0 |
| | $10^{-6}$ | 10.2 | – | 100 (0.01%) |
| IAA | $10^{-5}$ | 8.7 | + + | 300 (0.03%) |
| | $10^{-6}$ | 9.4 | + + | 214 (0.02%) |
| NAA | $10^{-6}$ | 8.5 | + + | 225 (0.02%) |
| | $10^{-7}$ | 9.5 | + + + | 280 (0.03%) |
| Kinetin | $10^{-5}$ | 9.0 | + | 198 (0.02%) |
| | $10^{-6}$ | 7.9 | + | 158 (0.015%) |
| Intact plant | – | – | – | 633 (0.063%) |

influence of IAA ($10^{-5}$ M) and NAA ($10^{-7}$ M). 2,4-D inhibited the formation of lobeline.

By means of thin layer chromatography, the author detected maximum of three in the callus and in the stage of rhizogenesis five alkaloids (among them lobelanine, lobeline, and lobelanidine were identified). According to Wysokinska, rhizogenesis had a positive influence on alkaloid production. The author did not observe the full organogenesis of the cultures. In tissue cultures the lobeline content was lower than in intact plants.

Later alkaloid production of callus, suspension, and organized cultures from *Lobelia inflata* were studied by Krajewska and Szöke in the Pharmacognosy Institute of the Semmelweis University of Medicine in Budapest.

*Material and Method.* The callus tissues induced from the leaves of sterile seedlings were raised on MS (Murashige and Skoog 1962) medium with minerals and vitamins in the dark (Fig. 5). The regulators were added to the culture medium in various concentrations, the alkaloid precursor amino acids (phenylalanine, lysine) at $10^{-4}$ M.

The growth data of the callus tissues were determined after Maróti (1976) and Szöke (1979). Statistical evaluation of the results (Sváb 1981): by the Bartlett test, Student test at $P = 5\%$ and $P = 1\%$ level. The total alkaloid content in the cultures was determined photometrically (Mahmoud and El-Masry 1980), while for the lobeline content thin layer chromatography was followed by a densitometric determination at 249 nm (type: CS 920 Shimadzu High Speed TLC Scanner).

The selective analytical thin layer chromatography separation of the alkaloids of tissue cultures was carried out by the application of the Prisma model (Nyiredy et al. 1985; Krajewska, 1986).

*Results.* From callus tissues of leaf origin kept on a MS medium containing 1 mg/l kinetin and 1 mg/l 2,4-D (AMS-2), a vigorously growing tissue line could

**Fig. 5.** Callus tissue of leaf origin of *Lobelia inflata* (on AMS-2 culture medium)

be produced. The growth of the cultures reached 95 mg daily, and they retained their capacity for alkaloid biosynthesis.

### 2.1.1 *Growth Dynamics of Callus Tissues*

Callus cultures reached maximum weight and alkaloid content after 8 weeks of incubation, so it is then that they should be collected for chemical analysis (Table 3). Between the 6th and the 8th week the growth of the cultures is still intensive (about 50% increase in fresh weight). When the cultures were kept on hormone-free culture medium (S), their growth intensity decreased considerably.

### 2.1.2 *Effect of New Synthetic Regulators on Biomass and Alkaloid Production by Callus Tissues* (Krajewska et al. 1987)

The Sz/11 (1,4-dihydroxy-phthalazine-biguanidine salt) regulator added at a concentration of 1 mg/l considerably increased the growth of the cultures compared to the S culture medium, while at higher concentrations (10 and 50 mg/l) it caused significant growth inhibition (Fig. 6). On the other hand, in cultures showing the lowest rate of growth (50 mg/l Sz/11) the alkaloid content (%) was higher than in the vigorously growing ones containing 1 mg/l Sz/11, or in those raised either on S or on AMS-2 culture medium. However, as regards the alkaloid production (per dry weight of one culture) the callus tissues kept on a culture medium containing 1 mg/l Sz/11 are considered better.

The regulator Sz/28 (N-isopropyl-benzimidazolium-chloride) also stimulated the growth of the callus tissue, although an increase in its concentration – similarly to Sz/11 – resulted in significant growth inhibition. The daily growth of the cultures was reduced to almost half (Fig. 6).

**Table 3.** Growth dynamics of *Lobelia inflata* callus raised on AMS-2 culture medium

| Incubation period (week) | Fr. wt. of callus (g) X | $s_{\bar{x}}$ | n | Dry wt. of callus (g) | Dry matter content (%) | Growth value Fr. wt. | Growth value Dry wt. | Daily increase in fresh weight (mg/day) |
|---|---|---|---|---|---|---|---|---|
| 0 | 0.3152 | 0.0237 | [10] | 0.0123 | 3.91 | – | – | – |
| 1 | 0.5088 | 0.0258 | [10] | 0.0295 | 5.80 | 0.61 | 0.13 | 27.65 |
| 2 | 0.5820 | 0.0415 | [10] | 0.0344 | 5.92 | 0.84 | 1.79 | 10.45 |
| 3 | 0.6443 | 0.0238 | [10] | 0.0356 | 5.53 | 1.04 | 1.89 | 8.90 |
| 4 | 0.8661 | 0.0821 | [10] | 0.0456 | 5.27 | 1.74 | 2.70 | 31.68 |
| 5 | 1.6223 | 0.1551 | [10] | 0.0786 | 4.85 | 4.14 | 5.38 | 108.02 |
| 6 | 1.8825 | 0.1876 | [10] | 0.0790 | 4.50 | 4.97 | 5.41 | 37.17 |
| 7 | 2.1246 | 0.1078 | [10] | 0.0807 | 3.80 | 5.74 | 5.55 | 34.58 |
| 8 | 2.5005 | 0.0876 | [10] | 0.0900 | 3.60 | 6.93 | 6.30 | 53.70 |
| 9 | 2.5978 | 0.0814 | [10] | 0.0909 | 3.50 | 7.24 | 6.37 | 13.90 |
| 10 | 2.6181 | 0.1026 | [10] | 0.0911 | 3.48 | 7.30 | 6.39 | 2.90 |

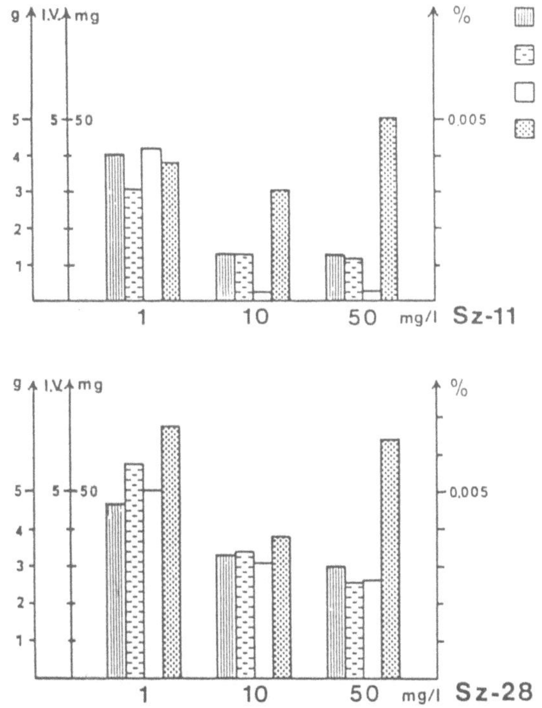


**Fig. 6.** Effects of regulators on growth and alkaloid content of *Lobelia inflata* callus tissue. *1* Fresh weight (g); *2* Increased value on dry weight basis; *3* daily growth (mg); *4* alkaloid content (%). (Krajewska et al. 1987)

An analysis of the percentage total alkaloid content of the cultures reveals that in response to 1 mg/l Sz/28 it increased compared to both the callus tissues grown on S- or AMS-2 basic culture media and those kept on culture media containing 1 mg/l kinetin or Sz/11.

### *2.1.3 Effect of Alkaloid Precursor Amino Acids on Alkaloid Production of Callus Tissues*

In order to further increase the alkaloid production of the cultures, phenylalanine and lysine were added to the culture medium, considering that they are the precursor amino acids of the piperidine base alkaloids that characterize the genus *Lobelia* (Donovan et al. 1975). However, the callus tissues grown on the basic culture medium containing 1 mg/l kinetin and 1 mg/l 2,4-D responded with a significantly reduced growth and alkaloid production to either lysine or phenylalanine (Fig. 7).

Considering that in *Datura innoxia* callus tissues the Sz/28 promoted the incorporation of phenylalanine into alkaloids (Szöke et al. 1984) we added new synthetic regulators instead of kinetin to the culture medium in the presence of the precursor amino acids in *Lobelia* callus also.

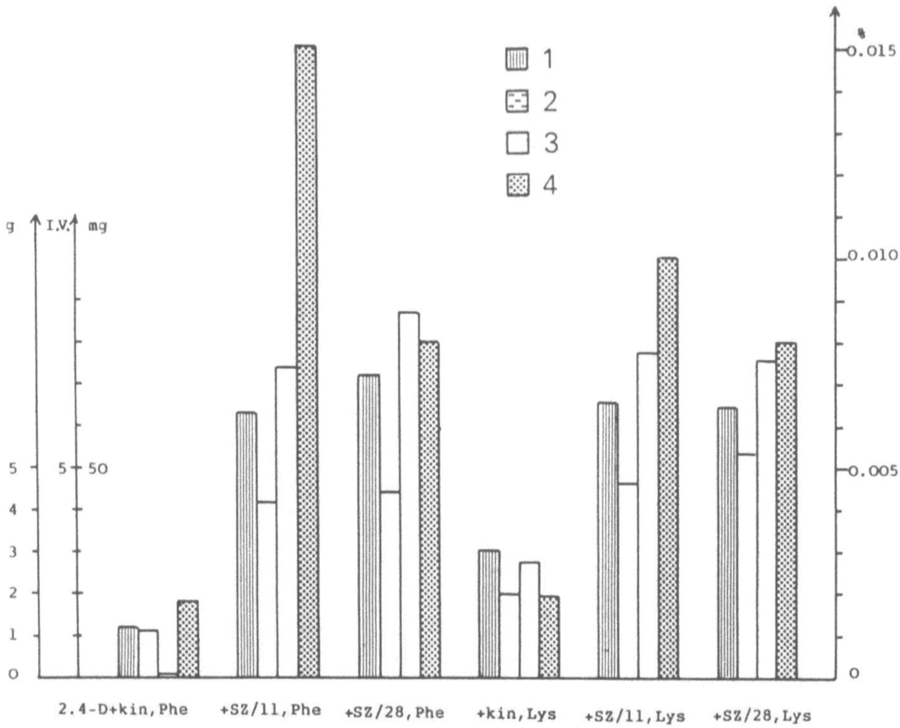


**Fig. 7.** Effects of regulators and precursor amino acids on the growth and alkaloid content of *Lobelia inflata* callus tissue. One mg/l 2,4-D, 1 mg/l cytokinin-like regulator and $10^{-4}$ M amino acid were added to the culture medium (see Fig. 6). (Krajewska et al. 1987)

Both regulators, combined with the amino acids, increased the biomass two- to threefold (compared to kinetin), and increased the alkaloid content still more. On a culture medium containing Sz/11 and 2,4-D, as well as $10^{-4}$ M phenylalanine, the total alkaloid content was 7.5-fold, but since the growth of the cultures is also more intensive, the alkaloid production (mg/culture) is already nearly 20-fold compared to the culture medium of similar composition but containing kinetin (Fig. 7).

On the other hand, for the biosynthesis of lobeline, the effect of Sz/28 is very favorable; the lobeline content is also highest in callus tissues raised on a culture medium containing phenylalanine (Table 4).

Both on basic culture medium and on one containing a regulator, callus tissues synthesize extremely large amounts of a so far unknown component whose quantity is reduced to minimum if the culture medium contains alkaloid precursor amino acids (Fig. 8). This molecule is supposed to be an important intermediary compound of biosynthesis. On the basis of its NMR spectrum, the compound is saturated and contains aromatic molecule parts which can be connected with lobeline, that is, it can be regarded as a lobeline derivative. By mass spectrometry its molecular weight is 485 (Figs. 9, 10).

**Table 4.** Changes in the lobeline content of callus tissues in response to regulators and precursor amino acids. (Krajewska et al. 1987)

| Culture medium | Lobeline content (%) |
|---|---|
| AMS-2 | 0.002 |
| S | 0.002 |
| Kin-1 | 0.001 |
| Sz/11-1 | 0.002 |
| Sz/28-1 | 0.002 |
| Sz/55-1 | 0.002 |
| 2,4-D + Sz/11 + Phe | 0.004 |
| 2,4-D + Sz/11 + Lys | 0.004 |
| 2,4-D + Sz/28 + Phe | 0.006 |
| 2,4-D + Sz/28 + Lys | 0.005 |

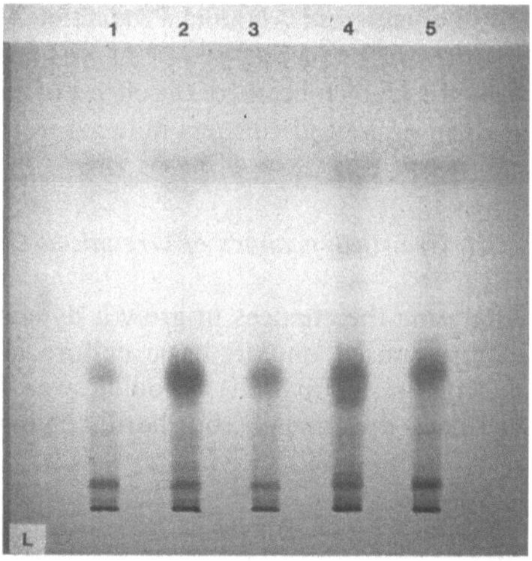


**Fig. 8.** Thin layer chromatography of alkaloids of *Lobelia inflata* callus tissue in $UV_{254}$ light. Developing mixture: tetrahydrofuran-acetone-diethylamine-hexane 1. 1:1. 1:2. 8:15

Thus in callus tissues of leaf origin from *Lobelia inflata* biomass, alkaloid, and lobeline production are all equally optimal if the new synthetic regulators (Sz/11 and Sz/28) are applied in combination with alkaloid precursor amino acids (phenylalanine, lysine).

## 2.2 Alkaloid Production in Suspension Cultures

Cell suspension cultures were raised in Braun-Melsungen Biostat-S type fermentor on basic culture medium (AMS-2), and on a culture medium containing phenylalanine + lysine as supplement. It was found that in the suspension cultures, compared with the basic culture medium, the alkaloid content rose twofold in response to the alkaloid precursor amino acids (Table 5).

Table 5. Total alkaloid content in the callus and cell suspension cultures

| Medium | Culture | Total alkaloid content (%) |
|---|---|---|
| AMS-2 | Callus | 0.003 |
| AMS-2 + Phe + Lys | Callus | 0.003 |
| AMS-2 | Suspension | 0.003 |
| AMS-2 + Phe + Lys | Suspension | 0.006 |

The cultures also synthesize very large quantities of the 485 molecular weight compound produced in the callus tissues (Figs. 9, 10).

## 2.3 Biomass- and Alkaloid Production by Organized Cultures

Analysis is given below of the effects of new type regulators and precursor amino acids in organized cultures induced from callus of hypocotyl origin (Fig. 11).

### *2.3.1 Growth Dynamics of Organized Cultures* (Krajewska and Szöke 1989)

Analyzing the changes in growth dynamics and alkaloid content of organized cultures raised on MS basic culture medium (2 mg/l IAA and 0.2 mg/l kin) during the incubation period, it was found that up to the 8th week they increased dynamically, but hardly changed afterwards (Fig. 12).

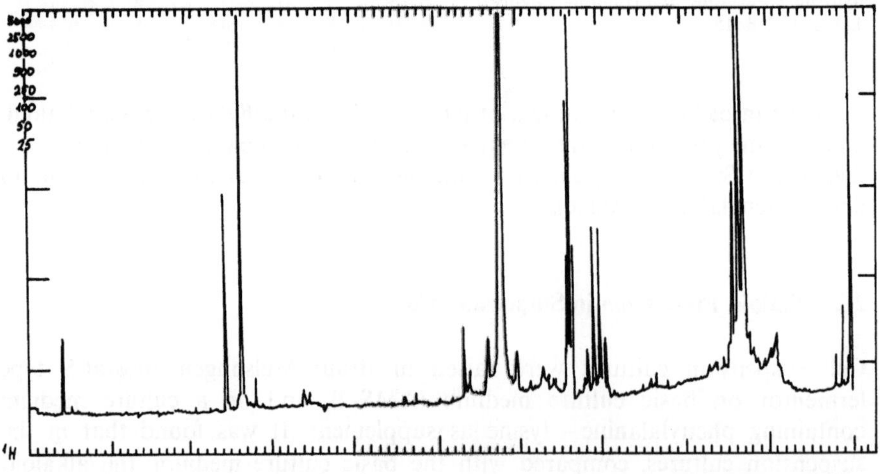

Fig. 9. $^1$H-NHR spectrum of the main component of callus and suspension culture, respectively

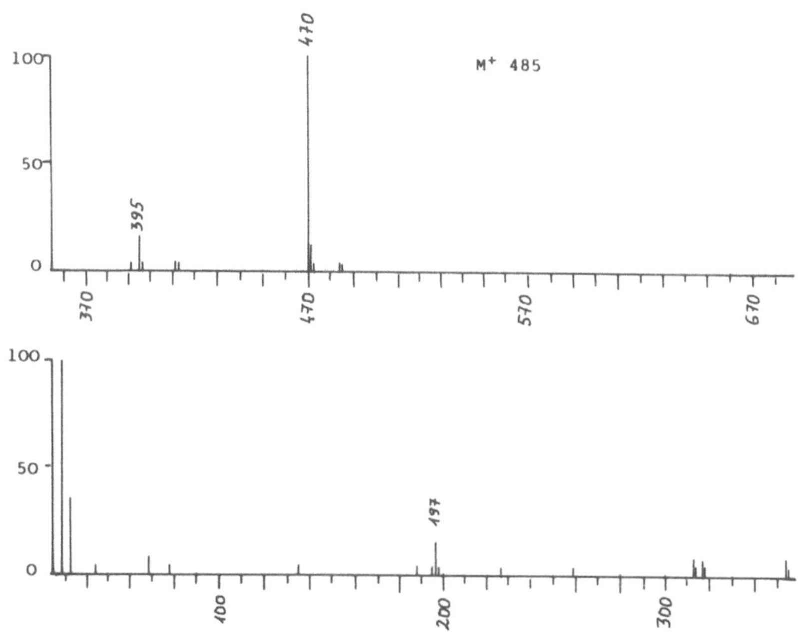


**Fig. 10.** Mass spectrum of the main component of callus and suspension culture

**Fig. 11.** Organized culture of *Lobelia inflata*

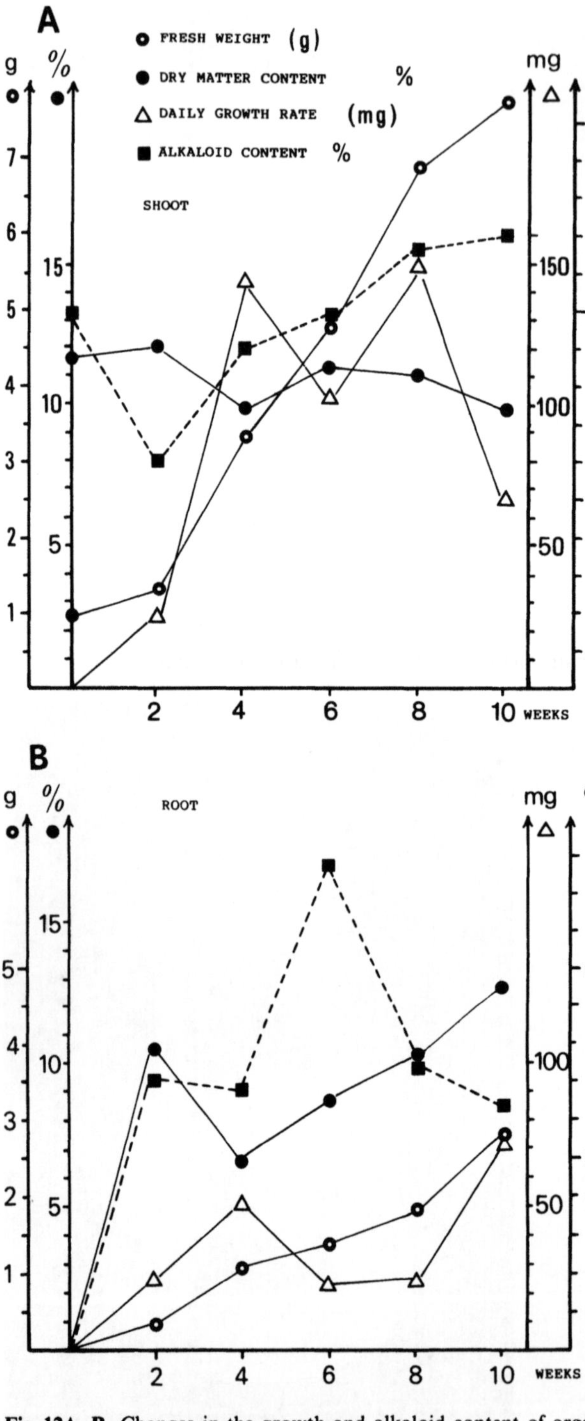


**Fig. 12A, B.** Changes in the growth and alkaloid content of organized culture (**A** shoot; **B** root) during the culturing period. (Krajewska and Szöke 1989)

The total alkaloid content in the cultures is lower than in the intact plant (MS: shoot 0.232%, root 0.147%; intact: shoot 0.395%, root 0.30%), but substantially exceeds that of the callus tissue (0.015%). This is in agreement with the opinion that there is a close correlation between the biosynthesis of secondary substances and the morphogenesis of the plant (Table 6).

From the cultures the total alkaloid components were isolated by preparative chromatography, and their UV-, IR-, mass- and NMR spectra determined with the purpose of studying their structure (Figs. 13, 14, 15).

Both the proton ($^1$H-NMR) and the carbon ($^{13}$C-NMR) spectra unequivocally prove that the substances examined can be regarded as homogeneous, and as different from one another. Both in the saturated and in the aromatic part they show different characters which, however, can be connected with lobeline; they can thus be regarded as lobeline derivatives. In the shoot we identified lobeline, norlobeline, and lobelanidine.

To study the unknown alkaloids of tissue cultures it was necessary to take up the $^1$H and $^{13}$C-NMR spectrum of lobeline, the basic molecule (Fig. 15), considering that in the literature no data on them are available.

### *2.3.2 Effects of Alkaloid Precursor Amino Acids on Alkaloid Production by Organized Cultures* (Szöke et al. 1992, 1993c)

The analysis of the endogenous amino acid level of organized cultures revealed that among the precursor amino acids important for the alkaloid synthesis lysine was present in the shoot of the organized culture in a quantity more than twice the amount in the intact plant, while the quantity of phenylalanine was less than in the intact herb (Table 7).

With phenylalanine and lysine added to the culture medium the free amino acid content considerably increases in the shoots of the cultures, while the growth of the plants is slightly reduced. The alkaloid production by the herb of the cultures is also somewhat lower as a consequence of amino acid application (similarly to callus tissues on basic culture medium). From the point of view of alkaloid formation, the culture medium with IAA + Kin + Phe proved to be the best combination. The total alkaloid content of the shoot is 0.215%, that of the root 0.164%; the latter value is higher than that of the cultures raised on basic culture medium (Fig. 16).

The result considered most important is the seven fold increase in lobeline content of the shoot in response to phenylalanine, although even lysine causes a fourfold increase in it.

To complete the experiment, after incubation with radioactive phenylalanine and lysine, respectively, the scintillation method was used to measure, in the shoot and root of the cultures, the incorporation of the isotopes into the total alkaloid fraction, then – following separation and isolation by thin layer chromatography – into the individual alkaloids.

The radioactive amino acids (DL-3-phenylalanine-1-$^{14}$C and lysine-6-$^3$H) increase the alkaloid biosynthesis better in the root than in the shoot of the organized cultures.

**Table 6.** Trend of biomass formation and alkaloid production in various organs of organized *Lobelia inflata* cultures raised in hormone-free and IAA + Kin 2/0.2 culture medium (in comparison to intact plant)

| Culture medium | Organ | Fresh weight (g) | | Dry weight (g) | Dry matter content (%) | Increased value | | Total alkaloid content (%) | Labeline content % | Alkaloid production mg/culture |
|---|---|---|---|---|---|---|---|---|---|---|
| | | X | $s_{\bar{x}}$ | | | By fresh weight | By dry weight | | | |
| S | Shoot | 5.42 | 0.327 | 0.5435 | 10.02 | 6.70 | 5.65 | 0.186 | 0.021 | 1.01 |
| | Root | 1.53 | 0.265 | 0.1637 | 10.50 | – | – | 0.150 | 0.062 | 0.24 |
| MS | Shoot | 6.88 | 0.309 | 0.7651 | 11.12 | 8.78 | 8.36 | 0.232 | 0.007 | 1.77 |
| IAA + Kin 2/0.2 | Root | 1.77 | 0.117 | 0.1814 | 10.25 | – | – | 0.147 | 0.010 | 0.27 |
| Intact plant | Shoot | | | | – | | | 0.395 | 0.047 | – |
| | Root | | | | | | | 0.300 | 0.173 | – |

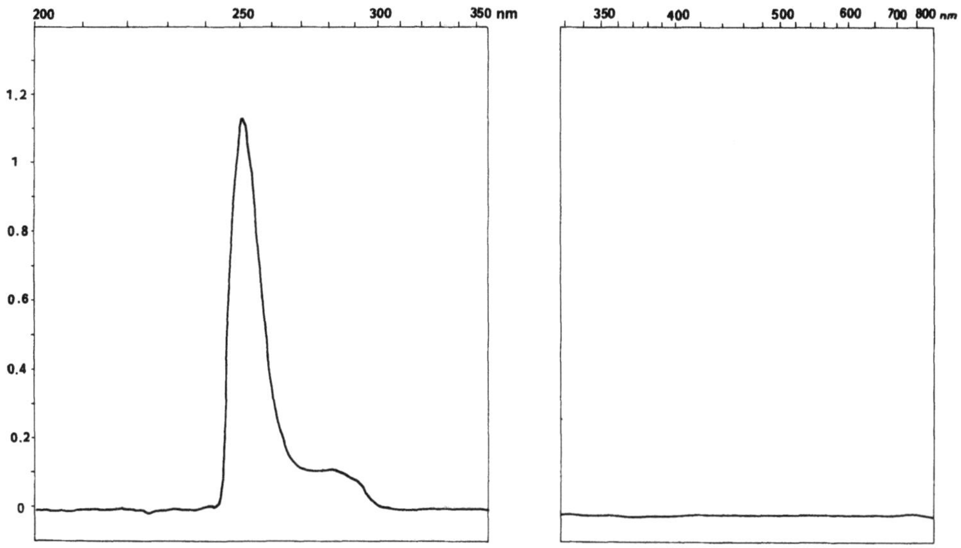


**Fig. 13.** UV spectrum of lobeline

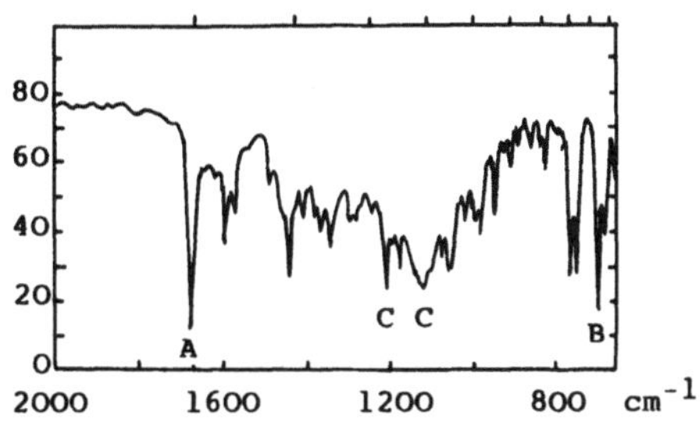


**Fig. 14.** IR spectrum of lobeline (KBr)

The results of scintillation counting clearly show that in the course of incubation the plantlets absorb and incorporate into the alkaloids considerable quantities of amino acids.

In the Dragendorff positive spots detected by thin layer chromatography, 66% of the activity of the purified total fraction is incorporated in the case of incubation with lysine, and only 22% on incubation with phenylalanine. It can thus be established that the organized cultures incorporate lysine in a greater measure than phenylalanine into their alkaloids, so the former can be regarded as a more important alkaloid precursor for the cultures (Fig. 17).

### *2.3.3 Growth and Alkaloid Production of Organized Cultures as Influenced by Regulators and Precursor Amino Acids*

The influence of regulators added one by one to the culture medium at various concentrations was studied (Table 8). Two of the compounds examined – Sz/11

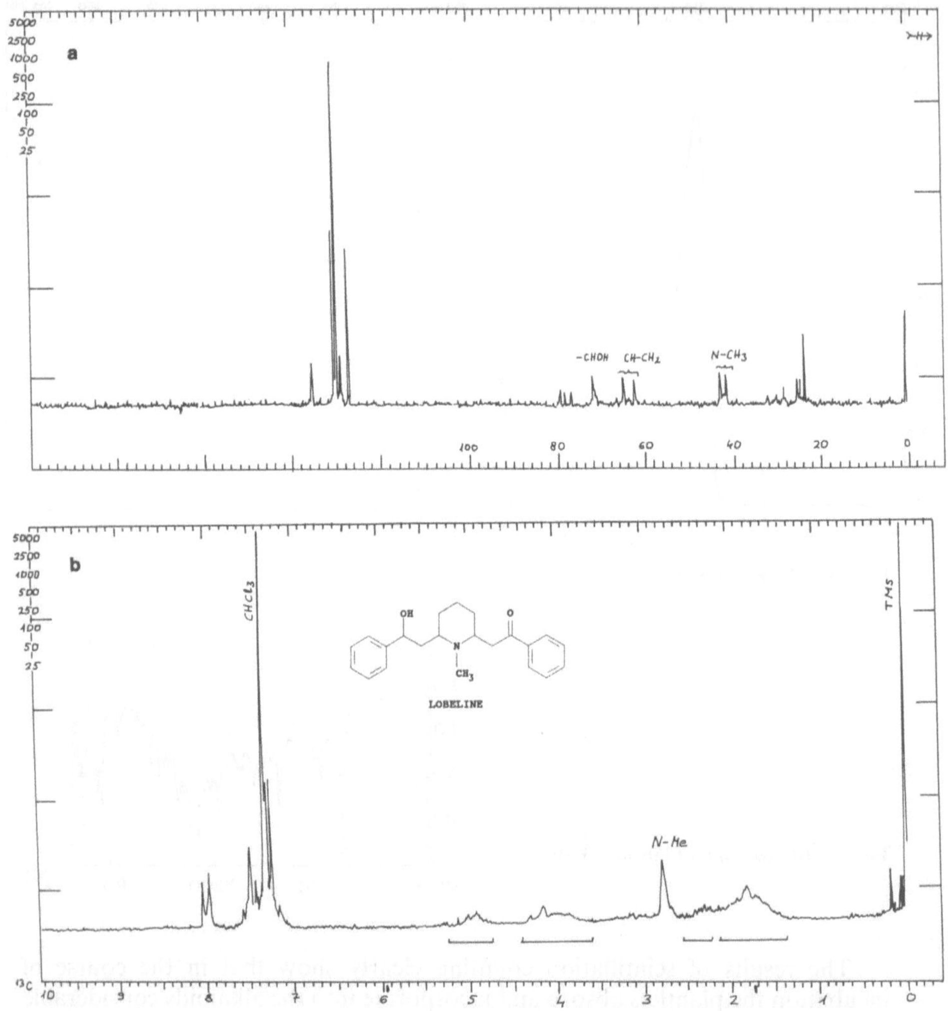


**Fig. 15. a** $^{13}$C-NMR spectrum of lobeline; **b** $^{1}$H-NMR spectrum of lobeline. (Szöke et al. 1992)

and Sz/28 – proved to have a cytokinin-like effect, therefore the influence of a conventionally used cytokinin was also studied.

With a phthalazine derivative examined for the effect on organized cultures of *L. inflata* of the Sz/11 regulator (1,4-dihydroxy-phthalazine-biguanidine salt), it was found that a so-called generative type culture developed. Compared to the kinetin, it stimulated the growth of the herb significantly. With an increase in concentration, however, the growth of the cultures was reduced by it (similar to the kinetin), while their alkaloid- and lobeline content increased.

The other phthalazine derivative – the compound Sz/41 – when applied at a concentration of 1 mg/l, stimulated the development of the herb compared to

**Table 7.** Changes in the free amino acid content in organized *Lobelia inflata* cultures in response to regulators and alkaloid precursor amino acids. (Krajewska and Szöke 1989)

| Culture medium | Part of org. culture | Amino acid content of cultures, $\mu$M/g | | | | | | | | | | | | | |
|---|---|---|---|---|---|---|---|---|---|---|---|---|---|---|---|
| | | Lys | His· | Arg | Asp | Ser | Glu | Pro | Gly | Ala | Val | Ile | Leu | Tyr | Phe |
| IAA + Kin 2/0.2 | Shoot | 2.227 | + | 11.544 | 6.115 | 29.940 | 3.203 | 0.686 | 0.763 | 2.242 | 1.342 | 0.442 | 1.174 | 0.214 | 1.464 |
| IAA + Kin 2/0.2 | Root | 1.313 | 1.838 | 49.875 | 7.088 | 112.088 | 14.438 | 2.888 | 1.313 | 23.100 | 3.150 | 0.525 | 1.575 | – | 1.575 |
| IAA + Kin + Phe + Lys | Shoot | 6.710 | 6.406 | 23.790 | 26.840 | 221.110 | 22.570 | 7.016 | 3.966 | 16.470 | 7.320 | 2.440 | 3.356 | + | 5.490 |
| IAA + Kin + Phe | Shoot | 4.118 | 4.738 | 13.726 | 16.012 | 106.140 | 12.048 | 4.420 | 1.982 | 10.060 | 5.338 | 1.526 | 1.830 | 0.610 | 2.288 |
| IAA + Kin + Lys | Shoot | 5.186 | 6.710 | 21.660 | 24.400 | 170.496 | 18.606 | 6.100 | 3.060 | 11.860 | 6.400 | 1.526 | 2.746 | 0.610 | 3.660 |
| S(hormone-free) | Shoot | 2.625 | 2.625 | 94.500 | 11.025 | 200.813 | 15.488 | 3.413 | 1.313 | 11.288 | 3.413 | 1.050 | 1.050 | 1.313 | 3.150 |
| Sz/11-10 | Shoot | 1.838 | 3.150 | 74.603 | 9.450 | 181.913 | 9.713 | 3.413 | 1.313 | 6.563 | 2.100 | 1.050 | 1.575 | 1.838 | 5.675 |
| Sz/28-10 | Shoot | 2.625 | 2.625 | 105.788 | 16.800 | 224.175 | 15.750 | 5.250 | 2.363 | 11.025 | 4.463 | 1.575 | 2.888 | 1.838 | 5.775 |
| Sz/55-10 | Shoot | 2.906 | 3.395 | 150.332 | 16.003 | 324.912 | 7.759 | 4.849 | 2.425 | 12.609 | 5.819 | 1.455 | 2.425 | 3.395 | 6.789 |
| Kin-10 | Shoot | 5.484 | 5.484 | 60.724 | 34.472 | 159.055 | 25.071 | 5.484 | 3.527 | 10.968 | 7.444 | 1.567 | 2.743 | 0.784 | 2.743 |
| Intact herb | – | 0.916 | 0.610 | 1.220 | 3.660 | 8.646 | 4.210 | + | 0.229 | 4.576 | 2.135 | 1.220 | 2.136 | 0.904 | 1.830 |

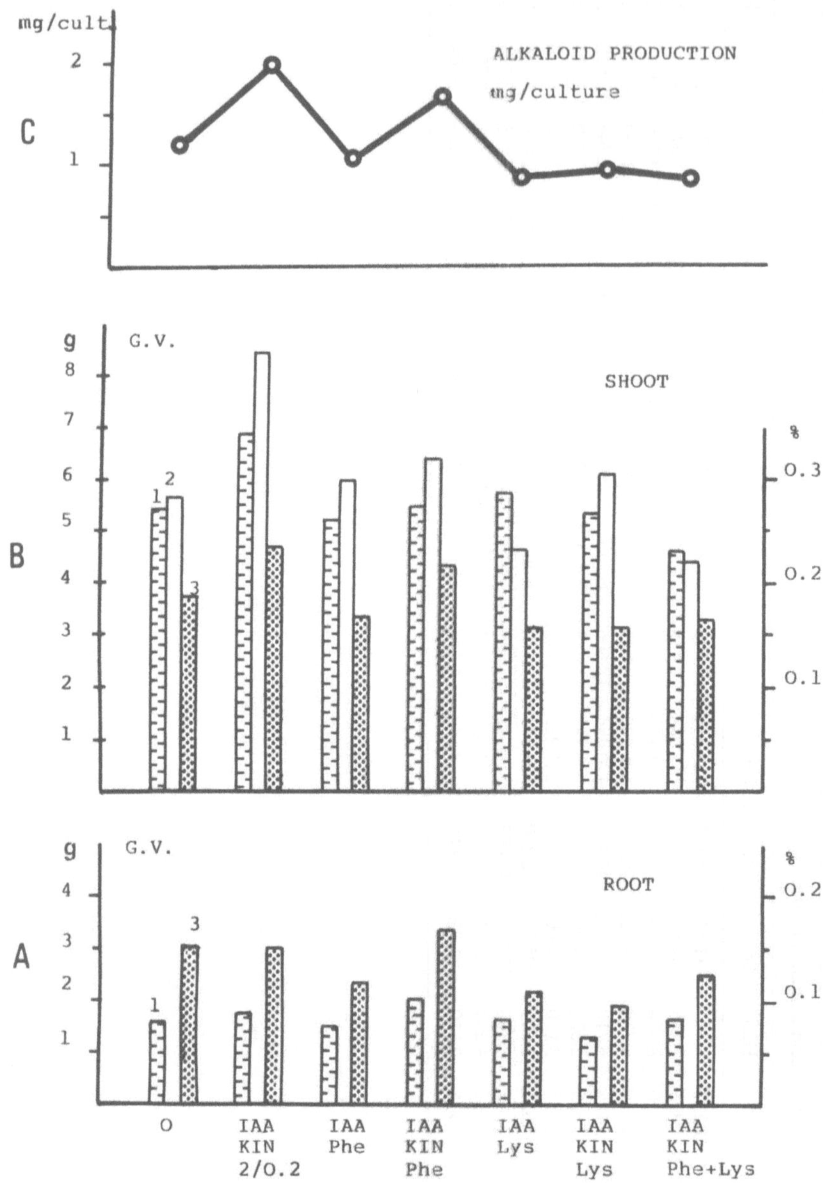


**Fig. 16.** Effect of regulators (Kinetin, IAA) and precursor amino acids on the biomass and alkaloid formation of organized *L. inflata* cultures. **A, B** *1* Fresh weight (g); *2* growth value (G.V.) by dry weight; *3* total alkaloid content (%). **C** Alkaloid production (mg/culture).

kinetin, but inhibited root development. At this concentration its effect agreed, in essentials, with the effect of Sz/11, in respect to both biomass production and total alkaloid content, but the lobeline content was twice that in the former case.

A benzimidazole derivative, Sz/28 (N-isopropyl-benzimidazolium-chloride), also stimulated shoot growth compared to kinetin. With an increase in concentration, biomass production, as in callus tissues, decreased, but the total alkaloid content and lobeline content considerably increased. The alkaloid production of the cultures was better than that under the influence of kinetin.

The hydroquinone derivative Sz/6r had an extremely positive effect on the biomass production of the shoot. In comparison to kinetin, it increased the fresh weight of the shoot twofold. The total alkaloid content remained unchanged, but alkaloid production redoubled. The lobeline content also increased remarkably. With increasing concentration, this compound did not inhibit the growth of the shoot as much as the other compounds examined.

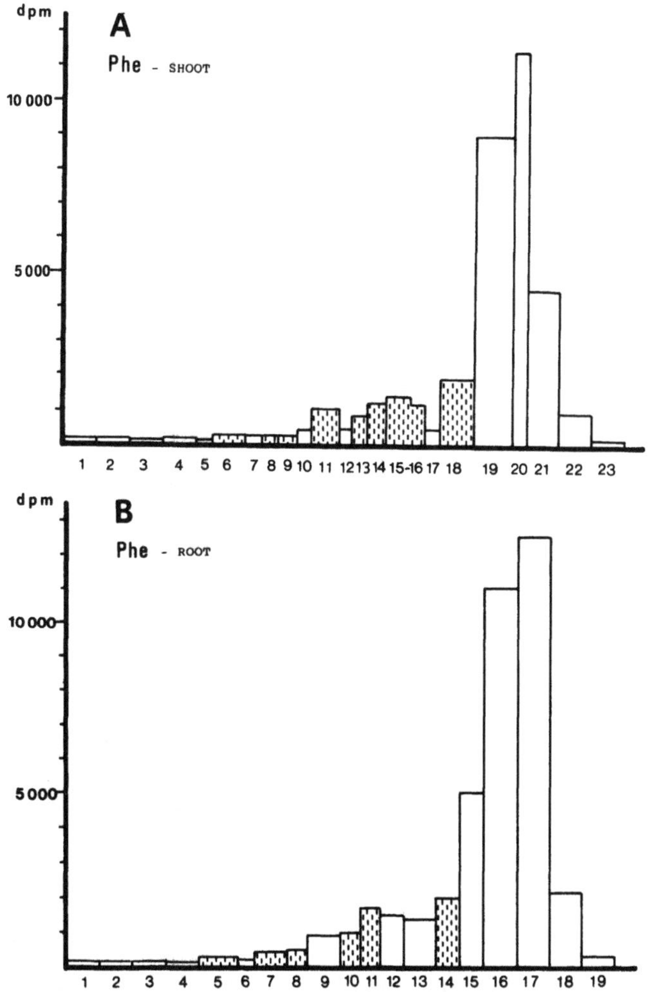


**Fig. 17A–D.** Incorporation of radioactive phenylalanine and lysine in the alkaloids of the shoot (**A**, **C**) and root (**B**, **D**) of organized *L. inflata* cultures, on the basis of thin layer chromatography. The *hatched parts* indicate the activity of the Dragendorff positive spots. (Szöke et al. 1992)

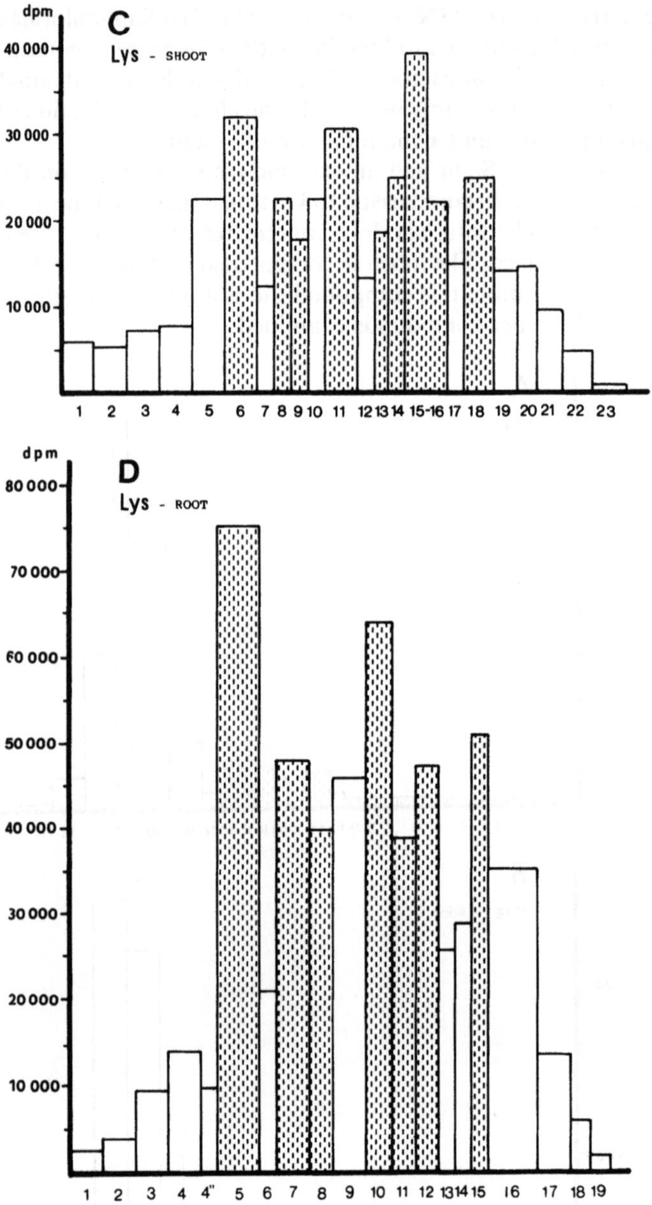


Fig. 17C, D

In response to 50 mg/l Sz/6r the alkaloid content of the roots became very high – 0.176%. The lobeline content also increased parallel with increases in the concentration. At the same time, the quantity of the characteristic root component decreased.

**Table 8.** Effect of regulators of various concentration on the biomass- and alkaloid formation of shoot in organized *Lobelia inflata* cultures. (Krajewska et al. 1986)

| Culture medium | Fresh weight (g) | | Dry weight (g) | Dry matter content (%) | Growth value | | Total alkaloid content (%) | Alkaloid production mg/culture | Lobeline content (%) |
|---|---|---|---|---|---|---|---|---|---|
| | X | $s_{\bar{x}}$ | | | By fr. wt. | By dry wt. | | | |
| S | 5.42 | 0.327 | 0.5435 | 10.02 | 6.71 | 5.65 | 0.186 | 1.01 | 0.021 |
| Kin-1 | 3.64 | 0.444 | 0.3105 | 8.53 | 4.64 | 2.80 | 0.142 | 0.44 | 0.026 |
| Kin-10 | 1.72 | 0.143 | 0.3115 | 18.30 | 1.42 | 2.81 | 0.073 | 0.23 | 0.052 |
| Kin-50 | – | – | – | – | – | – | – | – | – |
| Sz/11-1 | 4.50 | 0.150 | 0.5981 | 13.29 | 5.41 | 6.32 | 0.122 | 0.73 | 0.026 |
| Sz/11-10 | 3.92 | 0.308 | 0.4543 | 11.59 | 4.57 | 4.56 | 0.148 | 0.67 | 0.039 |
| Sz/11-50 | 1.01 | 0.075 | 0.2436 | 24.12 | 0.44 | 1.98 | 0.149 | 0.36 | 0.037 |
| Sz/41-1 | 4.85 | 0.413 | 0.3468 | 7.15 | 5.90 | 3.24 | 0.120 | 0.42 | 0.053 |
| Sz/41-10 | – | – | – | – | – | – | – | – | – |
| Sz/41-50 | – | – | – | – | – | – | – | – | – |
| Sz/28-1 | 4.99 | 0.429 | 0.6257 | 12.54 | 6.10 | 6.66 | 0.155 | 0.97 | 0.016 |
| Sz/28-10 | 4.66 | 0.350 | 0.6384 | 13.70 | 5.65 | 6.81 | 0.165 | 1.05 | 0.023 |
| Sz/28-50 | 3.03 | 0.360 | 0.4060 | 13.40 | 3.31 | 3.97 | 0.251 | 1.02 | 0.025 |
| Sz/55-1 | 6.78 | 0.589 | 0.7343 | 10.83 | 8.64 | 7.99 | 0.153 | 1.12 | 0.030 |
| Sz/55-10 | 4.72 | 0.551 | 0.4833 | 10.24 | 5.71 | 4.92 | 0.129 | 0.62 | – |
| Sz/55-50 | – | – | – | – | – | – | – | – | – |
| Sz/6r-1 | 6.83 | 0.513 | 0.6120 | 8.96 | 8.71 | 6.49 | 0.136 | 0.83 | 0.041 |
| Sz/6r-10 | 7.03 | 0.759 | 0.6376 | 9.07 | 8.99 | 6.81 | 0.092 | 0.59 | – |
| Sz/6r-50 | 5.44 | 0.423 | 0.5125 | 9.44 | 6.72 | 5.27 | 0.071 | 0.36 | 0.042 |

A benztriazole derivative (Sz/55), like Sz/6r, highly stimulated biomass production in the shoot, except at higher concentrations, when it caused inhibition.

At a concentration of 1 mg/l, the fresh weight of the shoot was significantly higher, while its total alkaloid content was somewhat lower than in cultures raised on hormone-free medium; however, the lobeline content redoubled under the influence of Sz/55 compared to cultures raised on hormone-free medium.

The highest alkaloid production in the shoot (1,12 mg/culture) was obtained with this regulator.

The growth intensity of the root was similar to that obtained with Sz/28, but its alkaloid content (0.19%) was essentially higher. Both alkaloid production and lobeline content were highest in the root under the influence of this regulator. With an increase in concentration (to 10 mg/l), lobeline content further increased (0.08%). The amount of the so-called root component, that we isolated for the purpose of structural examinations, also increased to a remarkable extent. On the basis of NMR spectra this compound is probably identical with one of the components isolated from the shoot (7). Considering the saturated and aromatic molecule parts, it can also be regarded as a lobeline derivative (Fig. 18).

To sum up, it can be seen that in comparison with kinetin, as a conventionally applied cytokinin, at 1 mg/l concentration all new type regulators

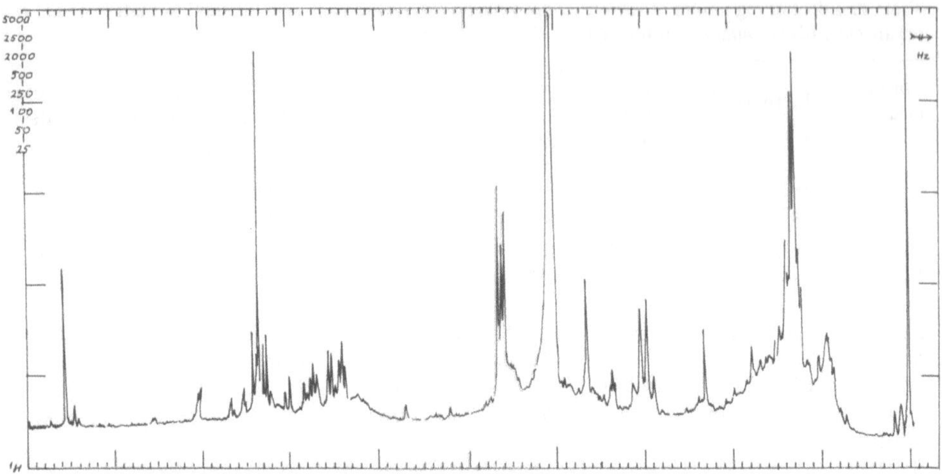

**Fig. 18.** $^1$H-NMR spectrum of the component "7"

stimulated the growth of the shoot (Sz/28, Sz/55, Sz/6r) significantly. The total alkaloid content was increased by Sz/28 and Sz/55, while slightly reduced by the phthalazine derivatives (Sz/11 and Sz/41). The lobeline content, at the same time, was highest in response to Sz/41.

The growth and alkaloid content of the root was favorably influenced by the regulator Sz/55 alone. At a concentration of 10 mg/l, all regulators, except the compound Sz/41, significantly increased biomass production of the shoot compared to kinetin. The total alkaloid content was also more favorably influenced by any of the regulators than by kinetin, although the highest lobeline content was attained in response to 10 mg/l kinetin. At a higher concentration (50 mg/l) the cultures hardly grew, if at all.

*Effect of Regulators on the Amino Acid Composition of the Cultures.* Of the precursor amino acids important for alkaloid biosynthesis, the amount of lysine is increased by kinetin, while the amount of phenylalanine increased in response to Sz/28 and Sz/55.

Considering that we originally produced organized cultures of *Lobelia* on IAA + kin 2/0.2 basic culture medium where they displayed vigorous growth, and even their alkaloid production was very good, we thought it necessary to study the effect of kinetin and of the two regulators already being manufactured (Sz/11 and Sz/28) also in the presence of IAA (2 mg/l) (Fig. 19).

With the application of kinetin, the total alkaloid content (%) and alkaloid production gradually decrease parallel with the increase in concentration, while the lobeline content increases threefold.

If besides IAA, Sz/11 is added to the culture medium, its increasing concentration causes a gradual decrease in the fresh weight of the shoot. However, the alkaloid content of shoot and root is the highest (shoot: 0.267%) in response to 10 mg/l. The alkaloid production substantially exceeds that of cultures raised with a similar concentration of kinetin.

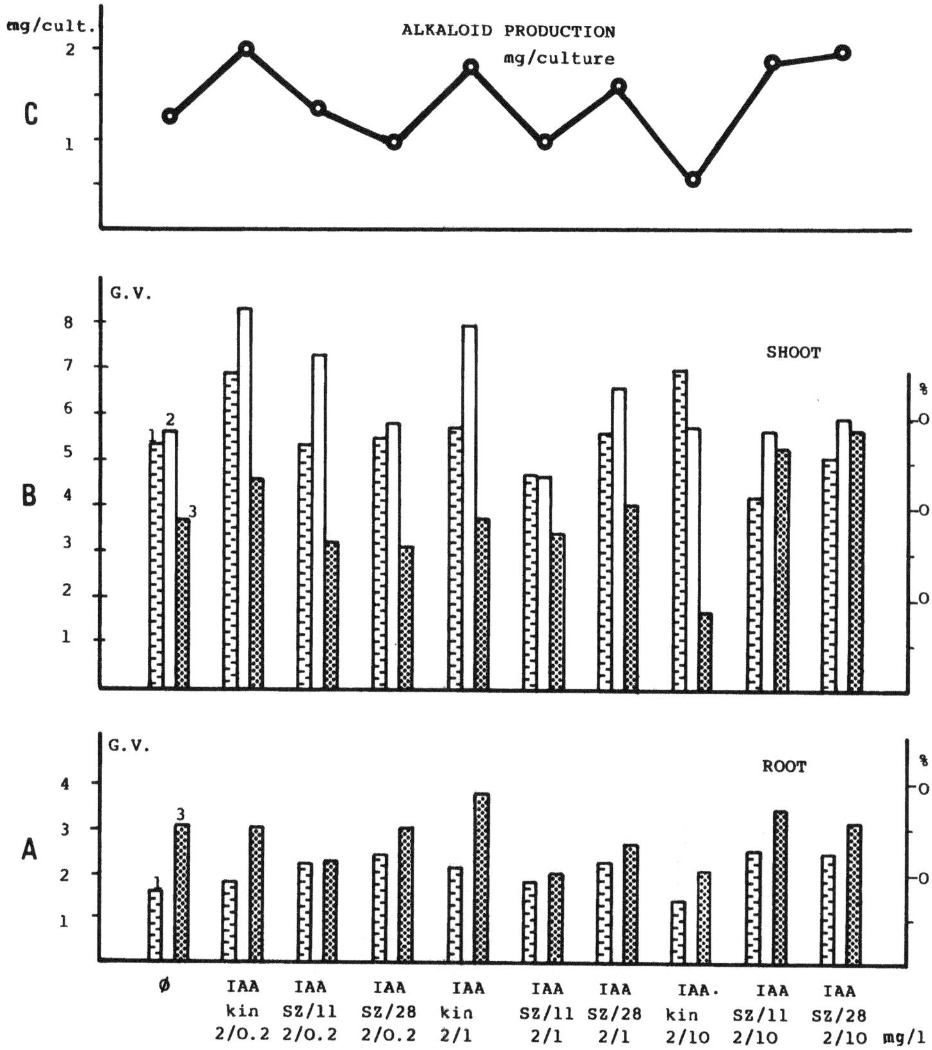


**Fig. 19.** Effect of IAA and cytokinin-like regulators on the biomass and alkaloid formation of organized *L. inflata* cultures. **A, B** *1* fresh weight (g); *2* growth value (G.V.) by dry weight; *3* total alkaloid content (%). **C** Alkaloid production (mg/culture)

With a joint application of IAA and Sz/28, the cultures develop well at all three concentrations. Parallel with an increase in the concentration, the alkaloid content gradually rises. In response to 10 mg/l the lobeline content is three times as high (0.41%) as when Sz/11 is applied.

Thus, it can be established that the alkaloid content of organized cultures is essentially higher if the cultures are raised on a medium containing both IAA and some new type cytokinin-like regulator (compared with cultures raised on a medium containing a new type regulator by itself). Therefore, to further influence the alkaloid content, the medium containing 1 mg/l IAA and 1 mg/l of a new type regulator (Sz/11 or Sz/28) was completed also with alkaloid precursor amino acids (phenylalanine, lysine) (Figs. 20, 21).

When the compound Sz/11 is applied, the biomass and total alkaloid production by the shoot and root of the cultures will be optimal if phenylalanine is added to the medium.

If the alkaloid precursor amino acids are added to the culture medium containing 2 mg/l IAA beside Sz/28, lysine has an optimal effect significant at

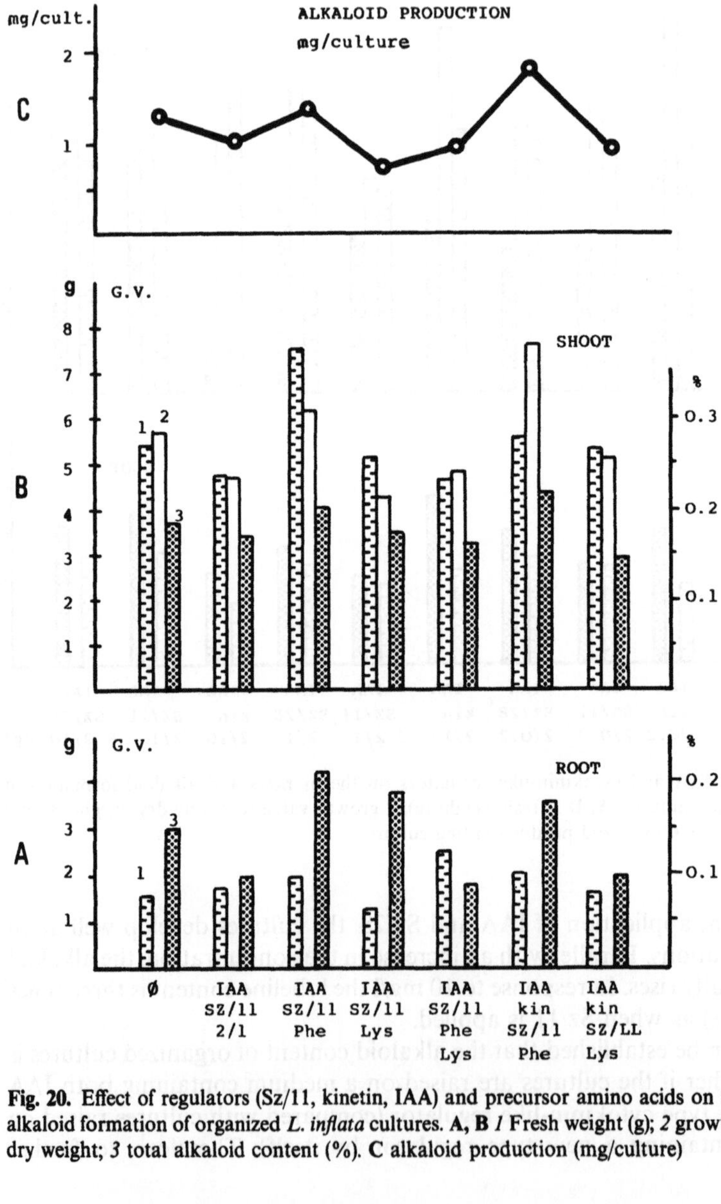


**Fig. 20.** Effect of regulators (Sz/11, kinetin, IAA) and precursor amino acids on the biomass- and alkaloid formation of organized *L. inflata* cultures. **A, B** *1* Fresh weight (g); *2* growth value (G.V.) by dry weight; *3* total alkaloid content (%). **C** alkaloid production (mg/culture)

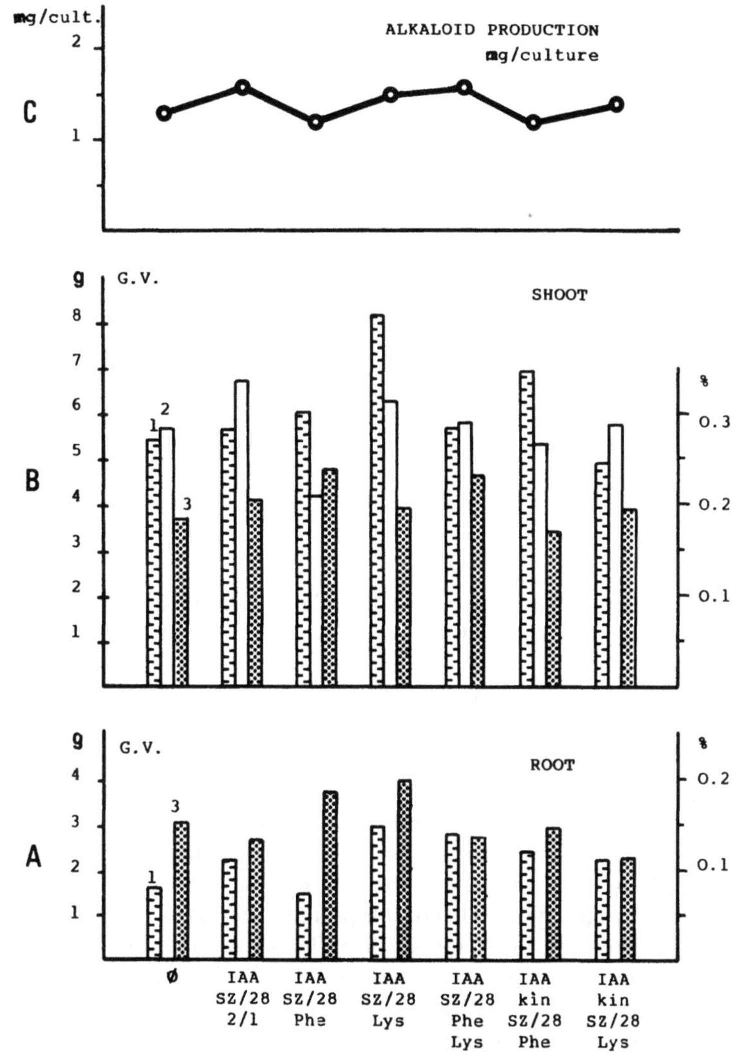


**Fig. 21.** Effect of regulators (IAA 2 mg/l + Sz/28 1 mg) and amino acids (phenylalanine or lysine) on the development of organized *L. inflata* cultures. **A, B** *1* Fresh weight (g); *2* growth value (G.V.) by dry weight; *3* total alkaloid content (%). **C** Alkaloid production (mg/culture)

$p = 1\%$ and $p = 5\%$ on biomass production by the shoot and root of the cultures (8.12, and 2.90 g, respectively), as opposed to the application of Sz/11, when the effect of phenylalanine was outstanding. In response to lysine, the lobeline content of the herb also shows a considerable increase (0.085%), while with the application of phenylalanine it is only half. It should be noted that the regulator Sz/11 with either amino acid induces lobeline biosynthesis to a lesser extent than the combination of Sz/28 and lysine.

The optimal effect of lysine applied in combination with the regulator Sz/28 is explained by the fact that under the influence of Sz/28 added by itself to the culture medium, the endogenous phenylalanine level of the cultures rises. It is thus supposed that an external application of the other alkaloid precursor – lysine – is required for the biosynthesis of lobeline.

Finally, it can be established that the biomass and alkaloid production of both shoot and root are stimulated by the combined application of new type regulators (besides IAA) and alkaloid precursor amino acids.

If cultures having a different organization level (callus and organized culture) are compared for the combined effect of new types of regulators and amino acids, the following conclusions can be drawn:

- the regulator Sz/11 combined with phenylalanine considerably increases the total alkaloid content in both types of culture,
- under the influence of Sz/28, the lobeline content increases first (in organized cultures in response to lysine, in callus tissues as a result of phenylalanine application).

### 2.4 Increasing the Propagation Rate in Vegetative Micropropagation
(Pomázi and Szöke 1989; Szöke et al. 1992, 1993b)

In plants propagated in vitro, the biosynthesis of leaf surface waxes and of fatty acids found primarily in cell membranes suffers damage, with the consequence that test tube explants are not able to tolerate fluctuation of temperature, and owing to membrane injury, dry out however high the relative humidity of the environment, and a considerable proportion of them die.

In order to increase the efficiency of propagation of explants in the course of in vitro micropropagation, the cultures were treated with a chloro-acetyl-hexamethylene type (TI-35) compound (Mészáros 1986; Pomázi and Szöke 1989; Szöke et al. 1992). Under the influence of this compound, the composition of the wax components and membrane lipids becomes similar to that of intact plants raised in the field, that is, the wax layer thickens, the amount of acids in it decreases, while the proportion of primary alcohols increases, which prevents the overpermeability of the leaf surface. As for the lipids, the proportion of unsaturated fatty acids increases, which makes the membrane more flexible. As a result of these changes the drought tolerance of the plants increases.

Organized cultures treated with the compound TI-35 tolerated transplantation stress well compared with the control group (the rate of survival changed according to a maximum curve). At a concentration of 5 mg/l the compound had a favorable effect on the survival of the explants, in fact, nearly 100% (Figs. 22, Table 9). The vulnerability of in vitro cultures can be traced back not only to a change in biochemical processes, but also transformation of the anatomical conditions influences the viability of the cultures. According to our experiments, there are differences in histological characteristics between in vitro and in vivo plants. In the case of greenhouse plants, the mesophyll is more differentiated, the palisade layer and the spongy parenchyma are well separated.

**Fig. 22.** Organized cultures of *Lobelia inflata* after transplantation. (Szöke et al. 1992).

**Table 9.** Percentage survival of organized cultures of *Lobelia inflata* after transplantation. (Pomázi and Szöke 1989)

| Treatment | | TI-35 (mg/l) | | | |
|---|---|---|---|---|---|
| Culture | Control | 2 | 5 | 10 | 20 |
| Explant | 96 | 105 | 128 | 80 | 128 |
| Explant surviving | 0 | 52 | 120 | 30 | 45 |
| Survival % | 0 | 49 | 94 | 37 | 35 |

The result of comparison of stomata on excoriated pieces of epidermis may partly explain the differences in survival rate observed on transplantation. It is known from published data (Wandle et al. 1979, 1983) that the stomata of plants raised in vitro are larger and fewer than those of plants grown in vivo, which has been confirmed by our own measurements. By comparing the untreated plants cultivated sterile (control) with those grown in greenhouse (control) for the number of stomata on the abaxial epidermis, it can be seen that in the course of transplantation the number of stomata redoubles, but at the same time their size slightly decreases. In the organized cultures treated with TI-35, the number of stomata also increased significantly. This is a favorable tendency considering that this numerical increase makes it possible for the plants to react quickly to fluctuation in humidity in the very instant of transplantation. On the abaxial epidermis of the treated plants, the number of stomata – after transplantation – exceeds by far even in vivo that of the control group. The number of stomata per $mm^2$ increases by 100 on an average, in the same way as in the control group (Fig. 23).

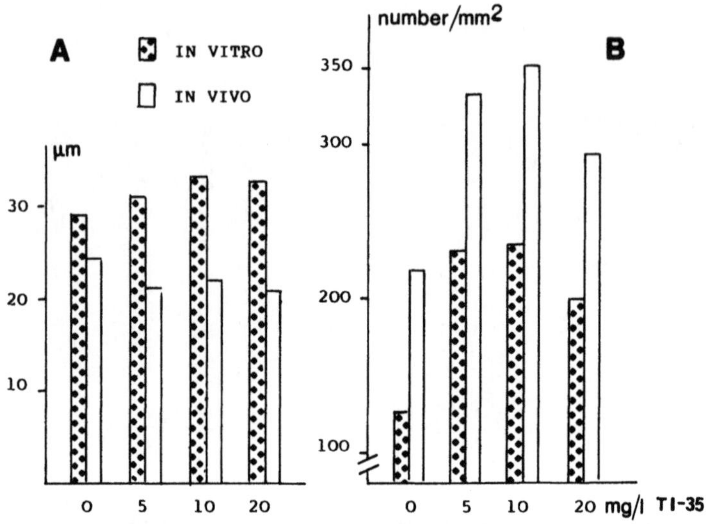


**Fig. 23A,B.** Size of stoma (**A**) and number of stomata (**B**) on the abaxial leaf epidermis of organized cultures of *Lobelia inflata.* (Pomázi and Szöke 1989)

As regards the stoma size, no such unambiguous correlations were found as in the case of stoma number. As a consequence of transplantation, the size of stoma was found to decrease significantly with both untreated and treated plants. Treatment with TI-35 had no well definable effect on the dimensional change of stomata; it caused only a slight increase in vitro and a slight decrease in vivo. On the basis of variance analysis the latter can be regarded as significant.

Considering that our experiments were carried out with medicinal plants, an important question, beyond increasing the rate of survival, was whether the compound TI-35 had a negative influence on the biomass and alkaloid production of the plants. The compound tested reduced to some extent the percentage total alkaloid content of the shoot both in vitro and in vivo, but since it increased biomass production, thus the alkaloid production on dry weight basis slightly increased. It should be noted that the percentage of alkaloid content of the shoot in plants raised under greenhouse conditions agreed, in essentials, with that of intact plants grown in the field. It is interesting, however, that the alkaloid content of the root was in each treatment very high, about four times that in the root of the intact plant (Table 10).

To sum up, our examinations confirm that the antidote-like compound TI-35 can be successfully used in the course of vegetative micropropagation, not only in the case of agricultural and horticultural crops, but also in the cultivation of *Lobelia inflata,* because it increases the rate of survival and has no disadvantageous effect on, but even slightly increases, the production of the special metabolic products.

**Table 10.** Changes in the alkaloid content of organized cultures of *Lobelia inflata* in response to treatment (before and after setting out). (Pomázi and Szöke 1989)

| TI-35/mg/l | | In vitro | | In vivo | |
|---|---|---|---|---|---|
| | | Total alkaloid content (%) | Alkaloid production mg/test tube | Total alkaloid content (%) | Alkaloid production mg/plant |
| 0/control | Shoot | 0.278 | 1.89 | 0.478 | 1.32 |
| | Root | 0.151 | 1.01 | 1.232 | 0.83 |
| 5 | Shoot | 0.206 | 1.58 | 0.425 | 1.66 |
| | Root | 0.163 | 1.11 | 1.630 | 0.43 |
| 10 | Shoot | 0.246 | 1.99 | 0.397 | 1.78 |
| | Root | 0.078 | 0.61 | 1.105 | 0.52 |
| 20 | Shoot | 0.200 | 1.62 | 0.403 | 1.43 |
| | Root | 0.068 | 0.57 | 1.159 | 1.05 |
| Intact | Shoot | | | 0.395 | |
| | Root | | | 0.308 | |

## 2.5 Hairy Root Cultures

Recently, the production of lobeline by the hairy roots of *L. inflata* transformed with *Agrobacterium rhizogenes* ATCC 15 834 has been reported (Yonemitsu et al. 1990).

Continuing our chemical research on this culture, three new polyacetilene compounds were isolated (Ishimaru et al. 1991, 1992).

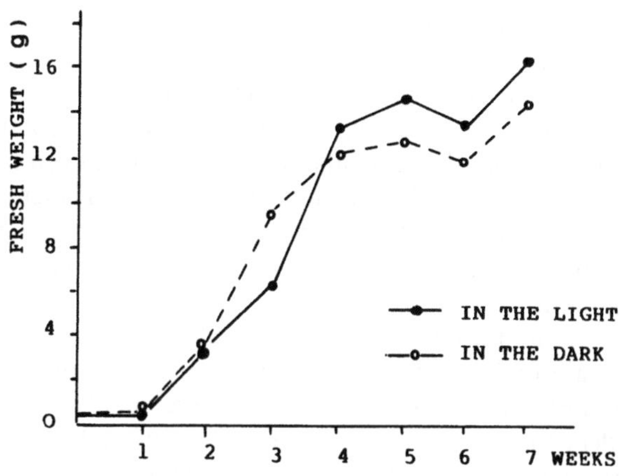


**Fig. 24.** Growth rate of the hairy roots from *L. inflata*. (Ishimaru et al. 1991)

Hairy roots were induced by direct infection with *A. rhizogenes* strain ATCC 15834 and maintained in hormone-free MS liquid medium. Ca. 0.4 g (fr. wt.) of the hairy roots were inoculated into MS liquid medium containing 3% sucrose (50 ml in 100 ml flask) and cultured at 25 °C on a rotary shaker at 100 rpm in the dark or light. The growth rate was determined periodically and the data shown as the mean of two determinations (Fig. 24).

From hairy roots (clone Li-A) cultured for 6 weeks in the dark on a rotary shaker (100 rpm), three polyacetylenes (lobetyol, lobetyolin, and lobetiolinin) were isolated, and their structure established on the basis of chemical and spectroscopic ($^1$H and $^{13}$C-NMR) evidence (Ishimaru et al. 1991, 1992). For details on transformation studies and production of secondary metabolites by hairy roots of *Lobelia* see Tanaka et al. (1993).

## 3 Summary and Conclusion

Alkaloid precursor amino acids and plant growth regulators affect not only the growth and differentiation of tissue cultures but also their secondary metabolism. Studies on the biomass and alkaloid production of in vitro and in vivo cultures and callus and cell suspension cultures from *Lobelia inflata* are summarized here. On the basis of qualitative and quantitative chromatographic results – isolation and spectroscopic ($^1$H-NMR, $^{13}$C-NMR, MS) identification of the major compounds, as well as comparison with intact plant and authentic substances – it was stated that the organized cultures have the ability to synthetize the characteristic alkaloids of the intact plant.

Among the more than 20 piperidine-base alkaloids contained in the shoot, special attention has been directed to the study of the content of the nicotine agonist α-lobeline because of its important role in medicine (e.g., stimulative effect on the respiratory organs).

The new cytokinin-like regulators (N-isopropyl-benzimidasolium-chloride and 1,4-dihydroxy-phtalazine-diguanidine-chloride) had a more positive influence on biomass and alkaloid content in the organized cultures than kinetin.

When we used regulators with alkaloid precursors (phenylalanine, lysine) in the medium, the total alkaloid content was even more increased in the case of phenylalanine; while the lobeline content was increased better on adding lysine.

In addition, a 95% survival rate was achieved during in vivo micropropagation of organized cultures using a chloracethyl-hexamethylene-type compound.

Thus, the compounds described above can be efficiently used in the future to attain higher biomass and active substance production in the pharmaceutical industry and field production of medicinal plants. From previous experience it is clear that we have to study the effect of such regulators on the biosynthesis of special plant compounds before making decisions about their use (yield increasers, herbicides, etc.) in the field cultivation of medicinal plants.

# 4 Protocol

Media and culture conditions
Callus tissues of leaf origin from *Lobelia inflata*, grown in the dark on MS medium containing 1 mg/l kin and 1 mg/l 2,4-D; 26 °C.
Alkaloid production: MS medium + 1 mg/l 2,4-D + 1 mg/l SZ-11 + $10^{-4}$ M phenylalanine
Lobeline production: MS medium + 1 mg/l 2,4-D + 1 mg/l SZ-28 + $10^{-4}$ M phenylalanine
Cell suspension cultures produced from callus tissues were raised in Braun-Melsungen Biostat S-type fermentor on MS liquid medium 1 mg/l kin + 1 mg/l 2,4-D + $10^{-4}$ phenylalanine and lysine.
Organized cultures raised on MS basic medium (2 mg/l IAA and 0.2 mg/l kin, in the light, 14-h photoperiod, 25 ± 2 °C. Culture periods: 6–8 weeks.
Alkaloid production (in the shoot) : MS medium + 1 mg/l IAA + 1 mg/l SZ-11 + $10^{-4}$ M phenylalanine.
Lobeline production (in the shoot): MS medium + 2 mg/l IAA + 1 mg/l SZ-28 + $10^{-4}$ M lysine
Micropropagation: MS medium 2 mg/l IAA + 0.2 mg/l kin + 5 mg/l TI-35.
Hairy roots: induced by direct infection with *Agrobacterium rhizogenes* ATCC 15834, growth in hormone-free MS liquid medium, for 6 weeks in the dark on rotary shaker (100 rpm), 25 °C.

*Acknowledgments.* The author wishes to acknowledge the useful information provided by Dr. Anna Krajewska (Poznan, Poland).

# References

Barker JK, Skelton RE, Cheng-Yu Ma (1979) Identification of drugs by high-pressure liquid chromatography with dual wavelength ultraviolet detection. J Chromatogr 168:417–427
Berger F (1954) Handbuch der Drogenkunde. Verlag für Medizinische Wissenschaften Wilhelm Maudrich, Wien, pp 299–312
Boit HG (1961) Ergebnisse der Alkaloid-Chemie bis 1960. Akademia-Verlag, Berlin, pp 125–153
Borkowski B (1973) Chromatografia cienkowarstwowa w analizie farmaceutycznej. Panstwowy Zaktad Wydawnictw Lekarskich, Warszawa
British Pharmaceutical Codex (1973) The Pharmaceutical Press, London, pp 270–271
Cleene M de, Ley J de (1981) The host range of infections hairy-root. Bot Rev 47:147–194
Daldrup T, Susanto F, Michalke P (1981) Combination of TLC, GLC (OV 1 and OV 17) and HPLC (RP 18) for a rapid detection of drugs, intoxicants and related compounds. Fresenius Z Anal Chem 308:413–427
El Masry S, El-Ghazool MG, Omar AA (1980) A simple TLC method for the separation and semiquantitative evaluation of alkaloids in liquid galenicals. Sci Pharm 48:330–334
El Sayed MA, Abdel Salam NA, Mohammed YA (1978) Spectrophotometric determination of emetine and lobeline by charge transfer complexation. Planta Med 34:430–436
Gedeon J, Gedeon S (1954) Die Alkaloide von *Lobelia nicotianaefolia* Heyne. Acta Helv 29:49–51
Heeger EF (1956) Handbuch des Arznei- und Gewürzpflanzenbaues Drogengewinnung. Deutscher Bauerverlag, Berlin
Hegnauer R (1966) Chemotaxonomie der Pflanzen Bd. 4. Birkhäuser Verlag, Basel, pp 404–414
Hlavackova Z (1972) Inzmiencivost'nektorych priznakov u lobelii odutlej (*L. inflata*). Cesk Farm 21:319–324
Ishimaru K, Yonemitsu H, Shimomura K (1991) Lobetyolin and lobetyol from hairy root culture of *Lobelia inflata*. Phytochemistry 30:2255–2257
Ishimaru K, Sadoshima S, Neera S, Koyama K, Takahashi K, Shimomura K (1992) A polyacetylene gentiobioside from hairy roots of *Lobelia inflata*. Phytochemistry 31:1577–1579
Kaczmarek F (1960) Mikrometoda fotometrycznego oznaczania lobeliny. Biul IRL 21:1–17
Kaczmarek F (1961) Badania chemiczne niektórych gatunków rodzaju *Lobelia*. Biul IRL 26:89–156

Kaczmarek F, Steinegger E (1958) Paper chromatography of alkaloids of *Lobelia*. Pharm Acta Helv 33:201–206

Keogh MF, O'Donovan DG (1970) Biosynthesis of lobeline. J Chem Soc 6:2470–2472

Koukol J, Conn EE (1961) The metabolism of aromatic compounds in higher plants. IV. Purification and properties of the phenylalanine deaminase of *Hordeum vulgare*. J Biol Chem 236:2692–2698

Krajewska A (1986) Ujtipusú regulátorok hatása a *Lobelia inflata* L. szövettenyészetekre. PhD Thesis, Budapest

Krajewska A, Szöke É (1989) The studies on regenerated cultures of *Lobelia inflata* L. Herba Pol 35, 4:171–178

Krajewska A, Szöke É, Petri G, Botz L, Szarvas T (1986) Effect of new synthetic regulators on the alkaloid production of organized cultures of *Lobelia inflata* L. 24th Hung Annu Meet Biochem, Gödöllö, p 114 (Abstr)

Krajewska A, Szöke É, Botz L, Szarvas T (1987) Effect of new synthetic regulators on biomass and alkaloid production by callus tissues of *Lobelia inflata* L. Acta Bot Hung 33:407–411

Krajewska A, Szöke É, Szarvas T (1988) Wplyw nowych syntetycznych cytokinin na hodowle tkankowe stroiczki rozdetej (*Lobelia inflata* L.) (The influence of the new synthetic cytokinins on *Lobelia inflata* L. tissue culture). Herba Pol 34:27–34

Krochmal A, Wilken L, Chien M (1972) Lobeline content of four Appalachian lobelias. Lloydia 35:303

List PH, Hörhammer L (1976) Hagers Handbuch der Pharmazeutischen Praxis, Vol V. Springer, Berlin Heidelberg New York, pp 551–566

Luckner M, Bessler O, Luckner R, Korn E (1968) Vorschlage für den Drogenteil des DAB7 *Herba Lobeliae*. Pharm Zentralhalle 107:28–39

Mahmoud ZF, El-Masry S (1980) Colorimetric determination of lobeline and total alkaloids in *Lobelia* and its preparation. Sci Pharm 48:365–368

Maróti M (1976) A növényi szövettenyésztés alapjai. Akadémiai Kiadó, Budapest

Mészáros A (1986) A növényi szövettenyésztés módszereinek alkalmazása a gerbera szaporitóanyag elöállitásában. PhD Thesis, Budapest

Mothes K, Schütte H (1963) Biosynthesis of alkaloids. Angew Chem Int Ed Engl 2:341

Munier R, Macheboeuf M (1949) Preparation of modified Dragendorff reagent. Bull Soc Chim Biol 31:1144

Murashige T, Skoog F (1962) A revised medium for rapid growth and bioassays with tabacco tissue cultures. Physiol Plant 15:473–497

Nyiredy Sz, Erdelmeier CAI, Meier B, Sticher O (1985) "PRISMA": Ein Modell zur Optimierung der mobilen Phase für Die Dünnschichtchromatographie, vorgestellt anhand verschiedener Naturstofftrennungen. Planta Med 51:241–246

O'Donovan DG, Forde T (1971) The biosynthesis of *Lobelia* alkaloids. Part II. The role of lobelanine in the biosynthesis of lobeline. J Chem Soc (C) 17:2889–2890

O'Donovan DG, Long DJ, Forde E, Geary P (1975) The biosynthesis of *Lobelia* alkaloids Part III. Intermediates in the biosynthesis of lobeline: biosynthesis of 8,10-diethyl-lobelidione. J Chem Soc Perkin Trans I 5:415–419

Pharmacopée française (1972) De la commission nationale de pharmacopée par l'ordre national des pharmaciens IX. Paris

Pomázi A, Szöke É (1989) Comparative leaf-anatomical studies of in vitro and in vivo cultured plants of *Lobelia inflata* L. Vth Symp of Hungarian Plant Anatomy. JATE Szeged, p 33 (Abstr)

Robinson MM, Pierson WG, Dorfman L, Lambert BF, Lucas RA (1964) The structure of lobinaline. Tetrahedron Lett 24:1513–1519

Schenk RU, Hildebrandt AC (1972) Medium and techniques for induction and growth of monocotyledonous and dicotyledonous plant cell cultures. Can J Bot 50:199–204

Schöpf C, Kauffmann Th, Berth P, Bundschuh W, Dummer G, Fett H, Habermehl G, Wieters E, Wust W (1957) The strongly hydrophilic minor alkaloids of *Lobelia inflata*. Ann Chem 608:88–127

Smogrovicova H, Spetkova O, Jindra A (1966) Biochemistry of *Lobelia* alkaloids. Abh Dsch Akad Wiss Berlin Kl Chem Geol Biol 3:147–149

Smogrovicova H, Jindra A, Kovács P (1972) Biosynthesis of alkaloids. XXVIII. L-phenylalanine-2-oxoglittarate aminotransferase from *Lobelia inflata* L. plants. Chem Zvesti 26:360–366

Steinegger E, Grütter H (1950) Lurenine, the new crystalline alkaloid from *Lobelia urens*. Pharm Acta Helv 25:276–277

Steinegger E, Ochsner F (1956a) Determination of lobeline of drugs and alkaloidal extracts. Pharm Acta Helv 31:65–72

Steinegger E, Ochsner F (1956b) The alkaloids of *Lobelia salicifolia*. Pharm Acta Helv 31:89–96

Sváb J (1981) Biometriai módszerek a kutatásban. Mezögazdasági Kiadó, Budapest

Szöke É, Kuzovkina IN, Verzár-Petri G, Szmirnov AM (1979) Növekedést befolyásoló anyagok hatása a kamilla (*Matricaria chamomilla* L.) kallusztenyészetek biomasszaképzésére [Effect of growth regulating substances on the biomass production of callus cultures of chamomile (*Matricaria chamomilla* L.)]. Herba Hung 18:41–57

Szöke É, Szarvas T, Pozsár BI, Verzár-Petri G (1984) Effect of a new synthetic cytokininlike regulator on the tissue culture, and alkaloid production of *Datura innoxia*. Acta Biochem Biophys Acad Sci Hung 19:62

Szöke É, Krajewska A, Neszmélyi A, Pomázi A, Mészáros A (1992) In vitro multiplication of *Lobelia inflata* plants and their alkaloid production. Planta Med 58:625–626

Szöke É, Krajewska A, Neszmélyi A (1993a) NMR characterization of alkaloids from *Lobelia inflata*. Planta Med 59:704

Szöke É, Neszmélyi A, Krajewska A, Szarvas T (1993b) Incorporation of radioactive precursors into the alkaloids of organized cultures from *Lobelia inflata*. Int Symp on Medicinal and aromatic plants, Tiberias, pp 65–66 (Abstr)

Szöke É, Pomázi A, Krajewska A, Neszmélyi A, Szarvas T (1993c) Incorporation of radioactive precursors into the alkaloids of organized cultures from *Lobelia inflata* and their vegetative micropropagation. Acta Hortic 344:444–461

Tanaka M, Yonemitsu H, Shimomura K, Ishimaru K, Mochida S, Endo T, Kaji A (1993) Transformation in *Lobelia inflata*. In: Bajaj YPS (ed) Biotechnology in agriculture and forestry, vol 22. Plant protoplasts and genetic engineering III. Springer, Berlin Heidelberg New York, pp 253–265

Teusher E (1979) Pharmakognosie Teil III. Akademie Verlag, Berlin

Vágujfalvi D (1993) Speciális növényi anyagcsere. Tankönyvkiadó, Budapest

Wagner H (1980) Pharmazeutische Biologie. Fischer, Stuttgart, pp 181–182

Waldi D (1961) Systematic analysis of alkaloids. J Chromatogr 6:61

Wandle K, Quinlow A, Simpkins I (1979) Abscisic acid and the regulation of water loss in plantlets of *Brassica oleracea* var. *botrytis* regenerated through apical meristem culture. Ann Bot 43:745–752

Wandle K, Dobbs EB, Short KC (1983) In vitro acclimatization of aseptically cultured plantlets to humidity. J Am Soc Hortic Sci 108:386–389

Wheals BB (1980) Isocratic multi-column high-performance liquid chromatography as a technique for qualitative analysis and its application to the characterization of basic drugs using an aqueous methanol solvent. J Chromatogr 187:65–85

Wieland H, Koschara W, Dane E, Renz I, Schwarze W, Linde W (1939) Über die Nebenalkaloide von *Lobelia inflata*. Liebigs Ann Chem 540:103–156

Wiewiórowski M, Skolik J (1959) Photometric determination of *Lupinus* alkaloids. Rocz Chemii 33:461–469

Wysokinska H (1977) Wystepowanie alkaloidów w hodowli tkankowej *L. inflata* L. Farm Pol 33:725–727

Yonemitsu H, Shimomura K, Satake M, Mochida S, Tanaka M, Endo T, Kaji A (1990) Lobeline production by hairy root culture of *Lobelia inflata* L. Plant Cell Rep 9:307–310

# XX *Medicago* Species (Alfalfa): In Vitro Culture and the Production of Secondary Metabolites

R. EDWARDS and A. D. PARRY[1]

## 1 Introduction

### 1.1 Distribution and Importance of *Medicago* Species

The genus *Medicago* (family Leguminosae) contains a number of species, which, following breeding efforts over 2000 years, have become the world's major forage legumes. Due to their ability to fix nitrogen, the various species have been cultivated for use as green-manuring agents and as a forage crop of high nutritional value to pigs, cattle, sheep, and poultry (Duke 1981). At various times, *Medicago* has also been used as a source of fiber for paper production and as a salad or vegetable garnish for human consumption, while the seeds have been extracted for edible oils and dye-stuffs. The most commonly encountered *Medicago* species are listed and briefly described in Table 1. Within the individual species there is considerable genetic variability, which has ensured the distribution of the plants in a wide variety of environments. This large genetic pool has allowed the plant breeders to incorporate desirable traits into commercial cultivars. An example of this has been the development of cold-hardy alfalfa cultivars by crossing *M. falcata* with *M. sativa* (Duke 1981).

Although seven *Medicago* species are described in Table 1, in terms of total area under cultivation, *M. sativa* is by far the predominant forage legume crop in the world today. Originally cultivated in Persia (Iran), *M. sativa* L., commonly referred to as alfalfa in the USA or lucerne in Europe, has been distributed through Europe, Asia, and the Americas by invading armies who planted the crop as forage for their horses (Stuteville and Erwin 1990). Although grown throughout Europe and Asia, it is in North America that alfalfa has become a major crop, initially in the southwestern states and later with the introduction of winter-hardy varieties into the northern states and Canada.

### 1.2 Propagation of *Medicago sativa*

Alfalfa is a cross-pollinating auto-tetraploid, and this gives rise to considerable genetic variability even within a given cultivar such that isogenic lines of alfalfa

[1] Department of Biological Sciences, University of Durham, South Road, Durham, DH1 3LE, United Kingdom

Biotechnology in Agriculture and Forestry, Vol. 28
Medicinal and Aromatic Plants VII (ed. by Y.P.S. Bajaj)

**Table 1.** Common names, distribution, and uses of *Medicago* species

| *Medicago* species | Common name | Description (cold tolerance +/−) | Distribution | Uses (recorded medical uses +/−) |
|---|---|---|---|---|
| *arabica* | Spotted/ southern burclover | Annual (−) | Europe, Pacific/ Southern America | Pasture (−) |
| *falcata* | Yellow flowered alfalfa, sickle medic | Perennial (+) | Europe, Asia, Northern America | Pasture, hay (−) |
| *lupulina* | Black medic, yellow trefoil | Annual/ perennial (+) | Europe, Asia, Africa (alpine) | Forage, green manure (+) |
| *orbicularis* | Button clover/ large disk medic | Annual (−) | Southern Europe Asia | Forage (−) |
| *polymorpha* | California/ toothed burclover | Annual (−) | Europe, Asia Pacific America | Forage, edible (+) |
| *sativa* | Alfalfa, lucerne | Perennial (−/+) | Europe, Asia America | Forage, hay, edible (+) |
| *scutellata* | Snail medic | Annual/ perennial (−) | Southern Europe, Asia | Forage (−) |

are not available. Alfalfa is either grown mixed with grasses or other legumes for pasture use or monocultured in rows or ridges on well-ploughed land. The seed is scarified or soaked in water prior to sowing, and may also be inoculated with competent strains of *Rhizobium meliloti* to promote nodulation. Monocultured "stands" of alfalfa are harvested by mowing after the first flowers have opened and the crop may then be cut again as growth is re-established throughout the season. In the USA, highest yields are obtained when the plants are grown under irrigation in hot southern areas. The harvested hay may either be baled, used in silage, or pelleted before use as animal fodder. For human consumption either the fresh foliage or, more usually, young sprouts are used.

A recent development in the agronomic use of the alfalfa has arisen from its potential for genetic transformation, which is unusual among the legumes. Both *Agrobacterium tumefaciens*- and *A. rhizogenes*-mediated transformations of alfalfa have been established, though subsequent out-breeding may be required to introduce the transgene into commercially useful varieties as the transformable cultivars are of limited use in the field (Nisbet and Webb 1990). Programmes to develop improved disease-resistance traits and produce compounds of commercial importance are already underway in transgenic plants. As will be discussed later, this development opens up a range of exciting possibilities for the production of secondary products in alfalfa.

### 1.3 Medicinal Uses of *Medicago* Species

The Greek herbalist Dioscorides, writing in the first century A.D., described the use of *M. sativa* "which being dried is mixed for its sweet savor's sake to salt sauces" and also discussed its merits as a salad vegetable but made no mention of its medical uses (Gunther 1934). Similarly, the *Medicago* genus in general does not seem to have been extensively used in folk medicine, though the antibacterial properties of aqueous extracts have been mentioned in several articles (Duke 1981). In particular, extracts from the leaves of *M. lupulina* and *M. sativa* have been used as aids to digestion, as laxatives, and have also been recommended for the treatment of peptic ulcers and for urinary and bowel problems. Alfalfa leaf preparations have been variously described as being estrogenic, diuretic, and useful in the treatment of scurvy. Seed extracts have apparently been used in India in poultices for boils, and a water infusion is reported to assist in weight gain. In view of the estrogenic coumestans and anti-inflamatory flavonoids and saponins present in this genus, many of these claims may have a chemical basis.

## 2 Secondary Plant Products in *Medicago*

In this section examples of the various classes of secondary plant products observed in *Medicago* are cited, together with associated biological functions of interest. It should be noted that the list of compounds observed cannot be said to be comprehensive, as of the genus only *M. sativa* has been examined in any detail. In addition, the identification of some compounds must be regarded as tentative due to the nature of the analytical techniques used.

### 2.1 Betaines

Stachydrine (**1**), 1-homostachydrine (**2**), and trigonelline (**3**) have been identified in the seeds of *M. sativa* (Harborne et al. 1971). These compounds are cited as having activity in increasing milk production and menstruation in animals (Duke 1981). Seeds were found to contain 8.6 nmol/seed and 16.4 nmol/seed of trigonelline and stachydrine, respectively (Phillips et al. 1992). Various alfalfa cultivars also contain stachydrine and homostachyridine in leaf tissues and the two related betaines glycine betaine (**4**) and hydroxyproline betaine (**5**) (Wood et al. 1991). However, trigonelline was not detected in aerial plant parts. The content of all of these betaines in immature leaves growing from defoliated plants varied considerably in the 15 alfalfa cultivars surveyed (stachyridine 2.4–11.3 μmol/g fresh weight; hydroxyproline betaine 0.2–10.1; homostachyridine 1.2–5.3; and glycine betaine 0.2–2.2). An inverse relationship between the levels of stachyridine and hydroxyproline betaine was demonstrated, and it was suggested that the two compounds had a common precursor (proline). In addition to their biological activities in grazing animals, betaines act as cytoplasmic osmolytes in plants, accumulating to high levels in response to foliar damage or exposure to high salt concentrations (Weretilnyk et al. 1989). Other studies have shown that these compounds are exuded from seeds and roots, and are important in the rhizosphere as osmoprotectants and as a carbon source for nodulating bacteria (Goldmann et al. 1991). Phillips et al. (1992) have reported that stachydrine and trigonelline are exuded from alfalfa seeds prior to germination and play a role in activating the *NodD2* protein but not the homologous *NodD1* protein in the nodulating bacteria *Rhizobium meliloti.*

In spite of the potentially high levels of betaines encountered in field-grown *Medicago*, little is known about the biosynthetic origins of these compounds. Glycine betaine is synthesized from choline and stachyridine from proline via N-methylproline, but the pathways leading to homostachyridine and hydroxyproline betaine remain to be elucidated. In alfalfa, maximal levels of these compounds are obtained in either seeds or new foliar growth arising from defoliated plants, or in plants irrigated with 0.2–0.3M NaCl. The extraction and analysis of these compounds has been described in several recent papers (Weretilnyk et al. 1989; Wood et al. 1991).

## 2.2 Saponins

The saponins are glycoside conjugates of triterpenes which have been isolated from both the roots and leaves of the following *Medicago* species: *sativa* (Massiot et al. 1988; Oleszek et al. 1990; Gorski et al. 1991; Massiot et al. 1991; Wymansimpson et al. 1991 and related earlier references), *hispida* (Mahato 1991), and *lupulina* (Oleszek et al. 1988). The aglycones of the saponins are termed the sapogenins. The saponins appear to be defence-related compounds with a bitter taste and detergent-like properties (Lewis and Elvin-Lewis 1977). Their biological activities as feeding deterrents in animals include the ability to lyse red blood cells and in this regard they are particularly toxic to fish. They also have antifungal properties when tested in bioassays (Oleszek et al. 1990) and recent studies in our laboratory have shown them to be highly toxic to sap-sucking insects, (Edwards unpubl. data). Medicinally some of the saponins have anti-inflamatory activities (Lewis and Elvin-Lewis 1977); however, the saponins of *Medicago* have not been exploited for this purpose. In general, the presence of these glucosides is considered to be an undesirable trait in a forage legume due to reported toxic effects in cattle. In alfalfa, saponins are more abundant in the roots than in the leaves and can consist of up to 20 related triterpene conjugates. The best studied of these compounds is the 3-*O*-glucopyranoside of medicagenic acid (**6**). The glucoside of medicagenic acid is also observed in the roots of *M. lupulina* (Oleszek et al. 1988). The recent studies in alfalfa roots by Oleszek et al. (1990) have identified six different sugar conjugates of medicagenic acid, two glycosides of hederagenin (**7**) and one glycoside of soyasapogenol (**8**). Sugar moieties included glucosides, glucuronides, and oligosacharides containing arabinose, rhamnose, and xylose. Interestingly, the 3-*O*-glucoside of medicagenic acid was up to 20 times more effective than the alternative glycosides of medicagenic acid at inhibiting the growth of *Trichoderma viride*, and was also more active than the conjugates of hederagenin and soyasapogenol. In addition to glycosides of medicagenic acid, glycosides of soyasapogenol and hederaginin have also been reported in other *Medicago* species. Mahato (1991) identified two complex oligosaccharide conjugates of soyasapogenol and hederageninin in the aerial parts of *M. hispida*.

The saponins can be isolated either from the roots or the leaves of *Medicago* and after extraction analyzed directly by HPLC using a refractive index detector, while after hydrolysis the aglycones can be determined by GC (Oleszek et al. 1990).

### 2.3 Chalcones, Flavones, and Flavanones

The flavonoids of *Medicago* are particularly interesting in view of both their wide variety of biological properties and the apparent elegance of the regulation of their metabolism. For convenience, the flavonoids discussed in the following two sections have been categorized as compounds with nondefence-related functions (chalcones **9**, flavones **10**, and flavanones **11**) and defence-related functions (isoflavonoid phytoalexins **12** and **13**, coumestans **14**).

In a comprehensive taxometric study, Classen et al. (1982) described the flavonoid content of healthy leaves of 47 species of *Medicago* complementing the earlier studies of Harborne (Harborne et al. 1971). Following acid hydrolysis, the flavonoids cyanidin (**15**) quercetin (3,3′,4′,5,7-pentahydroxyflavone), kaempferol (3,4′,5,7-tetrahydroxyflavone), luteolin (3′,4′,5,7-tetrahydroxyflavone, apigenin (4′5,7-trihydroxyflavone), chrysoeriol (4′,5,7-trihydroxy-3′-methoxyflavone), and tricin (4′,5,7-trihydroxy-3′,5′-dimethoxyflavone) were identified by two-dimensional TLC. However, none of the compounds was quantified with respect to the levels of the free or glycosylated forms in the leaves from each species. On the basis of their respective flavonoid contents, the relatedness of the species was established, with compounds such as apigenin present in all but one of the *Medicago*. Similarly, derivatives of quercetin and kaempferol were found in appreciable quantities in half of the species studied. Although anthocyanidins were found in 33 of the species tested, the proanthocyanidin condensed tannins were absent in all cases. This is consistent with the results of several studies which have surveyed forage legumes for tannins (Albrecht and Muck 1991). Tannins are desirable in silage production and in all instances *Medicago* species have tested negatively for proanthocyanidin content. The taxometric study of the *Medicago* also demonstrated that as plants mature the complexity of their flavonoid content increases. In seedlings of *M. sativa*, 9 flavonoids were observed, while in mature leaves 18 compounds were detected. A similar increase in complexity was also observed in the flavonoids of *M. lupulina*. In both instances, the flavonoids present in the seedlings were also seen in the mature tissue. In alfalfa seedlings infected with *Ascochyta imperfecta*, the glycosides of 7,4′-dihydroxyflavone, 7,3′,4′-trihydroxyflavone, and tricin were reported in both healthy and diseased leaves (Olah and Sherwood 1971). During infection, the dihydroxy- and trihydroxy-flavones were also observed as their aglycones.

Recent interest in the flavonoids and chalcones of the *Medicago* has been stimulated by the discovery of their involvement in the activation of the genes required for nodulation in the symbiotic bacteria *Rhizobium meliloti.* In *M. sativa*, the flavonoids luteolin and chryseriol are exuded from preformed stores in the seeds during imbibition and induce the *nod* genes in *Rhizobium* (Peters et al. 1986; Hartwig et al. 1990). In the case of luteolin, the flavonoid is released as the glucoside, which is then hydrolysed by an extracellular beta-glucosidase (Hartwig and Phillips 1991). Quercetin-3-*O*-galactoside has been determined as the dominant flavonoid exuded from the seed (Hartwig et al. 1991). Although quercetin glycoside and its aglycone do not activate *nod* genes directly, both compounds, together with luteolin, stimulate the growth of *Rhizobium*, but not the growth of bacterial or fungal pathogens. In contrast to the 3′,4′,5,7-substituted flavonoids exuded by the seeds, the roots of 3-day-old seedlings released the 5-deoxy *nod*-activators 4′,7-dihydroxyflavanone (**11**), 4′,7-dihydroxyflavone (**10**) and 4,4′-dihydroxy-2′-methoxychalcone (**9**), possibly in the form of the respective glucosides (Maxwell and Phillips 1990). The subtle differential regulation of the release of *nod*-inducing flavonoids by the seedlings is apparently mediated by the use of preformed stores of the flavonoids in the seeds and by the production of *de novo* synthesized compounds in the roots. The elegance of the system can be demonstrated by the selective 2′-*O*-methylation of the chalcone in the roots which channels flavonoid biosynthesis away from mainstream isoflavonoid production to the selective formation of a potent *nod*-activating factor (Dixon et al. 1992). The outcome of this regulation of flavonoid biosynthesis in *M. sativa* is that seedlings are able to chemically regulate the rhizosphere to produce conditions which favor their successful colonization by nodulating bacteria.

The flavonoids of *Medicago* do not appear to have been exploited for medical uses, though several flavonoid constituents have biological activities (Cody et al. 1986). For example, the flavonoids apigenin, luteolin, and quercetin have anti-inflammatory activities, but presumably are not present at pharmacologically active concentrations in plant preparations, or are of limited usefulness due to the undesirable toxic activities of other secondary products of the genus.

## 2.4 Isoflavonoid Phytoalexins and Related Metabolites

*Medicago* species represent an excellent system for the study of isoflavonoid phytoalexins, as they produce a relatively simple range of isoflavones (**12**), pterocarpans (**13**), and coumestans (**14**) following infection with fungal pathogens. Healthy leaves of *M. sativa* are reported to contain trace amounts of the isoflavonoids daidzein (4′,7-dihydroxyisoflavone), genistein (4′,5,7-trihydroxyflavone), biochanin A (5,7-dihydroxy-4′-methoxyisoflavone), and formononetin (7-hydroxy-4′-methoxyisoflavone) (Guggolz et al. 1961). Similarly, the glycosides of formononetin and daidzein have also been reported in healthy mature leaves (Olah and Sherwood 1971). The related metabolite coumestrol has been reported to accumulate in mature, but not young leaves, of several *Medicago* species (Classen et al. 1982). Livingston et al. (1961) reported that the foliage of *M. sativa* contained 56 mg coumestrol/kg fresh weight of tissue or 84 mg/kg dried material. Similarly, several varieties of *M. polymorpha* were found to contain on average 255 mg coumestrol/kg dry weight in the stems and 98 mg/kg in the pods (Barbetti and Fang 1991). However, our investigations support the observations of Harborne (Harborne et al. 1971) that these compounds are normally absent from leaves of uninfected *Medicago* plants but that they accumulate in response to a range of biotic or abiotic stresses which may be encountered under field conditions. HPLC analysis shows a single predominant metabolite present in the leaves of healthy mature and immature *M. sativa* plants (Fig. 1; Table 2).

In contrast to the leaves, significant quantities of isoflavonoids accumulate in healthy roots (Fig. 1). Formononetin glucoside was reported in the roots of *M. sativa* seedlings with the aglycone being exuded following exposure to copper salts (Maxwell and Phillips 1990). The 3-O-glucoside of medicarpin was first characterized in the roots of *M. sativa* by Sakagami et al. (1974). Our subsequent investigations in roots and cell cultures of alfalfa confirmed the presence of a 6″-O-malonyl-glucoside of medicarpin (Kessmann et al. 1990). A conjugate, originally identified as a malonylated glucoside of afrormosin (7-hydroxy-4′,6-dimethoxyisoflavone) or its structural isomer alfalone (6-hydroxy-4′,7-dimethoxyisoflavone), was found to be a major component in root extracts. However, our subsequent studies have shown that this conjugate is predominantly formononetin-7-O-glucoside-6″-O-malonate (Tiller et al. 1993). A glycoside of coumestrol, some of which was malonylated, was also observed as a metabolite in alfalfa roots (Tiller et al. 1993). In healthy roots, the conjugates of formononetin and medicarpin accumulate with age (Table 2). On the basis of the content of malonylated glucosides in the whole root of young plants, their accumulation is not affected by infection with nodulating *Rhizobium*. However, our studies have shown that in older plants (57d), uninoculated roots contain four times more formononetin glucoside malonate than nodulated roots (Tiller et al. 1993). In inoculated roots, the conjugates of medicarpin and formononetin

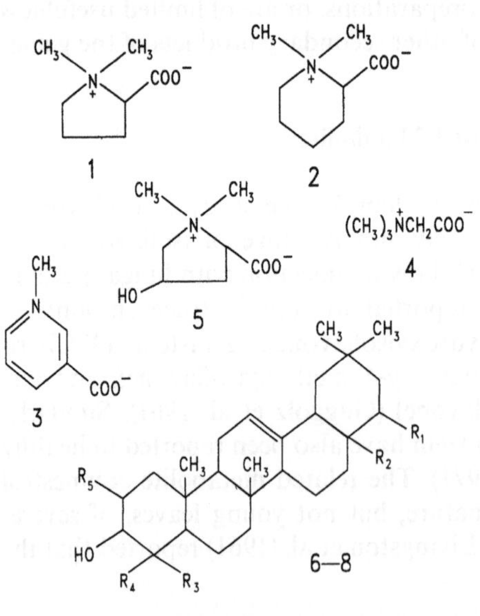


**Fig. 1.** **1** Stachydrine; **2** 1-homostachydrine; **3** trigonelline; **4** glycine betaine; **5** hydroxyproline betaine; **6** medicagenic acid (R1 H, R2 COOH, R3 $CH_3$, R4 COOH, R5 OH); **7** hederagenin (R1 H, R2 COOH, R3 $CH_3$, R4 $CH_2OH$, R5 H); **8** soyasapogenol (R1 OH, R2 $CH_3$, R3 $CH_2OH$, R4 $CH_3$, R5 H)

**Table 2.** Levels of coumestrol glucoside (CG), formononetin glucoside (FG) and malonylated glucoside (FGM), medicarpin (M) and its malonylated glucoside (MGM), in plants and cell cultures of *M. sativa*[a]

| Tissue | Age (days) | CG[b] | FG[b] (nmol/g f.wt.) | FGM[b] | MGM | M |
|---|---|---|---|---|---|---|
| Seedlings | | | | | | |
| Leaves | 10 | n.d. | n.d. | n.d. | n.d. | n.d. |
| Roots | 10 | n.d. | n.d. | 52.9 | 28.3 | n.d. |
| Leaves | 30 | n.d. | n.d. | 25.7 | 11.7 | 7.3 |
| Roots | 30 | n.d. | n.d. | 665.4 | 76.5 | n.d. |
| Leaves | 56 | n.d. | n.d. | 5.4 | 5.2 | 3.6 |
| Roots | 56 | n.d. | 51.0 | 918.0 | 123.2 | n.d. |
| Copper treated whole seedlings[c] | 10 | 53.5 | n.d. | 6.1 | 8.8 | 76.1 |
| Cell cultures | | | | | | |
| Control | 6 | n.d. | 17.5 | 7.5 | 36.1 | 2.0 |
| Elicited[d] | 6 | n.d. | 35.2 | 14.3 | 145.3 | 141.4 |

[a] Unpublished data from recent studies in the authors' laboratory.
[b] Aglycone not detected in any of the samples.
[c] 1 mM $CuCl_2$ for 3 d.
[d] Treated for 24 h with a yeast cell wall preparation. n.d. not detected.

are present at higher concentrations in the roots than in the nodules (Paiva et al. 1991). Interestingly, Dakora et al. (1993) have reported that alfalfa roots selectively exude these conjugates into the rhizosphere immediately following inoculation with *Rhizobium*. Formononetin glucoside malonate was shown to be a selective inducer of *nod*-genes in *Rhizobium* (Dakora et al. 1993).

*Medicago* species accumulate a range of phenolic phytoalexins in response to infection by bacteria or fungi. In *M. sativa* the major phytoalexins are the isoflavans (**16**) vestitol (6′,7-dihydroxy-4′-methoxyisoflavan and sativan (7-hydroxy-4′,6′-dimethoxyisoflavan) and the pterocarpan medicarpin (**13**) (Ingham 1982). In the majority of cases there is an inverse correlation between the extent of formation of these compounds and disease susceptibility following the initial infection (Dixon et al. 1992). For example, the ability of alfalfa to accumulate phytoalexins in stem tissue following infection with the wilt-causing pathogen *Verticillium albo-atrum* correlates with the resistance of the plant to the progression of disease symptoms (Khan and Milton 1978). In addition to medicarpin, sativan, and vestitol, the closely related compounds 4-methoxy medicarpin, 10-methoxy medicarpin, isosativan, 5′-methoxy sativan, 7-hydroxy-2′,3′,4′-trimethoxyisoflavan, and 7,5′-dihydroxy-2′,3′,4′-trimethoxyisoflavan have also been recently reported in the *Medicago* as trace constituents (Dewick 1988; Spencer et al. 1991). Several coumestans also accumulate in *M. sativa* during infection with fungi, most notably coumestrol (**14**), 9-O-methyl coumestrol, 4-O-methyl coumestrol, several dimethoxylated forms such as 3-hydroxy-8,9-dimethoxy coumestrol, wairol (3-hydroxy-7,9-dimethoxy coumestrol), and medicagol (**17**) (Olah and Sherwood 1971; Biggs and Shaw 1980; Dewick 1988). In alfalfa the accumulation of coumestans was observed to be greatest following exposure to pathogenic as opposed to non-pathogenic fungi (Sherwood et al. 1970).

The types and quantities of isoflavonoid phytoalexins accumulating following infection is dependent upon the pathogens used and the methods chosen for analysis. For example, when droplets of fungal spores of the corn pathogen *Helminthosporium turcicum* were applied to the alfalfa foliage and then subsequently extracted, the predominant phytoalexin was the pterocarpan medicarpin (Smith et al. 1971). This was not the case when alfalfa pathogens such as *Stemphylium botryosum* or *Leptosphaerulina briosiana* were used, apparently due to the metabolism of the phytoalexin by the fungi (Higgins and Millar 1969; Higgins 1972). Similarly, very different relative quantities of phytoalexins were obtained in alfalfa when foliar tissue rather than drop diffusates were analyzed (Higgins 1972). The potential shortcomings of using infected plant tissues as a source of isoflavonoid phytoalexins can be overcome either by using aseptic cell cultures (Edwards and Kessmann 1992) or by treating seedlings with abiotic elicitors such as UV-B irradiation or heavy metal salts (Dewick and Martin 1979a,b). Our own studies with seedlings of *M. sativa* treated with copper salts have shown the usefulness of this system (Fig. 2; Table 2). However, the abiotic elicitation of phytoalexins cannot be considered to accurately reflect isoflavonoid metabolism during infection. For example, our studies have shown that at copper concentrations of 50 $\mu$M, petioles accumulate the glycosides of isoflavonoid phytoalexins whereas at 500 $\mu$M the aglycones accumulate. This inhibition of glycosylation appears to be due to the extreme sensitivity of the medicarpin 3-O-glucosyl transferase to copper and this phenomenon may explain why phytoalexin glucosides have not been previously observed in elicitation studies using metal salts. The interest in the isoflavonoids of the *Medicago* has concentrated on their antifungal and estrogenic activities rather than on potential medical uses. The phytotoxic effects of phytoalexins such as medicarpin are also of continuing interest to crop scientists (Dornbos et al. 1990). Pterocarpan phytoalexins have activities toward fungal and nematode pathogens of man but their usefulness is limited by their adverse activities as respiratory inhibitors and hemolyzing agents (Cody et al. 1986). Interestingly, recent studies have shown that medicarpin is a potent inhibitor of leukotriene biosynthesis in mastocytoma cells, and appears to be the active component in the antiinflammatory Chinese herb *Dalbergia odorifera* (Jiangxiang) (Miller et al. 1989).

## 2.5 Other Secondary Products of Interest

*Medicago* species have excellent nutritional value and contain high concentrations of vitamins A, B, C, and D, and the vitamin A precursor beta-carotene (Duke 1981). The content of vitamin C declines following harvest, but vitamins D2 and D3 accumulate in alfalfa hay during exposure to UV-light as a result of the breakdown of sterol precursors, and these vitamins contribute to the antirachitic activity of alfalfa (Horst et al. 1984). In addition to the toxic secondary products already listed above, the amino acid derivative canavanine has been reported in several *Medicago* species (Duke 1981).

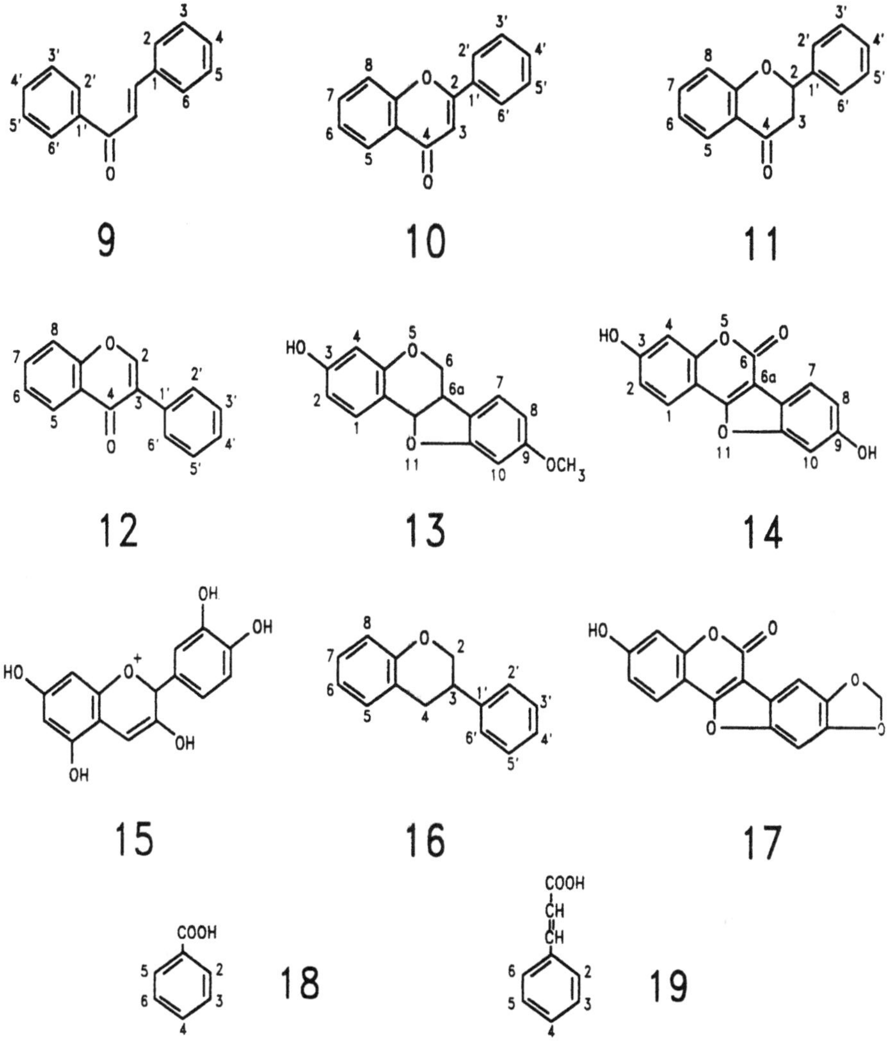


**Fig. 2.** **9** Chalcone skeleton; **10** flavone skeleton; **11** flavanone skeleton; **12** isoflavone skeleton; **13** medicarpin; **14** coumestrol; **15** cyanidin; **16** isoflavan skeleton; **17** medicagol; **18** benzoic acid; **19** cinnamic acid

# 3 In Vitro Approaches

## 3.1 Tissue Culture Studies

*Medicago* is one of the legumes on which extensive studies have been conducted on various aspects of protoplasts, cell and tissue culture, and the regeneration of plants. These aspects have been reviewed (see Gilmour et al. 1989; Arcioni et al.

1990; Nisbet and Webb 1990). In view of the commercial importance of *M. sativa*, most tissue culture studies have concentrated on this species, particularly as it is possible to regenerate whole plants from the embryogenic callus of certain genotypes (Miller 1987). Emphasis has been placed on screening cultivars for disease resistance traits and selecting for improved resistance arising through somaclonal variation, using *Phytophthora megasperma* or *Verticillium albo-atrum* as the pathogens for selection (Latunde-Dada and Lucas 1985; Miller 1987; Frame et al. 1991). The usefulness of the *in vitro* system stems from the retention of resistance traits through the culturing procedure such that resistant cultures are derived from resistant plants and susceptible cultures from susceptible plants. For screening purposes, either callus or suspension cultures can be used. Calli are inoculated with the pathogen, and resistance is assessed by monitoring hyphal development both inside and outside the callus (Latunde-Dada and Lucas 1985; Miller 1987). Suspension cultured cells are treated with filtrates from mycelial cultures of the fungus which contain toxins which inhibit cell growth and viability (Frame et al. 1991). Cells which are resistant to the toxic filtrates may then be selected and plants regenerated on Bio2y medium as described by Atanassov and Brown (1984). Using these selection methods, regenerants have been obtained possessing improved or modified resistance to *V. albo-atrum* (Latunde-Dada and Lucas 1983; Frame et al. 1991).

These in vitro infection studies have been particularly useful in determining the importance of isoflavonoid phytoalexins in the defence response of alfalfa. A good correlation could be established between the amount of medicarpin accumulating in calli infected with *V. albo-atrum* and the restriction to hyphal growth (Latunde-Dada and Lucas 1985). The resistance of the calli to the fungus could be abolished when the cells were grown on medium containing an inhibitor of medicarpin synthesis such as glyphosate. It was also reported that increased temperature enhanced medicarpin accumulation and reduced hyphal colonization (Latunde-Dada and Lucas 1986). In contrast to the studies with *Verticillium*, it has recently been reported that no correlation could be established between phenolic accumulation and susceptibility to *Fusarium oxysporum* in cell lines derived from resistant and susceptible lines of alfalfa (Cvikrova et al. 1992). However, the relationship between resistance and the presence of specific antifungal phytoalexins was not established.

### 3.2 Protocol for Tissue Culture

In *M. sativa*, callus cultures can be readily initiated from either embryos using the procedure described by Miller (1987) or by germinating seeds on callus-inducing medium as described by Latunde-Dada and Lucas (1985) or Jorrin and Dixon (1990). We used the following protocol to initiate cell lines for flavonoid metabolism studies.

1. Treat alfalfa seeds with concentrated sulfuric acid for 15 min and after thorough rinsing in sterile distilled water treat with a 1 mg/ml solution of mercuric chloride in 30% (v/v) ethanol for 30 min.

2. After extensive washing, the seeds are imbibed in sterile water for 2 h at 25 °C before distributing onto agar containing modified Schenk and Hildebrandt medium (SHM) prepared as described by Miller (1987).
3. The plates are incubated at 25 °C in the dark for 3–8 weeks with friable callus being removed from the germinating seeds at regular intervals. The callus should be collected together into lumps 10 mm in diameter on fresh SHM agar plates and incubated for a further 3–4 weeks at 25 °C in the dark with regular inspections to remove redifferentiating tissue.
4. Healthy friable callus may then be broken up with a spatula and transferred to a 100-ml Erlenmeyer flask containing 25 ml liquid SHM medium. After incubating at 25 °C in the dark on an orbital shaker (130 rpm) for 7–10 days, the finely divided cells are decanted off into a 250-ml Erlenmeyer flask containing 50 ml medium.
5. The suspension-cultured cells may now be subcultured every 10–14 days and may be used in secondary product biosynthesis studies 3–7 days after subculture. Cell wall elicitors derived from fungi are used to induce the accumulation of defense-related flavonoids (Edwards and Kessmann 1992).

### 3.3 Secondary Product Biosynthesis in Cell Cultures

Phenolic compounds and in particular the isoflavonoids appear to be the only secondary products of *Medicago* which have been the subject of study in cell cultures. In particular, the production of antifungal phytoalexins by elicitors of fungal origin in *M. sativa* is proving to be a useful model system to study the role of phenolic compounds in disease resistance in this species (Dixon et al. 1992).

When grown on modified Schenk and Hildebrandt medium, suspension cultured cells of *M. sativa* produce a range of compounds very similar to those observed in the roots of healthy plants (Fig. 3) with the 6″-O-malonyl glucosides of formononetin (**12**, FGM) and medicarpin (**13**, MGM) as the major constituents (Kessmann et al. 1990; Tiller et al. 1993). In unelicited cells, typical yields of these two compounds are in the range of 10 nmol/g fresh weight of cells for FGM and 50 nmol/g for MGM. In addition to these malonylated glucosides the non-malonylated glucoside of formononetin (FG) is also routinely observed (15 nmol/g). In view of the instability of the malonyl ester, at least some of the FG could arise from either chemical or enzymic hydrolysis of the FGM during storage and extraction. Interestingly, we have not observed the glucoside of medicarpin in cultures possibly because it appears to co-chromatograph with FGM (Fig. 3). Neither FG, FGM, or MGM are observed in the growth medium due to the presence of extracellular beta-glucosidases. Following treatment of alfalfa cell cultures with elicitors derived from fungal cell walls, analysis by HPLC has shown that the predominant elicited product is medicarpin, with concentrations of up to 400 nmol/g fresh weight of the phytoalexin being obtained (Dalkin et al. 1990). Medicarpin accumulates both in the cells and the medium, and in some instances the proportion of the total phytoalexin found extracellularly can exceed the amount of the compound found in the cells.

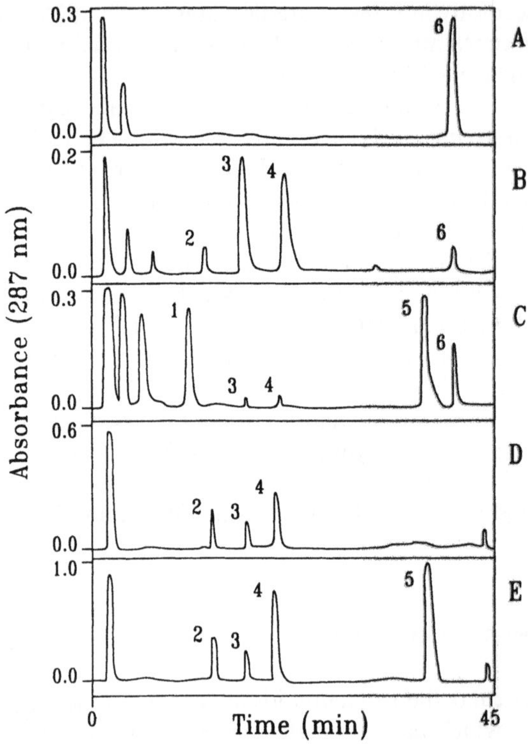


**Fig. 3A–E.** HPLC analysis of various *Medicago* extracts (equivalent fresh weights; see Edwards and Kessmann for details 1992). **A** 30-day-old leaf. **B** 30-day-old root. **C** Copper chloride (1 mM)-treated 30-day-old seedlings (whole). **D** Control cultured cells (6 days post subculturing). **E** Cells treated with yeast cell wall elicitor preparation. Compounds numbered are: **1** coumestrol glucoside; **2** formononetin glucoside; **3** (formononetin glucoside malonate) **4** (medicarpin glucoside malonate); **5** medicarpin; **6** CV, presently unidentified. (unpubl.)

Labeling studies have demonstrated that the majority of medicarpin is being formed *de novo* from phenylalanine (Kessmann et al. 1990), but we also have evidence that at least some of the phytoalexin is derived from the turnover of FGM or MGM. During elicitation, levels of MGM increase whereas levels of FG and FGM are largely unchanged (Table 2). We are currently studying the enzymes involved in the conjugation and turnover of these compounds in seedlings and cell cultures of *M. sativa* and have established that the regulation of the levels of these compounds is strictly regulated in both untreated and elicited tissue.

The nature of the other elicited secondary products in alfalfa cell cultures remains to be determined, but the range of phytoalexins produced does not accurately reflect the production of these compounds in plants. Gustine and Moyer (1982) reported that when callus cultures of *M. sativa* were treated with mercuric chloride, they synthesized medicarpin but did not accumulate the isoflavans vestitol and sativan which are major phytoalexins in planta. Our recent studies suggest that sativan is a minor metabolite in elicited alfalfa cell cultures, but that the cell lines are unable to make coumestrol (unpubl. data). Two recent papers have demonstrated how flavonoid biosynthesis may be usefully manipulated in alfalfa cultures. A naphthoquinone derived from *Penicillium* inhibited the growth of alfalfa callus and elicited the formation of antifungal phytoalexins (Kobayashi et al. 1987). Three elicited metabolites were identified

4′,7′-dihydroxyflavone, the isoflavone alfalone, and a beta-hydroxychalcone licodione. The chalcone showed pronounced antifungal activity toward *Pyricularia oryzae* and *Cladosporium herbaru* (Kobayashi et al. 1988). These authors speculated that the napthoquinone may have been acting as a selective inhibitor of flavonoid biosynthesis rather than as a conventional elicitor.

In addition to flavonoids, the hydroxybenzoic acids 4-hydroxybenzoic acid, vanillic acid (4-hydroxy-3-methoxybenzoic acid), salicylic acid (2-hydroxybenzoic acid), anisic acid (4-methoxybenzoic acid), syringic acid (4-hydroxy-3,5-dimethoxybenzoic acid), cinnamic acid, coumaric acid (4-hydroxycinnamic acid), caffeic acid (3,4-dihydroxycinnamic acid), ferulic acid (4-hydroxy-3-methoxycinnamic acid) (**18**, **19**) and sinapic acid (4-hydroxy-3,5-dimethoxycinnamic acid) have all been reported in varying amounts in alfalfa callus (Cvikrova et al. 1990). Levels varied depending on the age of the callus and ambient lighting, but in the majority of cases the compounds were present in their bound or conjugated forms. Levels of both benzoic acids and cinnamic acids increased in cultured cells following treatment with culture filtrates from *Fusarium oxysporum* (Cvikrova et al. 1992). In recent studies, we have determined several cinnamic acids and their glycoside conjugates in alfalfa cell cultures (Orr et al. 1993) and shown that the accumulation of the various acids during elicitation is dependent upon the type of elicitor used. In addition to these phenolic metabolites alfalfa cell cultures also contain saponins with the sapogenins medicagenic acid and soyasapogenol B identified as the major aglycones (Besson et al. 1989). These saponins are constitutive metabolites in cell cultures and do not accumulate following treatment with fungal elicitors (Edwards, unpubl. data).

### 3.4 Extraction and Analysis of Isoflavonoids from Cell Cultures

The analysis of isoflavonoid phytoalexins and their biosynthetic enzymes has been recently reviewed by the author (Edwards and Kessmann 1992). The reader is also referred to the comprehensive reviews on isoflavonoid analysis by Dewick (1988) Williams and Harborne (1989).

## 4 Conclusions and New Developments

*Medicago* species comprise a commercially important genus which contains an interesting range of biologically active secondary metabolites. In the past a good deal of attention has been given to the flavonoid content of these plants, but it is also clear that the saponins and betaines warrant further study. This is particularly true of the *in vitro* studies, which have concentrated on the isoflavonoids. The potential for gene manipulation of *Medicago* presents novel possibilities for enhancing the production of existing metabolites and of engineering whole new pathways of secondary product biosynthesis. It has been suggested that alfalfa could be engineered to produce alternative stereoisomers of pterocarpan phytoalexins using late biosynthetic enzymes from other legumes (Dixon et al. 1992). The positive isomer of medicarpin is more active as a fungicide than the negative isomeric form which is normally synthesized in alfalfa. It has been proposed to introduce the last two enzymes of pterocarpan biosynthesis from a species such as peanut which makes the positive isomers into alfalfa, thereby enhancing disease resistance. Alternative objectives of the genetic engineers may be to use antisense technology to inhibit coumestan biosynthesis and reduce the estrogenic potential of *Medicago* species, or to disable mainstream isoflavonoid biosynthesis and determine the effect of reducing phytoalexin accumulation on disease resistance. Alternative strategies might include the manipulation of genes encoding branch point enzymes such that medically useful flavonoids could be produced at much higher concentrations than those seen normally.

## References


Albrecht KA, Muck RE (1991) Proteolysis in ensiled forage legumes that vary in tannin concentration. Crop Sci 31:464–469

Arcioni S, Damiani F, Pezzotti M, Lupotto E (1990) Alfalfa, Lucerne (*Medicago* spp.). In: Bajaj YPS (ed) Biotechnology in agriculture and forestry, vol 10. Legumes and oilseed crops I. Springer, Berlin Heidelberg New York, pp 197–241

Atanassov A, Brown D (1984) Plant regeneration from suspension culture and mesophyll protoplasts of *Medicago sativa* L. Plant Cell Tissue Organ Cult 3:149–162

Barbetti MJ, Fang CS (1991) Relationship between phoma black stem severity and herbage and seed yield and coumestrol content in three *Medicago polymorpha* var. *Brevispina* cultivars. Aust J Agric Res 42:409–415

Besson V, Lavaud C, Massiot G, Le Men-Olivier L, Bianchi S, Flament P, Dattee Y (1989) Medicagenic acid and saponins in callus and suspension cultures of alfalfa. Phytochemistry 28:1379–1380

Biggs DR, Shaw GJ (1980) Wairol, a new coumestan from *Medicago sativa*. Phytochemistry 19:2801–2802

Classen D, Nozzolillo C, Small E (1982) A phenolic-taxometric study of *Medicago* (Leguminosae). Can J Bot 60:2477–2495

Cody V, Middleton E, Harborne JB (eds) (1986) Plant flavonoids in biology and medicine. Biochemical, pharmacological and structure-activity relationships. Progression in clinical and biological research, vol 213. Alan Liss, New York, pp 1–571

Cvikrova M, Meravy L, Hrubcova M, Eder J (1990) Dark-induced changes in the content of phenolic acids in callus cultures of alfalfa (*Medicago sativa*). Biol Plant 32:161–170

Cvikrova M, Binarova P, Eder J, Nedelnik J (1992) Accumulation of phenolic acids in filtrate-treated alfalfa cell cultures derived from genotypes with different susceptibility to *Fusarium oxysporum*. J Plant Physiol 140:21–27

Dalkin K, Edwards R, Edington B, Dixon RA (1990) Stress responses in alfalfa (*Medicago sativa* L.) 1. Induction of phenyl propanoid biosynthesis and hydrolytic enzymes in elicitor-treated cell suspension cultures. Plant Physiol 92:440–446

Dakora FD, Joseph CM, Phillips DA (1993) Alfalfa (*Medicago sativa* L.) root exudates contain isoflavonoids in the presence of *Rhizobium meliloti*. Plant Physiol 101:819–824

Dewick PM (1988) Isoflavonoids. In: Harborne JB (ed) The flavonoids. Chapman and Hall, London, pp 126–209

Dewick PM, Martin M (1979a) Biosynthesis of pterocarpan and isoflavan phytoalexins in *Medicago sativa*: the biochemical interconversion of pterocarpans and 2'-hydroxyisoflavans. Phytochemistry 18:591–596

Dewick PM, Martin M (1979b) Biosynthesis of pterocarpan, isoflavan and coumestan metabolites of *Medicago sativa*: chalcone, isoflavone and isoflavanone precursors. Phytochemistry 18:597–602

Dixon RA, Choudhary AD, Dalkin K, Edwards R, Fahrendorf T, Gowri G, Harrison MJ, Lamb CJ, Loake GL, Maxwell CA, Orr J, Paiva NL (1992) Molecular biology of stress-induced phenylpropanoid and isoflavonoid biosynthesis in alfalfa. In: Stafford HA, Ibrahim RK (eds) Phenolic metabolism in plants. Plenum Press, New York, pp 91–138

Dornbos DL, Spencer GF, Miller RW (1990) Medicarpin delays alfalfa seed germination and seedling growth. Crop Sci 30:162–166

Duke JA (1981) Handbook of legumes of world economic importance. Plenum Press, New York

Edwards R, Kessmann H (1992) Isoflavonoid phytoalexins and their biosynthetic enzymes. In: Gurr SJ, McPherson MJ, Bowles DJ (eds) Molecular plant pathology, vol 2. A practical approach. Oxford University Press, Oxford, pp 45–62

Frame B, Yu K-F, Christie BR, Pauls KP (1991) *In vitro* selection for resistance to *Verticillium* wilt in alfalfa (*Medicago sativa* L.) using a fungal culture filtrate. Physiol Mol Plant Pathol 39:325–340

Gilmour DM, Golds TJ, Davey MR (1989) *Medicago* protoplasts: Fusion, culture and plant regeneration. In: Bajaj YPS (ed) Biotechnology in agriculture and forestry, vol 8. Plant protoplasts and genetic engineering I. Springer, Berlin Heidelberg New York, pp 370–388

Goldmann A, Boivin C, Fleury V, Message B, Lecoeur L, Maille M, Tepfer D (1991) Betaine use by rhizosphere bacteria: genes essential for trigonelline, stachydrine, and carnitine catabolism in *Rhizobium meliloti* are located on *pSym* in the symbiotic region. Mol Plant Micro Interac 4:571–578

Gorski PM, Mierscm J, Plosynski M (1991) Production and biological activity of saponins and canavanine in alfalfa seedlings. J Chem Ecol 17:1135–1143

Guggolz J, Livingston AL, Bickoff EM (1961) Forage estrogens: detection of daidzein, formononetin, genistein and biochanin A in forages. J Agric Food Chem 9:330–332

Gunther RT (1934) The Greek herbal of Dioscorides. Oxford University Press, Oxford

Gustine DL, Moyer BG (1982) Retention of phytoalexin regulation in legume callus cultures. Plant Cell Tissue Organ Cult 1:255–263

Harborne JB, Boulter D, Turner BL (eds) (1971) Chemotaxonomy of the Leguminosae. Academic Press, London

Hartwig UA, Phillips DA (1991) Release and modification of *nod*-gene-inducing flavonoids from alfalfa seeds. Plant Physiol 95:804–807

Hartwig UA, Maxwell CA, Joseph CM, Phillips DA (1990) Chrysoeriol and luteolin released from alfalfa seeds induce *nod* genes in *Rhizobium meliloti*. Plant Physiol 92:116–122

Hartwig UA, Joseph CM, Phillips DA (1991) Flavonoids released naturally from alfalfa seeds enhance growth rate of *Rhizobium meliloti*. Plant Physiol 95:797–803

Higgins VJ (1972) Role of the phytoalexin medicarpin in three leaf spot diseases of alfalfa. Physiol Plant Pathol 2:289–300

Higgins VJ, Millar RL (1969) Comparative abilities of *Stemphylium botryosum* and *Helminthosporium turcium* to induce and degrade a phytoalexin from alfalfa. Phytopathology 59:1493–1499

Horst RL, Reinhardt TA, Russell JR, Napoli JL (1984) The isolation and identification of vitamin D2 and vitamin D3 from *Medicago sativa* (alfalfa plant). Arch Biochem Biophys 231:67–71

Ingham JL (1982) Phytoalexins from the Leguminosae. In: Bailey JA, Mansfield JW (eds) Phytoalexins. Blackie, Glasgow, pp 21–80

Jorrin J, Dixon RA (1990) Stress responses in alfalfa (*Medicago sativa* L.) II. Purification, characterization and induction of phenylalanine ammonia-lyase isoforms from elicitor-treated cell suspension cultures. Plant Physiol 92:447–455

Kessmann H, Edwards R, Geno PW, Dixon RA (1990) Stress responses in alfalfa (*Medicago sativa* L.) V. Constitutive and elicitor-induced accumulation of isoflavonoid conjugates in cell suspension cultures. Plant Physiol 94:227–232

Khan FZ, Milton JM (1978) Phytoalexin production and the resistance of lucerne (*Medicago sativa* L.) to *Verticillium albo-atrum*. Physiol Plant Pathol 13:215–221

Kobayashi A, Yata S, Hino T, Kawazu K (1987) A new fungal naphthoquinone which stimulates the production of antifungal compounds in alfalfa callus. Agric Biol Chem 51:2857–2860

Kobayashi A, Yata S, Kawazu K (1988) A B-hydroxychalcone and flavonoids from alfalfa callus stimulated by a fungal naphthoquinone PO-1. Agric Biol Chem 52:3223–3227

Latunde-Dada AO, Lucas JA (1983) Somaclonal variation and reaction to *Verticillium* wilt in *Medicago sativa* L. plants regenerated from protoplasts. Plant Sci Lett 32:205–211

Latunde-Dada AO, Lucas JA (1985) Involvement of the phytoalexin medicarpin in the differential response of callus lines of lucerne (*Medicago sativa*) to infection by *Verticillium albo-atrum*. Physiol Plant Pathol 26:31–42

Latunde-Dada AO, Lucas JA (1986) Influence of temperature on host resistance and fungal sensitivity to medicarpin in lucerne callus lines infected with *Verticillium albo-atrum*. Physiol Mol Plant Pathol 28:89–97

Lewis WH, Elvin-Lewis MPF (1977) Medical botany. Wiley, New York

Livingston AL, Bickoff EM, Guggolz J, Thompson CR (1961) Quantitative determination of coumestrol in fresh and dried alfalfa. Agric Food Chem 9:135–137

Mahato SB (1991) Triterpenoid saponins from *Medicago hispida*. Phytochemistry 30:3389–3393

Massiot G, Lavaud C, Guillaume D, Lemenolivier L (1988) Reinvestigation of the sapogenins and prosapogenins from alfalfa (*Medicago sativa*). J Agric Food Chem 36:902–909

Massiot G, Lavaud C, Besson V, Lemenolivier L, Vanbinst G (1991) Saponins from aerial parts of alfalfa (*Medicago sativa*). J Agric Food Chem 39:78–82

Maxwell CA, Phillips DA (1990) Concurrent synthesis and release of *nod*-gene-inducing flavonoids from alfalfa roots. Plant Physiol 93:1552–1558

Miller DK, Sadowski S, Han GQ, Joshua H (1989) Identification and isolation of medicarpin and a substituted benzofuran as potent leukotriene inhibitors in an anti-inflammatory Chinese herb. Prostaglandins Leukotrienes Essential Fatty Acids 38:137–143

Miller SA (1987) Tissue culture methods in phytopathology II-Fungi. In: Dixon RA (ed) Plant cell culture, a practical approach. IRL Press, Oxford, pp 215–229

Nisbet A, Webb B (1990) Transformation in legumes. In: Bajaj YPS (ed) Biotechnology in agriculture and forestry 10. Legumes and oilseed crops I. Springer, Berlin Heidelberg New York, pp 38–48

Olah AF, Sherwood RT (1971) Flavones, isoflavones and coumestans in alfalfa infected by *Ascochyta imperfecta*. Phytopathology 61:65–69

Oleszek W, Price KR, Fenwick GR (1988) Triterpene saponins from the roots of *Medicago lupulina* (black medic trefoil). J Sci Food Agric 43:289–297

Oleszek W, Price KR, Colquhoun IJ, Jurzysta M, Ploszynski M, Fenwick GR (1990) Isolation and identification of alfalfa (*Medicago sativa* L.) root saponins: their activity in relation to a fungal bioassay. J Agric Food Chem 38:1810–1817

Orr JD, Edwards R, Dixon RA (1993) Stress responses in alfalfa (*Medicago sativa* L.). XIV. Changes in the levels of phenylpropanoid pathway intermediates in relation to regulation of L-phenylalanine ammonia-lyase in elicitor-treated cell suspension cultures. Plant Physiol 101:847–856

Paiva NL, Edwards R, Sun Y, Hrazdina G, Dixon RA (1991) Stress responses in alfalfa (*Medicago sativa* L.) 11. Molecular cloning and expression of alfalfa isoflavone reductase, a key enzyme of isoflavonoid phytoalexin biosynthesis. Plant Mol Biol 17:653–667

Peters NK, Frost JW, Long SR (1986) A plant flavone, luteolin induces expression of *Rhizobium meliloti* nodulation genes. Science 233:977–980

Phillips DA, Joseph CM, Maxwell CA (1992) Trigonelline and stachydrine released from alfalfa seeds activate *Nod D2* protein in *Rhizobium meliloti*. Plant Physiol 99:1526–1531

Sakagami Y, Kumari S, Suzulii A (1974) Isolation and structure of medicarpin-B-D-glucoside from alfalfa. Agric Biol Chem 38:1031–1034

Sherwood RT, Olah AF, Oleson WH, Jones EE (1970) Effect of disease and injury on accumulation of a flavonoid estrogen, coumestrol, in alfalfa. Phytopathology 60:684–688

Smith DG, McInnes AG, Higgins VJ, Millar RL (1971) Nature of the phytoalexin produced by alfalfa in response to fungal infection. Physiol Plant Pathol 1:41–44

Spencer GF, Jones BE, Plattner RD, Barnekow DE, Brinen LS, Clardy J (1991) A pterocarpan and two isoflavans from alfalfa. Phytochemistry 30:4147–4149

Stuteville DL, Erwin DC (eds) (1990) Compendium of alfalfa diseases, 2nd edn. APS Press, St Paul

Tiller SA, Parry AD, Edwards R (1993) Isoflavonoid conjugates and their response to developmental change and abiotic stress in alfalfa (*Medicago sativa* L.). Acta Horticult (in press)

Weretilnyk EA, Bednarek S, McCue KF, Rhodes D, Hanson AD (1989) Comparative biochemical and immunological studies of the glycine betaine synthesis pathway in diverse families of dicotyledons. Planta 178:342–352
Williams CA, Harborne JB (1989) Isoflavonoids. In: Dey PM, Harborne JB (eds) Methods in plant biochemistry, vol 1. Academic Press, London, pp 421–449
Wood KV, Stringham KJ, Smith DL, Volonec JJ, Hendershot KL, Jackson KA, Rich PJ, Yang W-J, Rhodes D (1991) Betaines of alfalfa. Plant Physiol 96:892–897
Wymansimpson CL, Waller GR, Jurzysta M, McPherson JK, Young CC (1991) Biological activity and chemical isolation of root saponins of six cultivars of alfalfa (*Medicago sativa* L.). Plant Soil 135:83–94

# XXI *Papaver bracteatum* Lindley (Giant Scarlet Poppy): In Vitro Production of Benzophenanthridine Alkaloids

I. L. I. Hook, R. Lecky, and H. Sheridan[1]

## 1 Introduction

### 1.1 Botanical Characteristics

*Papaver bracteatum* Lindley [Papaveraceae, section Oxytona Bernh ( = *Macrantha* Fedde)], commonly known as the Giant Scarlet Poppy, is a robust perennial plant with unbranched flower stalks (70–120 cm), bearing erect bracteate buds expanding into conspicuous flowers having six purplish crimson petals, each usually with a black blotch longer than broad from base to mid-line. Leaves are long (up to 45 cm), pinnately dissected, and deeply incised, growing as a basal rosette. Capsules are open-pored, obovate-globose in shape, up to 3.5 cm long with 14–20 stigmas (Fig. 1); (Goldblatt 1974). The plant is almost totally self-incompatible requiring cross-pollination (usually by bees) for seed production. Fleshy tap roots (up to 1 cm in diameter) arise from a broad crown. They rarely branch, and grow downward 2 m or more into the soil. In temperate climates flowering occurs from mid-May to June. The plants enter dormancy

**Fig. 1.** A mature capsule of *Papaver bracteatum* Lind.

[1] Department of Pharmacognosy, School of Pharmacy, Trinity College, Dublin, Ireland

Biotechnology in Agriculture and Forestry, Vol. 28
Medicinal and Aromatic Plants VII (ed. by Y.P.S. Bajaj)

once the capsules are mature and new rosettes of leaves are visible above ground from August onwards. These remain during the following winter, with new flowering stems arising in the spring. Although the plant is native to the Caucasus, Iran, and Turkey, growing in mountainous regions between 2000 and 2500 m, it is easily cultivated throughout Europe, as well as in Israel (Milo et al. 1988) and the USA (Davis 1982).

In addition to *Papaver bracteatum*, the section Oxytona includes *P. orientale* L. and *P. pseudo-orientale* Fedde. These species resemble each other in many ways (Goldblatt 1974; Nyman and Bruhn 1979) and can produce interspecific hybrids (Milo et al. 1988). Many oriental poppy cultivars have been developed from *P. bracteatum* (Corrigan and Martyn 1981), with the result that misidentifications have occurred and earlier literature reports on alkaloid content must be interpreted with caution. Plants of the genus *Papaver* characteristically produce a milky white latex within articulated laticifers, which anastomose when mature. In both *P. bracteatum* and the structurally similar *P. somniferum*, no laticifers can be seen in nongerminated seeds (Nessler and Mahlberg 1978). They can first be observed after germination when the radicle has protruded about 2 mm. Laticifers differentiate shortly after the first appearance of the sieve elements, but before that of the tracheary elements and at the level of root hairs they are well developed (Rush et al. 1985). A large number enter the capsule in association with the phloem, but no laticifers appear to extend into the ovules (Fairbairn and Williamson 1978). Latex of both *P. somniferum* and *P. bracteatum* contains abundant small vesicles with which the morphinan alkaloids are biosynthetically associated. These "alkaloidal vesicles" can be separated from the latex by ultra-centrifugation (Kutchan et al. 1985).

## 1.2 Secondary Metabolites and Their Commercial Value

A large number of mainly isoquinoline alkaloids have been isolated from plant parts of *P. bracteatum*. These can be classified into the structural types shown in Fig. 2 and Table 1. Thebaine is the major alkaloid present in *P. bracteatum*. However, at least three chemical races of the plant are known to exist (Phillipson 1983; Preininger 1986): one containing thebaine together with orientalidine; a second with thebaine and alpinigenine, and a third with almost pure thebaine. The thebaine content of plants produced from different seed populations also varies and a number of high thebaine yielding strains have been selected over the years, e.g., UNB-2, ARYA-I, ARYA-II. Although small amounts of thebaine are present in all plant organs of *P. bracteatum*, the highest content is found in the capsules some 4 to 6 weeks after flowering when concentrations up to 3% (of dry wt.) have been reported (Fairbairn and Helliwell 1977; Davis 1982). If flower bud formation is prevented, the thebaine content remains high in the roots, where up to 0.9% of the alkaloid has been recorded (Milo et al. 1988).

Thebaine is the only alkaloid in *P. bracteatum* with commercial value. Although it has pharmacological activities similar to those of morphine, it is not used clinically as it is only a weak analgesic, but a more potent narcotic. Its importance is due to its usefulness as the starting material in the semi-synthesis of a number of analgesics (such as codeine, hydrocodone, oxycodone, nalbuphine, naltrexone, oxymorphone, thebacon) and of naloxone, a narcotic antagonist. The world usage of thebaine is currently low (not more than 5 tonnes per annum) and industrial requirements are satisfied by opium or concentrate of poppy (*P. somniferum*) straw, from which thebaine can be isolated as a by-product. The current price for thebaine on the world market is US $350 per kilo (MacFarlan Smith Ltd. pers. comm. 1992).

Many of the minor alkaloids of *P. bracteatum* also have biological activities (Kettenes-van den Bosch et al. 1981). Codeine, for example, has well-known pharmacological effects but is either present in very low concentrations ( < 0.004%) or absent. Isothebaine, like thebaine, has narcotic and analgesic

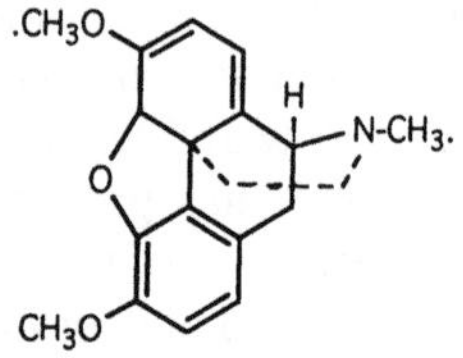

1

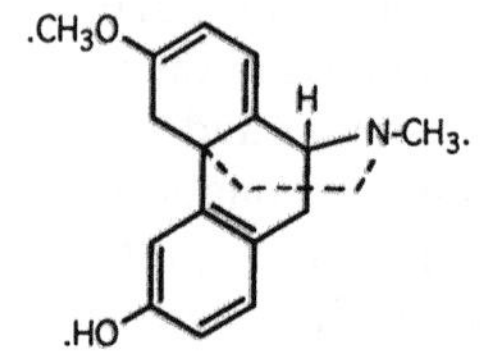

2

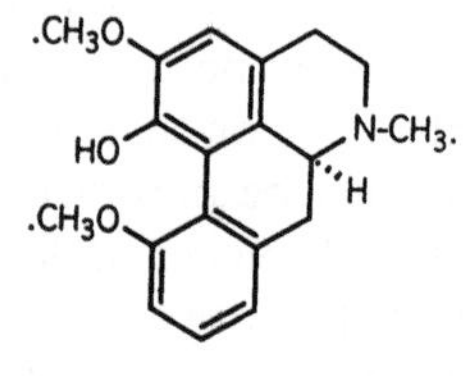

3

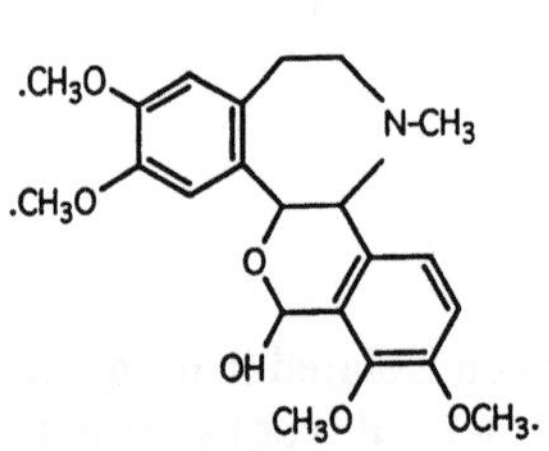

4

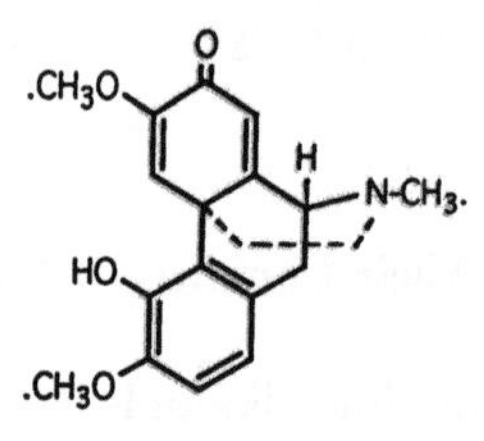

5

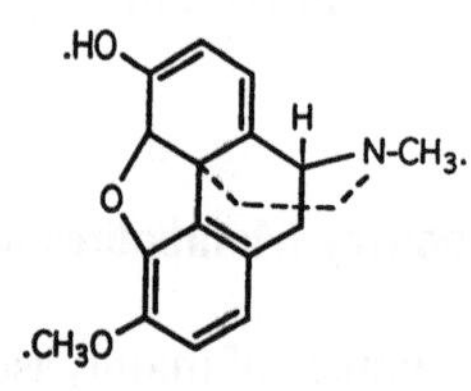

6

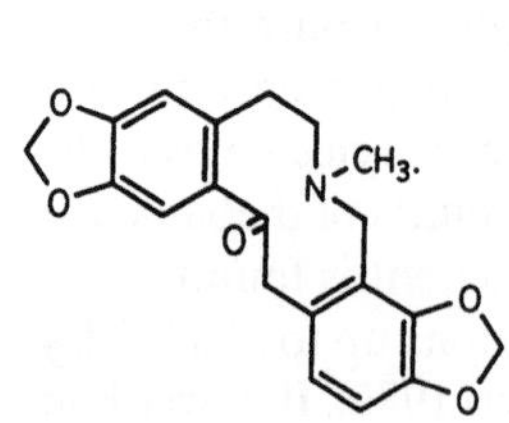


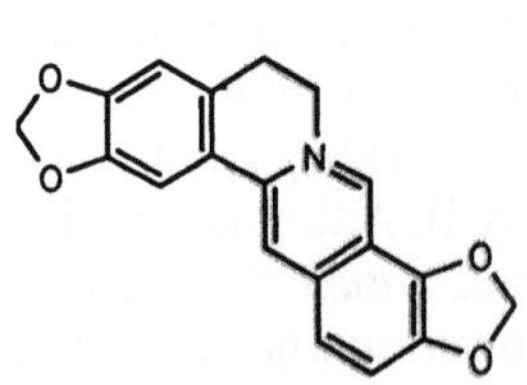


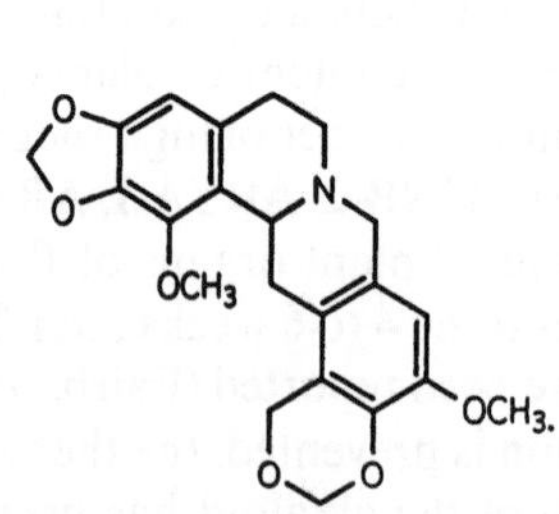


a 8 9 10

**b**

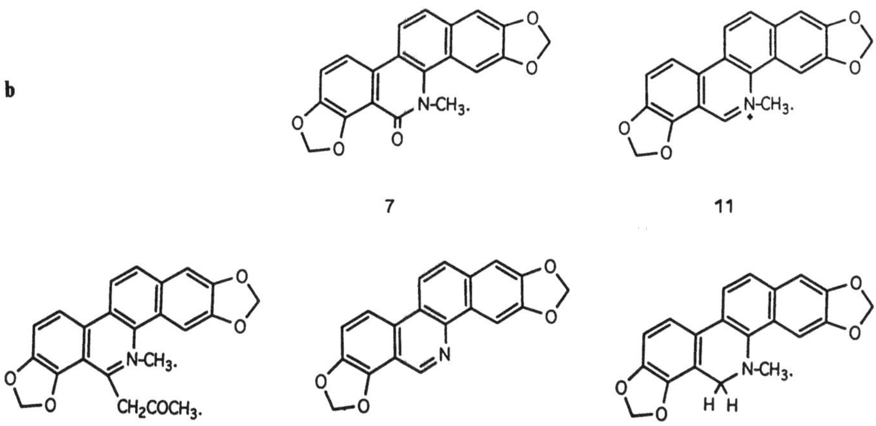


7 11

12 13 14

**Fig. 2. a** Isoquinoline alkaloids isolated from *Papaver bracteatum* plants (see Table 1). **b** Benzophenanthridine alkaloids isolated from *Papaver* cell cultures. **7** Oxysanguinarine; **11** sanguinarine; **12** 6-acetonylsanguinarine; **13** nor-sanguinarine; **14** dihydrosanguinarine

**Table 1.** Alkaloids isolated from *Papaver bracteatum* plants. (Kettenes-van den Bosch et al. 1981; Phillipson 1983; Preininger 1986)

| Alkaloid group | Example (see Fig. 2) |
|---|---|
| Protoberberine | Tetrahydropalmatine |
| | Coptisine (**9**) |
| | ( − )-Oreophiline |
| | ( − )-Orientalidine (**10**) |
| Rhoeadine/Papaverrubine | Alpinine |
| | Epialpinine |
| | ( + )-Alpinigenine (**4**) |
| | Papaverrubines B,C,D,E |
| Protopine | Muramine |
| | Protopine (**8**) |
| Benzophenanthridine | Oxysanguinarine (**7**) |
| (Pro)aporphine | ( − )-Oreintalinone |
| | ( + )-Isothebaine (**3**) |
| | ( − )-Nuciferine |
| | ( + )-Bracteoline |
| | Floripavidine |
| (Pro)-morphinan | ( + )-Salutaridine (**5**) |
| | Neopine |
| | Codeine (**6**) |
| | Oripavine (**2**) |
| | 14-β-Hydroxycodeine |
| | 14-β-Hydroxycodeinone |
| | α-Thebaine-N-oxide |
| | β-Thebaine-N-oxide |
| | Thebaine (**1**) |

activities, but is too toxic to be of value. (−)-Nuciferine and tetrahydropalmatine are CNS depressants, while protopine has sedative effects as well as bactericidal properties. Few biological studies, however, have been carried out with most minor alkaloids present in *P. bracteatum*.

## 2 In Vitro Approaches to Alkaloid Production

### 2.1 Review of Tissue Culture and Thebaine Production

Because of the commercial importance of thebaine as the starting material used in the synthesis of a number of valuable analgesics, several research groups have examined the potential of plant cell cultures derived from *P. bracteatum*, as an in vitro source of this alkaloid (see review by Kamo and Mahlberg 1988). Results of many years' work are shown in Table 2. From this it is apparent that only in a few instances has thebaine been produced and then not in good yields or on a sustained basis. The production of morphinan alkaloids by poppy cell cultures (*P. bracteatum* and *P. somniferum*) is known to be associated with cell differentiation and morphogenetic processes (Kutchan et al. 1983; Constabel 1985; Kamo and Mahlberg 1988]. Alkaloid synthesis is presumed to depend on laticifer formation (Kutchan et al. 1985; Rush et al. 1985), but neither undifferentiated callus nor suspension cultures derived from poppies have so far been shown to develop these specialized structures. Although laticifer-like cells were demonstrated in cultures of *P. bracteatum* upon removal of hormones from the medium (Kutchan et al. 1983), the lack of typical laticifers with vesicular contents is still presumed to be the explanation for the absence of morphinan alkaloids in routinely grown, undifferentiated poppy cultures.

### 2.2 Formation of Benzophenanthridine Alkaloids by Cell Cultures

Chemotaxonomic evaluation of the Papaveraceae has shown that protoberberine, protopine, and benzophenanthridine alkaloids are the most widespread, occurring with few exceptions in all genera of this family (Preininger 1986). The benzophenanthridines are present as minor metabolites in intact plants of the genus *Papaver* (Fig. 2b). It is known that they are synthesized at loci different to the laticifers associated with morphinan alkaloids (Kutchan et al. 1985). Table 2 shows the benzophenanthridines produced by undifferentiated cell cultures of *Papaver*. Some of these occur as major secondary metabolites in cell cultures of *P. bracteatum* (Lockwood 1981; Kutchan et al. 1983; Hook et al. 1988).

### 2.3 Biological Activity and Medicinal Value of Benzophenanthridine Alkaloids

Extracts of plants containing these alkaloids have well-documented uses in folk medicine for the treatment of a variety of ailments (Duke 1985). Numerous

**Table 2.** Media and culture conditions reported for the growth of *P. bracteatum* cell cultures CM: coconut milk; MS: Murashige and Skoog; 2,4-D: 2,4-dichlorophenoxyacetic acid, NAA: naphthalene acetic acid; 1AA: indole acetic acid; Kin: Kinetin

| Medium | Auxin mg/l | Cytokinin mg/l | Other supplements | Culture type | Other conditions | Alkaloids present | Reference |
|---|---|---|---|---|---|---|---|
| MS (no glycine) | 2,4-D(1) | Kin(0.1) | CM(7%) | Callus | 26°, Dark | Sanguinarine<br>Nor-sanguinarine<br>Dihydrosanguinarine<br>Oxysanguinarine<br>Chelirubine<br>Magnoflorine<br>Protopine | Ikutsu et al. (1974) |
| MS | 2,4-D(1) | Kin(0.1) | Polypeptone<br>CM | Suspension | 26° Dark<br>3% Sucrose | Not investigated | Kamimura and Akutsu (1976) |
| MS | NAA(1) | Kin(0.1) | CM (10%) | Callus suspension | 3% Sucrose | Thebaine<br>Protopine<br>Stylopine | Kamimura and Nishikawa (1976) |
| MS | 2,4-D(1) | Kin(0.1) | CM(10%) | Callus | 26° Dark<br>3% Sucrose | Thebaine | Kamimura et al. (1976) |
| MS | 2,4-D(0.1) | | | Suspension | 24°C<br>24-h light | Thebaine | Shafiee et al. (1976) |
| MS | 2,4-D(1) | Kin(0–0.1) | Various | Callus | | Sanguinarine<br>Orientalidine | Lockwood (1981) |
| MS | IAA(1) | | | Root suspension | 15-h light | Thebaine | Zito and Staba (1982) |
| MS | 2,4-D(10) | Kin(1) | Vit C (10 mg/l) | suspension | 12-h light | Sanguinarine<br>orientalidine | Lockwood (1984) |
| MS | 2,4-D(0.22)<br>NAA(0.186) | | | Suspension | Continuous dim light | Dihydrosanguinarine<br>sanguinarine | Hook et al. (1988) |
| MS | 2,4-D(0.22)<br>NAA(0.186) | | Copper | Suspension | 18-h light | Dihydrosanguinarine<br>sanguinarine | Lecky et al. (1992) |

summaries (with extensive bibliographies) dealing with the biological and pharmacological activities of the major quaternary benzophenanthridines have also been published in recent years (Shamma 1972; Simanek 1985; Simeon et al. 1989). Sanguinarine, the major quaternary benzophenanthridine produced by poppy cell cultures, has, for example, been found to interact with many enzyme systems and to have nematocidal, antifungal, antiprotozoal, and antibacterial activities (Mitscher et al. 1978; Simeon et al. 1989). Plants containing quaternary benzophenanthridines (e.g., species of *Fagara*, *Sanguinaria*, *Chelidonium*) have traditional uses in the treatment of warts, papillomas, etc. Isolated alkaloids were thought to have potentially useful anticancer activities at low concentrations (e.g., sanguinarine $< 1$ $\mu$g/ml) (Sethi 1985), as they interact with DNA. However, after extensive screening, these cytotoxic and antitumor effects were found to be of limited clinical interest. Antiinflammatory activity has also been ascribed to sanguinarine, which, together with its antimicrobial effects, suggested a usefulness for this alkaloid in the treatment of oral inflammations, e.g., gingivitis (Lenfeld et al. 1981). Sanguinarine has in addition been shown to have potential value as a substantive, human, dental antiplaque agent (Southard et al. 1984) and a number of toothpastes and mouthwashes containing it have been the subject of various patents (Anonymous 1988) and on sale in the USA since 1984. A minimum inhibitory concentration of 8–32 $\mu$g/ml against a variety of oral aerobic and anaerobic microorganisms has been reported (Southard et al. 1984). The inclusion of sanguinarine in various oral healthcare products is, however, not without controversy (Anonymous 1990), as the safety of this alkaloid on a long-term basis has not been established (Shamma 1972; Bowman and Rand 1980; Simeon et al. 1989). Dihydrosanguinarine and other sanguinarine derivatives, though not as active as the parent alkaloid, are not devoid of biological activity, showing anti-enzymatic (Simeon et al. 1989) and antibacterial effects; a minimum inhibitory concentration of between 140 and 150 $\mu$g/ml having been recorded for dihydrosanguinarine (Lecky 1992).

## 3 Production of Sanguinarine and Dihydrosanguinarine by Cell Cultures

### 3.1 Establishment of High-Alkaloid Yielding Cultures

To ensure authenticity of initial plant material, cell cultures were established from seeds derived from mature plants grown under controlled garden conditions. These were in turn developed from Arya II seeds obtained from UN sources in Geneva. Mature plants showed botanical characteristics typical of *P. bracteatum* (Sect. 1.1) and dry capsules on extraction yielded 0.70% of thebaine as the major alkaloid (Hook et al. 1988). Callus cultures were originally established in 1984 from surface sterilized seeds plated onto an agar-solidified MS salt medium (Murashige and Skoog 1962) containing 2,4-D (0.22 mg/l), NAA (0.186 mg/l), glycine (2 mg/l), nicotinic acid (0.5 mg/l), pyridoxine HCl

(0.5 mg/l), mesoinositol (200 mg/l), thiamine HCl (0.5 mg/l), and sucrose (30 g/l). Cultures grew as beige-colored aggregates and were subcultured every 4 to 6 weeks onto fresh medium.

Suspension cultures were initiated by transferring ninth generation callus pieces into liquid modified MS medium (100 ml in 250-ml conical flasks). Cultures were grown under continuous dim fluorescent light, in a culture room at 25 °C ( ± 2 °C), on an orbital shaker rotating at 90 rpm. Cell suspensions were subcultured at monthly intervals using standard procedurers. Cells were harvested by filtration through muslin and dried at < 40 °C in a fan-assisted oven.

### 3.2 Extraction of Alkaloids from Cell Cultures

*Extraction for Thebaine.* All callus and cell suspension material initially developed was extracted for thebaine with cold, 5% aqueous acetic acid using the recommended UN method (Wu and Dobberstein 1976). TLC using precoated silica gel plates (60 $F_{254}$, Merck) with solvent system SI: ethyl acetate-methanol–conc. ammonia (17:3:1) indicated the absence of thebaine (Rf 0.53). This was confirmed by HPLC carried out on a Spherisorb 10 ODS, 250 × 4.6 mm column; solvent system (15%) aq. ammonium carbonate (0.3%) in methanol (85%); flow rate 2 ml/min, with detection by UV at 285 nm. (Hook et al. 1988).

*Extraction for Benzophenanthridines.* Dried cell culture material was refluxed twice with ethyl acetate, followed by a third reflux with methanol. TLC on precoated silica gel plates with a solvent system SII: petroleum ether–ethyl acetate–acetone (8:2:1) showed the presence of a single major alkaloid visualized by its intense blue fluorescence under 366 nm. Methanol extracts showed additional minor alkaloidal constituents. Bulked ethyl acetate extracts were subjected to column chromatography on a silica gel column and eluted with a gradient of toluene–ethyl acetate (99:1) → petroleum ether–ethyl acetate–acetone (8:2:1). From the separated fractions, dihydrosanguinarine (DHS) was subsequently identified as the major alkaloid (SI Rf: 0.96; SII Rf: 0.91). Also identified as being present were sanguinarine (SI Rf: 0.96; SII Rf: 0.23) and in minor concentration, 6-acetonylsanguinarine. Column chromatography of the bulked methanol extracts also yielded dihydrosanguinarine, sanguinarine, 6-acetonylsanguinarine, and in addition a small amount of oxysanguinarine. All the isolated alkaloids were identified by spectroscopic and physical means, as well as by comparison with literature data (Lecky 1992). The presence of dihydrosanguinarine as the major metabolite produced by these cell cultures of *P. bracteatum* was confirmed by fluorescence microscopy, where blue fluorescent crystals were clearly visible within cultured cells (Fig. 3).

### 3.3 Yields of Dihydrosanguinarine from Cell Suspension Cultures

Cell suspension cultures were developed from a high dihydrosanguinarine-yielding callus (strain NB). Three cell lines were separated (June, 1986), all of

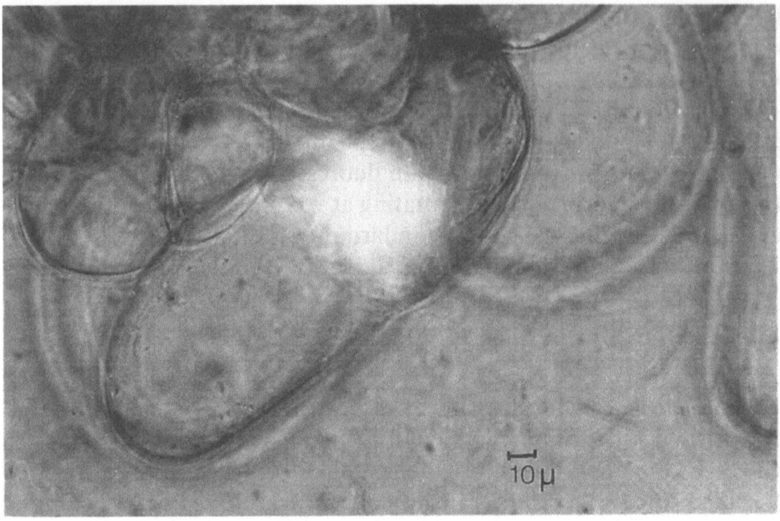


**Fig. 3.** Cultured cells of *P. bracteatum* showing crystalline intracellular deposits of dihydrosanguinarine visualized with fluorescence microscopy. (Hook et al. 1988)

which were initially similar in biomass yields (12.5 to 13.0 g dry wt./l) and percentage DHS (1.30 to 2.04% of dry cell wt.). All cell suspensions were grown as batch cultures for 28 days, under the conditions previously indicated (Sect. 3.1). Alkaloids were extracted as indicated above and quantified by HPLC, using a Spherisorb 10 ODS, 250 × 4.6 mm column, a solvent system of (25%) aq. ammonium carbonate (0.3%) in methanol (75%), a flow rate of 3 ml/min, with detection by UV at 285 nm (Hook et al. 1988).

*Short-Term Time Course Analysis.* Figure 4 shows the pattern of growth and alkaloid accumulation over a 28-day growth cycle. Growth (as dry wt.) reached a maximum on day 17 and then declined. Growth of the suspension cultures was accompanied by a decrease in sucrose content of the medium and by a concomitant increase in alkaloid (DHS) production. The maximum alkaloid content (day 21) coincided with the disappearance of sucrose from the medium. Dihydrosanguinarine synthesis thus appears to be growth-associated in this cell culture (Hook et al. 1988).

*Long-Term Growth Analyses.* The productive stability of two cell lines (NB2 and NB3) of *P. bracteatum* cell suspensions was assessed over a 3-year period. Biomass remained reasonably constant for both cell lines (NB2: 14.47 g dry wt./l, S.D. ± 1.58; NB3: 13.74 g dry wt./l, S.D. ± 2.44). Alkaloid production, however, was erratic especially during the first 18 months of culture (Fig. 5a,b), possibly due to a genetic instability in the cultures, and markedly decreased during the period of evaluation (NB2: % DHS decreased from 0.83 to 0.29; NB3: % DHS decreased from 2.55 to 0.73) (Lecky et al. 1992). This instability of

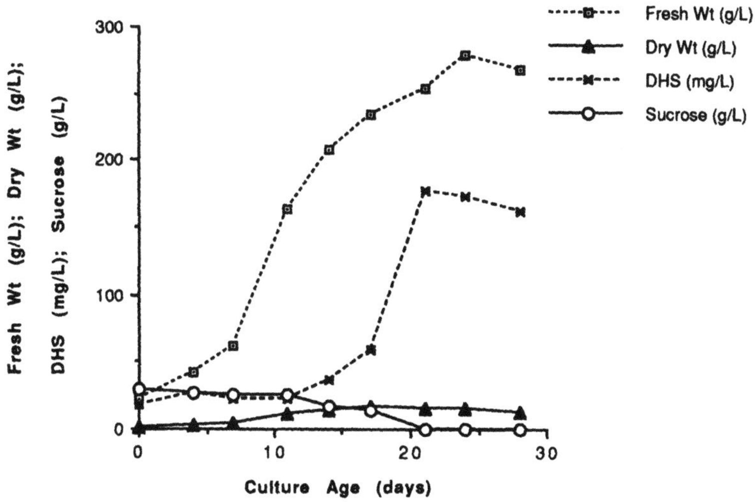


**Fig. 4.** Growth curve of *P. bracteatum* (cell line NB3) in batch suspension culture (50 ml medium in 100-ml flasks; results are the mean of three replicates). (Hook et al. 1988)

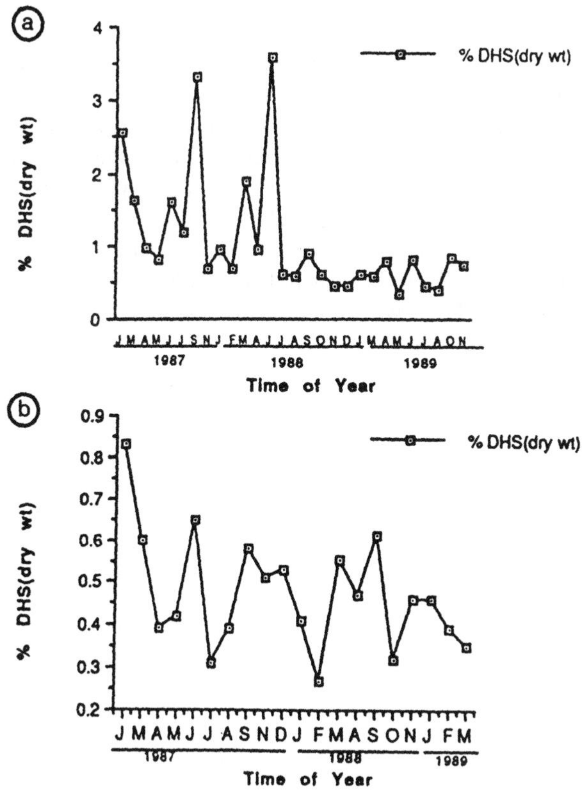


**Fig. 5a, b.** Alkaloid production by suspension cultures of *P. bracteatum* during a 3-year period of serial subculture (**a**) cell line NB3; (**b**) cell line NB2 (100 ml medium in 250-ml flasks). (Lecky et al. 1992)

secondary metabolite formation in nonselected cell cultures is a common phenomenon in plant tissue cultures (Morris 1986).

## 3.4 Enhancement of Alkaloid Production in Nonselected Cell Cultures

Newly established cultures of *P. bracteatum* (cell line NB3) were initially found to produce the alkaloid dihydrosanguinarine in yields frequently in excess of 2.55% (of dry cell wt.). This productivity was gradually lost on repeated subculture and in order to increase the production of DHS, a number of enhancement procedures were assessed.

### *3.4.1 Medium Modifications*

The concentration of several constituents of the culture medium were modified and their effects on cell growth and alkaloid formation evaluated (Lecky 1992).

*Sucrose* concentration in the medium was formulated to range from 0 to 7.5%. The results showed no significant increase in biomass production when sucrose levels were increased from 3% (control) to 7.5%. Highest dihydrosanguinarine levels were consistently produced with a 3% sucrose concentration (Fig. 6). This was in agreement with the results of Kamimura and Akutsu (1976).

*β-Cyclodextrin* was included in the culture medium at concentrations of 0.1, 0.5, and 1%. Results (Fig. 7) showed that the presence of this complexing agent during the growth cycle reduced alkaloid production without affecting biomass yields. The causes of these effects were not further investigated.

*Inorganic macroelements* present in the culture medium are known to affect secondary metabolite formation. Replacement of the macronutrient $CaCl_2$ by KCl or $NH_4Cl$, without altering ionic concentrations in the medium, yielded

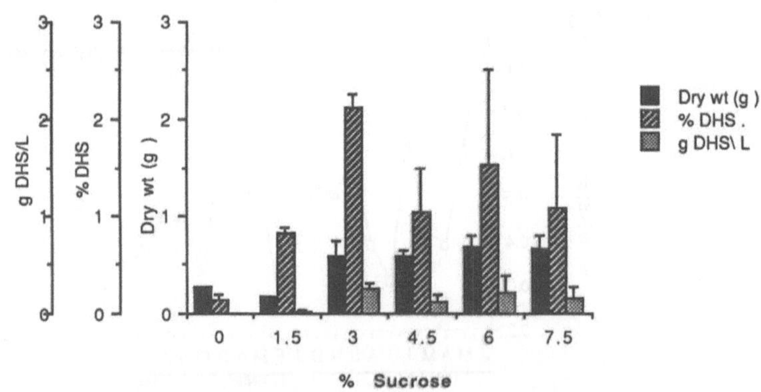


**Fig. 6.** Effect of sucrose concentration in culture medium on growth and alkaloid production by suspension cultures of *P. bracteatum* (cell line NB2). Dry wt. g/50 ml medium; values shown are the mean of three experiments; ± SD (Lecky 1992)

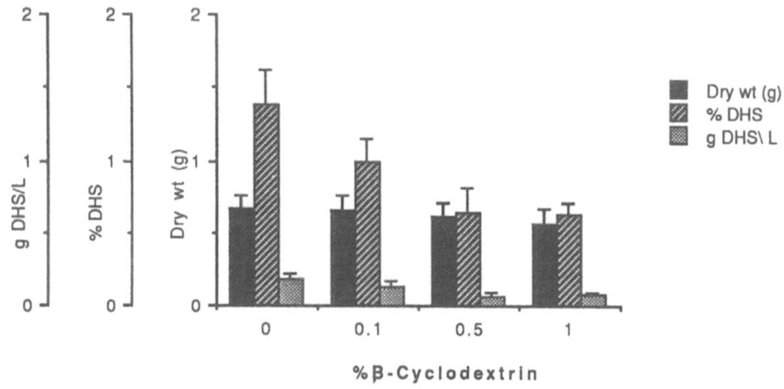


**Fig. 7.** Effect of β-cyclodextrin in medium on growth and alkaloid production of cell cultures of *P. bracteatum* (cell line NB2). Dry wt. g/50 ml medium; values shown are the mean of three experiments; ± SD (Lecky 1992)

results shown in Fig. 8. Only ammonium chloride produced an enhanced DHS yield, possibly due to an alteration in the nitrate: ammonium ratio in the medium.

*Inorganic microelements* were examined at several concentrations – half, twice, and four times those normally present in the modified MS medium used as control. The vitamins thiamine HCl and pyridoxine HCl were evaluated, as were the microminerals cobalt, manganese, zinc, and copper. Although all these constituents are known to be involved in a variety of enzyme reactions, their exact requirements for plant cell culture growth and secondary metabolite production are uncertain (DiCosmo and Towers 1983). Copper was the only

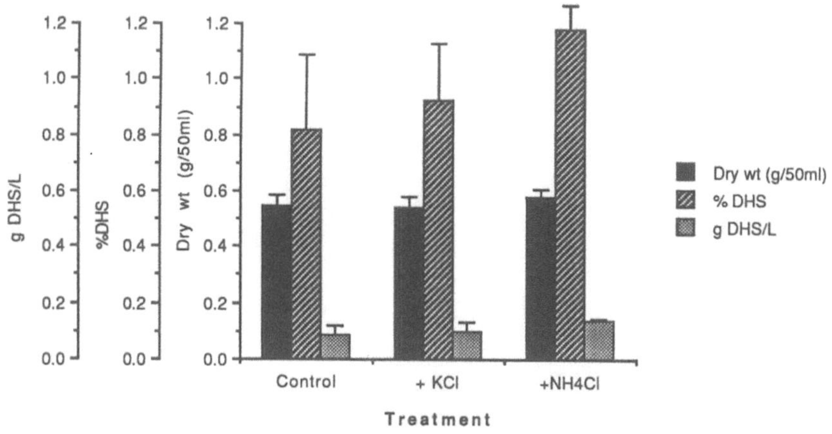


**Fig. 8.** Effect on growth and alkaloid production in suspension cultures of *P. bracteatum* (cell line NB3) by changes in medium of inorganic macroelements. Dry wt g/50 ml medium; values shown are the mean of four experiments; ± SD. (Lecky 1992)

microconstituent where increased concentrations in the nutrient medium significantly enhanced alkaloid yields (% DHS) without affecting growth (dry wt.). Doubling the copper concentration in the medium at the beginning of the growth cycle (day 0) produced a four fold increase in DHS yield at harvest (Fig. 9). For these enhanced alkaloid yields copper must be present at the beginning of the growth cycle or during the period of active cell division and growth (up to day 14). Once the culture is in its stationary phase (after day 21), additional copper does not affect alkaloid production (Lecky et al. 1992). The addition of metal ions to plant cell cultures has proved useful for the induction of secondary metabolite formation and has also been referred to as "abiotic elicitation" (Robins and Rhodes 1988).

### 3.4.2 *Immobilization*

The production of secondary metabolites has been associated with slow-growing or compact cells. Physically restraining the culture cells within or on a fixed support (i.e., immobilized) can encourage cell aggregation and contact-induced cytodifferentiation, which in turn can encourage increased metabolite formation (Rosevear and Lambe 1985; Hall et al. 1988). Numerous systems of cell immobilization have been investigated (Lindsey and Yeoman 1983). The most commonly used for plant cells cultures is entrapment in calcium alginate beads. Polyurethane foam has only rarely been used as an immobilizing agent (Corchete and Yeoman 1989), while no researchers have previously employed chemical wood pulp. For both agents, 0.5 g of washed and dried material was

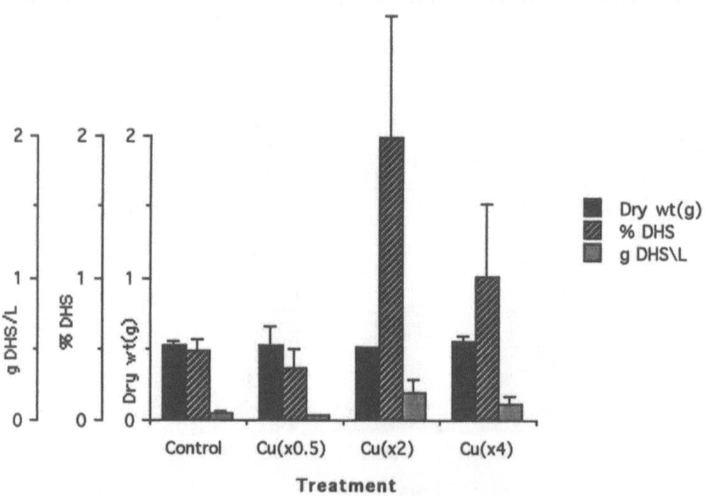


**Fig. 9.** Effect on growth and alkaloid production in *P. bracteatum* (cell line NB3) suspension cultures by various concentrations of copper in the culture medium. Dry wt g/50 ml medium; values shown are the mean of four experiments; ± SD. (Lecky 1992)

added per 50 ml of medium in a 100-ml conical flask and sterilized in situ prior to subculture of the cells. The suspension cultures were grown for 28 days, harvested, dried, and assayed as previously indicated. Results, shown in Fig. 10, indicate that both polyurethane foam and wood pulp enhanced alkaloid production without significantly increasing biomass yields. Wood pulp proved to be superior in all experiments, resulting in almost a doubling of dihydrosanguinarine yields (mg/l) when compared to the controls. Examination of the cultured cells within the polyurethane foam with a scanning electron microscope showed small cell aggregates within the pores, indicative of true immobilization (Fig. 11). The wood pulp on shaking became tightly compressed into small spheres which on sectioning were found not to have entrapped any suspension cells. In this case, the increased alkaloid yields can only be explained by some chemical interaction between the cellulosic cell walls of the wood pulp and the cells and/or enzyme systems of the cultured cells.

### 3.4.3 Biotic Elicitations

Enhancements of secondary metabolite production in plant cell cultures through inoculations with microbial preparations (elicitors) has proved successful in many instances (Robins and Rhodes 1988). A range of bacterial and fungal preparations, both living and dead, were examined for their ability to enhance alkaloid formation in cell cultures of *P. bracteatum* (cell line NB3). Three fungal species – *Alternaria longipes*, a *Penicillium*, and *Cyathus buleri*, as well as the bacterium *Escherichia coli* were evaluated. Cores (1 cm × 4 mm) of actively growing fungus, removed from the growing mycelial edge of the agar plate or 0.2 ml of bacterial culture, were used as inoculum for each flask. "Dead" samples

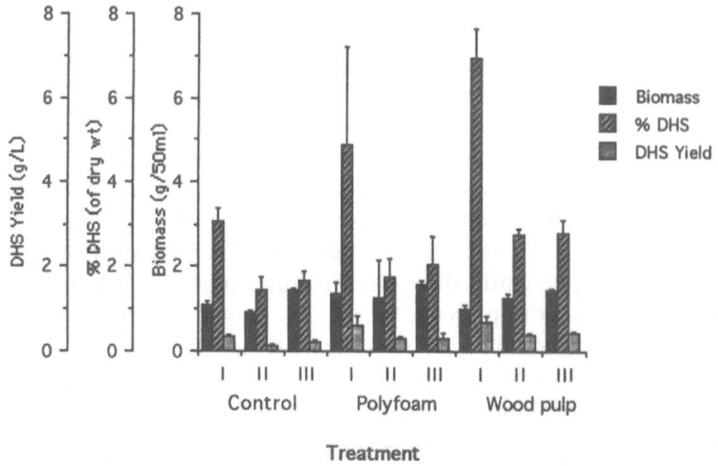


**Fig. 10.** Effect on growth and alkaloid production in suspension cultures of *P. bracteatum* (cell line NB3) by polyurethane foam and wood pulp as immobilizing agents. Dry wt g/50 ml medium; values shown are the mean of three experiments; ± SD. (Lecky 1992)

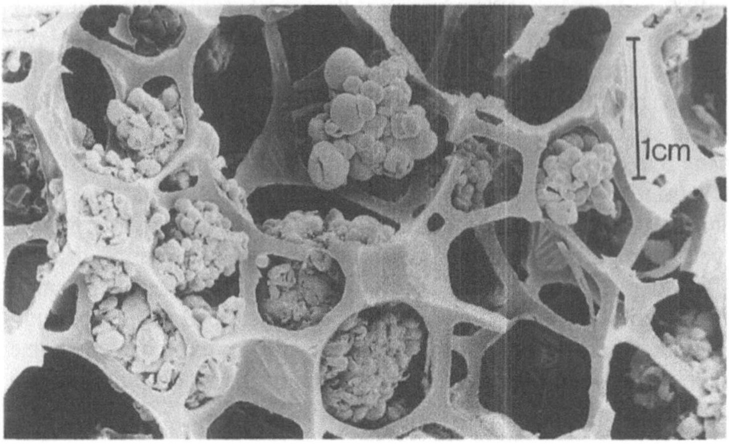


**Fig. 11.** Scanning electron micrograph of suspension cultures of *P. bracteatum* immobilized in polyurethane foam. (I. Hook and C. Reid, Electron Microscope Unit, Trinity College, Dublin)

were heat sterilized prior to addition to the cell cultures. All elicitor samples were added to the suspension cultures, growing in 50 ml of medium in 100 ml flasks, on day 14 of the growth cycle. Cells were harvested on day 28, dried and assayed.

Two experiments were carried out comparing the elicitation potential of living and dead microbes (Fig. 12a,b). From these it is obvious that living elicitors produced a variable enhancement in dihydrosanguinarine yields. Sanguinarine production was, however, increased (Fig. 12a) especially in the presence of the pathogenic fungus *Alternaria longipes*. In this instance, sanguinarine can be classified as a true phytoalexin, its enhanced production probably due to the activation by the fungus of the enzyme dihydrobenzophenanthridine oxidase which catalyzes the conversion of DHS to sanguinarine (Arakawa et al. 1992). The greatest increases in dihydrosanguinarine yields were produced by dead microbial elicitors, *A. longipes* being most consistently successful (an individual value of 10.3% DHS vs. 3.1% for the control having been recorded). These results are in agreement with those of Cline and Coscia (1988, 1989), Eilert and Constabel (1985, 1986), and Tyler et al. (1988), who showed increased production of benzophenanthridines when cultures of *P. bracteatum* and *P. somniferum* were grown in the presence of various fungal elicitors. The enhancement of alkaloid yields could result from the increased activities of enzymes involved in benzophenanthridine synthesis, such as tyrosine decarboxylase (Collinge and Brodelius 1989) and protopine-6-hydroxylase (Tanahashi and Zenk 1988). Another experiment combining elicitation with dead *A. longipes* and immobilization (Sect. 3.4.2) showed that additive increases in DHS production could be obtained by a combination of procedures. Figure 13 shows that cells immobilized with wood pulp and elicited with dead *A. longipes* resulted in a four fold increase in alkaloid yield.

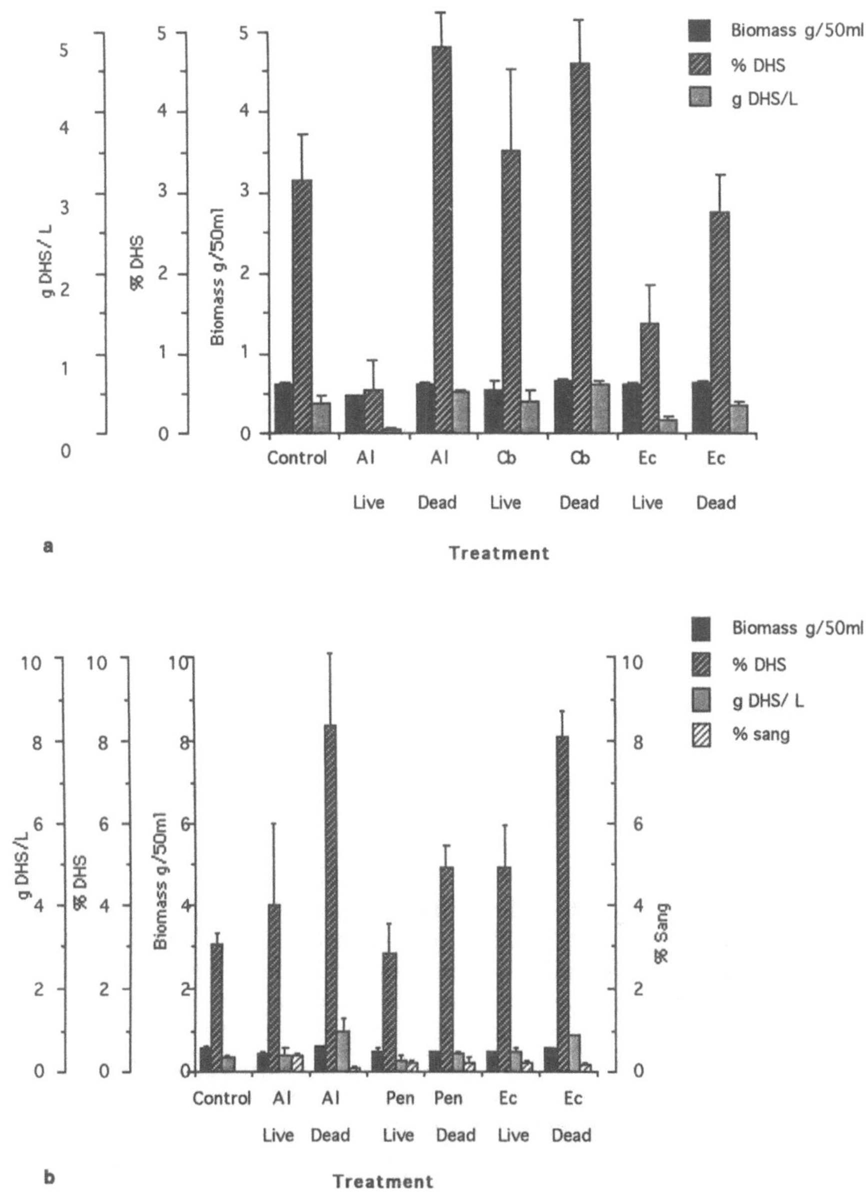


**Fig. 12a, b.** Effect on growth and alkaloid production in suspension cultures of *P. bracteatum* (cell line NB3) by various living and dead microbial elicitors. Dry wt g/50 ml; values shown are the mean of four experiments; ± SD. (Lecky 1992); *Al* = *Alternaria longipes*; *Pen* = *Penicillium*; *Ec* = *E. coli*; *Cb* = *Cyathus buleri*

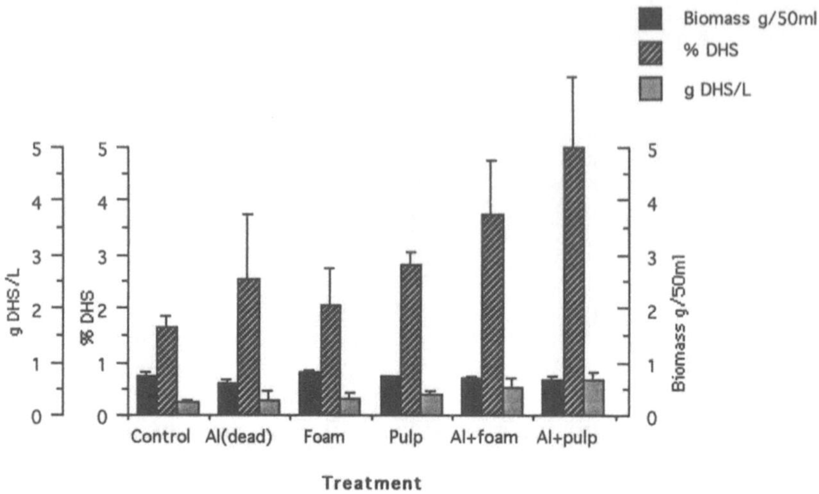


**Fig. 13.** Effect on growth and alkaloid production in suspension cultures of *P. bracteatum* (cell line NB3) by a combination of immobilization and elicitation with *A. longipes* (dead). Dry wt g/50 ml medium; values shown are the mean of four experiments; ± SD. (Lecky 1992). *Foam* Polyurethane foam; *pulp* chemical wood pulp

# 4 Summary and Conclusions

Callus and suspension cultures developed from seeds of *Papaver bracteatum* Arya II were found not to produce thebaine, the commercially valuable alkaloid characteristic of the intact parent plants. Instead, the secondary metabolites were identified as the benzophenanthridine alkaloids, dihydrosanguinarine and sanguinarine. High alkaloid-yielding cell lines were separated and found initially to produce in excess of 2.5% dihydrosanguinarine. On repeated subculture without selection, the alkaloid-producing capacity of the cultures was gradually lost. However, metabolite yields were found amenable to enhancement by manipulations such as (1) addition of copper salts to the culture medium; (2) immobilization in polyurethane foam and on wood pulp; (3) elicitation with fungal and bacterial preparations and a combination of (2) and (3). The commercial importance of dihydrosanguinarine is not obvious, although it has potential, due to its facile chemical conversion to sanguinarine – an alkaloid which is of current interest for the preparation of various antibacterial/antiplaque oral health care products.

# 5 Protocol

For the development of tissue cultures of *P. bracteatum* (Arya II), seeds were the best source material as they were easy to surface sterilize. The modified MS medium (specified in Sect. 3.1) was satisfactory for continued biomass and benzophenanthridine production with the high alkaloid-yielding culture investigated. Maximum alkaloid formation resulted after 3 weeks of culture under the growth conditions indicated. Enhancement of alkaloid (dihydrosanguinarine) production could be induced by increasing the copper content of the medium to 0.1 mg/l. Also, for increased alkaloid production, immobilization of culture cells in polyurethane foam and chemical wood pulp was found to be of value. The biotic elicitation of 14-day-old cultures with the plant pathogen *Alternaria longipes* also significantly enhanced alkaloid yields, but the technique requires refinement. Additional yields resulted from a combination of procedures – elicitation with immobilization.

# 4 References


Anonymous (1988) Extraction of benzophenanthridine alkaloids from plants (US Patent 4,767,861) through Chem Abstr 109(25): 226951S

Anonymous (1990) *Sanguinaria* extract safety. Washington Insight (newsletter for natural product scientists), 6 June: p 15

Arakawa H, Clark WG, Psenak M, Coscia CJ (1992) Purification and characterization of dihydrobenzophenanthridine oxidase from elicited *Sanguinaria canadensis* cell cultures. Arch Biochem Biophys 299(1): 1–7

Bowman WC, Rand MJ (1980) Textbook of pharmacology, 2nd edn. Blackwell, Oxford, pp 43, 49

Cline CD, Coscia CJ (1988) Stimulation of sanguinarine production by combined fungal elicitation and hormonal deprivation in cell suspension cultures of *Papaver bracteatum*. Plant Physiol 86: 161–165

Cline CD, Coscia CJ (1989) Ultrastructural changes associated with the accumulation and secretion of sanguinarine in *Papaver bracteatum* suspension cultures treated with fungal elicitor. Planta 178: 303–314

Collinge MA, Brodelius PE (1989) Dynamics of benzophenanthridine alkaloid production in suspension cultures of *Eschscholtzia californica* after treatment with yeast elicitor. Phytochemistry 28(4): 1101–1104

Constabel F (1985) Morphinan alkaloids from plant cell cultures. In: Phillipson JD, Roberts MF, Zenk MH (eds) The chemistry and biology of isoquinoline alkaloids. Springer, Berlin Heidelberg New York, pp 257–264

Corchete P, Yeoman MM (1989) Biotransformation of ( – )-codeinone to ( – )-codeine by *Papaver somniferum* cells immobilized in reticulate polyurethane foam. Plant Cell Rep 8: 128–131

Corrigan D, Martyn E (1981) The thebaine content of ornamental poppies belonging to the *Papaver* section *Oxytona*. Planta Med 42: 45–49

Davis AM (1982) Experiments in growing *Papaver bracteaum* Lindl. US Dep Agric, Agric Rev Manuals, Western Ser 25 (April)

DiCosmo F, Towers GHN (1983) Phytochemical adaptations to stress. In: Timmermann BN, Steelink C, Loewus FA (eds) Recent advances in phytochemistry, vol 18. Plenum Press, New York, pp 97–175

Duke JA (1985) CRC handbook of medicinal herbs. CRC Press, Boca Raton, pp 424–425

Eilert U, Constabel F (1985) Ultrastructure of *Papaver somniferum* cells cultured in vitro and treated with fungal homogenate eliciting alkaloid production. Protoplasma 128: 38–42

Eilert U, Constabel F (1986) Elicitation of sanguinarine accumulation in *Papaver somniferum* cells by fungal homogenates an induction process. J Plant Physiol 125: 167–172

Fairbairn JW, Helliwell K (1977) *Papaver bracteatum* Lindley: thebaine content in relation to plant development. J Pharm Pharmacol 29: 65–69

Fairbairn JW, Williamson EM (1978) Anatomical studies on *Papaver bracteatum*. Lindl. Planta Med 33: 34–45

Goldblatt P (1974) Biosystematic studies in *Papaver* section Oxytona. Ann Mo Bot Gard 61(2): 264–296

Hall RD, Holden MA, Yeoman MM (1988) Immobilization of higher plant cells. In: Bajaj YPS (ed) Biotechnology in agriculture and forestry vol 4. Medicinal and aromatic plants I. Springer, Berlin Heidelberg New York, pp 136–156

Hook I, Sheridan H, Wilson G (1988) Alkaloids of cell cultures derived from strains of *Papaver bracteatum.* Phytochemistry 27:2137–2141

Ikutsu A, Syono K, Furuya T (1974) Alkaloids of callus tissues and redifferentiated plantlets in the Papaveraceae. Phytochemistry 13:2175–2179

Kamimura S, Akutsu M (1976) Cultural conditions on growth of the cell culture of *Papaver bracteatum.* Agric Biol Chem 40(5):899–906

Kamimura S, Nishikawa M (1976) Growth and alkaloid production in the cultured cells of *Papaver bracteatum.* Agric Biol Chem 40(5):907–911

Kamimura S, Akutsu M, Nishikawa M (1976) Formation of thebaine in the suspension culture of *Papaver bracteatum.* Agric Biol Chem 40(5):913–919

Kamo KK, Mahlberg PG (1988) Morphinan alkaloids: biosynthesis in plant (*Papaver* sp.) tissue cultures. In: Bajaj YPS (ed) Biotechnology in agriculture and forestry, vol 4. Medicinal and aromatic plants I. Springer, Berlin Heidelberg New York, pp 251–263

Kettenes van den Bosch JJ, Salemink CA, Khan I (1981) Biological activity of the alkaloids of *Papaver bracteatum.* J Ethnopharmacol 3:21–38

Kutchan TM, Ayabe S, Kruegar RJ, Coscia EM, Coscia CJ (1983) Cytodifferentiation and alkaloid accumulation in cultured cells of *Papaver bracteatum.* Plant Cell Rep 2:281–284

Kutchan TM, Ayabe S, Coscia CJ (1985) Cytodifferentiation and *Papaver* alkaloid accumulation. In: Phillipson JD, Roberts MF, Zenk MH (eds) The chemistry and biology of isoquinoline alkaloids. Springer, Berlin Heidelberg New York, pp 281–294

Lecky R (1992) Biotechnological evaluation of *Papaver bracteatum* cell cultures. PhD Thesis Trinity College, Dublin

Lecky R, Hook I, Sheridan H (1992) Enhancement of dihydrosanguinarine production in suspension cultures of *Papaver bracteatum* I. Medium modifications. J Nat Prod 55(10):1513–1517

Lenfeld J, Koutil M, Marvsalek E, Slavik J, Preininger V, Simanek V (1981) Anti-inflammatory benzophenanthridine alkaloids from *Chelidonium majus.* Planta Med 43(2):161–165

Lindsey K, Yeoman MM (1983) Novel experimental systems for studying the production of secondary metabolites by plant tissue cultures. In: Mantell SH, Smith H (eds) Plant biotechnology. Cambridge University Press, Cambridge, pp 39–66

Lockwood B (1981) Orientalidine and isothebaine from cell cultures of *Papaver bracteatum.* Phytochemistry 20:1463–1464

Lockwood B (1984) Alkaloids of cell suspensions derived from four *Papaver* species and the effect of temperature stress. Z Pflanzenphysiol 114S:361–363

Milo J, Levy A, Palevich D (1988) HPLC analysis of the alkaloid spectrum in the roots and capsules of the species and hybrids of *Papaver* section Oxytona. J Chromatogr 452:563–570

Mitscher L, Park Y, Clark D, Clark III GW (1978) Antimicrobial agents from higher plants. An investigation of *Hunnemannia fumariaefolia* pseudoalcoholates of sanguinarine and chelerythrine. J Nat Prod 41:145–150

Morris P (1986) Long term stability of alkaloid productivity in cell suspension cultures of *Catharenthus roseus.* In: Morris P, Scragg AH, Stafford A, Fowler MW (eds) Secondary metabolism in plant cell culture. Cambridge University Press, Cambridge, pp 257–262

Murashige T, Skoog F (1962) A revised medium for rapid growth and bioassays with tobacco tissue cultures. Physiol Plant 15:473–497

Nessler CL, Mahlberg PG (1978) Laticifer ultrastructure and differentiation in seedlings of *Papaver bracteatum* Lindl., Population Arya II (Papaveraceae). Am J Bot 65(9):978–983

Nyman U, Bruhn JG (1979) *Papaver bracteatum* – a summary of current knowledge. Planta Med 35(2):97–117

Phillipson JD (1983) Infraspecific variation and alkaloids of *Papaver* species. Plant Med 48:187–192

Preininger V (1986) Chemotaxonomy of Papaveraceae and Fumariaceae. In: Brossi A (ed) The alkaloids, chemistry and pharmacology, vol 29. Academic Press, London, pp 1–99

Robins RJ, Rhodes MJC (eds) (1988) Manipulating secondary metabolism in culture. Cambridge University Press, Cambridge UK

Rosevear A, Lambe CA (1985) Immobilized plant cells. In: Fiechter A (ed) Advances in biochemical engineering/biotechnology 31. Plant cell culture. Springer, Berlin Heidelberg New York, pp 37–58

Rush MD, Kutchan TM, Coscia CJ (1985) Correlation of the appearance of morphinan alkaloids and laticifer cells in germinating *Papaver bracteatum* seedlings. Plant Cell Rep 4:237–240

Sethi ML (1985) Comparison of inhibition of reverse transcriptase and antileukemic activities exhibited by protoberberine and benzophenanthridine alkaloids and structure-activity relationships. Phytochemistry 24(3):447–454

Shafiee A, Lalezari I, Yassa N (1976) Thebaine in tissue culture of *Papaver bracteatum* Lindl. population Arya II. Lloydia 39:380–381

Shamma M (1972) The benzophenanthridines. In: Shamma M (ed) The isoquinoline alkaloids: chemistry and pharmacology. Academic Press, New York, pp 315–343

Simanek V (1985) Benzophenanthridine alkaloids. In: Brossi A (ed) The alkaloids, vol 26. Academic Press, New York, pp 185–234

Simeon S, Rios JL, Villar A (1989) Pharmacological activities of benzophenanthridine and phenanthrene alkaloids. Pharmazie 44(9):593–599

Southard GL, Boulware RT, Wahlborn DR, Groznik WJ, Thorne EE, Yankell SL (1984) Sanguinarine, a new antiplaque agent: retention and plaque specificity. JADA 108:338–341

Tanahashi T, Zenk MH (1988) One-step enzymatic synthesis of dihydrosanguinarine from protopine. Tetrahedron Lett 29(44):5625–5628

Tyler RT, Eilert U, Rijnders CO, Roewer IA, Kurz WG (1988) Semi-continuous production of sanguinarine and dihydrosanguinarine by *Papaver somniferum* L. cell suspension cultures treated with fungal homogenate. Plant Cell Rep 7:410–413

Wu F-F, Dobberstein RH (1976) Scientific research on *Papaver* species as sources of codeine, morphine and thebaine. United Nations Division of Narcotic Drugs; Document ST/SOA/SER, No 27 J/27

Zito SW, Staba EJ (1982) Thebaine from root cultures of *Papaver bracteatum*. Planta Med 45:53–54

# XXII *Phytolacca americana* L. (Pokeweed): In Vitro Production of Betacyanins and Medicinal Compounds

G. Petit-Paly, F. Andreu, J. C. Chénieux, and M. Rideau[1]

## 1 General Account

*Phytolacca americana* (Phytolaccaceae), in addition to serving as an occasional food, is a medicinal plant listed officially in the United States and in the French and Japanese Pharmacopeia. The plant has been used in folk medicine as a diuretic, purgative, antiscorbutic, and antisyphilitic agent (Fournier 1948). The roots are reputed in Korea to treat edema and rheumatism (Woo and Kang 1976a). At present, *P. americana* is used in some French homeopathic preparations to alleviate influenza, acute amygdalis, quinsy, mammary and rheumatic pains, and chronic pharyngitis (Voisin 1980; Nétien et al. 1986). The plant and its tissue cultures have been investigated as a source of saponins, betalains, mitogens, and antiviral proteins.

### 1.1 Botanical Traits and Classification

The genus *Phytolacca* belongs to the family Phytolaccaceae, order Centrospermae (or Caryophyllales), suborder Chenopodiineae (which includes ten families accumulating the pigments betalains). This suborder is closely related to the anthocyanin suborder Caryophyllineae. Both suborders are derived from a common ancestor, probably preadapted for $C_4$ photosynthesis, which had evolved ring-like inclusions composed of proteinaceous filaments, contained in the sieve-element plastids in all the Centrospermae families (Mabry 1980). About 35 species are assigned to the genus (Usher 1974).

*Phytolacca americana* L. (synonym: *P. decandra* L.; common names: English: pokeweed, pokesallet, pokeberry, pigeon berry, American scarlet berry; French: raisin d'Amérique; German: Kermesbeere) is a perennial herb, somewhat woody at the base, growing to a height of 1–3 m. Stems are subdichotomously branched, often red, with decurrent ridges from the leaf bases. Leaves are ovate-lanceolated, petiolated. Spikes of greenish white flowers turn reddish in fruits. Fruits are depressed-globose, dark purple juicy berries, about 10 mm in diameter (Fig. 1); they are dispersed by birds (Mc Donnell et al. 1984). *P. americana* originated in North America and was introduced into numerous countries in

[1] Plant Biotechnology Laboratory, EA-1370, Faculty of Pharmacy, University of Tours, 31 Avenue Monge, 37200 Tours, France

Biotechnology in Agriculture and Forestry, Vol. 28
Medicinal and Aromatic Plants VII (ed. by Y.P.S. Bajaj)

**Fig. 1.** *Phytolacca americana* L.

Europe (Sauer 1952). Thereafter, the species was spread to other continents and is now established in South Africa, Japan, Australia, New Zealand, Argentina, etc. (Edwards et al. 1988), so that it is now on a way to become an ubiquitous weed.

Optimal conditions for germination of *P. americana* seeds have been investigated (Edwards et al. 1988 and literature therein): alternating temperatures, light, and nitrate are three factors of importance to obtain appreciable seed germination; soaking seeds in concentrated $H_2SO_4$ prior to sowing increases rates of germination. Juice of pokeweed fruits and extracts of freshly harvested mature leaves stems or immature fruits are inhibitory for seed germination; in contrast, extracts of juvenile leaves have no effect on seed germination.

*P. americana* has long been considered poisonous for humans and animals (Fournier 1948). The more toxic organs are roots, stems and leaves being less toxic (Harding and Arena 1969); however, poisoning of horses and bovids was reported after consumption of pokeweed leaves (Jean-Blain and Grisward 1973). There are conflicting opinions about the toxicity of berries, some traditional sources indicating medicinal values when fermented (Wigginton 1973, reported in Edwards et al. 1988). The juice was formerly used as a dye to color jam and wine (Usher 1974). Yet, as this practice seemed to be hazardous, the cultivation of the plant was forbidden in Portugal (Le Maout and Decaisne 1876). Pokeberries were reported to be toxic to cattle, horses, and turkey poults (Barnett 1975), despite the fact that many wild birds are known to consume these fruits. Ingestion of berries by cows produced gastrointestinal disturbances and eventually death caused by paralysis of the respiratory organs (Hardin 1966, cited by Barnett 1975). Peripheral blood plasmocytosis was found in children exposed to pokeberries (Barker et al. 1966). Presence of toxins has been described by Driver and Francis (1979b). On the other hand, other reports claimed that the seeds are extremely toxic when ingested. Despite the toxicity of all parts of the plant, young stems and leaves seem to be edible if cooked: several authors claimed that *P. americana* sprouts were eaten locally in the United States or in France as a vegetable (Bois 1927 and references therein).

## 1.2 Chemical Constituents and Their Pharmacological Properties

### *1.2.1 Saponins*

All *Phytolacca* species are rich in saponins which have interesting biological activities: one of the most promising plants with molluscicidal properties is *Phytolacca dodecandra* l'Herit., from Ethiopia (Kloos and McCullough 1982; Parkhurst et al. 1989). The roots of *P. esculenta* have been used in traditional Chinese medicine for the treatment of tumors, edema, and bronchitis; they contain several saponins (Yang-hua 1992); one of them (esculentoside H) showed considerable activity in inducing immune interferon and tumor necrosis factor (Yang-Hua and Chu-Lu 1989). Fungistatic saponins have been isolated in *P. octandra* (Moreno and Rodriguez 1981).

*P. americana* also contains saponins. Ahmed et al. (1949) showed that an acidic water-soluble saponin (isolated from the roots) was toxic for rats. Stout et al. (1964) named phytolaccatoxin the toxic constituent and gave the structure of its aglycone phytolaccagenin. Later, several saponins with strong anti-inflammatory activity (Woo and Shin 1976) were isolated from roots (Table 1). Saponins were found to be distributed in all tissues (Woo and Kang 1976a), phytolaccoside E being the major saponin (Woo et al. 1976b). Hydrolysis of the total saponin mixture yielded four sapogenins (Fig. 2), jaligonic acid (**1**), its 3-O-methyl ester phytolaccagenin (**2**), esculentic acid (**3**), and its 3-O-methyl ester phytolaccagenic acid (also named phytolaccinic acid) (**4**). The fresh berry juice contains about 0.6% of a mixture of saponins, the hydrolysis of which afforded jaligonic acid (**1**), phytolaccagenin (**2**) and phytolaccagenic acid (**4**) (Johnson and Shimizu 1974).

Presence of esculentic acid as well as a minor aglycone (acinosolic acid methyl ester = pokeberry genin) also occurred in the hydrolysate (Kang and Woo 1980). Thin layer chromatography examination showed the presence of free acinosolic acid in the juice. *Phytolacca* saponins and phytolaccagenin exhibit almost the same potent anti-inflammatory activity as that of cortisone acetate (Woo and Shin 1976). At 15–30 mg/kg, crude saponin showed 50% inhibition of carrageenin-induced paw edema in rats (Woo et al. 1976b).

**Table 1.** Saponins isolated from *P. americana* roots

| Phytolaccoside | | Reference |
|---|---|---|
| A | 3-O-β-D-xylopyranosyl-esculentic acid 3-O methyl ester | Woo et al. (1978) |
| B | 3-O-β-D-glucopyranosyl-jaligonic acid 3-O methyl ester | Woo and Kang (1976a); Suga et al. (1978) |
| D | 3-O-[β-D-glucopyranosyl (1 → 4)-β-D-xylopyranosyl] esculentic acid 3O methyl ester | Woo et al. (1978) |
| $D_2$ | 3-O-[β-D-glucopyranosyl (1 → 2)-β-D-xylopyranosyl] esculentic acid 3O methyl ester | Kang and Woo (1987) |
| E | 3-O-[β-D-glucopyranosyl (1 → 4)-β-D-xylopyranosyl] jaligonic acid 3O methyl ester | Woo et al. (1978); Suga et al. (1978) |
| F | 3-O-[α-L-rhamnopyranosyl (1 → 2)-β-D-glucopyranosyl (1 → 2) β-D-xylopyranosyl] esculentic acid-3-O methyl ester | Kang and Woo (1987) |
| G | 3-O-β-D-xylopyranosyl-jaligonic acid | Woo and Kang (1977); Suga et al. (1978) |

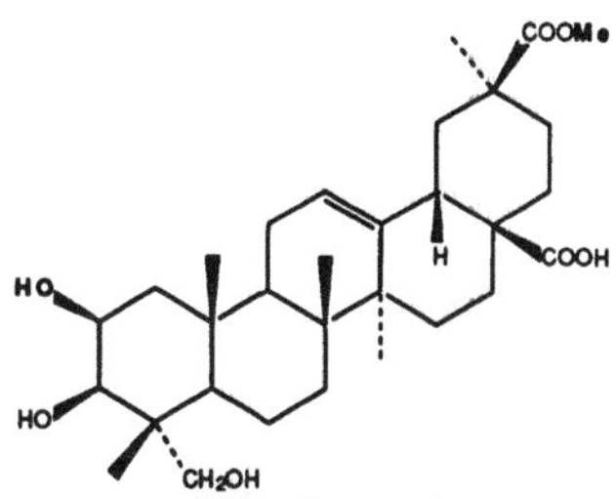


2 Phytolaccagenin
(jaligonic acid 3 O methylester)

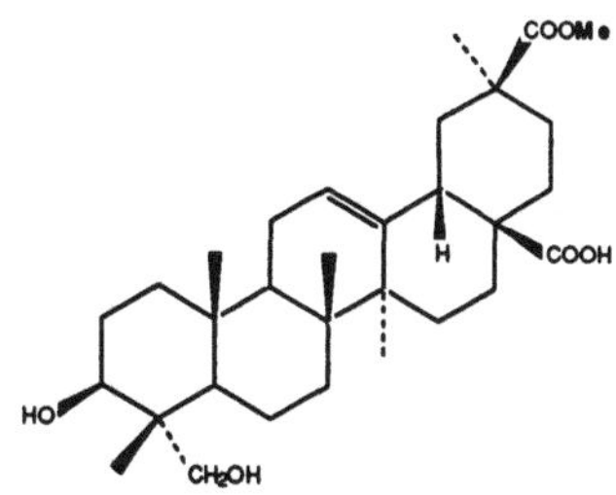


4 Phytolaccagenic acid
(esculentic acid 3 O methylester)
(phytolaccinic acid)

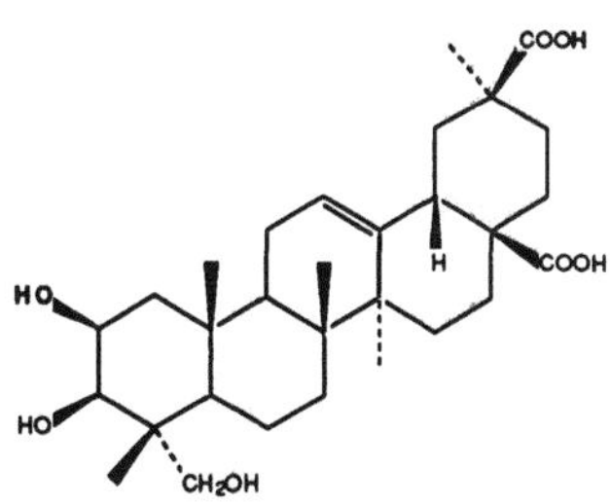


1 jaligonic acid

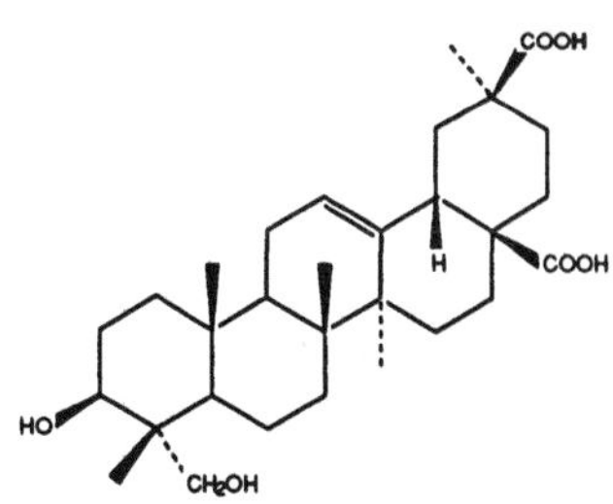


3 esculentic acid

**Fig. 2.** Sapogenins obtained after hydrolysis of *Phytolacca americana* root saponins

On the basis of thin layer chromatography, Burke and Le Quesne (1971) postulated the presence of free oleanic acid in root extracts and considered that phytolaccagenin most likely arises biogenetically from this compound. The authors also obtained 3-acetyloleanic acid from light petroleum extracts of seeds. Woo (1974) reported on the presence of a great amount of jaligonic acid in both ether and methanol extracts of roots. The terpenoid 3-acetylaleuritolic acid, which was previously known only from *Aleurites montana* (Euphorbiaceae), has been isolated in the seeds of *P. americana* (Woo and Wagner 1977).

### *1.2.2 Betalains*

The betalains comprise two groups of water-soluble vacuolar pigments, the red-violet betacyanins and the yellow-orange betaxanthins, occurring in the species of the suborder Chenopodiinae. Betacyanins are characterized by glycoside structure; betaxanthins do not contain sugar in their molecule. These compounds are polyfunctional, acting as pollinator attractant and as defense against viral or fungal infections (Mabry 1980).

The major red-violet pigment of *P. americana* (named phytolaccanin) was isolated by Wyler and Dreiding (1961) and was proved to be identical to betanin (**5**), the pigment of red beet (*Beta vulgaris*) (Fig. 3). Small amounts of three other pigments (probably identical to isobetanin (**7**) and to prebetanin and isoprebetanin, i.e., the sulfate monoesters of betanin and isobetanin, respectively) are also present, but account for less than 5% of the total red pigment (Wyler and Dreiding 1961). Biochemical investigations on the biosynthetic pathways leading to betanin have been carried out in whole plants, in tissue slices, and in suspension cultures of red beet (Mabry 1980; Zrÿd et al. 1982). From incorporation experiments with [$^{14}$C] tyrosine and [$^{14}$C] DOPA, the authors concluded that tyrosine is better incorporated into the upper part of the molecule (via cyclo-DOPA glucoside) than DOPA. The latter (which is formed from tyrosine) is efficiently incorporated into betalamic acid (which gives the lower part of the molecule of betanin): Fig. 4. On the other hand, four betacyanin-decolorizing enzymes catalyze the degradation of betacyanin, first to betanidin (**6**), then to yet unidentified yellow products. Two enzymes, $BD_1$ and $BDE_2$, are acidic and two others, $BDE_3$ and $BDE_4$, are basic. The enzyme activities are present in leaf,

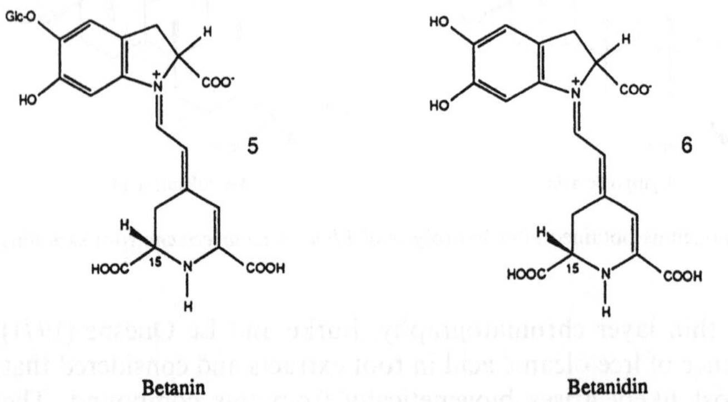


**Fig. 3.** Structures of betanin (**5**) and of betanidin (**6**) ; (isobetanin: C-15 epimer of betanin)

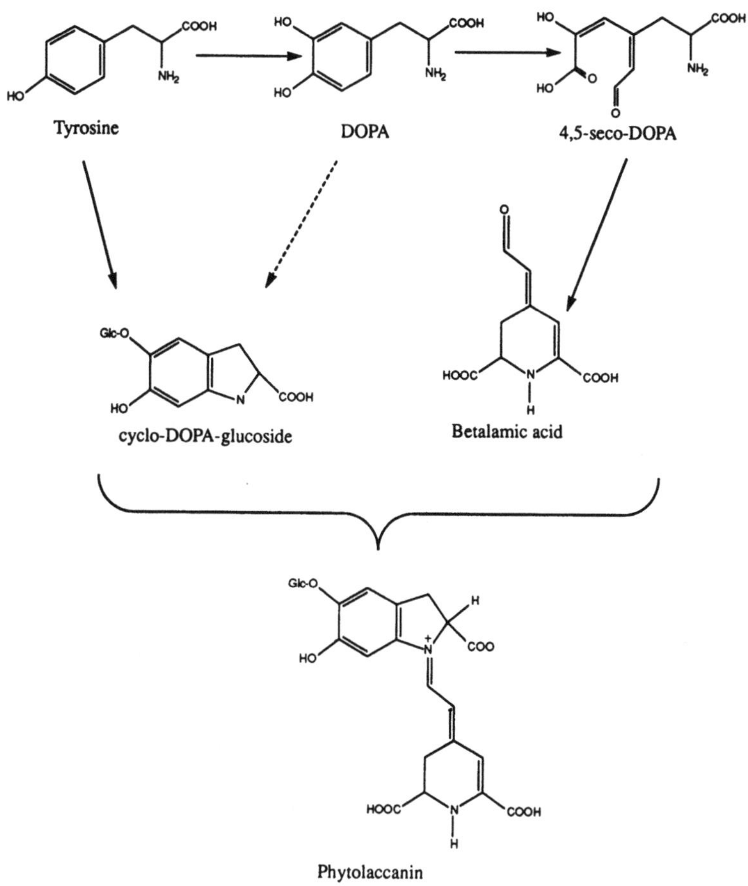


**Fig. 4.** The pathway of betalain synthesis

stem, seed, and root extracts (Kumon et al. 1990a), the highest specific activities for $BDE_2$ being found in the epidermidis of the red parts of the stem. The enzymes from *P. americana* apparently differ from those of *Amaranthus tricolor*, since in the latter species, betacyanin is degraded into cyclo-DOPA and betalamic acid (Elliott et al. 1983).

As for other betalains (Schütte and Liebisch 1985), the biosynthesis of phytolaccanin is controlled by several factors. Light induces or promotes the accumulation of the pigment in peeled epidermidis from the stems of *P. americana*. Kinetin enhances the pigment accumulation whereas abscisic acid or gibberellic acid are inhibitory. Moreover, some substances in the stem cortex may inhibit the accumulation of phytolaccanin (Kumon et al. 1990a,b).

The need for pigments from natural sources to replace artificial ones has been raised in the food industry, and the applicability of phytolaccanin as a food colorant has been investigated. The feasibility of this pigment as a natural colorant for gelatin and pectin gels was demonstrated by Driver and Francis

(1979a), but the colorant is rather unstable. Moreover, it is necessary to remove the toxic saponins before incorporation in foods (Driver and Francis 1979b).

### *1.2.3 Antiviral Proteins*

That *P. americana* contains virus inhibitors has been recognized for nearly 70 years (Duggar and Amstrong 1925). Pokeweed extracts were shown to inhibit viral transmission in plants (Wyatt and Sheperd 1969), and the replication of influenza virus (Tomlinson et al. 1974) and poliovirus (Ussery et al. 1977) in animal cells. The biological properties are due to several isoforms of a protein referred to as PAP (pokeweed antiviral protein). The first one (PAP) was previously reported to be a basic protein having a molecular weight of about 13 kDa (Wyatt and Schepherd 1969). Later, Irvin (1975) showed that the protein consists of a single polypeptide chain of 29 kDa. PAP was purified from leaves harvested in early spring from young plants. Another antiviral protein (PAP II), immunologically distinct from PAP despite similar amino acid composition (MM, 30 kDa), was purified from leaves harvested in summer (Irvin et al. 1980). A third antiviral isoform (31 kDa) (referred to as PAP-S) was isolated from seeds (Barbieri et al. 1982). Serum anti-PAP gave a weak reaction with PAP-S. Last of all, an isoform of PAP, PAP-R (MM, 30 kDa) was isolated from roots (Bolognesi et al. 1990).

All $PAP_s$ belong to the type 1 of the so-called ribosome-inactivating proteins ($RIP_s$) differing from (although evolutionarily related to) the ricin-like toxins. The latter have an A chain which enzymatically attacks ribosomes and a B chain which binds cell surfaces through interaction with sugars; the former consist of enzymes with activity like the above-mentioned A chain (Ready et al. 1984; Stirpe and Barbieri 1986).

$PAP_s$, like other $RIP_s$, are RNA N-glycosidases which depurinate the major ribosomal RNA, cleaving the N-glycosidic bond corresponding to adenine-4324 of rat liver 28S rRNA (Endo and Tsurugi 1988). They block ribosomes and elongation factor 1-dependent reactions, including both GTPase activity and enzyme-binding of aminoacyl-tRNA to ribosomes. They also inhibit the formation of the elongation factor 2-GTP-ribosome complex, with consequent arrest of protein synthesis (Jiménez and Vàsquez 1985, review; Endo et al. 1987; Endo and Tsurugi 1988). $PAP_s$ also depurinate 25S rRNA from plants, including ribosomes of *P. americana* (Prestle et al. 1992). PAP-R depurinates more efficiently the housefly ribosomal RNA than does PAP-S. It was found that PAP-R and PAP-S release 1 and 4 mol of adenine respectively/mol of ribosomes, in spite of the fact that they have similar inhibitory activity with a rabbit reticulocyte lysate system: contrary to most $RIP_s$, PAP-R is able to depurinate ribosomes at more than one site (Barbieri et al. 1992). A cDNA clone encoding the anti-viral protein (PAP) was recently isolated and characterized. Southern blot analysis indicated that PAP is encoded by a multigene family (Lin et al. 1991).

$PAP_s$ are nowadays of great interest since it was proved that these proteins may inhibit HIV-1-replication in primary $CD4^+$ cells (Zarling et al. 1990).

Preparation of PAP conjugated to monoclonal antibodies might also be useful for treatment of some types of cancer (Chu et al. 1990; Myers et al. 1991).

### *1.2.4 Lectins*

Several lectins with mitogenic and hemagglutinating properties have been characterized in root and leaf extracts of *P. americana*. Other lectins from berries, seeds, pulp, and stems had mitogenic activity and slight hemagglutinating properties (Goldstein and Poretz 1986). These lectins appear to be the causative agents for poisoning of children after the ingestion of *P. americana* berries, or after systemic exposure of cuts in the hand to pulp of berries (Liener 1986). Five mitogens (Pa-1, Pa-2, Pa-3, Pa-4 and Pa-5) were purified from roots by affinity chromatography. The pokeweed mitogen (PWM = Pa-1; GlcNAc-β-1,4 GlcNAc) is available commercially. PWM activates B and T lymphocytes, reacts only with lipoproteins, and is routinely used in immunology. It has been isolated from roots and possesses hemagglutinating, leucohemagglutinating, and mitogenic activities (Yokohama et al. 1978). The other mitogens from roots are active only for T-lymphocytes, of which Pa-2 exhibits the strongest activity (Yokohama et al. 1978). It could be used as marker to study pathologic structures in canine mammary glands (Castagnaro and Canese 1990, 1991), and to detect fungi in histologic sections of human tissues (Karayannopoulou et al. 1988).

### *1.2.5 Miscellaneous*

The flavonoids isoquercitrin and astragalin occurred in leaves of *P. americana*; cyclitol was found in the berries (Hegnauer 1969). Two new neolignans, americanol and isoamericanol, having an enhancing effect on choline acetyltransferase activity, were recently isolated in the seeds (Fukuyama et al. 1992). *P. americana* also contains α-spinasterol, its glucoside and acylated glucoside (Woo et al. 1976a), $\Delta^7$-stigmastenol, α-spinasteryl-D-glucoside, Δ-stigmastenyl-D-glucoside, 6′-palmityl-α-spinasteryl-D-glucoside, and 6′-palmityl-$\Delta^7$-stigmastenyl-D-glucoside (Woo 1974).

# 2 In Vitro Culture Studies

## 2.1 Establishment of Callus and Suspension Cultures

Tissue cultures of *P. americana* (Fig. 5A) can be easily raised from explants of stems or leaves harvested from whole plants or seedlings. Various media, such as MS (Murashige and Skoog 1962), LM (Linsmaier and Skoog 1965), B5 (Gamborg et al. 1968), or SH (Schenk and Hildebrandt 1972) may be used.

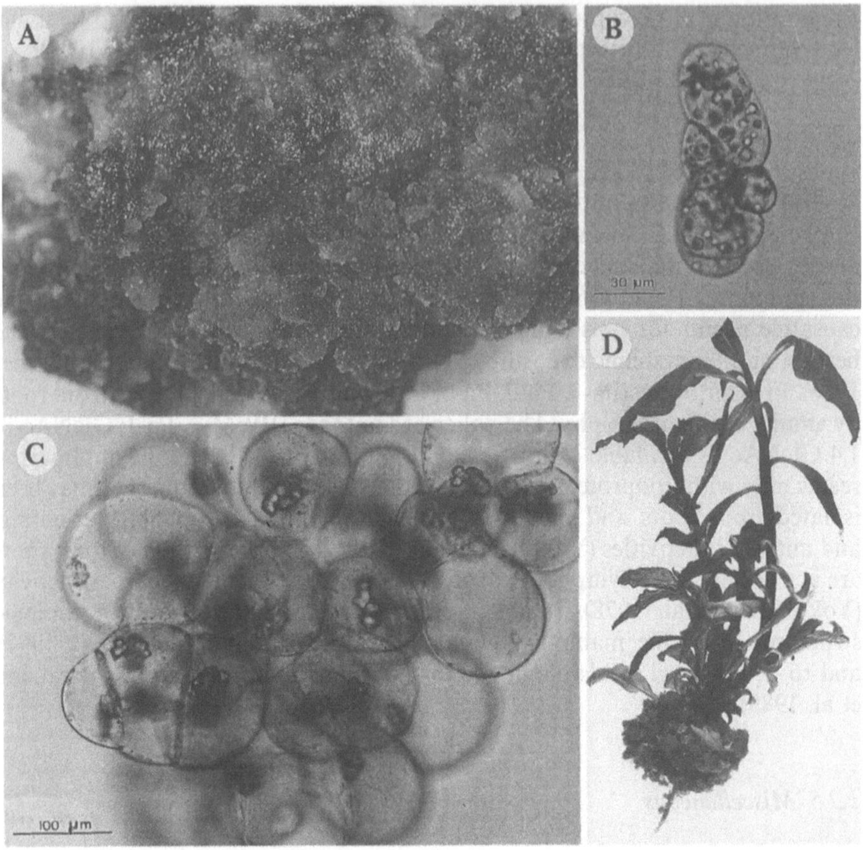


**Fig. 5A–D.** In vitro culture of *P. americana.* **A** Callus culture. **B** Suspension cultures (PA-87-St) derived from stem. **C** Suspension culture (PA-87-L) derived from leaves. **D** Shoot cultures. The culture medium of PA-87-St and PA-87-L lines is composed of MS medium plus 20 g/l sucrose and 1 mg/l 2,4-D. The shoot cultures are grown on an MS agar medium containing 30 g/l glucose and 1 mg/l benzyladenine. All the cultures are maintained in light. Note the starch granules in cell suspensions

Development of callus occurs in the presence of various phytohormones (Table 2). Schenk and Hildebrandt (1972) reported that callus growth was slightly better in light than in dark. In our opinion, the presence of naphthaleneacetic acid (NAA) and glucose favors obtaining dense chlorophyllous calli, whereas the presence of 2,4-dichlorophenoxyacetic acid (2,4-D) and sucrose favors friable betacyanin-producing calli from which it is easy to obtain suspension cultures (Fig. 5B,C). A modified medium containing twofold concentrations of the ordinary MS medium, 60 g/l sucrose, and 1 mg/l 2,4-D was used by Misawa et al. (1975a): the time required for obtaining maximum cell amounts was longer than that required in an ordinary medium, but the dry weight was about two times higher

**Table 2.** Cultural conditions used for growing in vitro cultures of *Phytolacca americana*

| Source of explants | Initiation medium | Callus/cells | Culture conditions | Response | Reference |
|---|---|---|---|---|---|
| Stem | LS + 2,4-D 1 mg/l | Callus | 27 °C, dark | Antimicrobial substances | Lin and Mathes (1973) |
| Stem | MS + 2,4-D 1 mg/l 1 g/l yeast extract[a] | Callus and cell suspension | 28 °C, 2000 lx | Plant virus inhibitors | Misawa et al. (1975a, b) |
| Leaf | B5 + NAA 1 mg/l + kinetin 1 mg/l + 2,4-D 0.2 mg/l | Callus and cell suspensions (lines 39 and 42) | 28 °C, dark | Ribosome-inactivating proteins | Barbieri et al. (1989) |
| ? | LS | Callus | 25 °C, 900 lx | Betacyanins | Ekiert and Dohnal (1984) |
| Stem | MS + 2,4-D 1 mg/l | Cell suspensions | 27 °C, dark | Betacyanins | Sakuta et al. (1986) |
| Stem[b] | MS + NAA 1 mg/l + kinetin 0.1 mg/l | Callus line PA-89-St | 24 °C, 1000 lx | | Petit, Andreu, Chénieux, Rideau (unpubl.) |
| Stem[b] | MS + 2,4-D 1 mg/l + kinetin 0.1 mg/l[a] | Callus and cell suspensions line PA-87-St | 24 °C, 1000 lx | Betacyanins | Petit, Andreu, Chénieux, Rideau (unpubl.) |
| Leaf[b] | MS + 2,4-D 1 mg/l + kinetin 0.1 mg/l[a] | Callus and cell suspensions line PA-87-L | 24 °C, 1000 lx | Betacyanins | Petit, Andreu, Chénieux, Rideau (unpubl.) |

Abbreviations: $B_5$: Gamborg et al. (1968) medium; 2,4-D: 2,4-dichlorophenoxy acetic acid; LS: Linsmaier and Skoog (1965) medium; MS: Murashige and Skoog (1962) medium; NAA: Naphthaleneacetic acid. (a): additives suppressed in the maintenance medium. (b): from 7-day-old seedlings.

than in the latter. This modified medium was selected for assays of production of a plant virus inhibitor.

*P. americana* shoot cultures (Fig. 5D) can be maintained under continuous light (3000 lx) at 24 °C on MS agar medium containing 30 g/l glucose and 1 mg/l benzylaminopurine.

## 2.2 Production of Antibiotic and Antiviral Compounds

Antimicrobial activity was detected in the secretions of *Phytolacca americana* explants. The expressed juice from callus also resulted in an inhibition of *Lemna* growth (Lin and Mathes 1973). Misawa et al. (1974, 1975) isolated an antibiotic substance from *P. americana* cultured cells and their media. The compound was active against *Bacillus subtilis* and *Streptococcus faecalis*. Its molecular mass was 976 daltons. Structural determination was, however, not given.

*P. americana* callus and suspensions were also found to possess inhibitory activities against tobacco mosaic virus infection on tobacco leaves. Therefore, Misawa et al. (1975a,b; Misawa 1977) investigated the production of plant virus inhibitor in *P. americana* cell cultures. The inhibitor accumulation appeared to accompany cell growth. A production medium derived from MS medium was selected. The active principles were purified by CM-cellulose column chromatography and ampholine electrofocusing. They were identified as five proteins, the molecular masses of which were from 11 to 31 kDa (Misawa 1977).

A ribosome-inactivating protein was found in considerable amounts in one cell line obtained from leaf tissue (Barbieri et al. 1989). The optimal conditions of production were the following: B5 medium supplemented with NAA (1 mg/l), kinetin (1 mg/l), and 2,4-D (1.2 mg/l), under light, i.e., conditions ensuring optimal growth. The protein (referred to as PAP-C) appears to be very similar, although not identical to one of the forms of PAP purified from leaves. Its partial N-terminal sequence is identical to that of PAP. It inhibits protein synthesis by a rabbit reticulocyte lysate with an $IC_{50}$ of 2.0 ng/ml. The inhibitor is basic, with molecular mass of 29 kDa. These results resemble those of Thomsen et al. (1991). The authors purified two new ribosome-inhibiting proteins (with molecular masses of 30 and 31 kDa) from in vitro cultures of *P. dodecandra*.

## 2.3 Production of Betalains

Betalain production by plant cell cultures has been investigated with several models, such as *Amaranthus tricolor* (Bianco-Colomas and Hugues 1990), *Basella alba* (Cyumel 1989), *Portulaca grandiflora* (Endress 1981; Schröder and Böhm 1984; Kishima et al. 1991), *Beta vulgaris* (Constabel and Haala 1968; Bokern et al. 1991a; Girod and Zrÿd 1991) and *Chenopodium rubrum* (Berlin et al. 1986; Bokern et al. 1991b). Most of these studies in vitro have evidenced that light, precursors (feeding of tyrosine), and phytohormone ratio play a major role in the regulation of betacyanin biosynthesis. However, it may be recalled that despite genetic similarity, different lines of one species may show distinct differences with regard to betacyanin accumulation. For example, light was shown to induce pigment formation in some lines of *Portulaca grandiflora* (Böhm et al. 1991; Kishima et al. 1991), but decreased the betacyanin content in other lines (Endress et al. 1984).

Regarding *Phytolacca americana*, first investigations on the betacyanin accumulation in cell suspensions in vitro are those of Misawa et al. (1973) and Komatsu et al. (1975). These authors showed that the pigment could be accumulated in relatively large amounts, but concluded that difficulties remained for economical production. Ekiert and Dohnal (1984) also reported the accumulation of betacyanin in callus cultures of different *Phytolacca* species (including *P. americana*) grown under light on MS agar medium. Since 1986, several investigations of the Komamine's group (Tohoku University, Sendai, Japan) were conducted on a cell line of *P. americana* maintained in an MS liquid medium containing 88 mM sucrose and 5 μM 2,4-D. The cells had the property of accumulating betacyanin in the dark. They did not excrete the pigment in the medium. The larger cell aggregates accumulated higher pigment contents than did the smaller cell aggregates (Sakuta et al. 1986). In 1987, our group obtained cell lines from stem and leaf explants excised from young seedlings of *P. americana* (Table 2). These lines differed from the above-mentioned line investigated by the Komamine's group since the cells accumulated betacyanin only in light. The pigment production disappeared either after transferring the cultures in the dark or after subculturing the cells in a medium containing glucose and NAA.

### *2.3.1 Relation Between Growth and Betacyanin Accumulation*

Contrary to the accumulation pattern described for most secondary metabolites (including the pigments anthocyanins), maximum betacyanin accumulation occurred during the logarithmic phase of the growth cycle. Evidence that betacyanin accumulation is positively correlated with cell proliferation was provided by experiments using factors regulating cell division: increasing the phosphate concentrations in the media enhanced both cell division and betacyanin accumulation. At concentrations greater than 1.25 mM, phosphate addition had no effect on growth and pigment accumulation (Sakuta et al. 1986); inhibiting the cell growth by 50 μM aphidocolin (an inhibitor of DNA synthesis) also suppressed the accumulation of betacyanin. Removing aphidocolin from the cells by washing restored cell division and betacyanin accumulation. Similar results were obtained with cycloheximide (Hirose et al. 1990).

In spite of the close correlation between cell division and betacyanin accumulation, optimal conditions for growth (expressed on cell number basis) appeared as somewhat different from those which gave optimal betacyanin accumulation. Increasing the sucrose concentrations in the media enhanced both cell number and pigment accumulation. Maximum betacyanin content was observed at 88 mM sucrose (on cell number basis) and at 175 mM sucrose (on fresh weight basis). Maximum cell number occurred at 234 mM sucrose. Cell size was affected by both sucrose concentration and water potential (Sakuta et al. 1987a). Increasing total nitrogen concentration (initial $NH_4^+ : NO_3^-$ ratio of 1:2) in the medium also enhanced cell number (optimum concentrations in total nitrogen: 5–10 mM) and betacyanin production (optimum concentrations: 40–80 mM on cell number basis; 10–80 mM on fresh weight basis). Increasing ammonium increased growth but decreased betacyanin accumulation; increasing nitrate had no effect on growth and increased betacyanin accumulation (Sakuta et al. 1987b).

### 2.3.2 *Effects of Plant Growth Regulators*

Sakuta et al. (1991) found that 2,4-D stimulated the betacyanin content per cell (optimal concentration: 5 μM). This may be due to an increase in free tyrosine level (i.e., the precursor of betacyanin) in the cells. The auxin also enhanced the cell growth (optimal concentration: 1 μM).

Contrary to the results obtained by Hirose et al. (1990), Petit and Andreu (unpubl.; Table 3) found that removing 2,4-D from the medium for one passage

**Table 3.** Effects of plant growth regulators and precursor on growth of suspension cultures of *Phytolacca americana* and on their betacyanin contents

| Cell line | 2,4-D | Other additives | Fresh weight (g/flask) 7 | 14 | Betacyanin content (nmol/g fr. wt.) 7 | 14 |
|---|---|---|---|---|---|---|
| | | 0 | 5.4 ± 0.8 | 12.04 ± 0.5 | 6.6 (100) | 8.2 (100) |
| PA-87-L | | 1 mM Tyr | 6.9 ± 0.7 | 12.5 ± 0.1 | 6.5 (100) | 10.6 (129) |
| | Present | 5 μM Z | 6.2 ± 0.5 | 12.3 ± 0.6 | 3.4 (51) | 5.4 (66) |
| | | 1 mM Tyr + 5 μM Z | 8.4 ± 1.3 | 13.0 ± 0.2 | 3.0 (45) | 7.5 (41) |
| | | 0 | 6.0 ± 0.8 | 11.6 ± 0.3 | 7.2 (109) | 10.0 (122) |
| | | 1 mM Tyr | 6.3 ± 0.7 | 11.7 ± 0.2 | 7.1 (107) | 11.3 (138) |
| | Absent | 5 μM Z | 5.4 ± 0.6 | 10.3 ± 0.5 | 8.5 (129) | 11.9 (145) |
| | | 1 mM Tyr + 5 μM Z | 5.6 ± 0.2 | 9.7 ± 0.6 | 8.5 (129) | 15.3 (186) |
| | | 0 | 4.2 ± 0.1 | 4.6 ± 0.1 | 23.3 (100) | 25.3 (100) |
| PA-87-St | | 1 mM Tyr | 3.7 ± 0.3 | 4.6 ± 0.2 | 38.9 (167) | 48.9 (193) |
| | Present | 5 μM Z | 4.2 ± 0.7 | 4.5 ± 0.3 | 10.0 (43) | 8.6 (34) |
| | | 1 mM Tyr + 5 μM Z | 3.2 ± 1.1 | 4.1 ± 0.3 | 9.6 (41) | 8.5 (33) |
| | | 0 | 4.9 ± 0.1 | 5.1 ± 0.1 | 48.5 (208) | 53.4 (211) |
| | | 1 mM Tyr | 4.7 ± 0.6 | 5.3 ± 0.2 | 79.3 (340) | 84.1 (332) |
| | Absent | 5 μM Z | 5.5 ± 0.3 | 5.1 ± 0.2 | 7.5 (32) | 11.0 (43) |
| | | 1 mM Tyr + 5 μM Z | 4.9 ± 0.2 | 6.1 ± 0.1 | 7.2 (31) | 11.4 (45) |

Abbreviations: 2,4-D: 2,4-dichlorophenoxyacetic acid; fr. wt.: fresh weight; Tyr: tyrosine; Z: zeatin. The cells were harvested at days 7 and 14. Betacyanin contents are expressed in nmol per g of fresh weight and (in parenthesis) in percent of the control, i.e., cells grown in their maintenance medium. (G. Petit Paly, F. Andreu, J.C. Chénieux, M. Rideau unpubl.).

enhanced the betacyanin production. The effect was more pronounced with PA-87-St cells. Moreover, adding 5 μM zeatin to the maintenance culture medium (containing 2,4-D) brings about a decrease in betacyanin production in the two lines without affecting the cell growth. This was a rather unexpected result, since cytokinins generally enhance the betacyanin production in the whole plants belonging to the Chenopodiinae. On the other hand, addition of zeatin to a 2,4-D-free medium had an enhancing effect on the betacyanin production in PA-87-L cells but decreased the betacyanin production in PA-87-St cells.

### 2.3.3 *Effects of Precursor Feeding*

Addition of the precursor tyrosine to the medium (final concentration 1 mM) increased the pigment formation in the proliferating cells, without any effects on growth (Hirose et al. 1990). This was also found after adding 1 mM tyrosine to the culture medium of the two lines PA-87-L and PA-87-St (Table 3). However, the line PA-87-St was more sensitive to tyrosine addition than was the line PA-87-L. Effect of tyrosine addition was apparently not dependent on the presence or absence of the auxin 2,4-D in the medium.

### 2.3.4 *Effects of Kanamycin*

While investigating the transformation of *P. americana* callus cultures with *Agrobacterium tumefaciens*, we added kanamycin to the culture medium in order to inhibit the growth of bacteria. During these experiments, it was found that kanamycin decreased the growth (expressed on a fresh weight basis) and induced (for the chlorophyllous line PA-89-St) or increased (for the betacyanin-accumulating line PA-87-St) pigment production (Fig. 6).

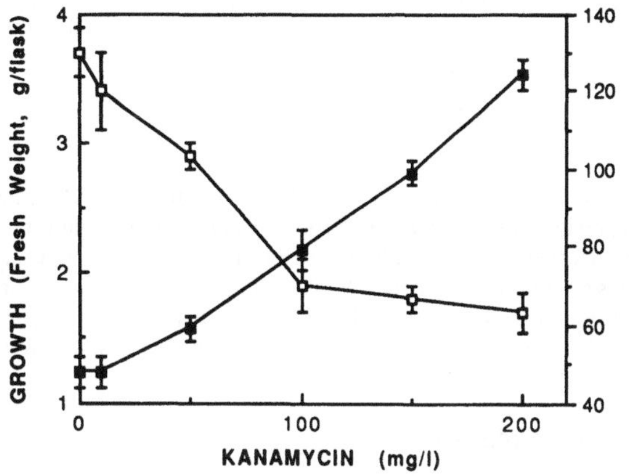


**Fig. 6.** Effect of kanamycin on growth of *P. americana* callus cultures (cell line PA-87-L) and on their betacyanin content. Pieces of callus tissue (1 g fresh weight) were grown for 21 days on MS medium containing 1 mg/l 2,4-D and various concentrations of kanamycin sulfate. (unpubl.)

### 2.4 Miscellaneous

With callus and cell suspension cultures of the related species *P. dodecandra*, it was shown that significant amounts of triterpenoid saponins are retained intracellularly in these cultures (Adams and Balandrin 1984). Tissue cultures of *P. americana* also produced saponins, but the pattern differs from that of the whole plant. Whereas phytolaccoside E is a major saponin in all parts of the plant, callus tissues from stems were found to accumulate phytolaccoside B as a major constituent (Woo and Kang 1976b).

Tissue cultures of *P. americana* were also recently used in physiological studies for investigating the uptake and accumulation of auxins in cells cultured in vitro (Sato et al. 1992).

## 3 Suggested Protocols

### 3.1 Induction of Primary Callus

Explants of stems or leaves were excised in March 1987 from 15-day-old seedlings germinated in light on Heller (1953) agar medium. Explants were cultivated under continuous white light (5000 lx), at 24 ± 1 °C, on Murashige and Skoog (1962) medium containing either 30 g/l glucose, 1 mg/l NAA, and 0.1 mg/l kinetin, or 20 g/l sucrose, 1 mg/l 2,4-D, and 0.1 mg/l kinetin. All media supported the appearance of primary callus.

### 3.2 Establishment of Callus Cultures

One stem-derived callus and one leaf-derived callus were maintained at 2-week intervals on MS medium supplemented with 20 g/l sucrose and 1 mg/l 2,4-D, at 24 ± 1 °C, under continuous white light (5000 lx). Two betacyanin-accumulating lines (referred to as PA-87-St and PA-87-L) were selected by subculturing the tissues on the same medium. Moreover, a chlorophyllous line (referred to as PA-89-St) was obtained after transferring in April 1989 1 g (fresh weight) of PA-87-St tissues on to MS medium supplemented with 30 g/l glucose and 1 mg/l NAA.

### 3.3 Cell Suspensions

Ten g of 2-week-old PA-87-St or PA-87-L callus was dissociated at 24°C by rotary shaking (100 rpm) in 250-ml Erlenmeyer flasks containing 50 ml of culture medium. The latter was composed of the MS medium plus 20 g/l sucrose and 1 mg/l 2,4-D. The cell suspensions were subcultured every 15 days at dilution of 1:10 (for PA-87-L line) or 1:5 (for PA-87-St line). The cells of the leaf-derived suspension were larger than those of the stem-derived suspension (Fig. 5). The two cell lines accumulated betacyanins in the light, but no pigment was biosynthesized in the dark.

### 3.4 Shoot Cultures

Shoots (3 cm high) were excised from green sterile seedlings and cultivated on MS agar medium containing 30 g/l glucose and 1 mg/l benzylaminopurine. Shoot cultures are maintained at 24 °C (3000 lx) and are subcultured every 4 weeks. Stems are intensively red, bearing green leaves.

### 3.5 Measurement of Betacyanins

Cells were harvested by filtration (nylon cloth, 30 $\mu$m) under partial vacuum, washed rapidly with cold distilled water, and frozen after fresh weight measurements. Four ml of 80% methanol were added to 500 mg frozen cells, followed by centrifugation at 10000 *g*. Betacyanin content was estimated by measuring absorbance at 535 nm ($\varepsilon_M = 5.66 \times 10^4$).

## 4 Conclusions and Perspectives

The studies on *Phytolacca americana* tissue cultures have shown that a number of economic and medicinal compounds may be produced in vitro.

From the pharmaceutical point of view, an important finding is that $RIP_s$ can be produced in cell suspensions. As *P. americana* is a very common species, easy to culture, it would be uneconomical to use cells grown in vitro for producing these proteins (Barbieri et al. 1989). However, cultures in vitro might be of interest since they are available throughout the year. Another important point is the fact that PAP-C slightly differs from the other $PAP_s$ isolated in the plants. It is well known that somaclonal variations often occur in tissue culture and therefore it may be expected that other forms of $PAP_s$, some of them having more active biological properties, exist in these tissues.

Another promising result is the accumulation of betalains in callus and suspension cultures. At present, the data available regarding the effect of environmental factors on these cultures represent only an initial step on the way to be explored. Increasing the accumulation of the pigment might be of interest with the aim of producing the dye on a commercial scale.

## References

Adams RP, Balandrin MF (1984) Application of biotechnological methods for the in vitro biosynthesis of bioactive saponins and propagation of endod. In: Lemma A, Heyneman D, Silangwa SM (eds) *Phytolacca dodecandra* (Endod). Tycooly, Dublin, pp 133–137

Ahmed ZF, Zufall CJ, Jenkins GL (1949) A contribution to the chemistry and toxicology of the root of *Phytolacca americana* L. J Am Pharm Assoc 38:443–448

Barbieri L, Aron GM, Irvin JD, Stirpe F (1982) Purification and partial characterization of another form of the antiviral protein from the seeds of *Phytolacca americana* L. (pokeweed). Biochem J 203:55–59

Barbieri L, Bolognesi A, Cenini P, Falasca AI, Minghetti A, Garofano L, Guicciardi A, Lappi D, Miller SP, Stirpe F (1989) Ribosome-inactivating proteins from plant cells in culture. Biochem J 257:801–807

Barbieri L, Ferreras JM, Barraco A, Ricci P, Stirpe F (1992) Some ribosome-inactivating proteins depurinate ribosomal RNA at multiple sites. Biochem J 286:1–4

Barker BE, Farnes P, La Marche PH (1966) Peripheral blood plasmocytosis following systemic exposure to *Phytolacca americana* (pokeweed). Pediatrics 38:490–493

Barnett BD (1975) Toxicity of pokeberries (fruit of *Phytolacca americana* Large) for turkey poults. Poult Sci 54:1215–1217

Berlin J, Sieg S, Strack D, Bokern M, Harms H (1986) Production of betalains by suspension cultures of *Chenopodium rubrum* L. Plant Cell Tissue Organ Cult 5:163–174

Bianco-Colomas J, Hugues M (1990) Establishment and characterization of a betacyanin producing cell line of *Amaranthus tricolor*: inductive effects of light and cytokinin. J Plant Physiol 136:734–739

Böhm H, Böhm L, Rink E (1991) Establishment and characterization of a betaxanthin-producing cell culture from *Portulaca grandiflora*. Plant Cell Tissue Organ Cult 26:75–82

Bois D (1927) Les plantes alimentaires chez tous les peuples et à travers les âges. 1. Phanérogames légumières. Lechevalier, Paris, pp 422–423

Bokern M, Heuer S, Wray V, Witte L, Macek T, Vanek T, Strack D (1991a) Ferulic acid conjugates and betacyanins from cell cultures of *Beta vulgaris*. Phytochemistry 30:3261–3265

Bokern M, Wray V, Strack D (1991b) Accumulation of phenolic acid conjugates and betacyanins, and changes in the activities of enzymes involved in feruloylglucose metabolism in cell-suspension cultures of *Chenopodium rubrum* L. Planta 184:261–270

Bolognesi A, Barbieri L, Abbondanza A, Falasca AI, Carnicelli D, Battelli MG, Stirpe F (1990) Purification and properties of new ribosome-inactivating proteins with RNA N-glycosidase activity. Biochim Biophys Acta 1087:293–302

Burke DE, Le Quesne PW (1971) 3-Acetyloleanic acid from *Phytolacca americana* seeds. Phytochemistry 10:3319–3320

Castagnaro M, Canese MG (1990) Lectin histochemical characteristics of the canine female mammary gland. Am J Vet Res 51:1815–1819

Castagnaro M, Canese MG (1991) Lectin histochemistry on squamous metaplasia in different epithelial tumors of dogs. Vet Pathol 28:8–15

Chu JY, Yang AD, Wang BM, Hu Z, Zhu XM, Zhang HJ, Qu JH, Luo LY, Guo R, Shi LR (1990) Monoclonal anti-human T cell antibody and PAP-s conjugate preparation and selective cytotoxic properties on leukemic cell. J Tongji Med Univ 10:15–18

Constabel F, Haala G (1968) Recherches sur la formation de pigments dans les tissus de betterave fourragère cultivés in vitro. In: Les cultures de tissus de plantes. CNRS, Paris, pp 223–229

Cyumel E (1989) *Basella alba* L.: in vitro culture and the production of betalains. In: Bajaj YPS (ed) Biotechnology in agriculture and forestry, vol 7. Medicinal and aromatic plants II. Springer, Berlin Heidelberg New York, pp 47–68

Driver MG, Francis FJ (1979a) Stability of phytolaccanin, betanin and FDαC red #2 in dessert gels. J Food Sci 44:518–520

Driver MG, Francis FJ (1979b) Purification of phytolaccanin (betanin) by removal of phytolaccatoxin from *Phytolacca americana*. J Food Sci 44:521–523

Duggar BM, Armstrong JK (1925) The effect of treating the virus of tobacco mosaic with the juices of various plants. Ann Mo Bot Gard 12:359–366

Edwards ME, Harris EM, Wagner FH, Cross MC, Miller GS (1988) Seed germination of American pokeweed (*Phytolacca americana*). I. Laboratory techniques and autotoxicity. Am J Bot 75:1794–1802

Ekiert H, Dohnal B (1984) Hodowla i wstepna analiza fitochemiczna tkanki kalusowej *Phytolacca* sp. Acta Pol Pharm 41:581–584

Elliott DC, Schultz CG, Cassar RA (1983) Betacyanin-decolourizing enzyme in *Amaranthus tricolor* seedlings. Phytochemistry 22:383–387

Endo Y, Tsurugi K (1988) The RNA N-glycosidase activity of ricin A-chain. J Biol Chem 263:8735–8739

Endo Y, Mitsui K, Motizuki M, Tsurugi K (1987) The mechanism of action of ricin and related toxic lectins on eukaryotic ribosomes. J Biol Chem 262:5908–5912

Endress R (1981) Conversion of phenylalanine into tyrosine by *Portulaca* callus. Plant Physiol 68:272–274

Endress R, Jäger A, Kreis W (1984) Catecholamine biosynthesis dependent on the dark in betacyanin-forming *Portulaca* callus. J Plant Physiol 115:291–295

Fournier P (1948) Le livre des plantes médicinales et vénéneuses de France, vol 3. Lechevalier, Paris, pp 211–213

Fukuyama Y, Hasegawa T, Toda M, Kodama M, Okazaki K (1992) Structures of americanol A and isoamericanol A having a neurotropic property from the seeds of *Phytolacca americana*. Chem Pharm Bull 40:252–254

Gamborg OL, Miller RA, Ojima K (1968) Nutrient requirements of soybean root cells. Exp Cell Res 50:151–158
Girod PA, Zrÿd JP (1991) Secondary metabolism in cultured red beet (*Beta vulgaris* L.) cells: differential regulation of betaxanthin and betacyanin biosynthesis. Plant Cell Tissue Organ Cult 25:1–12
Goldstein IJ, Poretz RD (1986) Isolation, physicochemical characterization and carbohydrate-binding specificity of lectins. In: Liener IE, Sharon N, Goldstein IJ (eds) The lectins, properties, functions and applications in biology and medicine. Academic Press, Orlando, pp 233–247
Harding JW, Arena JM (1969) Human poisoning from native and cultivated plants. Duke University Press, Durham, NC
Hegnauer R (1969) Chemotaxonomie der Pflanzen, Bd 5. Dicotyledoneae: Magnoliaceae–Quiinaceae. Birkhäuser, Basel, pp 305–311
Heller R (1953) Recherche sur la nutrition minérale des tissus végétaux cultivés in vitro. Ann Sci Nat Bot Biol Vég 14:1–223
Hirose M, Yamakawa T, Kodama T, Komamine A (1990) Accumulation of betacyanin in *Phytolacca americana* cells and of anthocyanin in *Vitis* sp. cells in relation to cell division in suspension cultures. Plant Cell Physiol 31:267–271
Irvin JD (1975) Purification and partial characterization of the antiviral protein from *Phytolacca americana* which inhibits eucaryotic protein synthesis. Arch Biochem Biophys 169:522–528
Irvin JD, Kelly T, Robertus JD (1980) Purification and properties of a second antiviral protein from *Phytolacca americana* which inactivates eucaryotic ribosomes. Arch Biochem Biophys 200:418–425
Jean-Blain C, Grisward M (1973) Plantes vénéneuses. Toxicologie. La Maison-Rustique, Paris, pp 52–53
Jiménez A, Vàsquez D (1985) Plant and fungal protein and glycoprotein toxins inhibiting eukaryote protein synthesis. Annu Rev Microbiol 39:649–672
Johnson A, Shimizu Y (1974) Phytolaccinic acid, a new triterpene from *Phytolacca americana*. Tetrahedron 30:2033–2036
Kang SS, Woo WS (1980) Triterpenes from the berries of *Phytolacca americana*. J Nat Prod 43:510–513
Kang SS, Woo WS (1987) Two new saponins from *Phytolacca americana*. Planta Med 53:338–340
Karayannopoulou G, Weiss J, Damjanov I (1988) Detection of fungi in tissue sections by lectin histochemistry. Arch Pathol Lab Med 112:746–748
Kishima Y, Nozaki K, Akashi R, Adachi T (1991) Light-inducible pigmentation in *Portulaca* callus; selection of a high betalain producing cell line. Plant Cell Rep 10:304–307
Kloos H, McCullough F (1982) Plant molluscicides. Planta Med 46:195–209
Komatsu K, Nozaki W, Takemusa M, Umemori S, Nakaminami M (1975) Production of pigment by plant tissue culture. Japan patent (open) 75-24494
Kumon K, Sasaki J, Sejima M, Takeuchi Y, Hayashi Y (1990a) Betacyanin-decolorizing enzymes from *Phytolacca americana*. Plant Cell Physiol 31:233–240
Kumon K, Sasaki J, Sejima M, Hayashi Y, Takeuchi Y (1990b) Effects of the cortex on light- and kinetin-induced accumulation of betacyanin in the epidermal tissue of *Phytolacca americana*. Plant Cell Physiol 31:391–393
Le Maout E, Decaisne J (1876) Traité général de botanique. Firmin-Didot, Paris, 468 pp
Liener IE (1986) Nutritional significance of lectins in the diet. In: Liener IE, Sharon N, Goldstein IJ (eds) The lectins, properties, functions and applications in biology and medicine. Academic Press, Orlando, pp 527–550
Lin A, Mathes C (1973) The in vitro secretion of growth regulators by isolated callus tissues. Am J Bot 60:34–41
Lin Q, Chen ZC, Antonin JF, White RF (1991) Isolation and characterization of a cDNA clone encoding the anti-viral protein from *Phytolacca americana*. Plant Mol Biol 17:609–614
Linsmaier EM, Skoog F (1965) Organic growth factor requirements of tobacco tissue cultures. Physiol Plant 18:100–127
Mabry TJ (1980) Betalains. In: Bell EA, Charlwood BV (eds) Secondary plant products. Springer, Berlin Heidelberg New York, pp 513–533
McDonnell MJ, Stiles EW, Cheplick GP, Armesto JJ (1984) Bird-dispersal of *Phytolacca americana* L. and the influence of fruit removal on subsequent fruit development. Am J Bot 71:895–901

Misawa M (1977) Production of natural substances by plant cell cultures described in Japanese patents. In:Barz W, Reinhard E, Zenk MH (eds) Plant tissue culture and its bio-technological application. Springer, Berlin Heidelberg New York, pp 17–26

Misawa M, Hayashi M, Nagano Y, Kawamoto T (1973) Production of a plant pigment, Japan patent (open) 73-6153

Misawa M, Hayashi M, Tanaka H, Keido K, Tomomasa M (1974) Production of plant virus inhibitor by *Phytolacca americana* suspension culture. Biotechnol Bioeng 17:1335–1347

Misawa M, Sakato K, Tanaka H, Hayashi M, Samejima H (1975) Production of physiologically active substances by plant cell suspension cultures. In: Street HE (ed) Tissue culture and plant science. Academic Press, London, pp 405–432

Moreno M, Rodriguez VM (1981) Yiamoloside B a fungistatic saponin of *Phytolacca octandra.* Phytochemistry 20:1446–1447

Murashige T, Skoog F (1962) A revised medium for rapid growth and bioassays with tobacco tissue cultures. Physiol Plant 15:473–497

Myers DE, Irvin JD, Smith RS, Kuebelbeck VM, Uckun FM (1991) Production of a pokeweed antiviral protein (PAP)-containing immunotoxin, B43-PAP, directed against the CD 19 human B lineage lymphoid differentiation antigen in highly purified form for human clinical trials. J Immunol Methods 136:221–238

Nétien G, Traisnel M, Vérain A (1986) Médicaments homéopathiques. Techniques et documentation. Lavoisier, Paris, pp 357–358

Parkhurst RM, Mthupha BM, Liang YS, Bruce JI, Lambert JDH, Collier TL, Apsimon JW, Wolde-Johannes L, Heath GE, Jones WO, Stobaens JK, Makhuber LP (1989) The molluscicidal activity of *Phytolacca dodecandra.* I. Location of the activating esterase. Biochem Biophys Res Commun 158:436–439

Prestle J, Schonfelder M, Adam G, Mundry KW (1992) Type I ribosome-inactivating proteins depurinate plant 25 S rRNA without species specificity. Nucleic Acids Res 20:3179–3182

Ready M, Wilson K, Piatak M, Robertus JD (1984) Ricin-like plant toxins are evolutionarily related to single-chain ribosome-inactivating proteins from *Phytolacca.* J Biol Chem 259:15252–15256

Sakuta M, Takagi T, Komamine A (1986) Growth related accumulation of betacyanin in suspension cultures of *Phytolacca americana* L. J Plant Physiol 125:337–343

Sakuta M, Takagi T, Komamine A (1987a) Effects of sucrose on betacyanin accumulation and growth in suspension cultures of *Phytolacca americana.* Physiol Plant 71:455–458

Sakuta M, Takagi T, Komamine A (1987b) Effects of nitrogen source on betacyanin accumulation and growth in suspension cultures of *Phytolacca americana.* Physiol Plant 71:459–463

Sakuta M, Hirano H, Komamine A (1991) Stimulation by 2,4-dichlorophenoxyacetic acid of betacyanin accumulation in suspension cultures of *Phytolacca americana.* Physiol Plant 83:154–158

Sato K, Iwakami S, Goda Y, Yamazaki T, Maitani T, Yoshihira K (1992) Uptake and accumulation of natural and synthetic auxins in cultured cells of plants: *Phytolacca americana* L., *Glycyrrhiza glabra* L. and *Rubia tinctorum* L. J Food Hyg Soc Jpn 33:157–163

Sauer JD (1952) A geography of pokeweed. Ann Mo Bot Gard 39:113–125

Schenk RU, Hildebrandt AC (1972) Medium and techniques for induction and growth of monocotyledonous and dicotyledonous plant cell cultures. Can J Bot 50:199–204

Schröder W, Böhm H (1984) Betacyanin concentrations in young cell cultures from *Portulaca grandiflora.* An analysis of variation. Plant Cell Rep 3:14–17

Schütte HR, Liebisch HW (1985) Alkaloids derived from tyrosine and phenylalanine. In:Mothes K, Schütte HR, Lückner M (eds) Biochemistry of alkaloids. VEB Deutscher Verlag, Berlin, pp 188–271

Stirpe F, Barbieri L (1986) Ribosome-inactivating proteins up to date. FEBS Lett 195:1–8

Stout GH, Malofsky BM, Stout VF (1964) Phytolaccagenin:a light atom X-ray structure proof using chemical information. J Am Chem Soc 86:957–958

Suga Y, Maruyama Y, Kawanishi S, Shoji J (1978) Studies on the constituents of phytolaccaceous plants. I. On the structures of phytolaccasaponin B, E and G from the roots of *Phytolacca americana* L. Chem Pharm Bull 26:520–525

Thomsen S, Hansen HS, Nyman U (1991) Ribosome-inhibiting proteins from *in vitro* cultures of *Phytolacca dodecandra.* Planta Med 57:232–236

Tomlinson JA, Walker VM, Flewett TM, Barclay GR (1974) The inhibition of infection by cucumber

mosaic virus and influenza virus by extracts from *Phytolacca americana*. J Gen Virol 22:225–232
Usher G (1974) A dictionary of plants used by man. Constable, London, 457 pp
Ussery DA, Irvin JD, Hardesty B (1977) Inhibition of poliovirus replication by a plant antiviral peptide. Ann NY Acad Sci 284:431–440
Voisin M (1980) Matière médicale du praticien homéopathe, Maloine, Paris, pp 953–958
Woo WS (1974) Steroids and pentacyclic triterpenoids from *Phytolacca americana*. Phytochemistry 13:2887–2889
Woo WS, Kang SS (1976a) Phytolaccoside B: triterpene glucoside from *Phytolacca americana*. Phytochemistry 15:1315–1317
Woo WS, Kang SS (1976b) Constituents of *Phytolacca* species. III. Components of overground parts and callus tissues. Soul Taehakkyo Saengyak Yonguso Opjukjip 15:111–113 (CA, 1978, 88:47504z)
Woo WS, Kang SS (1977) The structure of phytolaccoside G. Yakhak Hoe Chi 21:159–162 (CA, 1978, 88:191324h)
Woo WS, Shin Kh (1976) Antiinflammatory action of *Phytolacca* saponin. Soul Taehakkyo Saengyak Yonguso Opjukjip 15:90-6 (CA, 1978, 88:69156q)
Woo WS, Wagner H (1977) 3-Acetylaleuritolic acid from the seeds of *Phytolacca americana*. Phytochemistry 6:1845–1846
Woo WS, Chi HJ, Kang SS (1976a) Constituents of *Phytolacca* species. II. Comparative examination on constituents of the roots of *Phytolacca americana*, *P. esculenta* and *P. insularis*. Soul Taehakkyo Saengyak Yonguso Opjukjip 15:107–110 (CA, 1978, 88:85992q)
Woo WS, Shin KH, Kang SS (1976b) Constituents of *Phytolacca* species. I. Antiinflammatory saponins. Soul Taehakkyo Saengyak Yonguso Opjukjip 15:103–106 (CA, 1978, 88:146072p)
Woo WS, Kang SS, Wagner H, Seligmann O, Chari VN (1978) Triterpenoid saponins from the roots of *Phytolacca americana*. Planta Med 34:87–92
Wyatt SD, Sheperd RJ (1969) Isolation and characterization of a virus inhibitor from *Phytolacca americana*. Phytopathology 59:1787–1794
Wyler H, Dreiding AS (1961) Phytolaccanin, der Farbstoff der Kermesbeere (*Phytolacca decandra* L.) I. Mitteilung zur Kenntnis der Betacyane. Helv Chim Acta 44:249–257
Yang-hua Y (1992) Two new saponins from the roots of *Phytolacca esculenta*. Planta Med 58:99–101
Yang-hua Y, Chu-Lu W (1989) A new active saponin from *Phytolacca esculenta*. Planta Med 55:551–552
Yokohama K, Terao T, Osawa T (1978) Carbohydrate-binding specificity of pokeweed mitogens. Biochim Biophys Acta 538:384–396
Zarling JM, Moran PA, Haffar O, Sias J, Richman DD, Spina CA, Myers DE, Kuebelbeck V, Ledbetter JA, Uckun FM (1990) Inhibition of HIV replication by pokeweed antiviral protein targeted to $CD_4^+$ cells by monoclonal antibodies. Nature 347:92–95
Zrÿd JP, Bauer J, Wyler H, Lavandry P (1982) Pigment biosynthesis and precursor metabolism in red beet semi-continuous cell suspension cultures. In: Fujiwara A (ed) Plant tissue culture 1982. Maruzen, Tokyo, pp 387–388

# XXIII *Pueraria lobata* (Kudzu Vine): In Vitro Culture and the Production of Isoflavonoids

T. Hakamatsuka, Y. Ebizuka, and U. Sankawa[1]

## 1 Introduction

*Pueraria lobata* Ohwi ( = *P. thunbergiana* Benth.; = *P. hirsuta* Matsum.), also known as Kudzu, a member of the family Leguminosae, is a large creeping or climbing plant which sometimes grows to a length of 20 m. The leaves consist of three leaflets, the under surface being covered with silvery hair (Fig. 1). It bears reddish purple flowers in summer and fall. In Japan, Korea, and China, *P. lobata* grows wild in open places along roads, rivers, and forest

**Fig. 1.** *Pueraria lobata* Ohwi growing wild

[1] Faculty of Pharmaceutical Sciences, University of Tokyo, 7-3-1 Hongo, Bunkyo-ku, Tokyo 113, Japan

Biotechnology in Agriculture and Forestry, Vol. 28
Medicinal and Aromatic Plants VII (ed. by Y.P.S. Bajaj)

margins. Although the genus *Pueraria* includes 15 species and is widely distributed in the temperate and tropical regions of East Asia, it is the only species native to Japan. The dried roots are used under the name of Gegen (Kakkon in Japanese) as an important crude drug in Chinese traditional medicine, and Gegen Tao (Kakkon To in Japanese), a prescription containing Gegen, is a popular drug for the treatment of the common cold (Kinoshita 1982). The flowers are also used as a folk medicine against the effects of alcohol abuse. In addition to its medicinal use, starch of a good quality is manufactured from the roots of *P. lobata*, and used for cooking and baking. Raw roots yield about 20% starch. In the late 19th century, *P. lobata* was exported to the United States, where it was naturalized and called Kudzu Vine. The plant was largely used for erosion control, and the green leaves are sometimes fed to livestock (Kariyone and Koiso 1971; Horikawa 1972).

The main constituents of *P. lobata*, daidzein (**1**), daidzin (**2**), and puerarin (**3**), are isoflavonoids which are considered to be active principles in traditional medicine and are reported to stimulate cerebral and coronary blood circulation (Kinoshita 1982). Various types of isoflavonoids, flavonoids, and the related compounds were isolated from roots, stems, leaves, and flowers as well as from cell cultures of *P. lobata* (Table 1, Figs. 2, 3). Recently, the major isoflavone glucosides, daidzin (**2**), puerarin (**3**), and genistin (**17**), were proved to exist mostly as their 6"-*O*-malonyl esters in cell suspension cultures, fresh roots, and stems (Park et al. 1992). Besides these constitutive isoflavonoids, exogenous stimuli such as the treatment of leaves (Ingham 1982) and stems (Hakamatsuka et al. 1991a) with fungal elicitor or copper chloride induced the accumulation of phytoalexins of isoflavonoid structures, as shown in Fig. 4.

**Table 1.** Secondary metabolites identified in plants and tissue cultures of *Pueraria lobata* Ohwi[a]

| Plant part | Isoflavonoids | Flavonoids and related compounds | Reference |
|---|---|---|---|
| Roots | 1, 2, 3, 4, 5, 7, 8, 9, 10, 16, 19 | | Kinjo et al. (1987) |
| | 1, 2, 3, 4, 5, 6, 8, 10, 23 | | Ohshima et al. (1988) |
| | 2, 3, 17 | | Park et al. (1992) |
| | | 29, 30 | Kinjo et al. (1985) |
| Stems | 1, 2, 3, 4, 5, 9, 10, 16, 17, 18, 22, 28 | 28 | Kinjo et al. (1987) |
| Leaves | 2, 15, 17 | 25, 26, 27 | Kinjo et al. (1988) |
| Flowers | 1, 12, 16 | 24 | Kurihara and Kikuchi (1973) |
| | 15 | | Kurihara and Kikuchi (1975) |
| | 10, 13, 17, 20, 21 | | Kurihara and Kikuchi (1976) |
| | 14 | | Kubo et al. (1973) |
| | 11 | | Kubo et al. (1975) |
| | 15 | | Kinjo et al. (1988) |
| Callus | 1, 2, 3, 16, 22 | | Takeya and Itokawa (1982) |
| Suspension | 2, 3, 17, 18 | | Park et al. (1992) |

Structure and nomenclature of compounds listed above are shown in Figs. 2 and 3.
[a] Several saponins were isolated from flowers and leaves. (Kinjo et al. 1988).

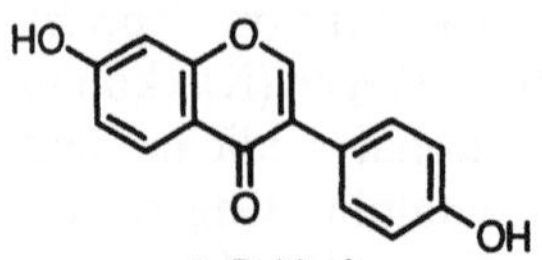
**1** Daidzein

**2** Daidzin

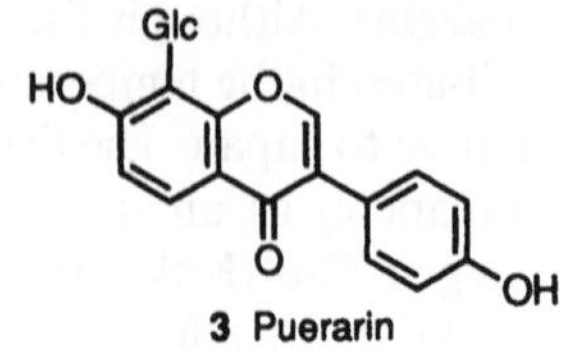
**3** Puerarin

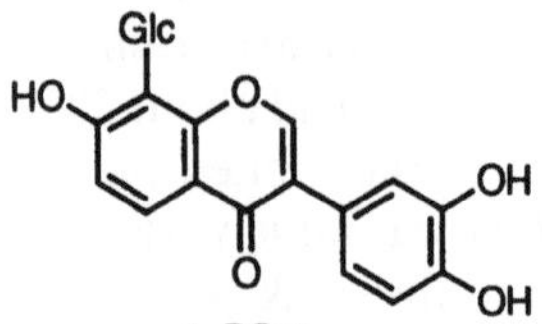
**4** PG-1

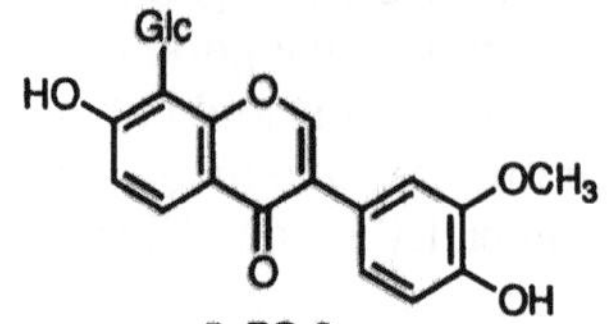
**5** PG-3

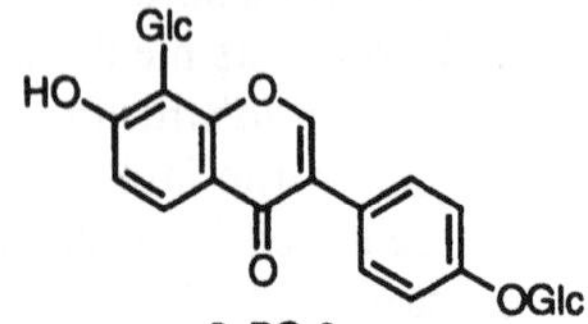
**6** PG-6

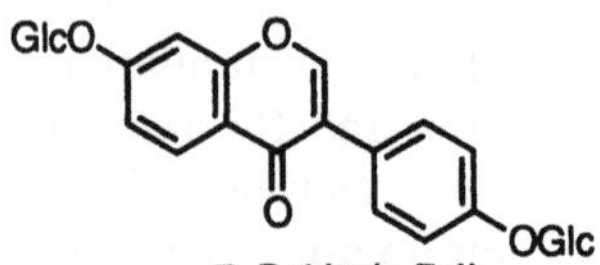
**7** Daidzein-7,4'-di-*O*-glucoside

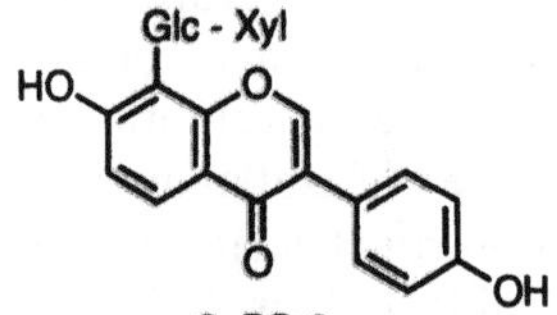
**8** PG-2

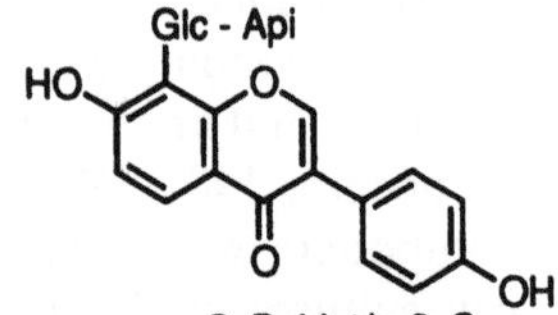
**9** Daidzein-8-*C*-apiosylglucoside

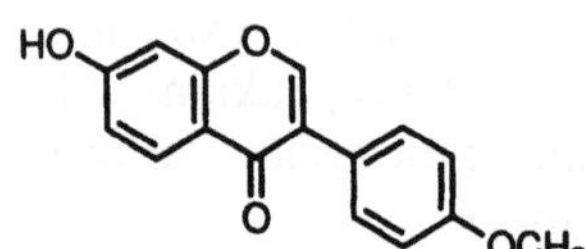
**10** Formononetin

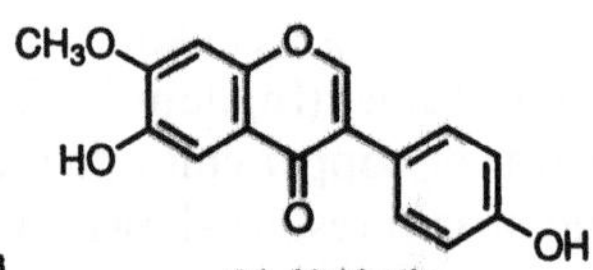
**11** Kakkatin

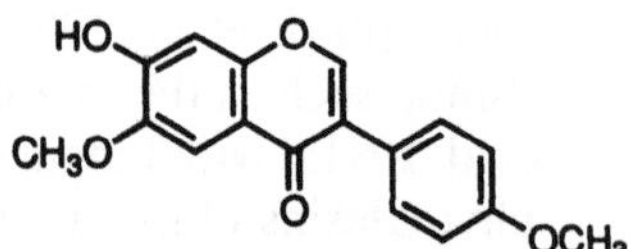
**12** Irisolidone

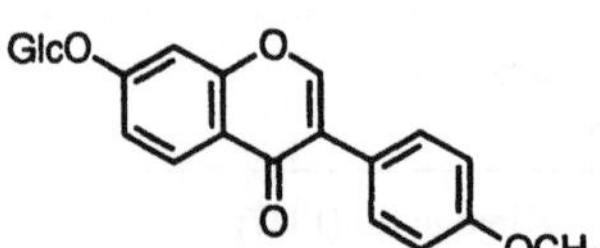
**13** Ononin

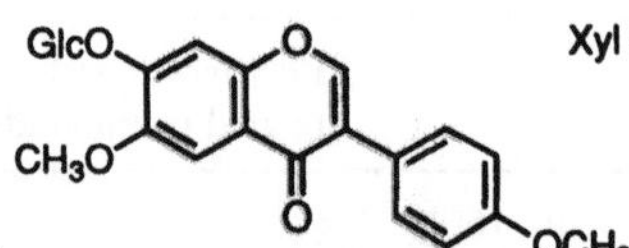
**14** Kakkalidone

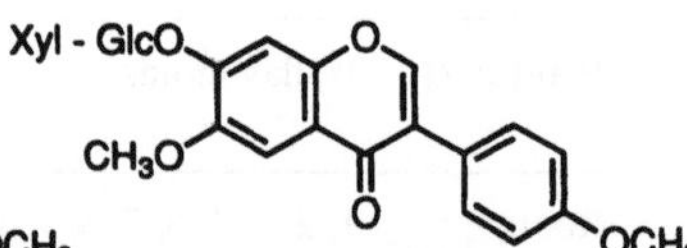
**15** Kakkalide

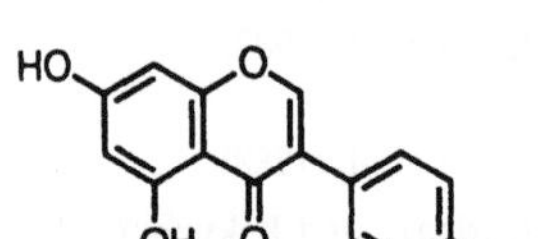
**16** Genistein

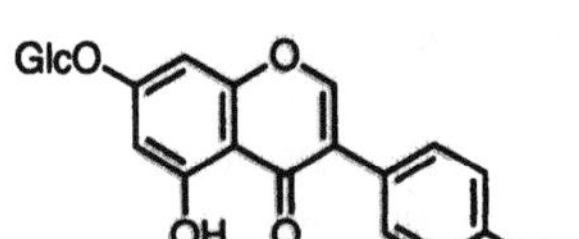
**17** Genistin

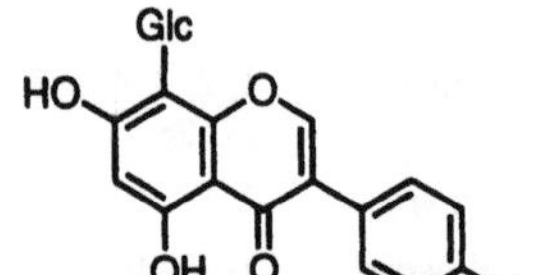
**18** Genistein-8-*C*-glucoside

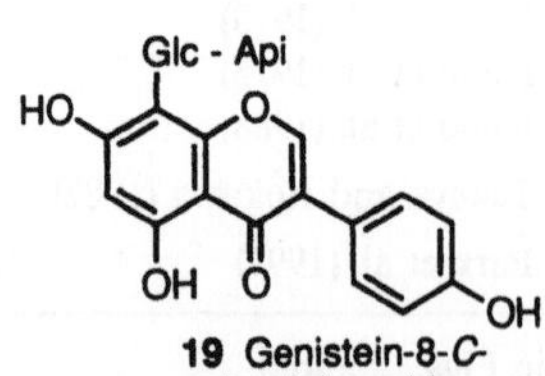
**19** Genistein-8-*C*-apiosylglucoside

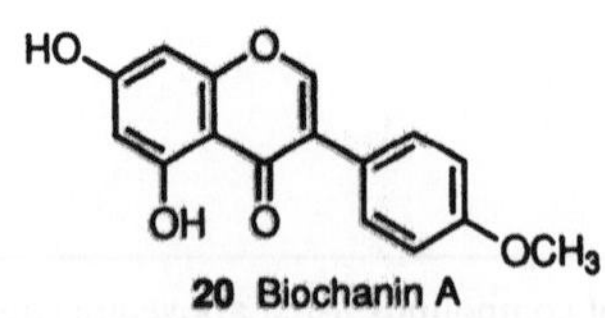
**20** Biochanin A

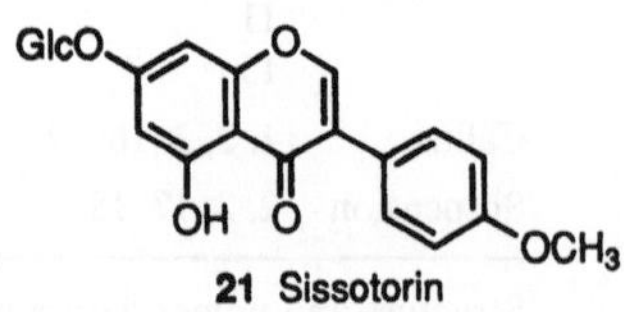
**21** Sissotorin

**Fig. 2.** Isoflavonoids from *Pueraria lobata* Ohwi. *Glc* Glucosyl; *Xyl* xylosyl; *Api* apiosyl

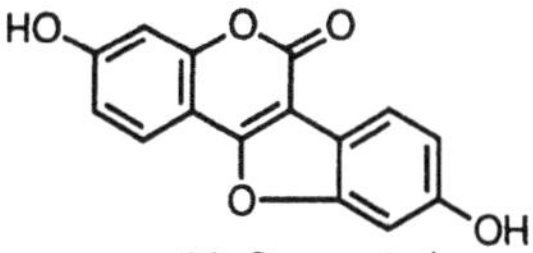
**22** Coumestrol

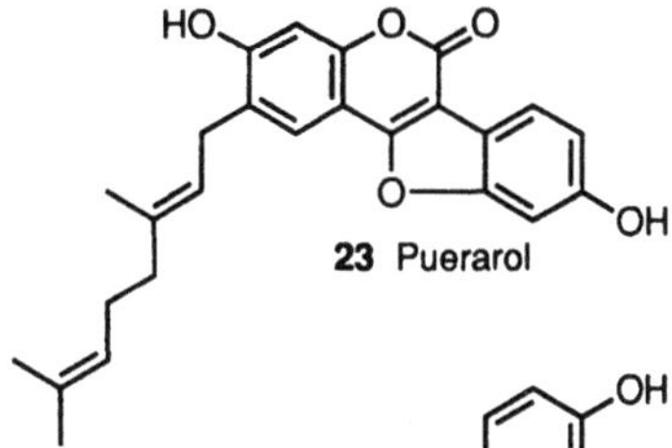
**23** Puerarol

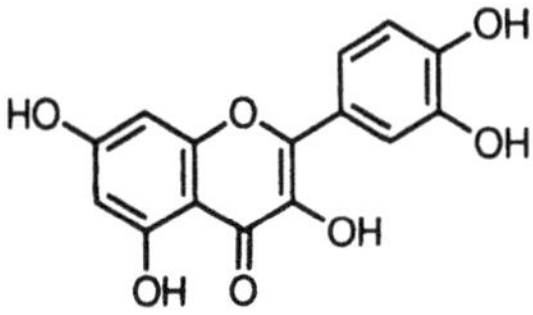
**24** Quercetin

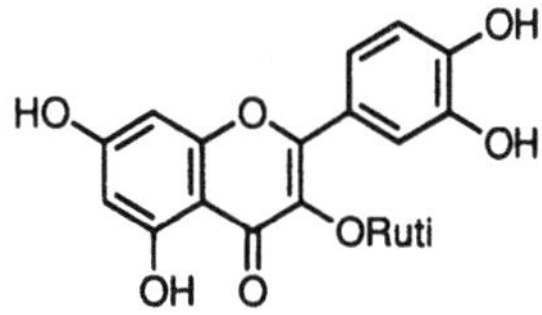
**25** Rutin

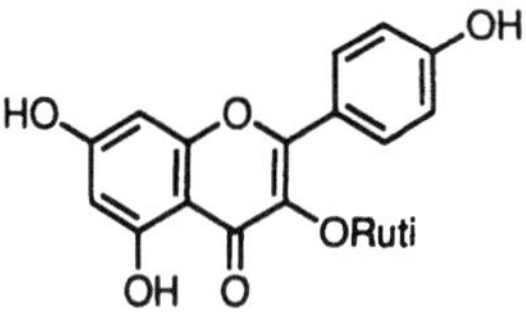
**26** Nicotiflorin

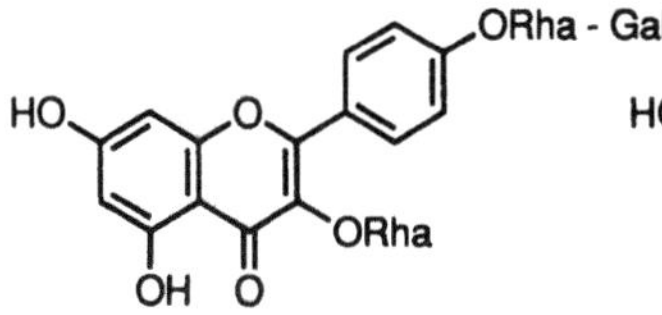
**27** Robinin

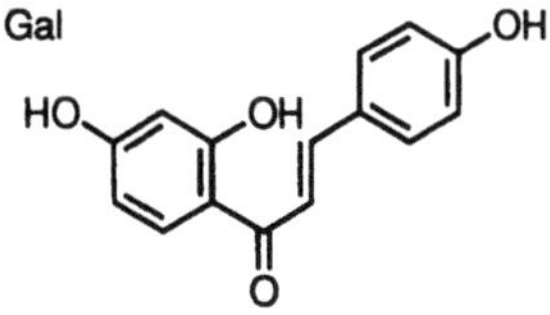
**28** Isoliquiritigenin

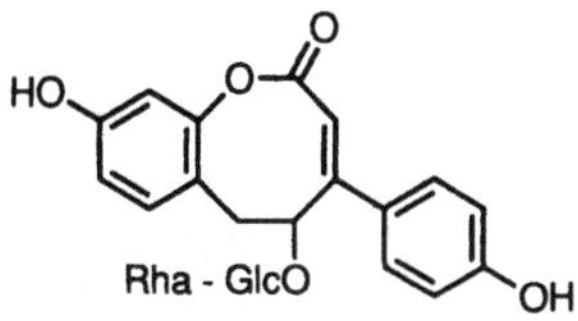
**29** Pueroside A

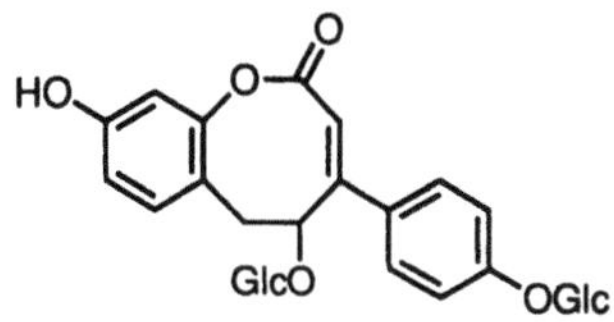
**30** Pueroside B

**Fig. 3.** Isoflavonoid, flavonoids, and related compounds from *Pueraria lobata* Ohwi. *Glc* Glucosyl; *Ruti* rutinosyl; *Rha* rhamnosyl; *Gal* galactosyl

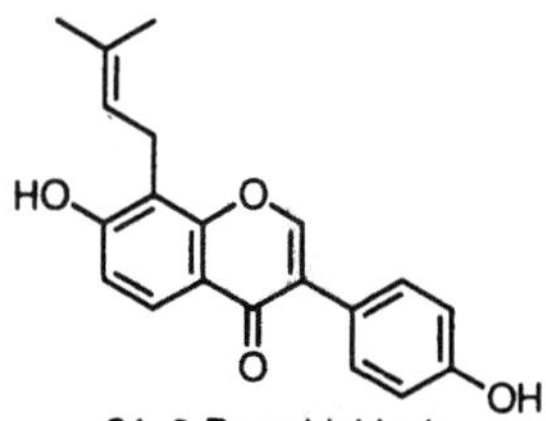
**31** 8-Prenyldaidzein

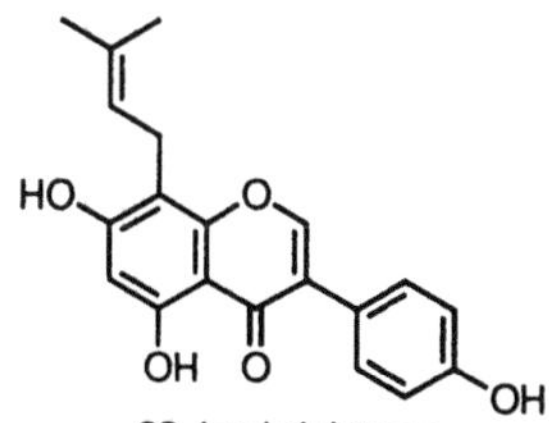
**32** Lupiwighteone

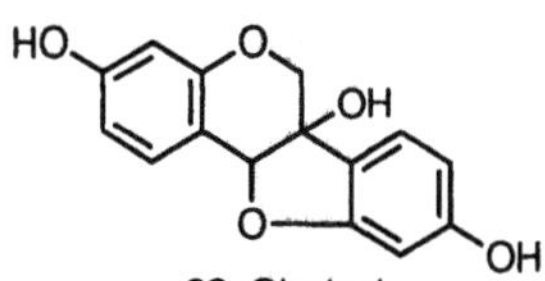
**33** Glycinol

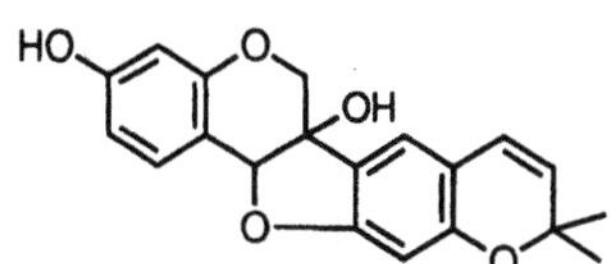
**34** Tuberosin

**Fig. 4.** Induced isoflavonoids from elicitor-treated leaves and stems of *P. lobata*. **33** and **34** were isolated from leaves. (Ingham 1982). **31, 32,** and **34** were isolated from stems. (Hakamatsuka et al. 1991a)

## 2 In Vitro Approaches

Takeya and Itokawa (1982) induced callus from the stem on MS basal medium (Murashige and Skoog 1962) containing 0.1 mg/l kinetin, 1 mg/l 2,4-dichlorophenoxyacetic acid (2,4-D), 3% sucrose, and 0.8% Difco Bacto agar. The callus tissues were subcultured on the same medium every 4 weeks at 26 °C in the dark. Suspension cultures were established from the callus cultures in MS medium containing 0.05 mg/l kin, 0.5 mg/l 2,4-D, and 3% sucrose on a rotary shaker (84 rpm) at 27 °C in the dark. The callus tissues produced puerarin (**3**), daidzin (**2**), daidzein (**1**), genistein (**16**), and coumestrol (**22**) which are also found in the mother plant (Table 1). The main products at the late stage of the callus cultures were daidzin (**2**) and puerarin (**3**) in the suspension culture and daidzin (**2**) in the static culture.

### 2.1 Callus Induction

Takeya and Itokawa (1982) induced callus from stem tissues, while the present authors have induced callus from leaf tissues and examined the effects of hormones on callus induction. The young leaves were soaked in tap water, sterilized with 20% sodium hypochlorite containing Tween 20 (0.05%) for 20 min, and washed three times with sterile distilled water. The sterilized leaves were cut into small segments (1 × 1 cm) and then placed on solidified agar (0.8%) in Petri dishes. After incubation for a week at 25 °C in the dark, surviving leaf segments were transferred to the induction medium consisting of MS salts, 3% sucrose, 0.8% agar, and hormones of various combinations (Table 2). After incubation for a further 2 weeks at 25 °C in the dark, callus induction from the cut surface was observed in some media (Fig. 5A). The MS medium supplemented with higher concentrations of 2,4-D (1 or 5 mg/l) effectively induced callus in combination with either cytokinin, benzylaminopurine (BAP), or kin. Higher concentrations of BAP also enhanced callus induction; however, the combin-

**Table 2.** Effects of hormones on callus induction from *P. lobata* leaf

| Cytokinin | (mg/l) | Auxin (mg/l) | | | | | |
|---|---|---|---|---|---|---|---|
| | | NAA | | | 2,4-D | | |
| | | 5 | 1 | 0.1 | 5 | 1 | 0.1 |
| BAP | 5 | + + | + + | − | + + | + + | + |
| | 1 | + + | + + | − | + + | + + | + |
| | 0.1 | − | − | − | + + | + | − |
| Kinetin | 5 | − | − | − | + + | + + | − |
| | 1 | − | − | − | + + | + | − |
| | 0.1 | − | − | − | + + | − | − |

+ + Callus was induced and grew well.
+ Callus was induced but failed to grow.
− Callus was not induced.

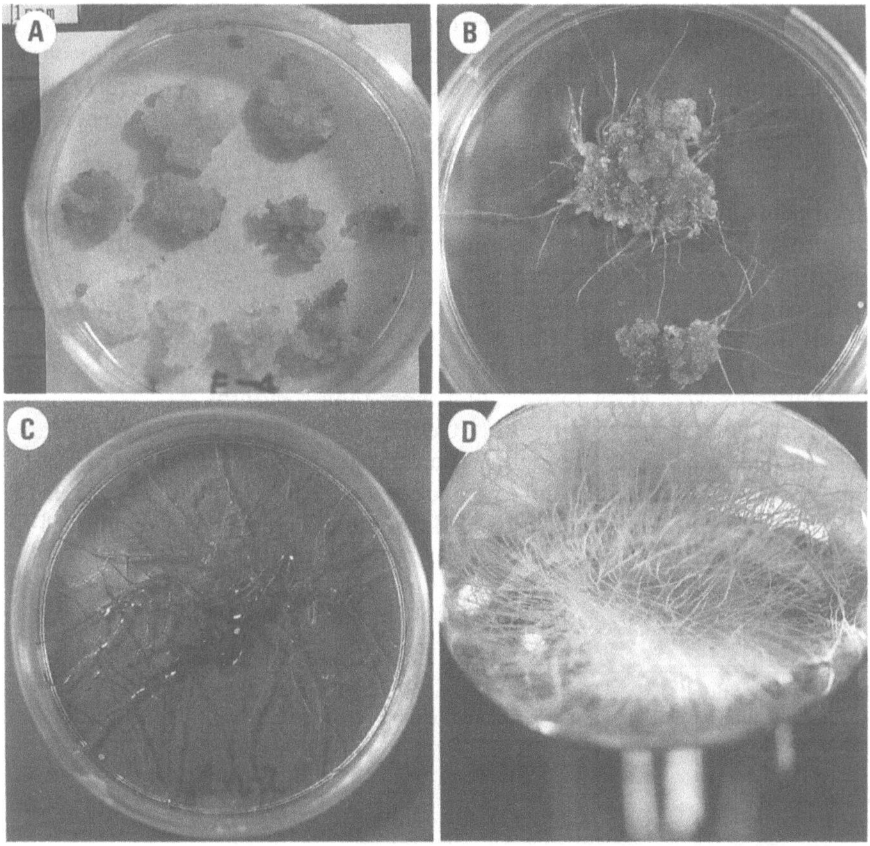


**Fig. 5A–D.** Callus cultures and root cultures of *P. lobata*. **A** Callus growing on MS medium containing NAA (1 mg/l), BAP (1 mg/l), 3% sucrose, and 0.8% agar after induction from leaf tissues on the same medium. **B** Adventitious root formation on MS medium containing 2,4-D (1 mg/l), BAP (0.01 mg/l), 3% sucrose, and 0.2% Gelrite from leaf callus cultured on MS medium containing 2,4-D (5 mg/l), kin (1 mg/l), 3% sucrose, and 0.8% agar. **C** Root cultures growing on MS medium containing NAA (1 mg/l), 3% sucrose, and 0.2% Gelrite, derived from root cut from propagated plantlets. **D** Root cultures from in vitro plantlet growing in MS liquid medium containing NAA (1 mg/l) and 3% sucrose, cultured on MS medium containing NAA (1 mg/l), 3% sucrose, and 0.2% Gelrite

ation of NAA and kin failed to induce callus. Combinations of relatively high concentrations of cytokinin and auxin were required for callus induction from leaves of *P. lobata*. Callus tissues induced from leaves were further used for root formation experiments.

## 2.2 Root Culture

Root cultures of *P. lobata* are expected to achieve a high production of isoflavonoids, because the isoflavonoid content of the roots is much higher than that of other plant parts. Furthermore, establishment of root culture is a prerequisite for biochemical studies on the tissue-specific expression of isoflavonoids. However, all attempts at regeneration of roots from callus cultures gifted by Takeya and Itokawa (1982) were unsuccessful, due to the loss of regenerating ability during long-term subculture (10 years from 1980).

Next, freshly induced callus derived from leaves was used for regeneration experiments. Leaf callus was transferred to root induction media which consisted of MS salts, 3% sucrose, 0.8% agar, high concentrations of auxins (5 or 1 mg/l), and low concentrations of cytokinins (0.1 or 0.01 mg/l). In all the combinations tested, however, root formation was not observed, and callus tissues turned brown and subsequently died within 1 month. Only in the medium consisting of 5 mg/l 2,4-D and 0.1 mg/l BAP, did callus tissues grow well, but roots were never formed after more than 4 months.

When agar was replaced by 0.2% Gelrite, adventitious roots were differentiated in some media after 2 months. The combination of higher concentrations (1 mg/l) of 2,4-D as an auxin and lower concentrations (0.1 or 0.01 mg/l) of BAP and kin as a cytokinin promoted root formation (Table 3). When only 2,4-D

**Table 3.** Effects of hormones on adventitious root formation from leaf callus of *P. lobata*[a]

| Hormone composition of callus induction medium from leaf[b] | | Hormone composition of root induction medium Auxin/Cytokinin (mg/l) | | | | |
|---|---|---|---|---|---|---|
| | | 2,4-D/BAP | | 2,4-D/kinetin | | 2,4-D only |
| Auxin/cytokinin (mg/1) | | 1/0.1 | 1/0.01 | 1/0.1 | 1/0.01 | 1 |
| NAA/BAP | 5.0/5.0 | – | + | + + | + | – |
| NAA/BAP | 5.0/1.0 | + | + + | + + | + + | – |
| NAA/BAP | 1.0/5.0 | + | + | + | + | – |
| NAA/BAP | 1.0/1.0 | + + | + + | C | + + | – |
| 2,4-D/BAP | 5.0/5.0 | + | – | C | + + | – |
| 2,4-D/BAP | 5.0/1.0 | + + | – | C | C | – |
| 2,4-D/BAP | 5.0/0.1 | – | – | – | + | – |
| 2,4-D/BAP | 1.0/5.0 | + + | – | + + | – | – |
| 2,4-D/BAP | 1.0/1.0 | – | + + | C | + + | – |
| 2,4-D/kin | 5.0/5.0 | + + | – | – | + + | – |
| 2,4-D/kin | 5.0/1.0 | + | + + | + | + | + + |
| 2,4-D/kin | 5.0/0.1 | + + | + | C | + | – |
| 2,4-D/kin | 1.0/5.0 | + | + | – | + | – |

[a] Induced leaf callus was transferred to MS medium supplemented with 3% sucrose, 0.2% Gelrite, and various concentrations of hormones at 25 °C in the dark. Observation of root formation after ca. 3 months' incubation.

[b] Every callus induced from leaves at different hormone concentrations (Table 2) was transferred to root induction media. This column shows the hormone composition of media in which leaf callus had been induced.

+ + Roots were formed and grew well. + Roots were formed but failed to grow. C Once roots were formed, callus was again induced from the surface of root tissues. – Roots were not formed.

**Table 4.** Effects of hormones on root cultures derived from plantlets of *P. lobata*

| | | Auxin (mg/l) | |
|---|---|---|---|
| Cytokinin (mg/l) | | NAA 1 | 2,4-D 1 |
| BAP | 0.1 | C | + C |
| | 0.01 | + + | C |
| Kinetin | 0.1 | + + | C |
| | 0.01 | + + | C |
| None | | + + | + C |

+ + Roots grew well. + C Once roots grew, callus was induced from the surface of root tissues. C Roots failed to grow and callus was induced from the surface of root tissues.

(1 mg/l) was added to the medium, root formation was not observed. The use of NAA as an auxin instead of 2,4-D inhibited root formation at any concentrations of cytokinins (data not shown). Even in the cases where roots were formed and grew, the callus tissues were induced again in some media from the surface of growing roots and consequently covered the whole root tissues.

Root cultures were also established from plantlets aseptically grown *in vitro*. The seeds were germinated on agar under continuous light and the growing roots were excised and transferred to MS medium solidified with 0.2% Gelrite (Table 4) (Fig. 5C). In contrast to the case of differentiation from leaf callus, NAA as an auxin was suitable for the maintenance of the root cultures, since the use of 2,4-D resulted in callus formation from root tissues. Medium supplemented with NAA (1 mg/l) and kin (0.1 mg/l) or NAA only (1 mg/l) gave the best growth rate for roots.

In liquid medium supplemented only with NAA (1 mg/l), both the adventitious roots from leaf callus and the root cultures from *in vitro* plantlets grew well on a rotary shaker (60 rpm) at 25 °C in the dark (Fig. 5D). When medium containing 2,4-D was used, the roots stopped growing, and callus formation was stimulated. Although the combination of NAA (1 mg/l) and low concentrations of cytokinin (BAP or kin) also gave good results, the use of NAA (1 mg/l) alone seemed better.

Compared with static callus and cell suspension cultures, root cultures produced isoflavonoids in much higher amounts. The contents of daidzin (**1**) and puerarin (**2**) in root cultures often reached three to four times those in the callus cultures (data not shown).

### 2.3 Cell Suspension Culture

The callus culture originally induced from stems by Takeya and Itokawa (1982) was transferred to cell suspension cultures which were maintained in MS liquid medium supplemented with kinetin (0.1 mg/l), 2,4-D (2 mg/l), and 3% sucrose on a rotary shaker at 28 °C in the dark. They were rotated at high speed (200 rpm), since the cells were easily aggregated at a low speed. Isoflavone glucosides, daidzin (**2**), puerarin (**3**), and genistin (**17**) were the main products of the suspension cultures and they mostly existed as 6″-*O*-malonyl esters (Park et al. 1992). This cell suspension culture was extensively used for biosynthetic studies of isoflavonoids, which are mentioned later.

## 2.4 Induction of Isoflavonoid Accumulation

Although cell suspension cultures of *P. lobata* produced an amount of isoflavonoids nearly equal to the mother plant (data not shown), the treatment of cultured cells with adequate materials led to much higher production of some isoflavonoids. Some of these cases are related to the defense responses in higher plants.

### *2.4.1 Induction by an Endogenous Elicitor*

The accumulation of phytoalexins against microbial attacks is one of the important defense responses in higher plants. Besides living pathogens, various materials are recognized as elicitors which can trigger phytoalexin accumulation. In some kinds of plants, their own cell wall components are known to play a part in phytoalexin induction, and these elicitors are called endogenous elicitors (Bailey 1980; Hahn et al. 1981).

When cell suspension cultures were treated with an endogenous elicitor at a final concentration of 0.05 mg/ml, which was prepared by the hydrolysis of *P. lobata* cell wall fraction with a fungal endopolygalacturonase, drastic changes in isoflavonoid metabolism were observed. The HPLC analysis of methanol extracts of suspension cultures treated with endogenous elicitor indicated that the major constitutive glucosides such as daidzin (**2**) and genistin (**17**) disappeared transiently within 5 h after treatment and then began to accumulate rapidly. After 48 h the amount of isoflavone glucosides sometimes reached more than fivefold that of untreated cells. The transient disappearance of major glucosides from the methanol-extractable fraction was partially due to the deposition to the cell walls to form nonextractable compounds (Park et al. unpubl.). Furthermore, the glucosides seemed to be degraded into their aglycones. In addition to the constitutive glucosides, three dimers of daidzein were accumulated (Fig. 6) (Hakamatsuka et al. 1992). Although one of these dimers has two chiral centers in its molecule, it was demonstrated to be optically inactive. Therefore, these dimers were formed by a nonspecific oxidative coupling reaction catalyzed by peroxidase-like enzyme(s).

Subsequently, a commercially available yeast extract (Bacto Yeast Extract from Difco) was found to have the same effects as the endogenous elicitor when it was added at a final concentration of 1 mg/ml.

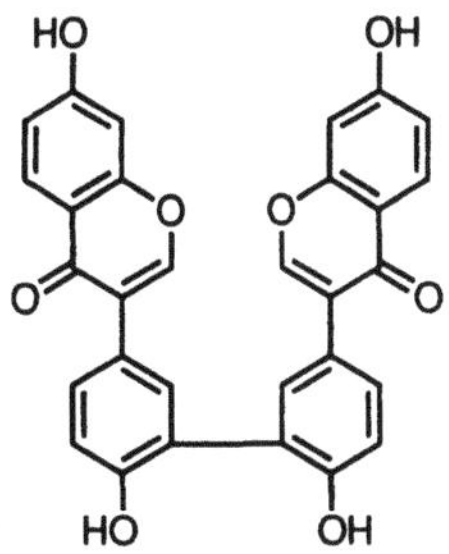

**35** Kudzuisoflavone A

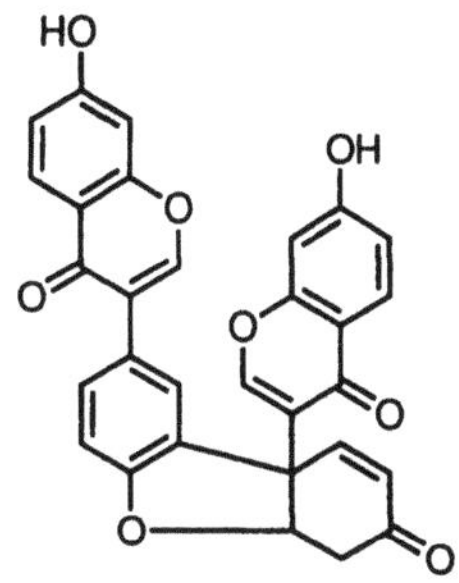

**36** Kudzuisoflavone B

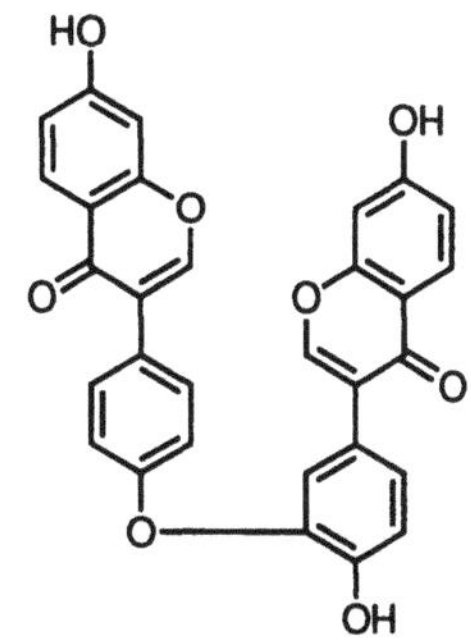

**37** Kudzuisoflavone C

**Fig. 6.** Induced isoflavone dimers from cell suspension cultures of *P. lobata* treated with the endogenous elicitor or yeast extract. Kudzuisoflavanone B (**36**) was isolated as a racemate

### 2.4.2 *Induction by a Fungal Elicitor*

Since specific pathogens for *P. lobata* had not been identified, glycoprotein elicitor from *Phytophthora megasperma* f. sp. *glycinea* (Keen and Legrand 1980), which is specific to soybean, was used for the induction experiments of isoflavonoid phytoalexins in cell suspension cultures of *P. lobata*. When glycoprotein elicitor from *Phytophthora megasperma* was added to the cell suspension cultures at a final concentration of 0.01 mg/ml, the treated cells turned dark brown within 10 h and stopped growing. This was quite a contrast to the endogenous elicitor or yeast extract-treated cells, whose color was light brown and whose growth was as good as in untreated cells. The HPLC analysis of methanol extracts of cells treated with glycoprotein elicitor indicated that pre-existing major isoflavone glycosides rapidly disappeared from the methanol-extractable fraction but these compounds were never induced again. Instead of isoflavone glucosides, a number of less polar compounds, including three dimers of daidzein described above, were accumulated (Fig. 7) (Shinkai et al. unpubl.).

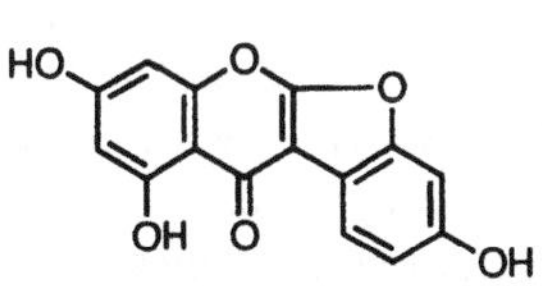

**38** Lupinalbin A

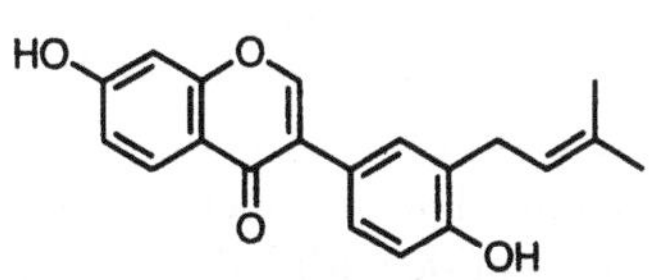

**39** Neobavaisoflavone

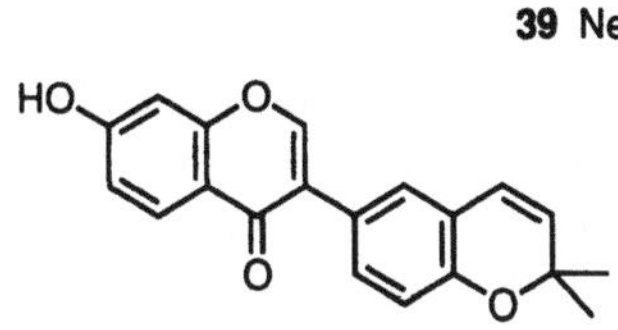

**40** Corylin

**Fig. 7.** Induced isoflavonoids from cell suspension cultures of *P. lobata* treated with the fungal elicitor or $CuCl_2$. In addition to the compounds illustrated here, daidzein dimers (**35**, **36**, **37**) and other isoflavonoids (**31**, **32**, **34**) were isolated from cell suspension cultures treated with fungal elicitor or $CuCl_2$

Among these compounds, tuberosin (**34**), 8-prenyldaidzein (**31**), and 8-prenylgenistein (lupiwighteone) (**32**) had been identified as phytoalexins in stems of *P. lobata* (Hakamatsuka et al. 1991a). Furthermore, the $CuCl_2$ treatment at a final concentration of 1 mM showed almost the same effects as treatment by fungal elicitor.

## 2.5 Biosynthetic Studies on Isoflavonoids

Enzymes involved in secondary metabolism are in general less active than those of primary metabolism. The activation of the target metabolic pathway is advantageous and often a prerequisite for biosynthetic studies of secondary metabolites. In this respect, cell suspension cultures of *P. lobata* treated with endogenous elicitor or yeast extract were convenient materials to investigate isoflavonoid biosynthesis. Since the preparation procedure of the endogenous elicitor is rather complicated (Hakamatsuka et al. 1990), easily available yeast extract has been used routinely for this purpose. Although the treatment of cell suspension cultures with glycoprotein elicitor or $CuCl_2$ induced the rapid accumulation of isoflavonoid phytoalexins, the response was so strong that the cells died soon after treatment because of their hypersensitive reaction. Therefore, these cells were not suitable for biosynthetic studies, in particular as the source of enzyme preparations.

The original plant and cell suspension cultures of *P. lobata* produce mainly 5-deoxy-type isoflavonoids, which lack the oxygen atom at C-5. Deoxychalcone synthase is an enzyme which catalyzes the formation of isoliquiritigenin, 5-deoxychalcone, from one molecule of *p*-coumaroyl CoA and three molecules of malonyl CoA in the presence of NADPH, and it was identified from cell suspension cultures of *P. lobata* (Hakamatsuka et al. 1988) (Fig. 8). Characterization of the deoxychalcone synthase indicated that this enzyme consisted of two separate enzymes, chalcone synthase, and co-acting reductase, and each polypeptide was purified to homogeneity (Hakamatsuka et al. unpubl.). The subunit of the chalcone synthase had a molecular weight of 43 kDa and its native form is dimeric. Two-dimensional electrophoresis of purified chalcone synthase indicated that it was composed of more than five isozymes with equal molecular weight. The reductase co-acting with chalcone synthase was a monomeric polypeptide with MW of 35 kDa and had at least two isozymes. Deoxychalcone synthase is a very unique and new-type polyketide synthesis enzyme system which affords a deoxy-type compound in the presence of NADPH and a hydroxy-type compound in the absence of NADPH. In the reactions of other polyketide synthetic enzymes such as 6-methylsalicylic acid synthase, the omission of NADPH from the reaction mixture results in the termination of chain elongation and the release of an incomplete intermediate.

Chalcone-flavanone isomerase catalyzes the interconversion of chalcone and flavanone (Fig. 8) and it was also purified from *P. lobata*. The enzyme was a monomer with MW 28 kDa, composed of two isozymes with equal molecular weight.

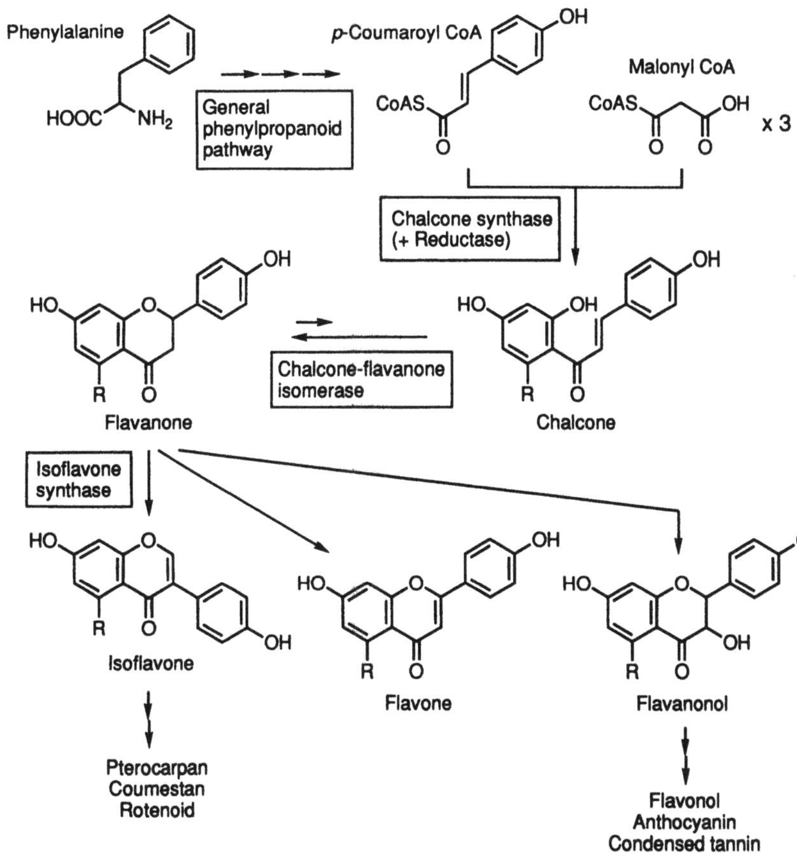


**Fig. 8.** Biosynthetic pathway of flavonoids and isoflavonoids. If the reductase acts with the chalcone synthase, *R* is H, and if not then *R* is OH

The biosynthetic course towards the flavanone, i.e., general phenylpropanoid pathway, construction of chalcone and its isomerization to flavanone, is common to both flavonoids and isoflavonoids (Fig. 8). The first reaction peculiar to isoflavonoid biosynthesis is the oxidative aryl migration catalyzed by isoflavone synthase and the activity of isoflavone synthase was detected in a microsomal preparation (Hakamatsuka et al. 1989, 1990). Since highly active chalcone-flavanone isomerase was present in the microsomal preparation, the transformation of both chalcone and flavanone into isoflavone was catalyzed in the presence of NADPH. Comparative experiments using [$^3$H] flavanone and [$^{14}$C] chalcone were then carried out and it was demonstrated that the true substrate for the isoflavone synthase was flavanone. When the microsomal preparation was carefully washed by repeated ultracentrifugation to remove the contaminating soluble enzymes, 2-hydroxyisoflavanone was isolated from the reaction mixture as a major product instead of isoflavone (Hashim et al. 1990).

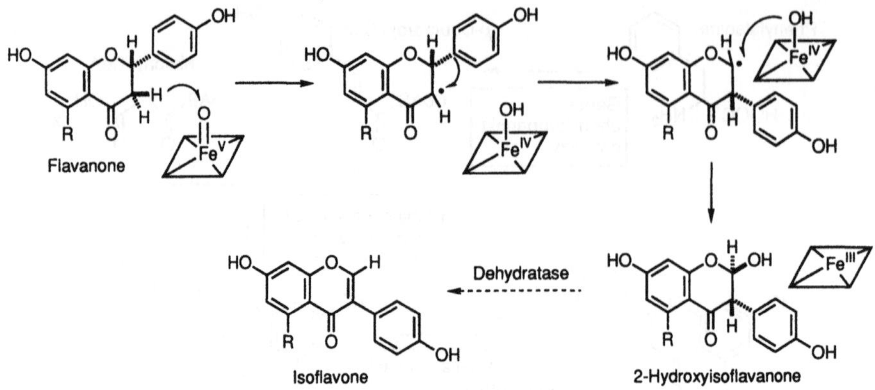


**Fig. 9.** Proposed reaction mechanisms for the oxidative aryl migration in isoflavone biosynthesis catalyzed by cytochrome P-450 enzyme system. The reactions illustrated by *solid arrows* are catalyzed by P-450. *Broken arrow* indicates the step catalyzed by a soluble dehydratase

The enzyme which catalyzed the conversion of flavanone to 2-hydroxyisoflavanone was proved to be a membrane-bound cytochrome P-450 enzyme system. P-450 and P-450:NADPH reductase were solubilized from the microsomal membrane and partially purified (Hakamatsuka et al. 1991b). With the partially purified enzymes the synthetic activity of 2-hydroxyisoflavanone was successfully reconstituted in the presence of plant lipids prepared from membrane fractions of *P. lobata*. The dehydration of 2-hydroxyisoflavanone to isoflavone was not a spontaneous reaction and was shown to be catalyzed by a specific enzyme which localized in a soluble enzyme fraction. Recently, the dehydratase was purified to homogeneity from *P. lobata* and was demonstrated to be a monomeric polypeptide with a mol.wt. of 38 kDa (Hakamatsuka et al. unpubl.). The reaction mechanism for oxidative aryl migration catalyzed by cytochrome P-450 was proposed as described in Fig. 9 (Hakamatsuka et al. 1991b). Hydroxylation associated with migration is a new category of cytochrome-P-450-mediated reactions.

## 3 Conclusions and Prospects

The tissue cultures of *P. lobata* grow fast and produce an amount of isoflavonoid compounds almost equal to the mother plant. This cell line maintains the ability to produce isoflavonoids, even after subculture for 10 years. The production of isoflavonoids in the cell cultures is drastically enhanced by treatment with various elicitors, and cells activated in this manner are a suitable experimental system for the studies of the regulation of genes and enzymes involved in the biosynthesis of isoflavonoids. Recently, in addition to the enzymatic studies mentioned above, cDNA clones encoding chalcone synthase (Nakajima et al.

1991), co-acting reductase (Shibuya et al. unpubl.), and chalcone-flavanone isomerase were isolated, sequenced, and expressed in *E. coli* as active enzymes. The enzymes were expressed in large amounts and detailed investigations on the stereochemistry of the enzymatic reaction and the characterization of active sites of the enzymes are now in progress.

Cell suspension cultures show different responses to the two types of stimuli, i.e., endogenous elicitor (and yeast extracts) and the fungal glycoprotein elicitor (and $CuCl_2$). The mechanisms of signal recognition by plant cells, and the regulation of differential expression of genes and enzymes are attractive fields in studies of the biosynthesis of natural products, and cell cultures of *P. lobata* would be a very suitable model system for these studies.

## 4 Protocol

*Best Explant.* Sterilized leaf segments (1 ×1 cm), Precultured in solidified agar (0.8%) for 1 week at 25 °C in the dark.

*Callus Induction Medium.* MS basal medium + 5 mg/l 2,4-D, 1–5 mg/l BAP or kinetin, 3% sucrose. The medium is solidified with 8 g/l agar.

*Rooting Medium.* MS basal medium + 1 mg/l 2,4-D, 0.01–0.1 mg/l BAP or kinetin, 3% sucrose. The medium is solidified with 2 g/l Gelrite.

*Acknowledgments.* The authors wish to thank Dr. H. Itokawa (Tokyo College of Pharmacy) for kindly supplying stem callus of *P. lobata*, and Dr. M. Yoshikawa (Kyoto Prefectural University) for a generous gift of a glycoprotein elicitor from *Phytophthora megasperma* f. sp. *glycinea*. Thanks are also due to Dr. H. Noguchi of Tokyo University for helpful discussions. Finally, contributions to this work from K. Shinkai, H. Terai, H. Murakami, and M. Yoshida are gratefully acknowledged.

## References

Bailey JA (1980) Constitutive elicitors from *Phaseolus vulgaris*; a possible cause of phytoalexin accumulation. Ann Phytopathol 12:395–402

Hahn MG, Darvill AG, Albersheim P (1981) Host-pathogen interactions. XIX. The endogenous elicitor, a fragment of a plant cell wall polysaccharide that elicits phytoalexin accumulation in soybeans. Plant Physiol 68:1161–1169

Hakamatsuka T, Noguchi H, Ebizuka Y, Sankawa U (1988) Deoxychalcone synthase from cell suspension cultures of *Pueraria lobata*. Chem Pharm Bull 36:4225–4228

Hakamatsuka T, Noguchi H, Ebizuka Y, Sankawa U (1989) Isoflavone synthase from cell suspension cultures of *Pueraria lobata*. Chem Pharm Bull 37:249–252

Hakamatsuka T, Noguchi H, Ebizuka Y, Sankawa U (1990) Isoflavone synthase from cell suspension cultures of *Pueraria lobata*. Chem Pharm Bull 38:1942–1945

Hakamatsuka T, Ebizuka Y, Sankawa U (1991a) Induced isoflavonoids from copper chloride-treated stems of *Pueraria lobata*. Phytochemistry 30:1481–1482

Hakamatsuka T, Hashim MF, Ebizuka Y, Sankawa U (1991b) P-450-dependent oxidative rearrangement in isoflavone biosynthesis: reconstitution of P-450 and NADPH: P-450 reductase. Tetrahedron 47:5969–5978

Hakamatsuka T, Shinkai K, Noguchi H, Ebizuka Y, Sankawa U (1992) Isoflavone dimers from yeast extract-treated cell suspension cultures of *Pueraria lobata*. Z Naturforsch 47c: 177–182
Hashim MF, Hakamatsuka T, Ebizuka Y, Sankawa U (1990) Reaction mechanism of oxidative rearrangement of flavanone in isoflavone biosynthesis. FEBS Lett 271: 219–222
Horikawa Y (1972) Atlas of the Japanese flora. Gakken, Tokyo, p 143
Ingham JL (1982) In: Bailey IA, Mansfield JW (eds) Phytoalexins. Blackie, Glasgow, pp 21–80
Kariyone T, Koiso R (1971) Altas of medicinal plants. Takeda Chemical Industries, Osaka, pp 101–102
Keen NT, Legrand M (1980) Surface glycoproteins: evidence that they may function as the race specific phytoalexin elicitors of *Phytophthora megasperma* f.sp. *glycinea*. Physiol Plant Pathol 17: 175–192
Kinjo J, Furusawa J, Nohara T (1985) Two novel aromatic glycosides, pueroside-A and -B, from Puerariae Radix. Tetrahedron Lett 26: 6101–6102
Kinjo J, Furusawa J, Baba J, Takeshita T, Yamasaki M, Nohara T (1987) Studies on the constituents of *Pueraria lobata*. III. Isoflavonoids and related compounds in the roots and the voluble stems. Chem Pharm Bull 35: 4846–4850
Kinjo J, Takeshita T, Abe Y, Terada N, Yamashita H, Yamasaki M, Takeuchi K, Murakami K, Tomimatsu T, Nohara T (1988) Studies on the constituents of *Pueraria lobata*. IV. Chemical constituents in the flowers and the leaves. Chem Pharm Bull 36: 1174–1179
Kinoshita T (1982) Chemistry of Puerariae Radix. Gendai Toyo Igaku 3: 58–62
Kubo M, Fujita K, Nishimura H, Naruto S, Namba K (1973) A new irisolidone-7-*O*-glucoside and tectoridin from *Pueraria* species. Phytochemistry 12: 2547–2548
Kubo M, Sasaki M, Namba K, Naruto S, Nishimura H (1975) Isolation of a new isoflavone from Chinese *Pueraria* flowers. Chem Pharm Bull 23: 2449–2451
Kurihara T, Kikuchi M (1973) Studies on the constituents of flowers. I. On the components of flower of *Pueraria thunberginana* Benth. Yakugaku Zasshi 93: 1201–1205
Kurihara T, Kikuchi M (1975) Studies on the constituents of flowers. V. On the components of flower of *Pueraria thunberginana* Benth. (2). Isolation of a new isoflavone glycoside. Yakugaku Zasshi 95: 1283–1285
Kurihara T, Kikuchi M (1976) Studies on the constituents of flowers. VI. On the components of flower of *Pueraria thunberginana* Benth. (3). Yakugaku Zasshi 96: 1486–1488
Murashige T, Skoog F (1962) A revised medium for rapid growth and bioassays with tobacco tissue culture. Physiol Plant 15: 473–497
Nakajima O, Akiyama T, Hakamatsuka T, Shibuya M, Noguchi H, Ebizuka Y, Sankawa U (1991) Isolation, sequence and bacterial expression of a cDNA for chalcone synthase from the cultured cells of *Pueraria lobata*. Chem Pharm Bull 39: 1911–1913
Ohshima Y, Okuyama T, Takahashi K, Takizawa T, Shibata S (1988) Isolation and high performance liquid chromatography (HPLC) of isoflavonoids from the *Pueraria lobata*. Planta Med 54: 250–254
Park H-H, Hakamatsuka T, Noguchi H, Sankawa U, Ebizuka Y (1992) Isoflavone glucosides exist as their 6″-*O*-malonyl esters in *Pueraria lobata* and its cell suspension cultures. Chem Pharm Bull 40: 1978–1980
Takeya K, Itokawa H (1982) Isoflavonoids and the other constituents in callus tissues of *Pueraria lobata*. Chem Pharm Bull 30: 1496–1499

# XXIV *Santalum album* L. (Sandalwood): In Vitro Culture and the Bioreactor Production of Secondary Metabolites

J. V. VALLURI[1]

## 1 Introduction

Sandalwood (*Santalum album* L.) is a semiparasitic, perennial tree belonging to the family Santalaceae. There are approximately 26 species of sandalwood whose natural distribution extends from India, reaching eastwards through the Malayan archipelago into drier areas of Australia and New Zealand, fanning out across various islands in the Pacific, and stopping just west of the coast of South America at the Juan Fernandez Islands (Morris 1982). Among the various species, *S. austro-caledonicum* Vieill, *S. freycinetianum* Gaud, *S. marchionense* skottsb, *S. yassi seem*, and *S. spicatum* are of economic importance. The most valuable sandalwood oil is distilled from the heartwood and roots of *Santalum album* L., confined to forests in Mysore (southern India). In India, sandalwood is found mainly in the Deccan Plateau and spreads over 300 miles from Dharwar in the north to Nilgiris in the south, and from Coorg in the west approximately 250 miles to Kuppam in the east (Venkatesan 1981).

Sandalwood is one of the oldest known perfumery materials famous for its scented oil and its use in carving. Approximately 2000 tons, representing over 75% of the total production of India, come from Mysore each year. The oil found in the root and heartwood has been acknowledged to be one of the most precious perfumery items from antiquity to modern times. The yellow brown wood is of great commercial importance and is used for carving and making curios of exquisite beauty. The essential oil of sandalwood consists of a mixture of two sesquiterpene alcohols and is obtained by low-pressure steam distillation, being then pale yellow in color, optically clear, and possessing the characteristic smooth fragrance of sandalwood.

India derives considerable revenue from the exportation of sandalwood and its products, with an overall share of more than 50% of the country's total exports (Sunderesh 1979). In 1974–1976, this amounted to 9 million US dollars, but now has fallen to 3 million US dollars. This dramatic decrease in production is attributed to lethal spike, a "yellows-type" disease caused by mycoplasma-like organisms (Dijkstra and Lee 1969; Hull et al. 1969; Varma et al. 1969). The disease is characterized by extreme reduction in the size of leaves and internodes. Spiked plants do not bear flowers or fruits; occasionally phylloid or abortive

[1] Department of Biological Sciences, Marshall University, Huntington, West Virginia 25701, USA

Biotechnology in Agriculture and Forestry, Vol. 28
Medicinal and Aromatic Plants VII (ed. by Y.P.S. Bajaj)

flowers are developed. Death occurs within 2–3 years after manifestation of the external symptoms. So far, progress in the study of spike disease has been very slow, and efforts to control or eliminate it have been unsuccessful. Temporary remission of disease symptoms occurs upon application of antibiotics belonging to the tetracycline group (Rao et al. 1975); a mixture of tetracycline and benylate applied by the girdling method gave the best results (Raychaudhari et al. 1972). However, the remission is only temporary, and repeated treatments are not economically justified for field application. Efforts to develop strains inherently resistant to the disease have not been successful (Venkatesh and Kedarnath 1964; Muniyappa et al. 1980).

## 2 In Vitro Approaches

Clonal propagation using tissue culture offers many potential advantages over conventional breeding techniques. The utility of the tissue culture approach as a means for clonal propagation of sandalwood has been well documented (Table 1). Srimathi and Sreenivasaya (1963) initiated the culture of sandalwood root tips, haustoria, and lateral buds. Rangaswamy and Rao (1963) demonstrated callus initiation and establishment from endosperm tissue of sandalwood on modified White's medium supplemented with 2 mg/l 2,4-D, 5 mg/l kinetin, and 0.25% yeast extract. Further, the embryogenic callus obtained on White's basal medium differentiated into embryoids and was subsequently used for plantlet regeneration (Rao 1965; Rao and Rangaswamy 1971). Regeneration of shoot buds in hypocotyl sections was accomplished on basal media supplemented with the auxins IAA, NAA, IBA, and NOA (1 mg/l) (Rao and Bapat 1978). Cytokinins stimulated and enhanced shoot bud formation far better than auxins. However, excised stem segments of mature plants did not respond to any hormonal treatment (Bapat and Rao 1979).

From the standpoint of tree improvement, regeneration of plantlets from embryogenic tissue and juvenile seedlings seldom retain the superior traits of a mature tree. The ability to culture mature tissue has some decided advantages. By selecting tissue from trees with superior traits, a genetic gain equal to one or two generations further than in the breeding program can be attained. Induction of callus and differentiation of embryoids and their subsequent development into plantlets from 20- to 25-year-old mature sandalwood trees was demonstrated by Laskhmi Sita et al. (1979).

We have been investigating novel bioreactor configurations that provide favorable conditions for the production of secondary metabolites from economically important plant cell cultures. Investigations were undertaken on sandalwood to exploit bioreactor technology for clonal propagation of disease-free species and to identify elite cell lines for secondary metabolite production. Previous reports on bioreactor culture of sandalwood suspension cultures have concentrated on the production of somatic embryos (Bapat et al. 1990) and encapsulation of embryogenic suspensions of sandalwood (Bapat and Rao

**Table 1.** In vitro culture of *Santalum album* L. (see also Lakshmi Sita 1986)

| Explant source | Response | Reference |
|---|---|---|
| 1. Seedling material | | |
| Root tips, haustoria lateral buds | Callus induction | Srimathi and Srinivasaya (1963) |
| Endosperm from dry seeds | Callus induction No endosperm proliferation in the absence of embryo | Rangaswamy and Rao (1963) |
| Embryo | Embryo callus from germinating seeds, differentiation | Rao (1965), Rao and Rangaswamy (1971) |
| Hypocotyl sections | Shoot bud regeneration | Rao and Bapat (1978), Rao and Raghava Ram (1983) |
| | Embryogenesis of hypocotyl callus | Bapat and Rao (1979) |
| | Plantlet formation, hypocotyl callus | Bapat and Rao (1984) |
| Nodal and internodal sections | Differentiation of multiple buds and somatic embryogenesis | Rao et al. (1984) |
| Leaf mesophyll, stem and hypocotyl callus | Isolation of protoplasts | Bapat et al. (1985) |
| Embryogenic cell suspensions | Plantlet formation | Rao and Ozias-Akins (1985) |
| 2. Mature trees (20–25-year-old) | | |
| Nodal and internodal sections of young shoots | Callus induction and differentiation of embryoids and their development into plantlets | Lakshmi Sita et al. (1979) |
| Suspension cultures | Induction of embryogenesis and plantlet formation | Lakshmi Sita et al. (1980a) |
| Endosperm (from fruits) | Callus induction and differentiation of embryoids into plantlets | Lakshmi Sita et al. (1980b) |
| | Establishment of plantlets in soil | Lakshmi Sita et al. (1982) |
| Leaf callus and suspension cells | Protoplast isolation | Lakshmi Sita and Shobha Rani (1983) |
| Cell suspensions | Bioreactor culture and production of phenolics | Valluri et al. (1991) |

1988). Our studies on bioreactor culture of sandalwood focus on the cultural factors that influence secondary metabolite production.

### 2.1 Establishment of Suspension Cultures

Photomixotrophic (PMT) and photoautotrophic (PAT) suspensions of higher plants are being increasingly employed to study secondary metabolism in bioreactors. Photomixotrophic sandalwood suspension cultures (SW-1) were developed by transferring calli to liquid MS medium modified with 20 g/l sucrose, 1 mg/l 2,4-D, and 1 mg/l BA, then culturing on an orbital shaker (120 rpm) at 25 °C, with a 16-h photoperiod using cool white fluorescent lighting (4–6 $W/m^2$). A second batch of sandalwood suspensions (SW-2) were developed from calli that were cultured under continuous cool white fluorescent lighting (12–15 $W/m^2$). Cell suspensions were subcultured at the late exponential phase (14 days) using an inoculum ratio of 1:8 (10 ml to 80 ml medium in 250-ml baffled flasks). We noticed that cells cultured in baffled flasks became more resistant to shear stresses caused by hydrodynamic or mechanical forces generated in a bioreactor. The shear tolerance of sandalwood suspensions may be improved by subculturing into baffled flasks, increasing shaker speed over time, and selecting for cells with higher growth rates. Both SW-1 and SW-2 cell lines were derived from the original calli and were maintained under the same culture regime, but at different photoperiods and light intensities. The behavior of many plant cell cultures is influenced by photoperiodicity, light intensity, and light quality (Seibert and Kadkade 1980). Increased photoperiod and the differences in light intensity during sandalwood cell culture seem to have altered the metabolic activity of the two cell lines. Ten to 14-day-old suspension cultures were used to inoculate the bioreactor. Inoculum volume was 350 ml, which was equivalent to 5 to 8 g fresh weight/l and 17.5% of reactor volume.

### 2.2 Bioreactor Culture of Sandalwood

Biomass and growth studies of sandalwood suspensions were conducted in a modified 2.5-l New Brunswick Bio-Flo II (New Brunswick Sci. Co., New Brunswick, NJ) mechanically agitated bioreactor (Fig. 1) with a 2.35-l working volume (Valluri et al. 1991). In culturing photosynthetic sandalwood cells in bioreactors, it is necessary to maintain optimum mixing and aeration without increasing the hydrodynamic and mechanical stress effects beyond tolerance levels. Reactor conditions were constant illumination (180–200 $\mu E/m^2/s$ measured at inside glass surface), 26 °C and an initial pH of 5.4. Agitation (60–100 rpm) of the bioreactor was provided by a "cell-lift" impeller which generated lower shear rates than the Rhuston-type impellers. Impeller speed between 60 and 100 rpm produced the best mixing without excessive foaming. A rapid reduction in biomass yields was observed with a pair of 5-cm flat blade impellers as rpm was increased beyond 60 rpm. The air flow rate was measured

**Fig. 1.** Bioreactor with cell-lift impeller used for culture of photomixotrophic sandalwood cell suspensions. The total working volume of the vessel is 2.5-l. Mixing is carried out by an incoming stream of compressed air that enters the vessel at the bottom

by flowmeter and expressed as volume of air per volume of medium per min (vvm). The gas composition was 21% $O_2$, 2% $CO_2$, and the bulk $N_2$. A modified sparger equipped with a fritted glass filter was used to aerate the medium at 0.025 to 0.5 vvm. Dissolved oxygen (DO) tension in the medium was measured by galvanic oxygen electrode (Phoenix instruments, Houston, TX). The initial volumetric oxygen transfer coefficient ($K_La$) of the bioreactor was determined dynamically as a function of the aeration rate using a gassing-out method to measure the increase or decrease in oxygen concentration after a step change in inlet gas composition. Data was continuously monitored and logged by Loggernaut Data Logging System (Cyborg Corporation, Newton, MA).

The choice of bioreactor design may be determined by the characteristics of the plant species (Vogelmann 1981). The stirred tank bioreactor used in this study contained a modified impeller system consisting of the cell-lift impeller

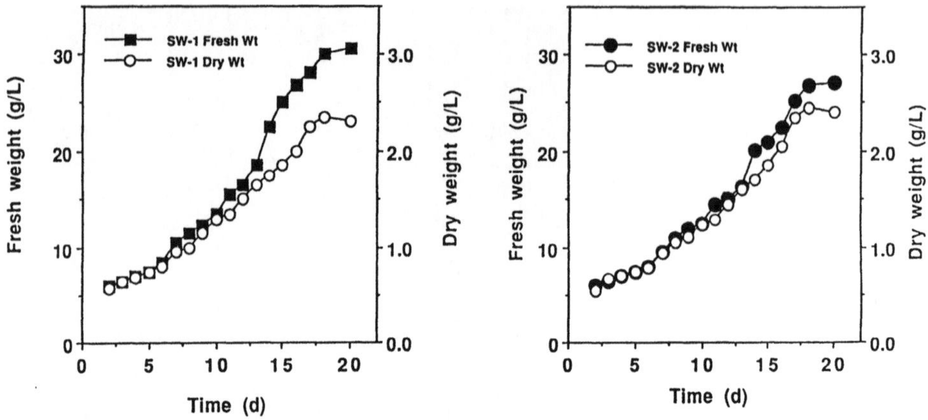


**Fig. 2.** Bioreactor culture of SW-1 and SW-2 sandalwood cell suspensions showing exponential growth phase of 11–12 days (*left*) and stationary growth phase of 3–5 days (*right*). (Valluri et al. 1991)

and was run in the batch mode. The lag phase typically observed in plant suspensions after inoculation was not readily apparent (Fig. 2). This may be due to the use of a relatively large inoculum of 350 ml which was equivalent to 17.5% of reactor volume. The exponential phase for SW-1 and SW-2 cell suspensions appeared to begin almost immediately and lasted for 11–12 days. The stationary phase lasted for 3 to 5 days, after which a rapid increase in dead cells was observed, as well as an increase in medium pH. During exponential growth, fresh weight Td was 5.8 days and the specific growth rate ($\mu$) was 0.12 day for SW-1 and SW-2 cell cultures. As cells entered the stationary phase, dry weight biomass yields levelled off, while fresh weights continued to increase very slightly. Microscopic observation revealed that the increase in fresh weight was due to cell swelling, probably due to uptake of water.

## 2.3 pH and Dissolved Oxygen During Bioreactor Culture

Optimal growth in plant cell cultures usually occurs in media with initial pH values in ranges 5–6. Typically, the buffering capacity of plant cell media is very low, since the usual concentration of phosphate is 1 to 5 mM. A characteristic U-shaped curve was observed for SW-1 and SW-2 cell suspensions cultured in the bioreactor (Fig. 3). The pH decreased from 5.5 to 3.5 during the exponential growth phase, and increased after day 11 as cells entered the stationary phase. When plant cells were cultured on $NH_4^+$ as the sole nitrogen source, the pH of the medium decreased as the ammonium was utilized. Conversely, the pH has been shown to rise when $NO_3^-$ is the sole nitrogen source (Treat 1988).

Monitoring of dissolved oxygen in the medium was used to suggest the physiological state of the cells. Prior to inoculation, the dissolved oxygen was measured to be a relative 98% at 0.3 vvm. This resulted in an initial $K_La$ of

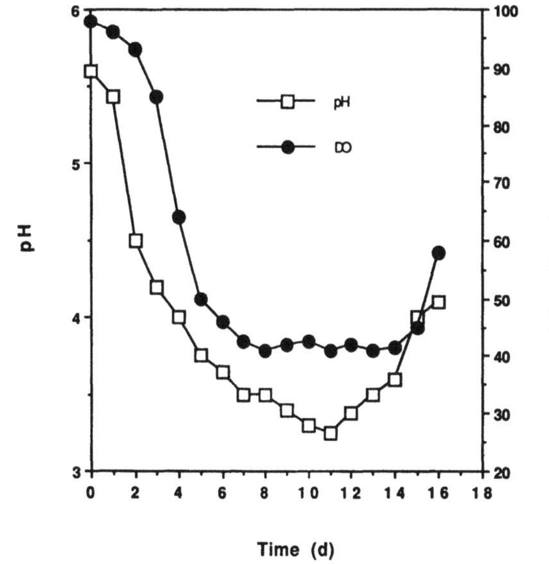


**Fig. 3.** Changes in pH and dissolved oxygen during bioreactor growth of SW-1 and SW-2 sandalwood cell suspensions. (Valluri et al. 1991)

20.9/h at 100 rpm. Plant cell cultures do not need the high $K_La$'s required for most aerobic microbial fermentation. An increase in the $K_La$ may be detrimental to the plant cell cultures due to removal of $CO_2$ (Smart and Fowler 1981). After inoculation, the dissolved oxygen dropped steadily over the next 8 days to near 40%, where it remained throughout exponential growth (Fig. 3). The onset of the stationary phase was characterized by an increase in the dissolved oxygen content of the medium, suggesting reduced demand and a change in the metabolic state of the cells. To optimize uptake of nutrients by cells and mixing of the medium, suspension cultures consisting of single cells would be ideal. The average cell aggregate size was greater in shake flasks than in the bioreactor. Shear generated at impeller tips and by air bubbles passing through liquid may have contributed to a reduction in cell aggregate size. Particle sizes in excess of 500 $\mu$m contained 15–30 cells. Cell aggregates in excess of 3–5 mm were observed during the latter stages of fermentation, typically on the order of five to nine clumps. The tendency for cell aggregates to increase in size during culture was typical of all impeller types. These larger aggregates appeared to be the result of sloughed-off debris from biofouling.

## 2.4 Aeration and Mixing

Both aeration and mixing constitute basic requirements of all mass growth systems. The supply of gaseous components can not only affect growth, but can induce product formation. Against the need to reduce shear in plant cell cultures there is the problem of providing effective mixing to maintain cells in suspension and achieve optimal productivity and growth. Since plant cells are larger than microbial cells, they tend to sediment rapidly and create conditions of inadequate mixing, causing stagnant or "dead" zones in the reactor. Relatively good mixing and low shear rates were accomplished during sandalwood cell culture because of using the cell-lift impeller. Another particularly troublesome phenomenon that was observed in the stationary phase was the cell deposition on the bioreactor wall above the fluid level. Aeration rates above 0.5 vvm when the impeller speed was 100 rpm also resulted in the deposition of a substantial amount of cells in a meringue that formed above the liquid on the vessel wall, drive shaft, and probes during the exponential phase. This accumulation in the head plate area was reduced by maintaining 0.3 vvm at 100 rpm. The narrow space between the baffles and the vessel proved to be ideal for the accumulation of larger cell aggregates. This buildup of biomass produced poor mixing. The removal of the baffles eliminated most fouling and mixing problems.

### 2.5 Essential Oils and Phenols

Very few tissue cultures from higher plants are capable of producing volatile oils in considerable quantities with components similar to those of the whole plant. The chemical composition of the volatile oil derived from the root and heartwood of sandalwood has been subject to numerous investigations. The chief constituent, amounting to more than 90% of the oil is santalol, a mixture of two primary sesquiterpene alcohols $C_{15}H_{24}O$, viz., α- and β-santalol, in which the α-form predominates (Sastry 1944). The structure of these sesquiterpene alcohols has been elucidated by Brunke and Tumbrink (1986). At present, our analytical investigations by TLC and GCMS of undifferentiated suspension cells cultured in the bioreactor have revealed no essential oil accumulation. However, we have noticed production of phenols in suspension cells of sandalwood over three growth periods.

Phenolic compounds were determined by the method of Swain and Hillis using gallic acid as a standard (1959). Phenolic content of the SW-1 cell line cultivated in the bioreactor was approximately 2.5 times that of the SW-2 cell line, 31.5 mg/l versus 12.5 mg/l, respectively. Phenolics began accumulating during the exponential phase and peaked during the stationary phase (Fig. 4). The bioreactor medium of the SW-1 and SW-2 cultures turned grayish brown after 14 and 6 days of growth, respectively. Phenolics from SW-1 and SW-2 cell lines were higher and appeared to accumulate faster in the bioreactor than in shake flasks (Fig. 4).

The differences in the amount of secondary metabolites produced in SW-1 and SW-2 sandalwood cultures seem to be influenced by photoperiodicity and light quality. The activities of many plant enzymes involved in the biosynthesis of flavones, flavonols, cinnamic acids, coumarins, lignins, chalcones, and anthocyanins are significantly influenced by light (Hahlbrock et al. 1980). For instance, the activities of enzymes in the flavonoid pathway in cultured cells of

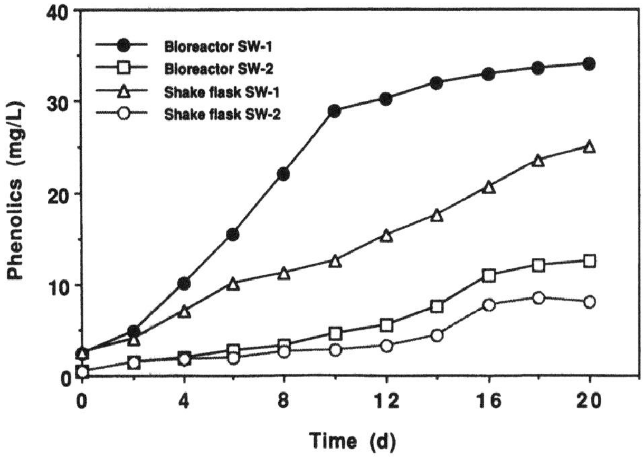


**Fig. 4.** Phenolic production of SW-1 and SW-2 sandalwood cell suspensions during the growth cycle. (Valluri et al. 1991)

parsley showed an increase when cells were exposed to light for 2 to 4 h. The effect of the light induction of phenylalanine ammonia lyase (PAL) and other flavonoid pathway enzymes in parsley cell cultures is shown by the degree to which secondary metabolites accumulate as a result of exposing cultures to continuous cool white fluorescent light. More than 20 flavone and flavonol glycosides are produced after such light treatment (Grisebach and Hahlbrock 1974). Flavone glycoside biosynthesis is most sensitive to UV light at wavelengths below 320 nm and subsequent post-UV irradiation in glycoside synthesis, as demonstrated by Wellmann (1975). Cool white light also stimulates the biosynthesis of numerous other secondary metabolites including steroidal sapogenins (e.g., diosgenin in tuber-derived callus and cell suspensions of *Dioscorea*) and steroidal alkaloids (e.g., solasodine and solamargine) in *Solanum* (Seibert and Kadkade 1980).

## 3 Summary and Conclusions

Sandalwood cell suspensions were developed and successfully cultured in a modified air-lift bioreactor. Accumulation of volatile oils was not detected; however, secondary metabolites in the form of phenolics were produced in the bioreactor environment. These studies have shown that the bioreactor environment, including factors such as inlet gas composition, lighting regimen, pH, dissolved oxygen, and impeller speed, plays an important role in cell growth and secondary metabolite production. Studies are underway to determine whether unorganized cultures of sandalwood contain enzymes crucial in the isoprenoid pathway. It is important to know the decisive biosynthetic enzymes of the secondary pathway to influence its expression or its regulation. The exploitation of variant cell strains and increased knowledge in regard to the biosynthesis and physiology of secondary metabolites will certainly improve the chance of using sandalwood cell cultures for future bioproduction of desired compounds.

## References

Bapat VA, Rao PS (1979) Somatic embryogenesis and plantlet formation in tissue cultures of sandalwood (*Santalum album* L.). Ann Bot **44**: 629–630

Bapat VA, Rao PS (1984) Regulatory factors for in vitro multiplication of sandalwood tree (*Santalum album* L.) I. Shoot bud regeneration and somatic embryogenesis in hypocotyl cultures. Proc Indian Acad Sci 93: 19–27

Bapat VA, Rao PS (1988) Sandalwood plantlets from synthetic seeds. Plant Cell Rep 7: 434–436

Bapat VA, Gill R, Rao PS (1985) Regeneration of somatic embryos and plantlets from stem callus protoplasts of sandalwood tree (*Santalum album* L.). Curr Sci 54: 978–982

Bapat VA, Fulzele DP, Heble MR, Rao PS (1990) Production of sandalwood somatic embryos in bioreactors. Curr Sci 59: 746–748

Brunke JE, Tumbrink L (1986) First total synthesis of spirosantalol, a sesquiterpene alchohol from east Indian sandalwood oil. Prog Essential Oil Res. Walter de Gruyter, Berlin, pp 321–327

Dijkstra J, Lee TS (1969) Presence of mycoplasma-like bodies in phloem of sandal affected with spike disease. Nature 224: 1121–1122

Grisebach H, Hahlbrock K (1974) Enzymology and regulation of flavonoid and lignin biosynthesis in plants and plant cell suspension cultures. In: Runeckles VC, Conn EE (eds) Metabolism and regulation of secondary plant products. Academic Press, New York, pp 21–52

Hahlbrock K, Schroder J, Vieregge J (1980) Enzyme regulation in parsley and soybean cell cultures. Adv Biochem Eng 18: 39–60

Hull G, Horne RW, Nayar RM (1969) Mycoplasma-like bodies associated with sandal spike disease. Nature 224: 1120–1122

Lakshmi Sita G (1986) Sandalwood (*Santalum album* L.). In: Bajaj YPS (ed) Biotechnology in agriculture and forestry, vol 1. Trees I. Springer, Berlin Heidelberg New York, pp 363–374

Lakshmi Sita G, Shobha Rani B (1983) Preliminary studies on isolation and culture of protoplasts from sandalwood. Experientia 45: 4–5

Lakshmi Sita G, Raghava Ram NV, Vaidyanathan CS (1979) Differentiation of embryoids and plantlets from shoot callus of sandalwood. Plant Sci Lett 15: 265–270

Lakshmi Sita G, Shobha J, Vaidyanathan CS (1980a) Regeneration of whole plants by embryogenesis from cell suspension cultures of sandalwood. Curr Sci 49: 196–198

Lakshmi Sita G, Raghava Ram NV, Vaidyanathan CS (1980b) Triploid plants from endosperm cultures of sandalwood by experimental embryogenesis. Plant Sci Lett 20: 63–69

Lakshmi Sita G, Vaidyanathan CS, Ramakrishnan T (1982) Applied aspects of tissue culture with special reference to tree improvement. Curr Sci 51: 88–92

Morris ET (1982) Romantic sandalwood – its history and use. Dragoco report. News letter from the perfume, cosmetics and toilet goods industry. DRAGOCO, Gerberding, Holzminden, pp 106–116

Muniyappa V, Vijaykumar N, Subba Rao M, Kushalappa V (1980) Studies on sandal spike disease in the forests of Karnataka State. KSTC, Bangalore, India

Rangaswamy NS, Rao PS (1963) Experimental studies on *Santalum album* L. Establishment of tissue cultures of endosperm. Phytomorphology 13: 450–451

Rao PS (1965) In vitro induction of embryo proliferation in *Santalum album* L. Phytomorphology 15: 175–179

Rao PS, Bapat VA (1978) Vegetative propagation of sandalwood plants through tissue culture. Can J Bot 56: 1153–1156

Rao PS, Ozias-Akins P (1985) Plant regeneration through somatic embryogenesis in protoplast cultures of sandalwood (*Santalum album* L.). Protoplasma 124: 80–86

Rao PS, Raghava Ram NV (1983) Propagation of sandalwood (*Santalum album* L.) using tissue and organ culture technique. In: Sen SK, Giles KL (eds) Plant cell culture in crop improvement. Plenum Press, New York, pp 119–124

Rao PS, Rangaswamy NS (1971) Morphogenetic studies in tissue culture of the parasite *Santalum album* L. Biol Plant 13: 200–206

Rao PS, Srimathi RS, Nag KC, Raychaudhari SP, Ghosh SK, Chenulu VV, Verma A (1975) Response of spike disease of sandal to mixed treatment with antibiotics and fungicides. Indian Natl Sci Acad 41: 340–342

Rao PS, Bapat VA, Mhatre M (1984) Regulatory factors for in vitro multiplication of sandalwood tree (*Santalum album* L.) II. Plant regeneration in nodal and internodal explants and occurrence of somaclonal variations in tissue culture raised plants. Proc Indian Natl Sci Acad 50: 196–202

Raychaudhari SP, Chenulu VV, Ghosh SK, Verma A, Rao PS, Srimathi RS, Nag KC (1972) Chemical control of spike disease of sandal. Curr Sci 44: 72–73

Sastry SG (1944) Oil of sandalwood east Indian. J Sci Ind Res 3: 173–187

Seibert M, Kadkade PG (1980) Environmental factors. A. Light. In: Staba JE (ed) Plant tissue culture as a source of biochemicals. CRC Press, Boca Raton, pp 123–142

Smart NJ, Fowler MW (1981) Effect of aeration on large scale culture of plants. Biotechnol Lett 3: 171–176

Srimathi RA, Sreenivasaya M (1963) Studies on the culture of root tips, haustoria and lateral buds of *Santalum album* L. In: Maheswari P, Rangaswamy NS (eds) Plant tissue organ culture. Delhi University Press, Delhi, pp 55–57

Sunderesh I (1979) Export potential of essential oils in the context of world trade. Seminar on Essential Oils, Bangalore, India

Swain T, Hillis WE (1959) The phenolic constituents of *Prunus domestica* I. The quantitative analysis of phenolic constituents. J Sci Food Agric 10: 63–68

Treat WJ (1988) Characterization of heterotrophic and photomixotrophic growth of callus and suspension cultures of *Pinus elliotti* and *Glycine max*. PhD Thesis, Texas A&M University, College Station, Texas

Valluri JV, Treat WJ, Soltes EJ (1991) Bioreactor culture of heterotrophic sandalwood (*Santalum album* L.) cell suspensions utilizing a cell-lift impeller. Plant Cell Rep 10: 366–370

Varma A, Chenulu VV, Raychaudri SP, Prakash N, Rao PS (1969) Mycoplasma-like bodies in tissue infected with sandal spike and brinjal little leaf. Indian Pathol 27: 289–291

Venkatesan KR (1981) A fresh look at the management of sandal. Proc 2nd All-India Sem, Salem, India

Venkatesh CS, Kedarnath S (1964) Breeding sandal for resistance to the spike disease. Indian For Bull 14: 243–250

Vogelmann H (1981) Aspects on scale-up and mass cultivation of plant tissue culture. In: Moo-Young M(ed) Advances in biotechnology, Pergamon Press, New York, pp 117–122

Wellmann E (1975) UV dose-dependent induction of enzyme related to flavonoid biosynthesis in cell suspension cultures of parsley. FEBS Lett 51: 105–107

# XXV *Santolina chamaecyparissus* L. (Cotton Lavender): In Vitro Culture and the Production of Lower Terpenoids

D. V. BANTHORPE[1]

## 1 General Account

### 1.1 Distribution, Secondary Metabolites and Importance

*Santolina* (L.) is a small genus of the large Asteraceae family of the Compositae, and is native to the lands of the western Mediterranean basin. Six main taxa and several subspecies and cultivars have been distinguished (Tutin et al. 1976; Valdes-Bermejo and Antonez 1981), but the most common, the most studied, and the only species cultivated as a herb outside its habitat is *Santolina chamaecyparrisus* L. (Syn: *S. incana* Lam.; cotton lavender). This plant is a hardy perennial dwarf herb (ca. 50 cm) – as indeed are all the members of the genus – found on wasteland and especially calcareous ground, and is highly resistant to drought. The narrow silvery green leaves (3 mm wide) are deeply lobed and the numerous branches of the stems each carry small yellow flower heads (Fig. 1). The whole plant is intensely aromatic and the epidermis of the leaves and tracts is covered with hairs and voluminous oil glands which show up well by electron microscopy (Derbessy et al. 1989). The blooming period in the UK is June to August.

*S. chamaecyparissus* is widely cultivated on a small scale as a medicinal herb throughout Europe and in adjacent parts of Asia and Africa. The infusion, essence, powder, and oil from the foliage, and to a lesser extent from the flower heads, are all widely used in folk and herbal medicine on account of their marked anthelmintic, antispasmodic, and emmenagogic properties, as well as being claimed to be efficacious for the treatment of eye infections, in stimulating the proliferation of scar tissue, and in relief of insect bites. The dried flower heads are used as an insect repellant, e.g., in bed linen. The essential oil also has found limited use in perfumery and as a flavoring for drinks (Grieve 1931; Guenther 1947; Stuart 1981; Chiej 1982). Despite such desirable properties, the plant has apparently never been cultivated on a large scale, although recent attempts to found a commercial venture in southern France have been appraised (Derbesy et al. 1989). One source of commercial oil (Oil of Santolina) is available (ex. Raphel-Carbonel, Vallauris, Alpes-Maritimes, France).

[1] Department of Chemistry, University College London, 20 Gordon Street, London WC1H 0AJ, United Kingdom

Biotechnology in Agriculture and Forestry, Vol. 28
Medicinal and Aromatic Plants VII (ed. by Y.P.S. Bajaj)

**Fig. 1.** *Santolina chamaecyparissus* L. (Courtesy of C.N. Banthorpe, University of Bristol, 1992)

These diverse medicinal claims, if valid, must result from the presence of pharmacologically-active secondary metabolites acting alone or in combination. Unfortunately, no screening or medicinal evaluation of either crude extracts or individual components has been reported. Nevertheless, detailed phytochemical analyses of *S. chamaecyparissus* have recently been made, although with very confusing results.

Most analyses were of foliage oils (yields 0.1 to 2% based on dry wt.; very variable) although a few studies have indicated that oil from the flowerheads is similar. Initial analyses using material from pooled specimens of unspecified parentage from both wild-type and cultivated (botanical garden) populations indicated that the steam-volatile oil of samples from the UK (Crombie et al. 1972; Allen et al. 1977); USA (Zaltov et al. 1964); Switzerland (Thomas and Willhalm 1964); Egypt (Aboutabl et al. 1987); and France (Derbessy et al. 1989) were broadly similar. The main components were monoterpenoids (up to 96% wt./wt.), with the balance being predominantly sesquiterpenoids. The interesting discovery was that the main component (45–95%) was artemisia ketone (**1**; Fig. 2), an irregular monoterpenoid previously found in a few *Artemisia* species that abrogates the Biogenetic Isoprene Rule in *not* being derived from geranyl pyrophosphate as parent, as are virtually all other mono- and higher terpenoids. This ketone was accompanied by lesser amounts of other irregular monoterpenoids: artemisia alcohol (**2**), yomogi alcohol (**3**), and santolina triene (**4**). Thus, *S. chamaecyparissus* is a member of a small group (ca. 20 members) of Compositae species that possess the ability to bypass the conventional route to monoterpenoids via geranyl pyrophosphate, and rather have available the enzymatic machinery to couple two $C_5$ moieties (3,3-dimethylallyl pyrophosphate) in a usually forbidden manner to generate the irregular skeletons shown. A recent very detailed study illustrates the situation so far described: GC-MS analyses of the oil from plants of French provenance revealed 146 components, of which 70 were identified. Of these, artemisia ketone (45%) and the regular monoterpenoid myrcene (5; 15%) were the main components. Headspace analyses of the exudate of the flower head indicated the same two compounds to predominate, but in reversed proportions (27; 42%). The balance of the oil was a nondescript mixture of monoterpenoids possessing the pinane, thujane, camphane, and menthane skeletons as well as acyclic compounds (Vernin 1991). These latter components were typical of oils of numerous herbs that have been investigated. Patterns of products similar to those observed in summer were found to exhibit

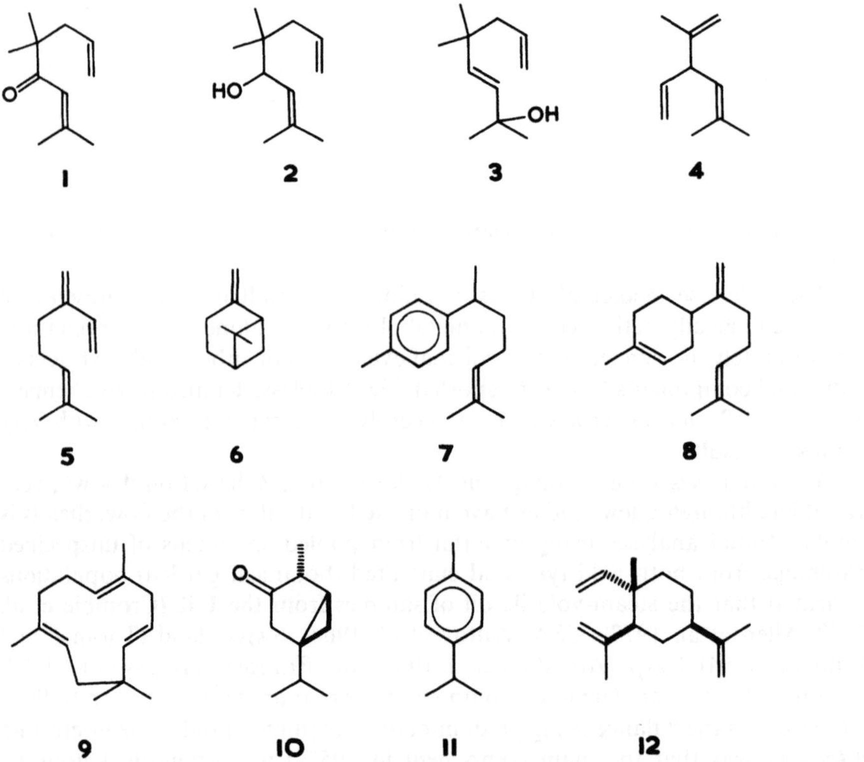


**Fig. 2.** Major mono- and sesqui-terpenoids isolated from *S. chamaecyparissus* L. (wild type and cultivars). **1** Artemisia ketone; **2** artemisia alcohol; **3** yomogi alcohol; **4** santolinatriene; **5** myrcene; **6** β-pinene; **7** ar-curcumene; **8** β-bisabolene; **9** α-humulene; **10** isothujone; **11** p-cymene; **12** β-elemene; **13** γ-cadinene; **14** α-cubebene; **15** α-copaene; **16** δ-cadinene; **17** α-ylangene; **18** spathulenol; **19** vulgarone B; **20** T-cadinol; **21** allo-aromadendrene; **22** α-muurolene; **23** elemol; **24** cubenol; **25** ledol; **26** β-phallandrene. N.B. Relative configurations are shown: the absolute stereochemistries of the naturally occurring compounds have rarely been reported (in many examples of monoterpenoids and sesquiterpenoids, both enantiomers of a compounds are known to occur in the same or in different plant species)

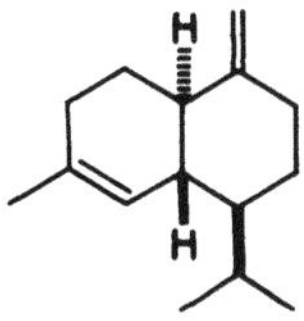
13

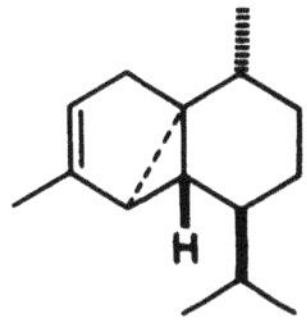
14

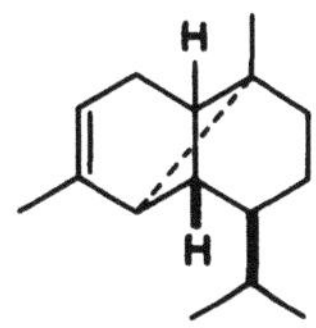
15

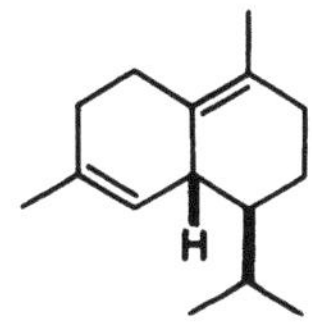
16

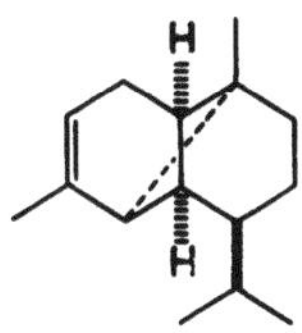
17

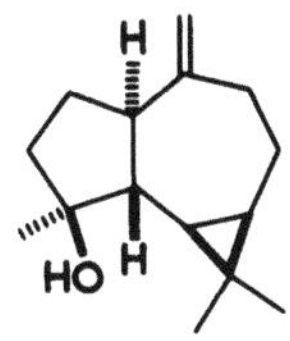
18

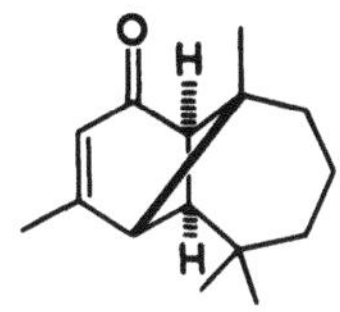
19

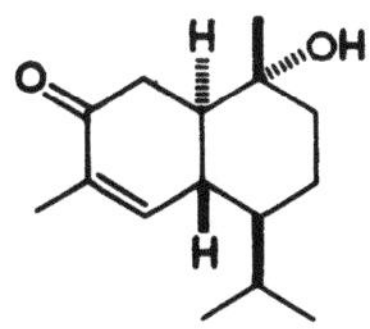
20

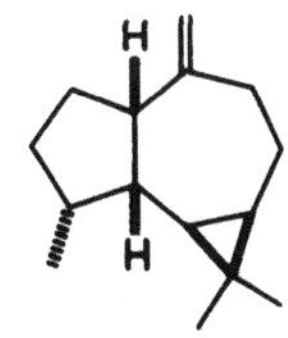
21

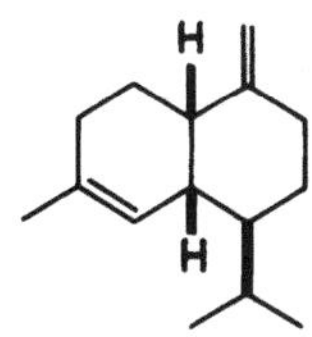
22

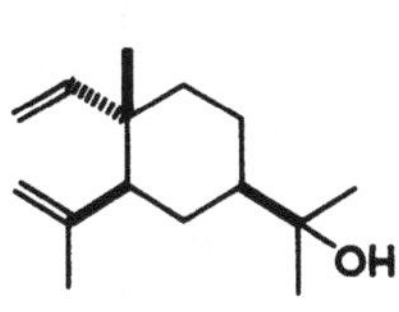
23

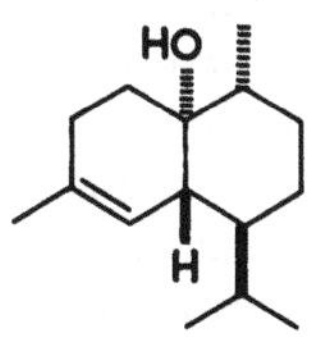
24

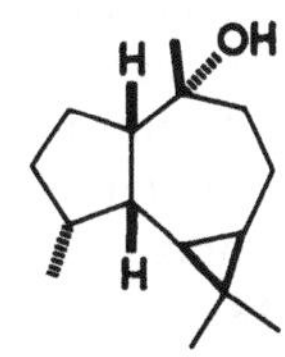
25

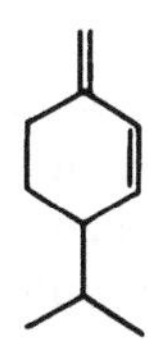
26

**Fig. 2.** (*Contd.*)

little seasonal variation or to depend on whether specimens were field-grown or maintained in the greenhouse (Allen et al. 1977).

In contrast, a line of nominally *S. chamaecyparissus* L. from a botanic garden in the UK yielded an oil comprising only 17% artemisia ketone but with a plethora of monoterpenoids including β-pinene (**6**) and the sesquiterpenoids ar-curcumene (**7**), β-bisabolene (**8**), and α-humulene (**9**) (Banthorpe et al. 1977); whereas samples from India contained mainly isothujone (**10**) and p-cymene (**11**) (Misra et al. 1981), and some from the former Soviet Central Asia contained predominately the sesquiterpenoids β-elemene (**12**), γ-cadinene (**13**), and β-bisabolene (**8**) (Ushakov et al. 1980). Subsequently, it was discovered that oils from the foliage of the botanically defined cultivars *S. chamaecyparissus* (L.) cv. *chamacyparissus* and cv. *neopolitana* grown in the UK differed completely from the (presumed) wild type in containing mainly sesquiterpenoids over the entire growing and winter seasons (the latter in the greenhouse; Baig et al. 1989b). Now the sesquiterpenoids comprised 52 to 65% of the oil (ca. 0.3% wt./wet wt.) and several well-known compounds were present: e.g., ar-curcumen (**7**), β-bisabolene (**8**), α-humulene (**9**), α-cubebene (**14**), γ-cadinene (**13**), α-copaene (**15**), δ-cadinene (**16**), and α-ylangene (**17**), together with lesser amounts of artemisia ketone and related irregular monoterpenoids. Of outstanding interest was the occurrence in both these cultivars of spathulenol (**18**, 21%) which is fairly uncommon and has been shown to display pronounced insecticidal properties (Hubert and Wiemer 1985), and the real rarity, vulgarone B (**19**, 22%). The latter has only once been reported (in *Tanacetum vulgare* L.; Uchio 1978) and was now shown to be a potent repeller of slugs. Dilute solutions (10 ppm aqueous) protected *Brassica* seedlings very effectively from predation by *Milax* and *Arion* species. Such effects were not shown by the sesquiterpenoids present in the oil from wild-type plants (Baig et al. 1989b).

Other recent studies have also revealed striking differences in oils obtained from different locations. Thus, wild-type plants from Spain yielded artemisia ketone together with T-cadinol (**20**) as the main components; whereas the designated "subspecies" *squarrosa*, *magnonica*, *incana*, and *tomentosa* yielded different patterns of either predominantly monoterpenoids (60% total) or sesquiterpenoids (30–61%). Amongst the main sesquiterpenoids were allo-aromadendrene (**21**), α-muurolene (**22**), elemol (**23**), cubenol (**24**), spathulenol (**18**), and ledol (**25**) (Villar et al. 1986; Perez-Alono and Velasco-Negueruela 1992). Elemol, here and elsewhere, may be an artifact of isolation (Banthorpe 1992).

Even with allowance for differences in cultivation and in geographical location, the situation as regards the terpenoid metabolites of *S. chamaecyparissus* is confused. The botanical status of the claimed "subspecies" is difficult to assess, and the terpenoid pattern may reflect the existence of chemical taxa, i.e., plant lines that are indistinguishable on morphological grounds but show different genetically determined patterns of secondary metabolism. Some 26 such chemotypes that breed true have been demonstrated for *Tanacetum vulgare* L. and less prolific situations occur with other species (Tetenyi 1970; Tetenyi et al. 1975). The phenomenon may be more widespread for herbaceous plants than has been suspected. Such taxa can only be distinguished by assay of individual plants or clones: use of pooled plant material obscures the situation.

The essential oils of several other *Santolina* species (none of which is widely distributed outside the Iberian peninsula) have also been recorded. Some species produce mainly monoterpenoids and others mainly sesquiterpenoids of the types that occur in the various cultivars of *S. chamaecyparissus* (Ushakov et al. 1976; Frazao et al. 1977; Perez-Alonso and Velasco-Negueruela 1988). *S. rosmarinifolia* L. and *S. oblongifolia* Lag. also produce acetylene derivatives and dihydrofurans derived from cyclization of nerolidol-5,8-oxides respectively (De Pascual et al. 1981, 1983).

### 1.2 Commercial Applications

No pharmacological screening has been carried out to check or validate the claims for medicinal significance for extracts from *S. chamaecyparissus.* In the absence of such data, the most obvious exploitable field is in horticulture/agriculture. Firstly, as a companion plant to deter insects, and secondly as a source of extracts or of vulgarone B for slug control. However, much more detailed studies are necessary to determine scope and limitations of the latter applications.

### 1.3 Extraction and Purification of Components

The volatile oil of *S. chamaecyparissus* is typically obtained by steam distillation or by extraction with solvents (Baig et al. 1989b; Derbessy et al. 1989). In the latter process, foliage (100 g) was frozen in liquid nitrogen, pulverized, and the cold powder extracted with n-$C_6H_{14}$ (500 $cm^3$) followed by $Et_2O$ (500 $cm^3$). The combined extracts were dried ($MgSO_4$) and reduced by rotary evaporation to ca. 10 $cm^3$. Losses during this procedure were low ( $< 5\%$). The components were separated by GC on Carbowax 20 M or Silicone SE-30 (20%; 3 m × 8 mm) with $N_2$ carrier (3.6 l/h) at 60 °C (5 min) followed by programming (2 °C/min) to 160 °C and then (6 °C/min) to 190 °C. Identifications were by Kovacs' indices, relative retention times and peak enhancement, and especially by GC-MS. These latter analytical studies utilized OV-101; BPI-methylsilicone, or Carbowax 20 M WCOT or SCOT capillary glass columns (12 to 50 m × 0.32 mm) run over 45–200 °C at 2 °C/min with He as carrier (0.6 l/h). Liquid chromatography on $Al_2O_3$ columns has also been used for separations, with eluants in the sequence: n-$C_6H_{14}$; n-$C_6H_{14}$: $C_6H_6$ (4:1); $C_6H_6$; $C_6H_6$-EtOAc (4:1); EtOAc; EtOAc-$CHCl_3$ (4:1). Fractions were assayed for purity by capillary GC and characterized by GC-MS and FT-IR and NMR spectroscopy.

## 2 Biotechnological Studies

Three sets of studies have relevance for possible biotechnological exploitation of *S. chamaecyparissus.*

### 2.1 Establishment of Cell-Free Extracts of Foliage that Sustain Biosynthesis of Terpenoids

A procedure using foliage of the herb was developed (Sect. 4.1) that produced cell-free extracts capable of the biosynthesis of irregular monoterpenoids (Banthorpe et al. 1977). The wild-type taxon used (see above) accumulated mainly artemisia ketone (**1** ca. 90%) in the extractable oil, together with the regular monoterpenoids myrcene (**5**) and β-phellandrene (**26**). Under optimized conditions the extracts converted the putative $C_5$-precursors isopentenyl, 3,3-dimethylallyl, and dimethylvinylcarbinyl pyrophosphates into mainly artemisia ketone (**1**) and also artemisia alcohol (**2**), as well as *trans-chrysanthemyl* alcohol (**27**) and lavandulol (**28**) in a combined yield of ca. 12% (see Fig. 3). The latter two irregular monoterpenoids have not been previously detected in the oils from any *Santolina* species, but **27** is believed to be closely related to the class of irregular monoterpenoids under consideration and **28**, which is common in lavender oils, is a scission product of **27** (Banthorpe et al. 1977). These cell-free extracts contained low levels of protein (ca. 0.1–0.5 mg/l), and attempts to fractionate the enzymic activities present in the extract led to loss of activity. This finding led to the exploration of cell cultures as a possible source of biomass for studies of enzyme extraction and purification.

### 2.2 Establishment of Cell Cultures

Only one study of tissue culture of *Santolina* species has been recorded (Baig et al. 1989a). Initial work used the "wild-type" line (see above) that produced mainly artemisia ketone. Explants from apical meristem, stem, leaf, petal, and root were set up on an initiation medium comprising Murashige and Skoogs'

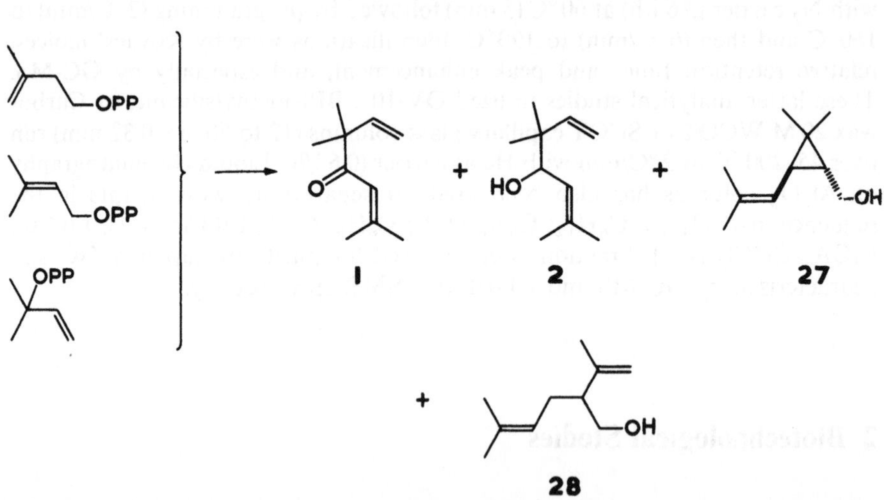


**Fig. 3.** Nonisoprenoid monoterpenoids produced by cell-free extracts of *S. chamacyparissus* using hemiterpenyl pyrophosphates ($C_5$-compounds) as substrates. *OPP* Pyrophosphate group; **27** chrysanthemyl alcohol; **28** lavandulol

standard formulation supplemented with phytohormones (Sect. 4.2). After 4 to 7 months, some 20 to 30% of explants had formed callus, and these were transferred to a maintenance medium which permitted more rapid growth with a subculture period of 23–25 days. After a settling-down period of four to six subcultures, the resulting callus appeared morphologically and biochemically stable for up to 2 years (when experiments were discontinued). This initial instability is a common phenomenon and presumably reflects the period when a more or less stable population is set up from the initial competing sublines present in the initial callus explant. The callus that emerged from this habituation process (under the chosen conditions) was white, friable, and appeared mainly composed of parenchyma-like cells with some tracheid formation and secondary thickening. Extraction and assay of callus over the period eight to ten cell cycles after transfer to maintenance medium revealed that only the callus derived from the apical meristem showed any ability to synthesize terpenoids de novo. A similar situation has been found for *Rosa* species: here only callus derived from apical meristem possessed the ability to synthesize and accumulate 2-phenylethanol (Banthorpe et al. 1988). Obviously, the source material for a cell culture can have a decisive influence on the secondary metabolism that is exhibited by the resulting cell line, presumably owing to the physiological and hormonal condition of the explant. This facet of production of secondary metabolites in tissue culture has been largely ignored by the botanists.

More recent studies (Banthorpe et al. unpubl. work) have utilized the cultivar Neopolitana, which produces sesquiterpenoids as the main component of its essential oil. Initiation of callus was carried out as above, but the maintenance medium and culture conditions were altered (Sect. 4.3). The success of initiation, the morphological properties, and the culture period of the callus line were similar to that previously described. Again, the lines derived from apical meristem were unique in accumulating terpenoids. (A distinction between synthesis and accumulation must be drawn: thus the observed accumulation is the balance between overall synthesis and degradation in vivo). The optimized conditions are outlined in Section 4.

Coarse cell suspensions (mainly clumps, ca. 2 mm in diameter) and fine suspensions (aggregates of up to 200 cells) were generated from the above lines of callus by conventional methods and were established on the appropriate MS medium plus supplements, but without addition of agar. Such suspensions were maintained on a gyrorotary shaker at 120 rpm and proliferated. Subcultures were carried out (10-day cycles) when the cell density was ca. $10^6$ cells/$cm^3$ and the culture was entering the stationary phase. Such lines were stable but produced no detectable (i.e., $< 10^{-6}$% wt./wt.) level of secondary metabolites at any stage of the cell cycle. Presumably all nutrients were now directed towards cell division and proliferation rather than being expended in secondary functions.

## 2.3 Products Obtained from Culture Lines

In the following, WT-callus refers to the line established from specimens of *S chamaecyparissus* L. of the wild type. This parent type produced mainly artemisia ketone accompanied by other monoterpenoids in its foliage. In contrast, N-callus refers to the line derived from *S. chamaecyparissus* L. cv. Neopolitana. Now the parent stock accumulates mainly sesquiterpenoids (cf. Sect. 1.1). Both callus lines were derived from explants of apical meristem. Clones of each line were bulked, extracted, and assayed by GC-MS at periods after 8 to 15 cell cycles following installation on the maintenance medium. Over this period, the lines were stable both morphologically and as to their secondary metabolism.

Six of the 11 lines of WT-callus yielded 0.01 to 0.03% (wt./wet wt.) of an oil which comprised only monoterpenoids and which corresponded at maximum to about 10% of the level present in foliage. These accumulations must represent de novo synthesis, as carry-over from the initial explant can be calculated to be quite negligible ($< 10^{-6}$% wt./wt.) after the given number of subcultures, and additionally as the composition of oils from callus and foliage were widely different. Indeed, the accumulations in callus are remarkably high (cf. Banthorpe et al. 1986), as levels in a largely undifferentiated culture are compared with those in the organs of synthesis and accumulation (foliage) of the parent. All productive cultures yielded the same two products, i.e., β-phellandrene (**26**) and myrcene (**5**) in approximately equal amounts (each ca. 0.01% wt./wt. at maximum). These are the main regular monoterpenoid components of the parent plant and, interestingly, are accumulated in culture in virtually the same yield (as % wt./wt.) as in the parent plant. The main component of the parent, artemisia ketone, or its relatives, could not be detected in extracts from calli, but *trans*-chrysanthemyl alcohol (**27**) occurred in low yields (ca. $10^{-4}$%) in most extracts. The balance of the extracts from calli (ca. 0.01% wt./wet wt. at maximum) consisted of 3,3-dimethylallyl alcohol, the pyrophosphate of which is almost certainly the immediate parent of artemisia ketone. It is significant that *trans*-chrysanthemyl alcohol is considered not to be an obligatory precursor for the ketone (Banthorpe et al. 1977).

N-callus was cultured under slightly different conditions and was selected for a (moderate) level of chlorophyll formation (see Sect. 4.2). Assays from this material (six lines) at the same stage of cultural history as the WT-callus gave a completely different and more promising pattern of products. The yield of oil (0.01–0.02% wt./wet wt.) was at maximum about 4% of that obtained from the field-grown plant and mainly comprised (> 96%) four sesquiterpenoids: α-himachalene (**29** Fig. 4), β-himachalene (**30**), himachalol (**31**), and ar-himachalene (**32**) in the ratio ca. 6:5:2:1. The balance was monoterpenoids of regular structure: (e.g., α-pinene; β-pinene; limonene). These findings are noteworthy, as relatively few culture lines have been reported to accumulate sesquiterpenoids, other than farnesol, which may well be a component of the route to steroids (Banthorpe et al. 1986; Banthorpe 1988). More significantly, this work indicates that these cultures contain part of the enzymic pathway to vulgarone B (see Sect. 1.1). This is illustrated in Fig. 4. The ion (**33**) derived from 2Z, 6E-farnesyl pyrophosphate – the latter is an ubiquitous precursor of sesquiterpenoids – can be hypothetically cyclized to yield **34**, which in turn could lead to vulgarone B (**19**). However, the inability to complete this pathway and the consequent accumulation of the biogenetic equivalent of **34** could directly lead to the

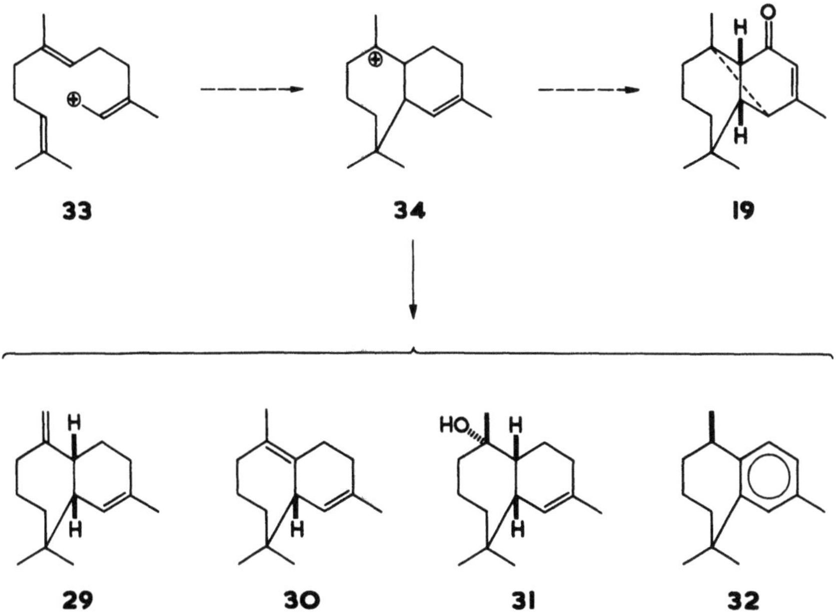


**Fig. 4.** Sesquiterpenoids accumulated by callus cultures of *S. chamaecyparissus* cv. Neopolitana. **19** Vulgarone B; **29** α-himachalene; **30** β-himachalene; **31** himachalol; **32** ar-himachalene; **33** ion derived from 2Z,6E-farnesyl pyrophosphate; **34** himachalane ion derived from cyclization of **33**

himachalane derivatives. Thus the necessary differentiation has been achieved in these culture lines to allow construction of a significant part of the pathway to the sesquipinane skeleton exemplified by **19**. These himachalane derivatives – or indeed any closely related structures – have never been reported to occur in any *Santolina* species.

## 2.4 Biotransformations by Cell Cultures

A brief study to explore the possibility of the above illustrated a very common shortcoming of the methodology of such studies (Banthorpe et al. unpubl. work). Cell suspension cultures prepared from WT-callus apparently gave low conversions of exogenous artemisia alcohol into its ketone, and of *trans*-chrysanthemyl alcohol into artemisia alcohol and lavandulol. Exogenously supplied compounds (5 mg/500 $cm^3$ medium) apparently showed 5–10% biotransformation after incubation for 24–48 h at 30 °C with a culture that had entered the stationary phase. Conventional controls suggested these were real effects, i.e., biotransformations, but the controls had neglected a crucial pH change. The initial suspension medium was made up to a nominal pH meter reading of 5.8, and as is usual in cell culture, was unbuffered. However, 9 days later, as the culture entered the stationary phase, the pH meter reading had dropped to 3.8! Use of controls of preconditioned media from which the cells had been spun off demonstrated that the apparent biotransformations could be attributed to the variation in pH. Thus, the enhanced acidity in the complex incubation (i.e., culture) medium caused both the predictable ring cleavage of *trans*-chrysanthemyl alcohol and also enhanced oxidation processes. Many purported "biotransformations" could have similar origins, as considerable alterations in pH over the cell cycle are commonly found in cell cultures (Brown and Banthorpe 1990).

## 3 Summary and Conclusions

Cell-free extracts from foliage of a wild-type line of *Santolina chamaecyparissus* have been prepared, that efficiently (up to 12% conversions) biosynthesized several irregular monoterpenoids from $C_5$-precursors. Callus cultures from the same line accumulated low yields of regular monoterpenoids (at maximum 10% of levels of total monoterpenoids in field-grown plants). In contrast, callus from the cultivar Neopolitana after ten cell cycles yielded four himachalane-type sesquiterpenoids (at max. 4% of levels of total sesquiterpenoids in field-grown plants) that are on the biosynthetic pathway to vulgarone B, the main component of the oil of the field-grown cultivar.

It seems unlikely that cell cultures could ever be developed on a commercial scale to produce economically any particular component of the oil of wild-type *S. chamaecyparissus* that proved to have medicinal value. However, any future demand for vulgarone B or spathulenol on a large scale could be deemed worthy of studies on genetic engineering. Appropriate gene transfers from appropriate *S. chamaecyparissus* cultivars to other plants or to microorganisms is feasible, and the first steps – the establishment of biomass from cultures that possess the ability for sesquiterpenoid biosynthesis – have been taken.

## 4 Protocols

### 4.1 Production of Cell-Free Extracts that Sustained Biosynthesis of Irregular Monoterpenoids (cf. Banthorpe et al. 1977)

a) Foliage (5 g) from *S. chamaecyparissus* L. (wild type) was pulverized in liquid $N_2$ and the cold powder was mixed with Polyclar AT (solid form; 2 g) and stirred into tris-$CH_3COOH$ buffer (0.1 M; 40 $cm^3$; pH 7.0) containing cysteine hydrochloride (10 mM), sodium diethyldithiocarbonate (10 mM), EDTA (Na salt,; 1 mM) and sucrose (0.25 M) at 4 °C.
b) After maintenance at 4 °C (1 h), the solution was filtered through glass wool and centrifuged (500 g; 10 min).
c) A portion of the supernatent (10 $cm^3$) was passed through Sephadex G-10 (45 × 25 cm) made up in the *tris*-$CH_3COOH$ buffer at 4 °C and eluted with water (50 $cm^3$ h). The excluded forerun (20 to 30 $cm^3$) contained the monoterpenoid synthetase activity. This fraction typically had protein concentration 0.06–0.13 mg $cm^3$.

d) An aliquot of this active fraction (1 $cm^3$) in sodium maleate-HCI buffer (0.1 M; 2 $cm^3$; pH 6.5) containing ATP, $NADP^+$, $NAD^+$, $MgCl_2$, $MnCl_2$, and dithiothreitol (5 mM each) and the $C_5$-pyrophosphate used as substrate (0.5 mg; $1 \times 10^6$ pm) was incubated at 30°C (2 h) onto the plateau region. After this period the reaction medium on extracted and assayed by GC-MS and with TLC-radiochemical ($2\pi$) scanning.

## 4.2 Production of Callus Cultures that Synthesized and Accumulated Monoterpenoids (Baig et al. 1989a)

a) Explants ($0.5 \times 0.2 \times 0.2$ cm) of apical meristem from young plants (1 year old) of *S. chamaecyparissus* L. (wild type) were surface sterilized by conventional methods and were cultured on MS standard medium supplemented with 2,4-D (2 mg/l), kinetin (0.3 g/l), and sucrose (30 g/l) that was made up in agar (Oxoid No. 2; 0.9% w/v) and taken initially to pH 5.5. The explants were maintained at constant illumination (ca. 600 lx; Thorn Growlux lamps; $\lambda_{max}$ 660 nm) and some 20 to 30% had formed sufficient callus for subculture after 4 to 7 months.
b) Subcultures were made at passage times of 23 to 25 days for 8 to 10 cycles before assays were made. The maintenance medium during this period of accumulation of biomass was as above except that the 2,4-D was replaced by NAA (2 mg/l). Subculture was made at the period of entry to the stationary phase when the mass of the original explant had been increased ca. 10- to 15-fold. Oil production was optimum with a light intensity of ca. 2000 lx and a photoperiod of 18 h at 28 °C followed by 6 h at 23 °C.

## 4.3 Production of Callus Cultures that Synthesized and Accumulated Sesquiterpenoids (Banthorpe et al. 1992)

These cultures were established from apical meristem of *S. chamaecyparrisus* L. cv. *neopolitana* by methods similar to the above. The differences were:
a) The MS medium was supplemented with BAP (0.5 mg/l) NAA (2 mg/l), gibberellin $A_3$ ($GA_3$; 10 mg/l), and sucrose (30 g/l). This medium had previously successfully been employed in inducing sesquiterpenoid accumulation in other species (Banthorpe et al. 1989). This particular medium was used both for initiation and for maintenance.
b) The culture conditions at the maintenance stage were: continuous illumination (ca. 600 lx; Philips warm white tubes, $\lambda_{max}$ 580 nm); 30 °C.
c) After the third subculture, selection was made to exercise green volumes of the callus. After a further seven to ten cycles, the resulting callus material was almost uniformly pigmented and the chlorophyll and (carotene + xanthophyll) concentrations were ca. 10 and 2% respectively of that in adult parent foliage. Increasing selection and culture did not increase the greening above this value.

*Acknowledgments.* The author is very grateful to Drs. M.A. Baig, G.P. Edwards, and Stephanie A. Branch, who carried out described experimental work, and to Dr. P. Yeo (Cambridge Botanical Garden) and the curators of the Chelsea Physic Garden, London, for providing the specimens of *S. chamaecyparissus*.

## References


Aboutabl EAA, Hammerschmidt F, Elazzoug AA (1987) Essential oil of *Santolina chamaecyparissus*. Sci Pharm 55:267–271 (Chem Abstr 108:226719)

Allen KG, Banthorpe DV, Charlwood BV, Voller CM (1977) Biosynthesis of artemisia ketone in higher plants. Phytochemistry 16:79–83

Baig MA, Banthorpe DV, Branch SA (1989a) Hemi- and mono-terpenoids from callus of *Santolina chamaecyparissus*. Fitoterapia 40:184–186

Baig MA, Banthorpe DV, Gutowski JA (1989b) Accumulation in cultivers of *Santolina chamaecyparissus* of a rare sesquiterpenoid with gastropod-repellant activity. Fitoterapia 40:373–375

Banthorpe DV (1988) Monoterpenes and sesquiterpenes In: Constabel F, Vasil IK (eds) Cell cultures and somatic cell genetics of plants, vol 5. Academic Press, San Diego, pp 143–159

Banthorpe DV (1992) classification of monoterpenoids and general procedures for their characterisation. In: Charlwood BV, Banthorpe DV (eds) Methods in plant biochemistry, vol 7. Academic press, London, pp 1–41

Banthorpe DV, Doonan SD, Gutowski JA (1977) Biosynthesis of irregular monoterpenes in extracts of higher plants. Phytochemistry 16:85–92

Banthorpe DV, Branch SA, Njar VCO, Osborne MG, Watson DG (1986) Ability of plant callus cultures to accumulate lower terpenoids. Phytochemistry 25:629–636

Banthorpe DV, Branch SA, Poots I, Fordham WD (1988) Accumulation of 2-phenylethanol by callus derived from leaf-bud of *Rosa damascena*. Phytochemistry 27:795–800

Banthorpe DV, Brooks CJW, Brown JT, Lappin GJ, Morris GS (1989) Synthesis and accumulation of polygodial by tissue cultures of *Polygonum hydropiper*. Phytochemistry 28:1631–1633

Brown GD, Banthorpe DV (1990) Growth and secondary metabolism of cell cultures in buffered media. Plant Sci 67:107–113

Chiej R (1982) Encyclopaedia of medicinal plants. Macdonald, London, p 278

Crombie L, Firth PA, Houghton RP, Whiting DA, Woods DK (1972) Cyclopropane-ring cleavage to yield artemisyl, santolinyl and lavandulyl compounds: model systems and biosynthetic studies. JCS Perkin I:642–646

De Pascual TJ, Gonzalez MS, De Dios MA, San Segundo JM, Vicente S, Bellido IS (1981) Essential oil of *Santolina rosmarinifolia*. Riv Ital EPPOS 63:355–356 (Chem Abstr 96:109953)

De Pascual TJ, Vicente S, Gonzalez MS, Bellido IS (1983) Nerolidol 5,8-oxides from essential oil of *Santolina oblongifolia*. Phytochemistry 22:2235–2238

Derbessy M, Touche J, Zola A (1989) Essential oil of *Santolina chamaecyparissus*. J Essent Oil Res I:269–275

Frazao S, Dominques A, Sousa B (1977) Studies on some oils of aromatic plants of the Portuguese Flora: *Santolina rosmarinifolia*. 7th Int Congr Essent Oils 7:169–171 (Chem Abstr 92:64521)

Grieve M (1931) A modern harbal. Jonathan Cape, London pp 451–453

Guenther E (1947) The essential oils, vol 6. Van Nostrand, New York, pp 221–223

Hubert TD, Wiemer DF (1985) Anti-repellant terpenoids from *Melampodium divaricatum*. Phytochemistry 24:1197–98

Misra LN, Siddiqui MS, Srivastrava SK, Dimri BP (1981) GC examination of an essential oil of *Santolina chamaecyparissus*. Perfume Flavor 6:25–27 (Chem Abstr 95:86130)

Perez-Alonso MJ, Velasco-Negueruela A (1988) Essential oils of four *Santolina* species. Flav Fragr J 3:37–42

Perez-Alonso MJ, Velasco-Negueruela A (1992) Essential oil components of *Santolina chamaecyparissus*. Flav Fragr J 7:87–91

Stuart M (1981) Encyclopaedia of herbs and herbalism. Orbis London pp 259–260

Tetenyi P (1970) Infraspecific chemical taxa of medicinal plants. Akad Kiado, Budapest

Tetenyi P, Kaposi P, Hethelyi E (1975) Variations in essential oil of *Tanacetum vulgare*. Phytochemistry 14:1539–1544

Thomas AF, Willhalm B (1964) Oil of *Santolina chamaecyparissus*. A new nonisoprenoid monoterpene hydrocarbon. Tetrahedron Lett:3775–3778

Tutin TG, Heywood VH, Burgess NA, Moore DM, Valentine DH, Walker SM, Webb DA (eds) (1976) Flora Europeaea, vol 4. Cambridge University Press, Cambridge, pp 144–145

Uchio Y (1978) Vulgarones A and B from *Tanacetum vulgare*. Tetrahedron 34:2893–2896
Ushakov VA, Muraveva D, Bakina LA (1976) Monoterpenoid compounds from essential oil of plants of the genus *Santolina*. Khim Prir Soedin:664–666
Ushakov VA, Muraveva DA, Bakina LA (1980) Sesquiterpenoids of the essential oil of the genus *Santolina*. Khim Prir Soedin:257–259
Valdes-Bermejo E, Antonez C (1981) *Santolina* species of the Iberian penisula. An Jard Bot Madr 38:127–140
Vernin G (1991) Volatile components of the oil of *Santolina chamaecyparissus*. J Essent Oil Res 3:49–53
Villar A, Giner RM, Rios JL (1986) Chemical composition of *Santolina chamaecyparissus* subspecies *squarrosa* essential oil. J Nat Prod 49:1143–1144
Zaltov LH, Brannon DR, Wecke JW (1964) On the occurrence of santolineniones. J Org Chem 29:2786

# XXVI *Sapium sebiferum* (Chinese Tallow): In Vitro Culture and the Production of Tannins and Other Phenolic Compounds

K. Ishimaru, H. Arakawa, and S. Neera[1]

## 1 Introduction

*Sapium sebiferum* (Euphorbiaceae), known as tallow tree (Fig. 1), originated in China and grows mainly in subtropical regions of the world. The tree usually grows up to ca. 15 m in height and is commonly used as a source of oil, which can be obtained from the seeds. The wax is extracted from the seeds by steaming without crushing, and it resembles cacao butter and is edible (Venketeswaran

**Fig. 1.** *Sapium sebiferum* grown at Saga University. (Photo by K. Ishimaru, May, 1991)

[1] Department of Applied Biological Sciences, Faculty of Agriculture, Saga University, 1 Honjo, Saga 840, Japan

Biotechnology in Agriculture and Forestry, Vol. 28
Medicinal and Aromatic Plants VII (ed. by Y.P.S. Bajaj)

1989). It is also used in the manufacture of candles, cosmetics, and soaps. The seed meal, after extraction of the oil, has a high content of protein and can be a valuable feed and fertilizer. This tree is also used as an indoor ornamental tree because the leaves turn beautiful autumn colors such as yellows, reds, and purples. The wood is also useful for biomass for making furniture, wood instruments, and similar products. Several approaches for mass (and rapid) propagation of this important biomass plant by tissue culture techniques have been undertaken (Venketeswaran and Gandhi 1982; Gandhi 1983; Kotwal et al. 1983; Mridula et al. 1983; Venketeswaran et al. 1983a, b, 1984). For details see Venketeswaran (1989).

Like other euphorbiaceous plants, it contains a high content of phenolic compounds, especially tannins (Nishioka 1983) in the leaves, barks, and stems as its main secondary metabolites. Recently, tannins have been reported to have an important medicinal anti-AIDS activity (Nishizawa et al. 1989; Nonaka et al. 1990) in addition to the well-known uses such as antidiarrhea, antivirus (Corthout et al. 1991), psychotropic (Ueki et al. 1985), and for improvement of nitrogen metabolism (Nagasawa et al. 1978, 1980; Shibutani et al. 1980, 1983).

In spite of some experiments on tissue cultures of this plant, there has been no report on the production of secondary metabolites in vitro. In this chapter, research on tissue cultures (callus and cell suspension cultures) of this plant used for production of its important secondary metabolites (phenolic constituents) is described (Neera and Ishimaru 1992; Neera et al. 1992).

## 2 In Vitro Culture Studies

Tissue culture techniques have new and valuable advantages for mass propagation as well as for the genetic improvement of forest trees (Bajaj 1986). Especially in this decade, tissue culture studies (callus, root, adventitious shoot, liquid suspension, and protoplast cultures) have been done for the purpose of rapid propagation of *S. sebiferum* (Chinese tallow) (see Venketeswaran 1989).

### 2.1 Establishment of Callus and Cell Suspension Cultures

For induction of callus, leaf segments of *S. sebiferum* collected in June (from young leaves) were surface sterilized (with 2% NaOCl) and placed aseptically on six types of MS solid media (Murashige and Skoog 1962) (containing 30 g/l sucrose, solidified with 2.5 g/l Gelrite) supplemented with various combinations of 2,4-dichlorophenoxyacetic acid (2,4-D), indole-3-acetic acid (IAA), and benzyladenine (BA) (Table 1). Of these six media, the medium with 1 mg/l 2,4-D and 0.1 mg/l BA proved to be the best for callus induction. After 2 months' culture, calli derived from leaf segments cultured on this medium were transferred to 19 types of MS media (10 ml medium/tube), A-I (combinations of 2,4-D and BA), J-R (combinations of BA and kinetin), and S [1 mg/l naphthaleneacetic acid

**Table 1.** Effects of growth regulators on callus formation on leaf segments of *Sapium sebiferum* cultured on MS solid media for 8 weeks. (Neera and Ishimaru 1992)

| 2,4-D (mg/l) | IAA (mg/l) | BA (mg/l) | Callus fr. wt.[a] (mg) |
|---|---|---|---|
| 0.1 | 0 | 0.1 | 448 |
| 1 | 0 | 0.1 | 954 |
| 0.1 | 0 | 0 | 101 |
| 1 | 0 | 0 | 44 |
| 0 | 3 | 0.1 | 471 |
| 0 | 3 | 0 | 0 |

[a] Average of five samples.

(NAA) – 0.1 mg/l BA] and cultured in the light (16 h photoperiod/day) or dark. The callus was also inoculated into MS liquid medium (50 ml medium per 100 ml flask) supplemented with 1 mg/l 2,4-D and 0.1 mg/l BA (medium T) and cultured in the dark to establish cell suspension culture.

The number of calli cultured on 18 media (A-R) for 5 weeks in the light or dark are shown in Fig. 2. The calli cultured on MS media with various combinations of 2,4-D and BA (media A-I) showed sufficient growth (over 60 mg per test tube) in the light except on medium C (Fig. 2a). When the calli were cultured in the dark with a high amount of BA (media G-I), their growth slightly decreased. Calli cultured on MS media with BA and kinetin (media J-R) in the light showed a growth level (Fig. 2b) almost similar to that of those cultured on media with 2,4-D and BA. However, on these media (J-R), the calli did not grow well in the dark. The level of growth of the calli on these media (J-R) under light was almost 1.5 to 3 times higher than in the dark.

The growth rates (0 to 8 week) of the calli (on media C, H and S) and cells (in medium T) are also shown in Fig. 3. The calli started to increase within 2 weeks and continued to grow until the end of the culture period (8 weeks). The growth rates of the calli cultured on C and S media were approximately three times greater than on H medium. The amount of cells cultured in liquid medium T increased from the beginning of culture and reached a maximum level at 5 weeks, when it remained unchanged.

## 2.2 Phenolic Constituents in the Plant and Cultured Cells

### 2.2.1 *Isolation of Phenolic Compounds*

To clarify the phenolic constituents of *S. sebiferum*, stems of the parent plants and calli cultured on media C and S were collected separately.

*Intact Plants.* Fresh stems (ca. 200 g, collected in February) were macerated and extracted with 80% aq. $Me_2CO$ at room temperature. The extract was

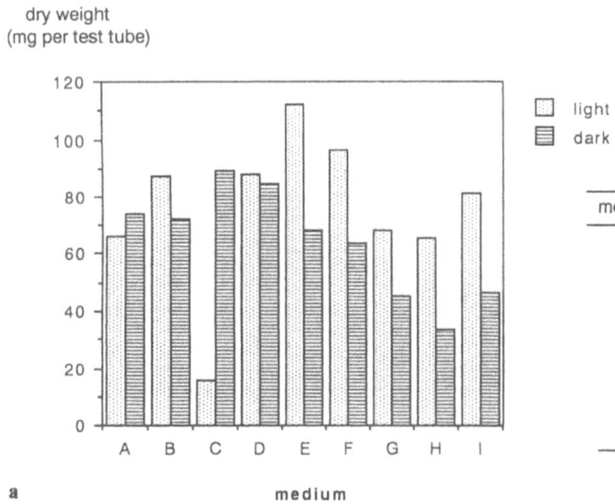


| medium | 2,4-D | BA (mg /l) |
|---|---|---|
| A | 0 | 0.5 |
| B | 0.1 | 0.5 |
| C | 0.5 | 0.5 |
| D | 0 | 1.0 |
| E | 0.1 | 1.0 |
| F | 0.5 | 1.0 |
| G | 0 | 3.0 |
| H | 0.1 | 3.0 |
| I | 0.5 | 3.0 |

a

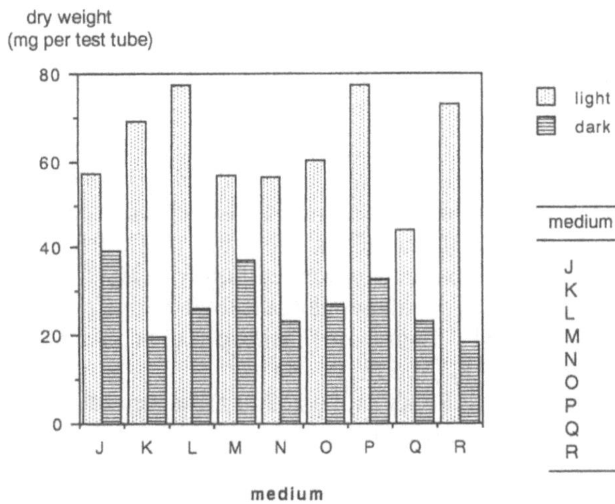


| medium | BA | kinetin (mg /l) |
|---|---|---|
| J | 0.1 | 0.5 |
| K | 0.5 | 0.5 |
| L | 1.0 | 0.5 |
| M | 0.1 | 1.0 |
| N | 0.5 | 1.0 |
| O | 1.0 | 1.0 |
| P | 0.1 | 2.0 |
| Q | 0.5 | 2.0 |
| R | 1.0 | 2.0 |

b

**Fig. 2a,b.** Growth of calli cultured on MS solid media for 5 weeks. (Neera et al. 1992). **a** With 2,4-D and BA. **b** With BA and kinetin

concentrated under reduced pressure to ca. 100 ml and subjected to column chromatography using Sephadex LH-20 (eluted with $H_2O$ containing an increasing amount of MeOH) to afford three fractions. Fraction (Fr) 1 was chromatographed over MCl get CHP-20P (stepwise elution with $H_2O$ and 30% MeOH) which gave β-glucogallin (Kashiwada et al. 1984) (**1**) (12.3 mg) and chlorogenic acid (**2**) (173.9 mg). Fr 2 was purified with Fuji gel ODS-G3 column (stepwise elution with $H_2O$ and 25% MeOH) to give tercatain (Tanaka et al. 1986) (**3**) (98.9 mg) and chebulagic acid (**4**) (Haddock et al. 1982) (180.2 mg). Fr 3 was chromatographed over Sephadex LH-20 with 60% EtOH to give geraniin

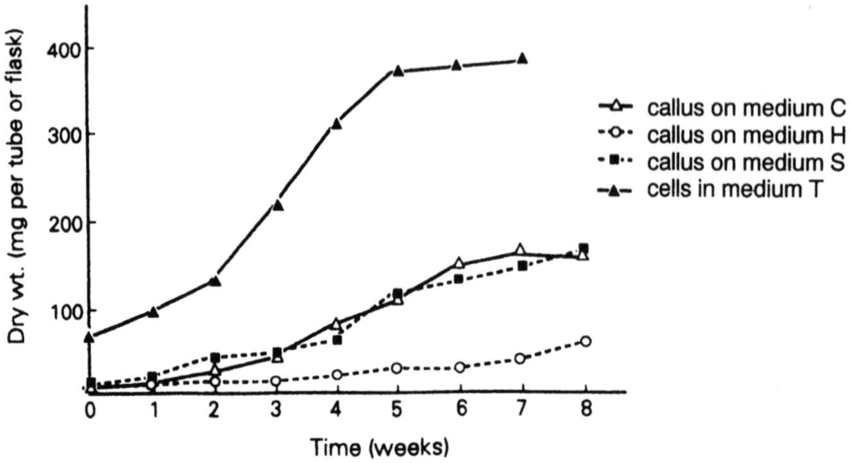


**Fig. 3.** Growth of callus and cell suspension cultures. (Neera and Ishimaru 1992)

(Okuda et al. 1982) (**5**) (419.9 mg), which was the main phenolic compound produced in this plant.

*Calli Cultured on Medium C.* Fresh calli (ca. 200 g), cultured on medium C for 6 weeks in the dark, were homogenized and extracted with 90% $Me_2CO$ at room temperature. The extract, after concentration to ca. 20 ml, was applied to Sephadex LH-20 column chromatography (elution with $H_2O$ containing increasing amount of MeOH) to give three fractions. Fr 1 was chromatographed over Sephadex LH-20 (with 80% EtOH) to afford gallic acid (**6**) (22.9 mg). Fr 2 was purified with Bondapak $C_{18}$ Porasil B (stepwise elution with $H_2O$ to MeOH) to give **5** (135.3 mg). Fr 3 was purified with Sephadex LH-20 (with 50% $Me_2CO$) to afford 1,2,3,4,6-penta-O-galloyl-β-D-glucose (Nishizawa et al. 1982) (**7**) (4.7 mg) and furosin (Saijo et al. 1989) (**8**) (38.5 mg).

*Calli Cultured on Medium S.* Fresh calli (ca. 200 g), harvested after 6 weeks culture on medium S in the dark, were homogenized and extracted as above to give three fractions. Fr 1 was separated with Sephadex LH-20 (with 80% EtOH) to give **6** (12.8 mg) and **1** (45.7 mg). Fr 2 was purified with Bondapak $C_{18}$ Porasil B (stepwise elution with $H_2O$ to MeOH) to give **3** (15 mg), **5** (254.5 mg), **8** (125.7 mg), and 1,2,3,6-tetra-O-galloyl-β-D-glucose (Nishizawa et al. 1983) (**9**) (4 mg). Fr 3 was separated with Sephadex LH-20 (with 80% EtOH) to give **7** (6.1 mg). All phenolic compounds (**1–9**), isolated here were identified by comparison of their physical and spectral (270 MHz $^1$H-nuclear magnetic resonance) data with those mentioned in the references. The chemical structures of **1–9** are shown in Fig. 4.

### 2.2.2 *Quantitative Determinations of Phenolic Constituents in Cultured Cells*

Lyophilized samples (callus or cell) were ground with pestle and mortar, and the powdered samples (20–60 mg) were extracted with 80% $Me_2CO$ (2 ml) for 15 h

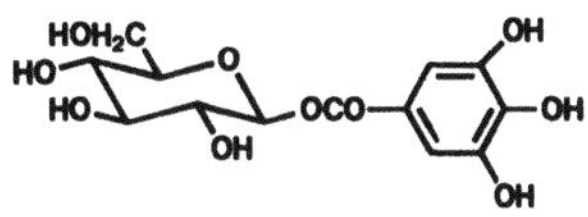

β-glucogallin (1)

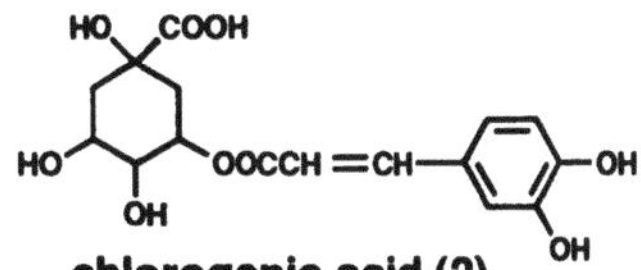

chlorogenic acid (2)

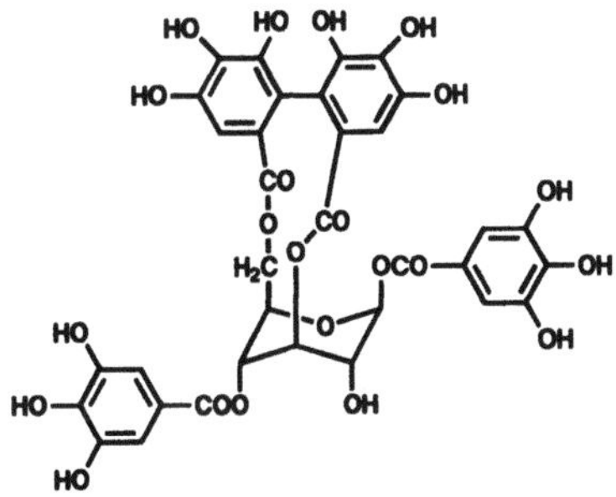

tercatain (3)

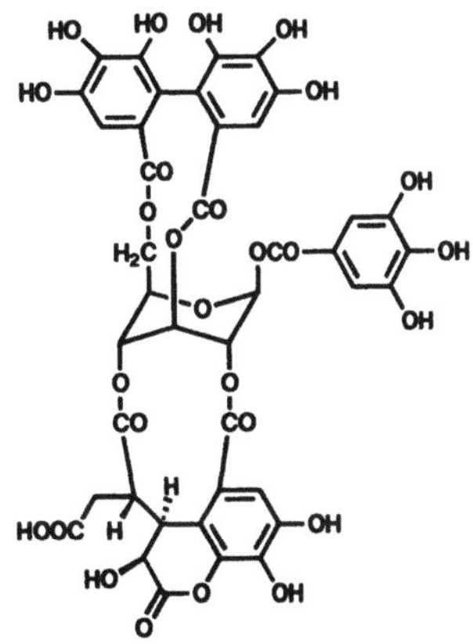

chebulagic acid (4)

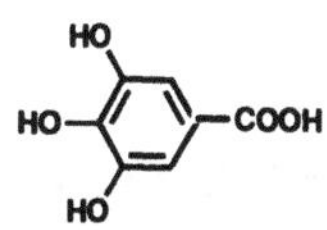

gallic acid (6)

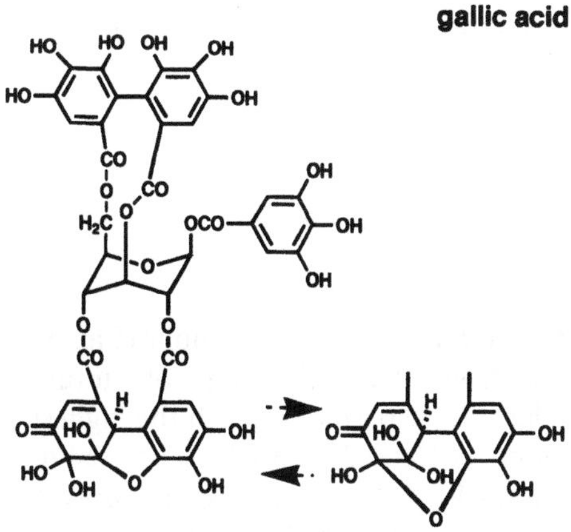

geraniin (5)

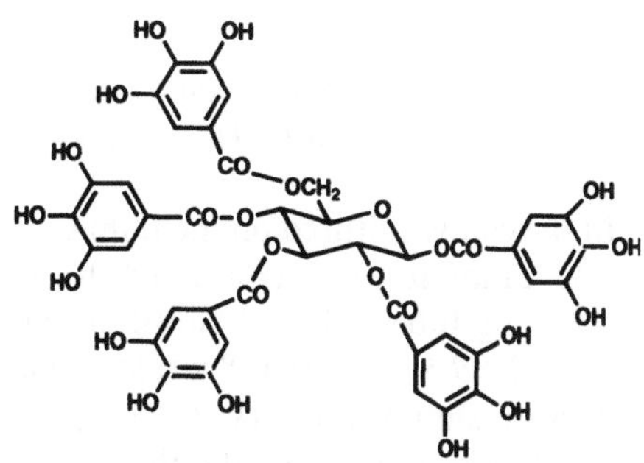

1,2,3,4,6-penta-*O*-galloyl-β-D-glucose (7)

**Fig. 4.** Chemical structures of phenolic constituents

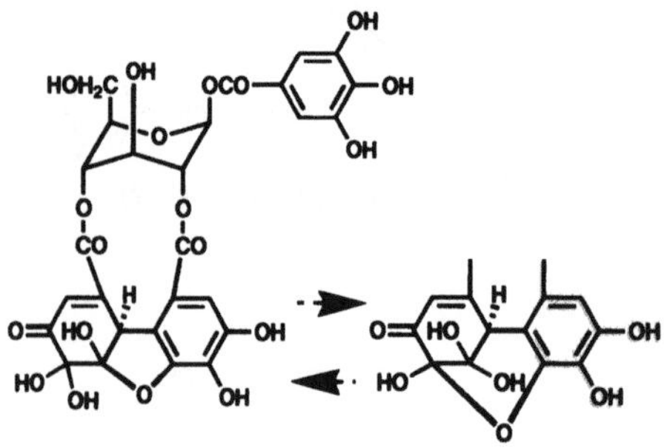

furosin (8)

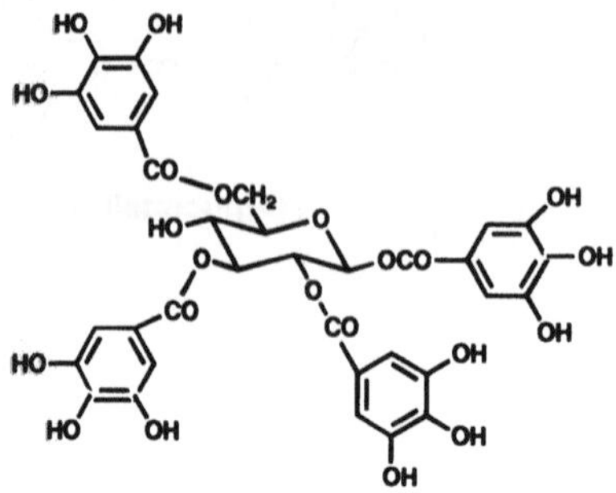

1,2,3,6-tetra-*O*-galloyl-β-D-glucose (9)

**Fig. 4.** (*Contd.*)

at room temperature. Each extraction, after filtration with a Millipore filter (0.45 μm), was injected (2–4 μl) to high performance liquid chromatography (HPLC, Shimadzu LC6A). HPLC conditions were as follows; column: Wakosil II 5C18 HG (4.6 mm i.d. × 160 mm), mobile phase: MeCN-50 mM $NaH_2PO_4$ (5:95 → 80:20, in 30 min), flow rate: 0.8 ml/min, column temperature: 40°C, detection: 280 nm (UV), $t_R$ (min): **6** (2.3), **1** (3.0), **2** (5.3), **5** (18.9), **4** (20.5), **3** (21.4), and **7** (22.8).

## 2.3 Production of Phenolic Compounds in Cultured Cells

The amounts of phenolic compounds in several cultures of *S. sebiferum* were determined by HPLC.

### *2.3.1 Time Course of Production of Phenolic Compounds in Calli Cultured on Media C, H, and S in the Dark*

In the callus cultured on medium C, compound **5** was the main phenolic (Fig. 5). The amount of **5** continued to increase from the early stage of culture (121.2 μg/tube at 1 week) until the end of the culture period to reach the highest level (2199.3 μg/tube at 8 weeks). The amounts of **1** and **3** gradually increased to their maximum levels (**1**: 265.3 μg/tube at 6 weeks, **3**: 160.0 μg/tube at 7 weeks), which were almost one-tenth the level in comparison with that of **5**. **2**, **4**, and **7** appeared only at the early stages (1 to 2 weeks) of this culture.

The callus cultured on medium H also produced high levels of **5** (520.8 μg/tube at 8 weeks) (Fig. 6), but the amount of **5** in this culture was almost one-quarter that cultured on medium C. The amounts of **1**, **3**, and **7** increased

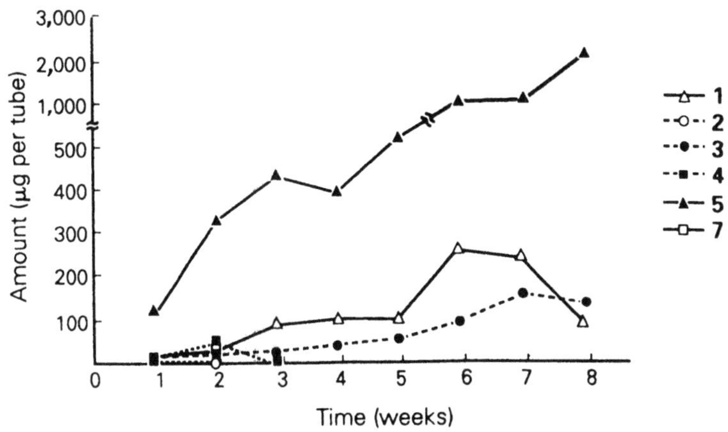


**Fig. 5.** Production of phenolic compounds in calli cultured on medium C. (Neera and Ishimaru 1992)

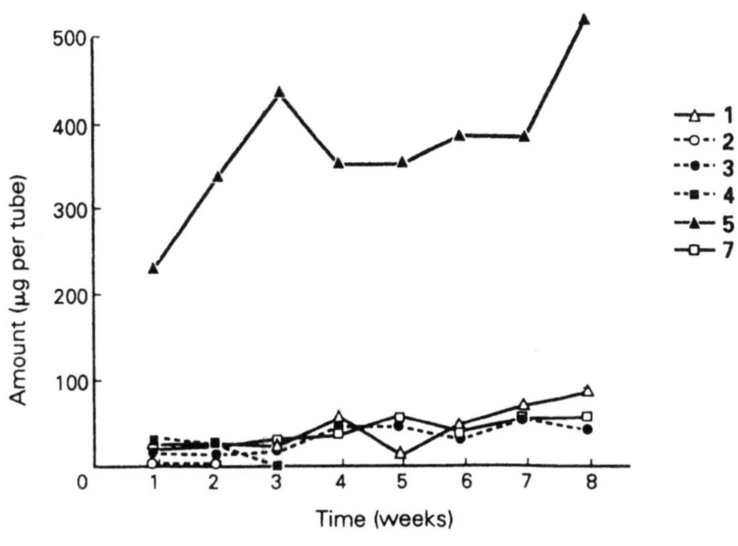


**Fig. 6.** Production of phenolic compounds in calli cultured on medium H. (Neera and Ishimaru 1992)

slightly throughout the culture time, but their levels were not very high (under 100 $\mu$g/tube). **2** and **4** appeared in low concentration only at the early stage in this culture.

In the callus cultured on medium S, the main product **5** increased rapidly after 2 weeks' culture and reached the highest level (3092.8 $\mu$g/tube) at 8 weeks (Fig. 7). The amounts of **1**, **3**, and **7** also gradually increased and reached over 150 $\mu$g per tube.

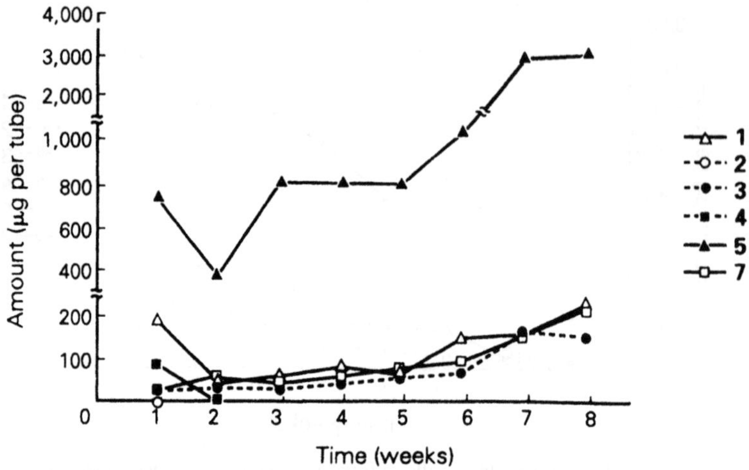


**Fig. 7.** Production of phenolic compounds in calli cultured on medium S. (Neera and Ishimaru 1992)

### *2.3.2 Time Course of Production of Phenolic Compounds in Cells Cultured in Liquid Medium T in the Dark*

In the cell suspension culture in medium T, phenolic constituents were similar to those produced in the callus cultures (Fig. 8). In this culture, as in the callus cultures, **5** was the main compound and reached over 3000 $\mu$g per flask at 7 weeks' culture. **1**, **3**, **4** and **7** were also detected with high levels in comparison with those in the callus cultures.

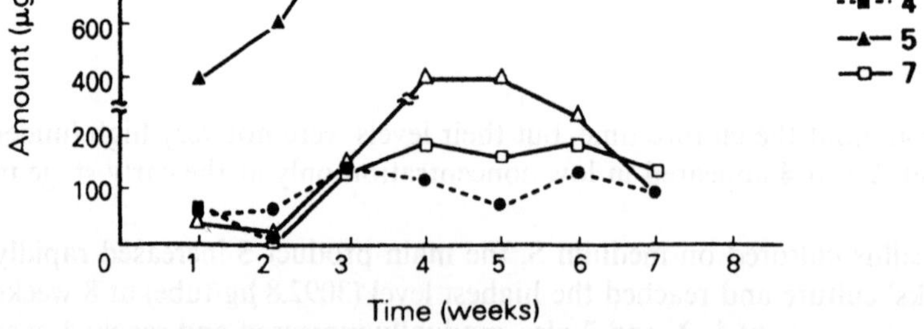


**Fig. 8.** Production of phenolic compounds in cell suspension culture. (Neera and Ishimaru 1992)

### *2.3.3 Production of Phenolic Compounds in the Calli Cultured on Media A-R in the Light or Dark*

The main compound produced in the calli cultured on these media was **5** both in the light (Figs. 9a, 10a) and dark (Figs. 9b, 10b). The average amount of phenolic compounds observed in calli cultured on media A-I (Fig. 9a, b) was a little larger than that of those cultured on media J-R (Fig. 10a and b). On these media (A-R) the illumination was less effective for the production of phenolic compounds in *S. sebiferum* calli.

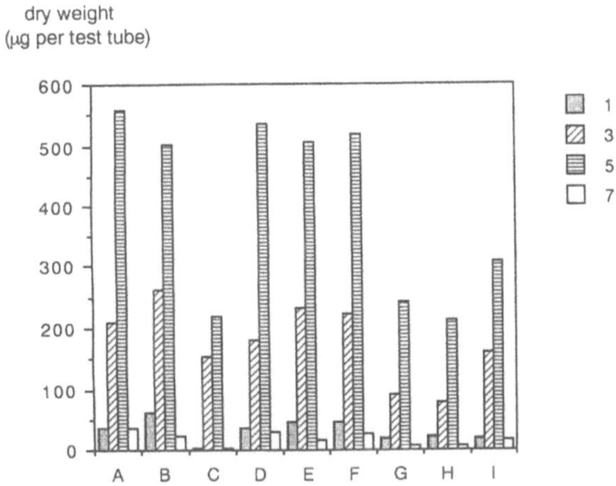


**a** medium

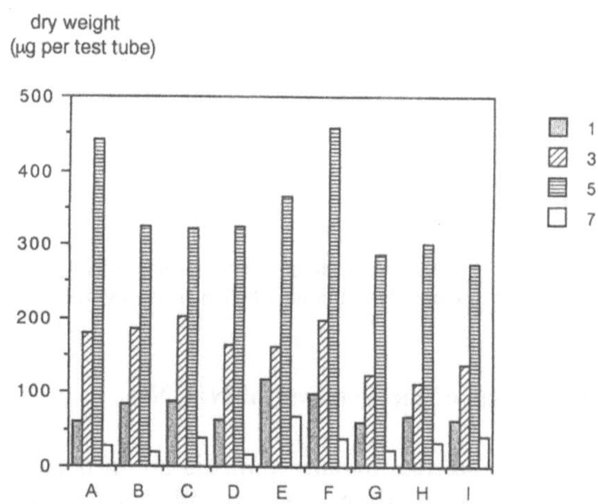


**b** medium

**Fig. 9a,b.** Production of phenolic compounds in calli cultured on MS solid media with 2,4-D and BA for 5 weeks. (Neera et al. 1992). **a** In the light. **b** In the dark

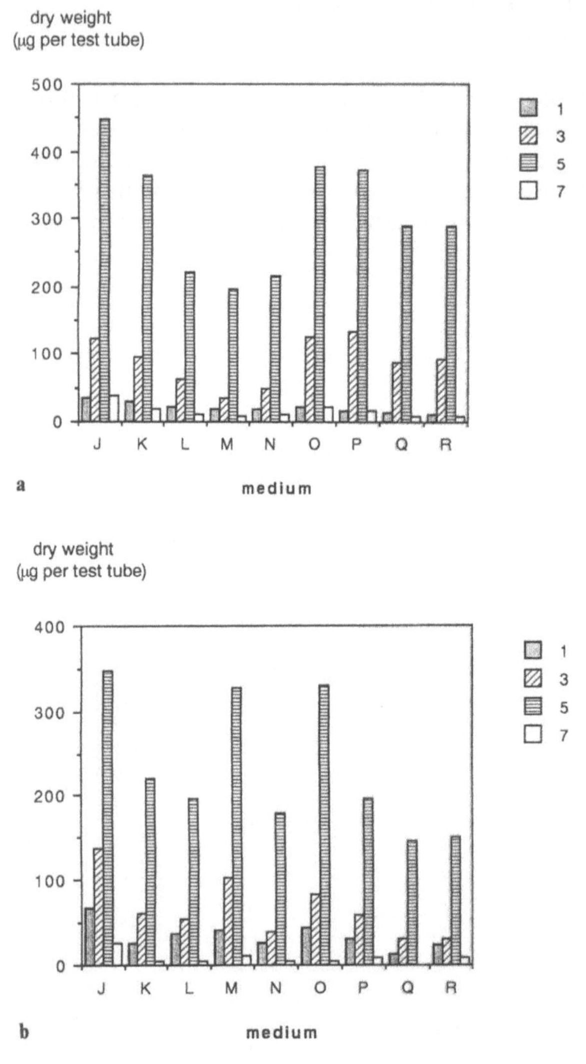


**Fig. 10a,b.** Production of phenolic compounds in calli cultured on MS solid media with BA and kinetin for 5 weeks (Neera et al. 1992). **a** In the light. **b** In the dark

## 2.4 Effects of the Removal of $NH_4NO_3$

Nitrogen sources in culture media are important factors for growth (Wetherell and Dougall 1976; Jeorge et al. 1988) and production of secondary metabolites such as tropane alkaloids (Deno and Uhara 1987), shikonin (Yazaki et al. 1987), tannin (Ishimaru and Shimomura 1991), etc. in plant tissue cultures.

The growth of calli cultured on media A-I without $NH_4NO_3$ (Fig. 11) was not so different from that of those cultured on media A-I containing $NH_4NO_3$

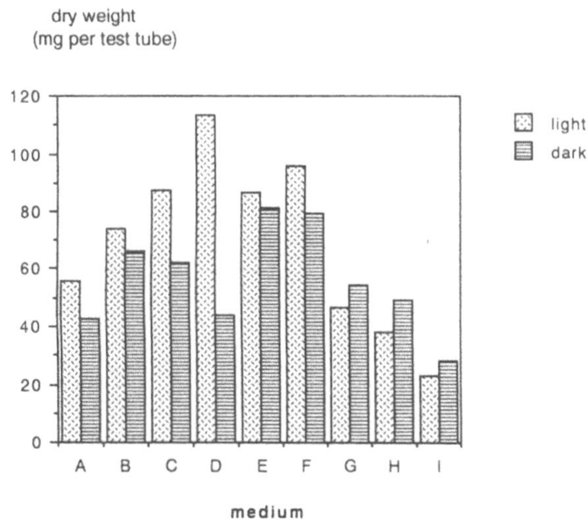


**Fig. 11.** Growth of calli cultured on MS solid media (minus $NH_4NO_3$) with 2,4-D and BA for 5 weeks. (Neera et al. 1992)

(Fig. 2a). The removal of $NH_4NO_3$ seemed to be less effective on the growth of the calli. In the media G-I (with 3 mg/l BA) under light, the removal of $NH_4NO_3$ was not suitable for growth.

On the other hand, the production of phenolic constituents (especially **5**) in the calli was remarkably enhanced by the removal of $NH_4NO_3$. Calli cultured on the media A-I without $NH_4NO_3$ produced a large amount of **5** [over 3000 $\mu$g per test tube in the light (Fig. 12a) and over 1900 $\mu$g per test tube in the dark (Fig. 12b)], which was almost 5–15 times that observed in calli cultured on media A-I containing $NH_4NO_3$ (Fig. 9a, b). In these media (minus $NH_4NO_3$), illumination seemed to be effective for the production of **5**.

## 2.5 Effects of Some Amino Acids

Some amino acids (glutamine, glycine, serine, etc.) are known to affect growth (Kamada and Harada 1984; Dodds and Roberts 1985; Jeorge et al. 1988) and secondary metabolism (Luckner et al. 1977; Jeorge et al. 1988) in plant tissue cultures. Therefore, we determined the effects of these amino acids on the growth and production of phenolic compounds in *S. sebiferum* calli. When the calli were cultured on MS media (A-I) with three amino acids (1 g/l glutamine, 50 mg/l glycine, and 50 mg/l serine), their growth level (Fig. 13) was almost similar to those cultured on media A-I without these amino acids (Fig. 2a). Therefore, for the growth of the calli, the addition of these amino acids seemed to be less effective.

The pattern of production of phenolics in calli cultured on these media (plus amino acids) (Fig. 14a, b) was different from that observed on media without

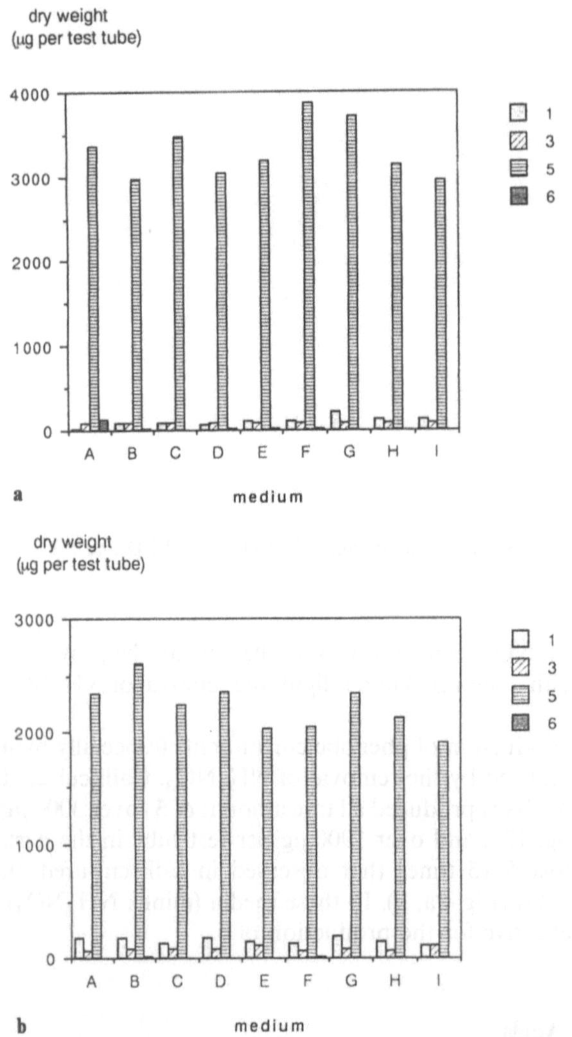


a

b

**Fig. 12a,b.** Production of phenolic compounds in calli cultured on MS solid media (minus $NH_4NO_3$) with 2,4-D and BA for 5 weeks. (Neera et al. 1992). **a** In the light. **b** In the dark

amino acids (Fig. 9a, b). In calli cultured on media containing these amino acids, increase in **5** was observed under some conditions [media C-E and G in the light (Fig. 14a) and D and G in the dark (Fig. 14b)]. For the production of the other phenolics (such as **3**, **5**, **7**, etc.), the addition of these amino acids to the medium was not effective. Compared to the effects of $NH_4NO_3$, these amino acids seemed to affect the production of phenolic constituents in *S. sebiferum* callus less strongly.

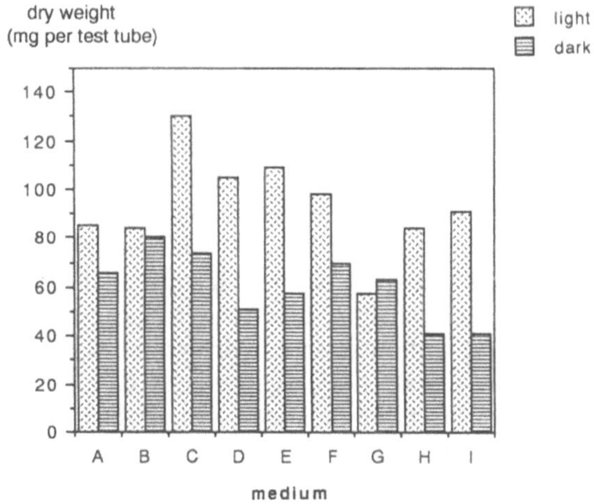


**Fig. 13.** Growth of calli cultured on MS solid media (plus amino acids) with 2,4-D and BA for 5 weeks. (Neera et al. 1992)

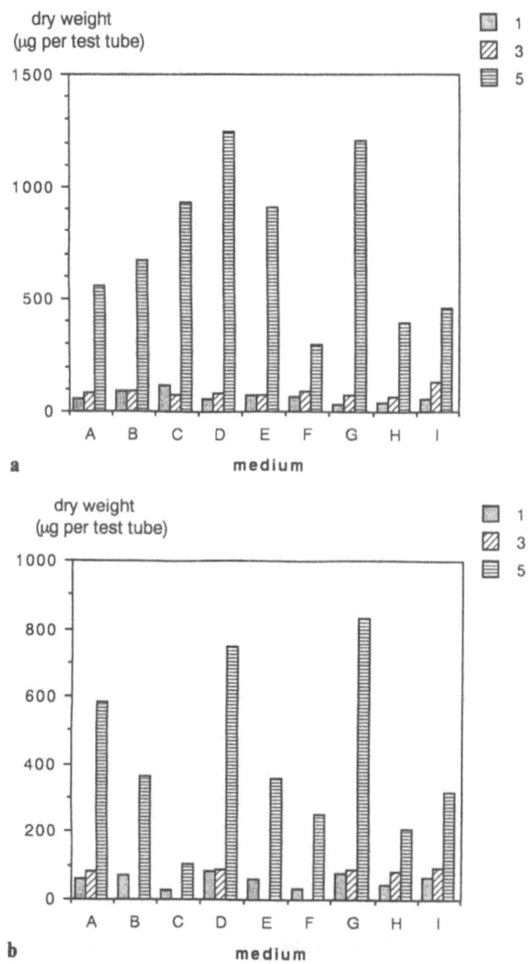


**Fig. 14a,b.** Production of phenolic compounds in calli cultured on MS solid media (plus amino acids) with 2,4-D and BA for 5 weeks. (Neera et al. 1992). **a** In the light. **b** In the dark

## 2.6 Effects of Casein Hydrolysate

Casein hydrolysate in combination with BA and kinetin affected the growth of callus cultures (Mridula et al. 1983). When *S. sebiferum* calli were cultured on MS media containing 100 mg/l casein hydrolysate with various combinations of BA and kinetin (media J-R), they showed sufficient growth, especially in the light (Fig. 15). The growth of the calli cultured on media M and N (containing casein hydrolysate) was more than twice as large as those cultured on the same media (M and N) without casein hydrolysate (Fig. 2b).

Mridula et al. (1983) reported that casein hydrolysate (with BA and kinetin) had some effects on secondary metabolism which caused organogenesis (plant regeneration) in *S. sebiferum* callus. We thereupon determined the effects of casein hydrolysate on the production of phenolics in calli (Fig. 16). Calli cultured on media J-R with casein hydrolysate produced a large amount of **5** in the light (Fig. 16a). Especially on the media J-L and P-R, the amount of **5** reached 1300–2600 μg per test tube, a level three to seven times larger than when cultured on media without casein hydrolysate (Fig. 10a). On media M-O the calli produced less **5** in the light, but the growth level of calli on these media (M-O) was fairly good (Fig. 15). A reversed relationship between the growth and production of **5** was observed in calli cultured on these media (plus casein hydrolysate) in the light.

## 2.7 Shoot Regeneration from Callus Cultures

Mridula et al. (1983) succeeded in the formation and elongation of multiple shoots from the leaf callus of this plant using a two-stage procedure; in the first

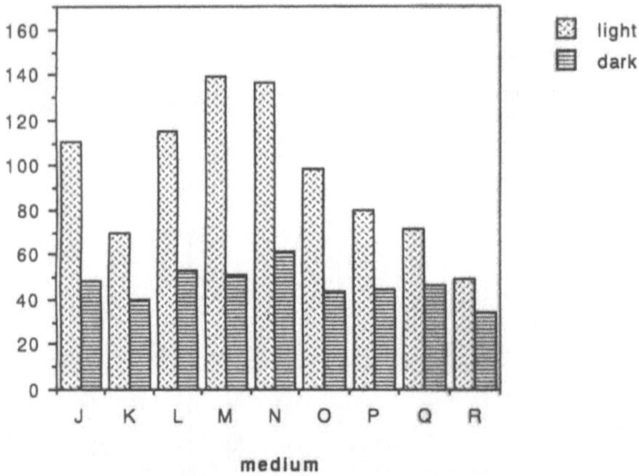


**Fig. 15.** Growth of calli cultured on MS solid media (plus casein hydrolysate) with BA and kinetin for 5 weeks. (Neera et al. 1992)

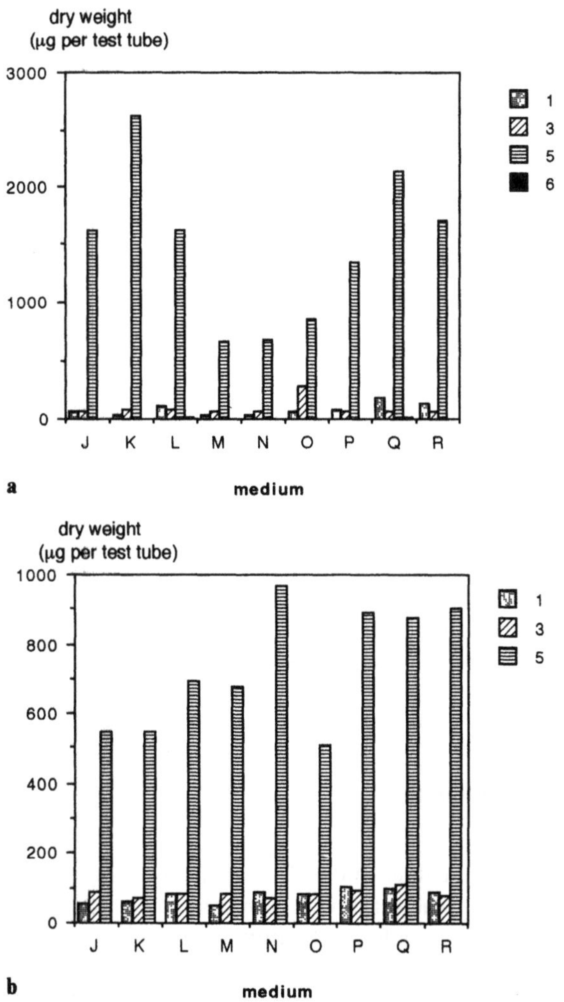


**Fig. 16a,b.** Production of phenolic compounds in calli cultured on MS solid media (plus casein hydrolysate) with BA and kinetin for 5 weeks. (Neera et al. 1992). **a** In the light. **b** In the dark

stage (shoot initiation), addition of casein hydrolysate with BA and kinetin was essential and in the second stage (shoot elongation) it was necessary to omit casein hydrolysate. In the course of chemical determination of secondary metabolites in *S. sebiferum* calli cultured on several media, MS medium supplemented with IAA and BA (without casein hydrolysate) was found to be more suitable for shoot regeneration. The shoots regenerated directly from callus cultured on MS solid medium containing 0.1 mg/l IAA and 1 mg/l BA (Fig. 17) elongated faster than those obtained from callus on MS medium with 1 mg/l BA, 0.5 mg/l kinetin (medium L) and 100 mg/l casein hydrolysate in 4 weeks' culture in the light. However, the growth rate of calli cultured on this medium

**Fig. 17.** Regenerated shoots of *Sapium sebiferum* from callus cultured on MS medium with 0.1 mg/l IAA and 1 mg/l BA. (Photo by K. Ishimaru, 1991)

(containing IAA and BA) was fairly low, which indicated this medium to be unsuitable for the production of secondary metabolites.

# 3 Conclusion

In both callus and cell suspension cultures of *Sapium sebiferum*, the main phenolic was geraniin and in some media these cultured cells were clarified to be usable for high production of geraniin. Especially, taking into account the volume of media (10 ml per tube for callus cultures and 50 ml per flask for cell suspension culture), callus cultures seem to be superior to cell suspension culture for the production of secondary metabolites. Among the media tested, MS medium without $NH_4NO_3$ containing 0.5 mg/l 2,4-D and 1 mg/l BA (medium F) was best for the production of geraniin (3900 $\mu$g per test tube in the light). The different responses of calli (to $NH_4NO_3$, amino acids, and casein hydrolysate) in the production of phenolics also suggest that these cultures might be applicable to biosynthetic study of hydrolyzable tannins.

# 4 Protocol

Leaf segments were sterilized, placed on MS solid media supplemented with various combinations of 2,4-D, IAA, and BA (Table 1), and cultured for 8 weeks in the dark at 25°C. Calli derived on the medium containing 1 mg/l 2,4-D and 0.1 mg/l BA (medium T) were cut and subcultured (on medium T) for over 1 year. The calli were also transferred to liquid medium T (50 ml medium/flask) and cell suspension culture was established (100 rpm, in the dark). A piece of fresh callus (0.3 g) was inoculated separately on various MS media (media A-S, media A-I without $NH_4NO_3$, media A-I with 1 g/l glutamine, 50 mg/l glycine and 50 mg/l serine, and media J-R with 100 mg/l casein hydrolysate, each 10 ml medium/tube) and cultured under dark or light (3000 lx, 16-h photoperiod per day) condition. The growth (amount of calli and cells) and the production of phenolics (quantified by HPLC) were determined.

## References


Bajaj YPS (1986) Biotechnology of tree improvement for rapid propagation and biomass energy production. In: Bajaj YPS (ed) Biotechnology in agriculture and forestry, vol 1. Trees I. Springer, Berlin Heidelberg New York, pp 1–23

Corthout J, Pieters LA, Claeys M, Berghe DAV, Vlietinck AJ (1991) Antiviral ellagitannins from *Spondias mombin*. Phytochemistry 30:1129–1130

Deno H, Uhara S (1987) Japan Patent 62-6675:347–351

Dodds JH, Roberts LW (1985) Production of secondary metabolites by cell cultures. In: Experiments in plant tissue culture. Cambridge University Press, Cambridge pp 35–53

Gandhi V (1983) Tissue culture of Chinese tallow – an euphorbiaceous tree. Curr Sci 52:363–365

Haddock EA, Gupta RK, Haslam E (1982) The metabolism of gallic acid and hexahydroxydiphenic acid in plants, Part 3. Esters of (*R*)- and (*S*)-hexahydroxydiphenic acid and dehydrohexahydroxydiphenic acid with D-glucopyranose ($^1C_4$ and related conformations). J Chem Soc Perkin Trans I 1982: 2535–2545

Ishimaru K, Shimomura K (1991) Tannin production in hairy root culture of *Geranium thunbergii*. Phytochemistry 30:825–828

Jeorge EE, Pattodk DJM, Jeorge HJ (1988) The constituents of culture media. In: Plant culture media commentary and analysis 2. Exegetics Limited. Edington, Westbury, Wilts, pp 338–369

Kamada H, Harada H (1984) Changes in endogenous amino acid compositions during somatic embryogenesis in *Daucus carota* L. Plant Cell Physiol 25:27–38

Kashiwada Y, Nonaka G, Nishioka I (1984) Tannins and related compounds 23. Rhubarb (4) : Isolation and structures of new classes of gallotannins. Chem Pharm Bull 32:3461–3470

Kotwal MM, Gupta PK, Mascarenhas AF (1983) Rapid multiplication of *Sapium sebiferum* Roxb. Plant Cell Tissue Organ Cult 2:133–139

Luckner M, Nover L, Bohm H (1977) Secondary metabolism in cell cultures of higher plants and problems of differentiation. In: Molecular biology, biochemistry and biophysics 23. Springer, Berlin Heidelberg New York, pp 104–119

Mridula MK, Gupta PK, Mascarenhas AF (1983) Rapid multiplication of *Sapium sebiferum* Roxb. by tissue culture. Plant Cell Tissue Organ Cult 2:133–139

Murashige T, Skoog F (1962) A revised medium for rapid growth and bioassays with tobacco tissue cultures. Physiol Plant 15:473–497

Nagasawa T, Shibutani S, Oura H (1978) Effect of kampo-prescriptions on rat serum constituents after the administration. Yakugaku Zasshi 98:1642–1650

Nagasawa T, Shibutani S, Oura H, Shoyama Y, Nishioka I (1980) Effect of extract from *Rhei rhizoma* on urea-nitrogen concentration in rat serum. Chem Pharm Bull 28:1736–1739

Neera S, Ishimaru K (1992) Tannin production in cell cultures of *Sapium sebiferum*. Phytochemistry 31:833–836

Neera S, Arakawa H, Ishimaru K (1992) Tannin production in *Sapium sebiferum* callus cultures. Phytochemistry 31:4143–4149

Nishioka I (1983) Chemistry and biological activities of tannins. Yakugaku Zasshi 103:125–142

Nishizawa M, Yamagishi T, Nonaka G, Nishioka I (1982) Tannins and related compounds. Part 5. Isolation and characterization of polygalloylglucoses from Chinese gallotannin. J Chem Soc Perkin Trans I 1982:2963–2968

Nishizawa M, Yamagishi T, Nonaka G, Nishioka I (1983) Tannins and related compounds. Part 9. Isolation and characterization of polygalloylglucoses from Turkish galls (*Quercus infectoria*). J Chem Soc Perkin Trans I 1983:961–965

Nishizawa M, Yamagishi T, Dutschman GE, Parker WB, Bodner AJ, Kilkuskie RE, Cheng Y-C, Lee K-H (1989) Isolation and characterization of four new tetragalloylquinic acids as a new class of HIV reverse transcriptase inhibitors from tannic acid. J Nat Prod 52:762–768

Nonaka G, Nishioka I, Nishizawa M, Yamagishi T, Kashiwada Y, Dutschman GE, Bodner AJ, Kilkuskie RE, Cheng Y-C, Lee K-H (1990) Inhibitory effects of tannins on HIV reverse transcriptase and HIV replication in H9 lymphocyte cells. J Nat Prod 53:587–595

Okuda T, Yoshida T, Hatano T (1982) Constituents of *Geranium thunbergii* Sieb. *et* Zucc. Part 12. Hydrated stereostructure and equilibration of geraniin, J Chem Soc Perkin Trans I 1982:9–14

Saijo R, Nonaka G, Nishioka I (1989) Tannins and related compounds. 84. Isolation and characterization of five new hydrolysable tannins from the bark of *Mallotus japonicus*. Chem Pharm Bull 37:2063–2070

Shibutani S, Nagasawa T, Yokozawa T, Oura H (1980) Effect of Rhubarb (*Rhei rhizoma*) extract on urea nitrogen and amino acid metabolism after the administration. Yakugaku Zasshi 100:434–442

Shibutani S, Nagasawa T, Oura H, Nonaka G, Nishioka I (1983) Mechanism of the blood urea nitrogen-decreasing activity of rhatannin from *Rhei rhizoma* in the rat. I. Chem Pharm Bull 31:2378–2385

Tanaka T, Nonaka G, Nishioka I (1986) Tannins and related compounds. 42. Isolation and characterization of four new hydrolysable tannins, terflavins A and B, tergallagin and tercatain from the leaves of *Terminalia catappa* L. Chem Pharm Bull 34:1039–1049

Ueki S, Nonaka G, Nishioka I, Fujiwara M (1985) Psychotropic activity of GAO and its active substances. J Med Pharm Soc WAKAN-YAKU 2:502–503

Venketeswaran S (1989) Chinese tallow (*Sapium sebiferum* Roxb.). In: Bajaj YPS (ed) Biotechnology in agriculture and forestry, vol 5. Trees II. Springer, Berlin Heidelberg New York, pp 342–358

Venketeswaran S, Gandhi V (1982) Mass propagation of forest trees for biomass production. Biomass 2:5–15

Venketeswaran S, Gandhi V, Romano EJ, Nagmani R (1983a) Mass propagation of selected trees for biomass by tissue culture. In: Proc NATO Inst Biomass utilization, Portugal. Plenum Press, New York, pp 191–204

Venketeswaran S, Gandhi V, Romano EJ, Nagmani R (1983b) Biomass energy production from forest trees by tissue culture. In: Proc Symp Energy from Biomass and wastes, vol III. Inst Gas Technol, Chicago, pp 69–99

Venketeswaran S, Nagmani R, Gandhi V (1984) Tissue culture propagation of selected tree genera, In: Proc BioEnergy '84 Conf, Gothenburg, Sweden. Elsevier, New York, pp 47–51

Wetherell DF, Dougall DK (1976) Sources of nitrogen supporting growth and embryogenesis in cultured wild carrot tissue. Physiol Plant 37:97–103

Yazaki K, Fukui H, Kikuma M, Tabata M (1987) Regulation of shikonin production by glutamine in *Lithospermum erythrorhizon* cell cultures. Plant Cell Rep 6:131–134

# XXVII *Tabebuia avellanedae* Lorentz ex Griseb. (Taheebo): In Vitro Culture and the Production of Naphthoquinones

S. UEDA[1]

## 1 General Account

### 1.1 Distribution and Importance of the Plant

Bignoniaceous plants of the genus *Tabebuia* grow throughout southern North America to South America. *Tabebuia avellanedae* Lorentz and Griseb. [synonym *T. heptaphylla* Vellosoi Toledo, *T. impetiginosa* Mart. ex DC. (Standl.), and *T. ipê* (Mart. ex Schum.) Standl.] is native to South America from Brazil to northern Argentina (Williams 1936; Hashimoto 1962; da Silva et al. 1977). The bark of the plant has been used as a diuretic and astringent. This plant has many colloquial names: Ipê, Ipê roxo, Ipê cavatan, Ipê comum, Ipê de São Paulo, Ipeuva, Aipê, Cavatan, Guiraiba, Lapacho, Pau d'Arco Roxo, Pau d'ArcoVelmelho, Peuva, Peuva roxa, Piuva, Queraiba, or Upeuva in Brazil; Lapacho or Lapacho negro in Argentina; Lapacho in Paraguay; Tayihú in Guarani; and Taheebo in ancient Inca (Hashimoto 1962; da Silva et al. 1977; Accorsi 1988). In an ancient language of the Incas, Quechua, which is still retained by the inhabitants of the sierras, *Tabebuia* species are called Tahuari (Williams 1936). *T. avellanedae* grows well in the semitropical regions of Brazil. The erect trunk of the tree with its spreading branches reaches a height of 30–50 m. The palmate compound leaves are composed of five to seven microphylls. The leaves are opposite, sharply toothed, and long pointed. The five-lobed rosy or purple showy blossoms, similar in shape to those of the genus *Catalpa*, open in mid- July to September (Fig. 1). The plants bear long suspended pods containing numerous seeds in September to October. The seeds are broadly winged. The timber is valuable for general construction and cabinetwork. The recent rediscovery of *Tabebuia* plants as effective drugs for cancer has made *T. avellanedae* an important medicinal resource (Hartwell 1968; Gottlieb and Mors 1980; M.M. Rao and Kingston 1982; Accorsi 1988; Ueda and Tokuda 1990; Zani et al. 1991; Saizarbitoria et al. 1992).

[1] Faculty of Pharmaceutical Sciences, Kyoto University, Sakyo-ku, Kyoto 606-01, Japan

**Fig. 1.** Herbarium specimen of *Tabebuia avellanedae* flowers. (Donated by Professor W.R. Accorsi)

## 1.2 Secondary Metabolites

From the heartwood of *T. avellanedae*, a group of naphthoquinones and the derivatives, lapachol (**1**), lapachol methyl ether (**2**), deoxylapachol (**3**), menaquinone-1 (**4**), β-lapachone (**5**), α-lapachone (**6**), dehydro-α-lapachone (**7**), and lapachenole (**8**); a group of anthraquinones, 2-methylanthraquinone (**10**), 2-hydroxymethylanthraquinone (**11**), 2-acetoxymethylanthraquinone (**12**), anthraquinone-2-aldehyde (**13**), anthraquinone-2-carboxylic acid (**14**), 1-hydroxyanthraquinone (**15**), 1-methoxyanthraquinoe (**16**), and 2-hydroxy-3-methylanthraquinone (**17**); as well as a new dimeric quinone, tabebuin (**9**) were isolated together with quercetin, and *o*- and *p*-hydroxybenzoic acids (Burnett and Thomson 1967a, b;

Figs. 2 and 3). Recently, from the inner bark of Argentine *T. avellanedae*, which was obtained from Kenneth Corwin (Dr. Meyer's Lapacho Co., Santa Monica, California), 5-hydroxy-2-(1-hydroxyethyl)-naphtho[2,3-*b*]-furan-4,9-dione (**18**), 8-hydroxy-2-(1-hydroxyethyl)naphtho[2,-3-*b*]-furan-4,9-dione (**19**), 2-acetyl-5-hydroxynaphtho[2,3-b]furan-4,9-dione (**20**), 2-acetyl-8-hydroxynaphtho [2,3-*b*]

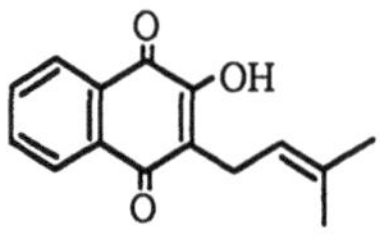

Lapachol (1)

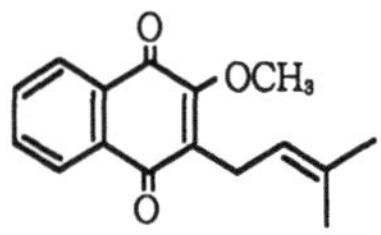

Lapachol methyl ether (2)

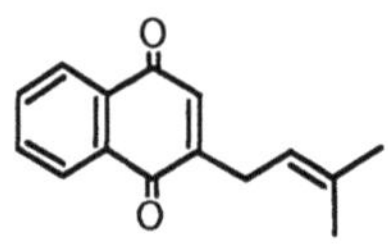

Deoxylapachol (3)

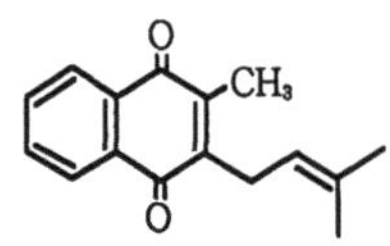

Menaquinone-1 (4)

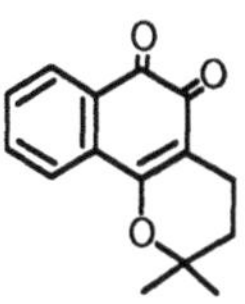

β-Lapachone (5)

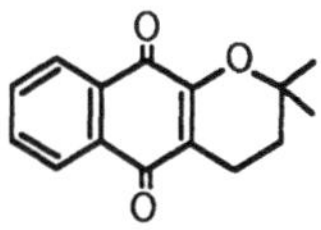

α-Lapachone (6)

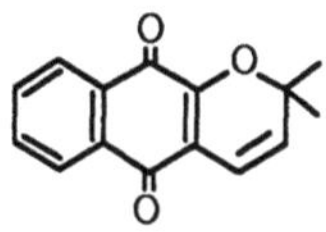

Dehydro-α-lapachone (7)

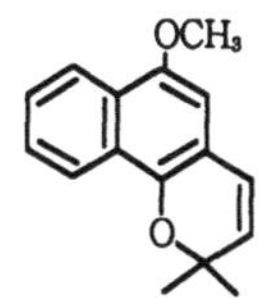

Lapachenole (8)

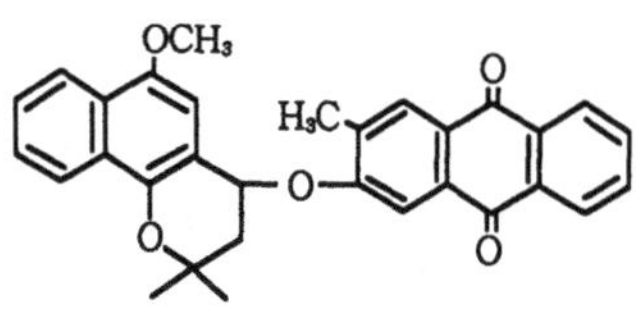

Tabebuin (9)

**Fig. 2.** Naphthoquinones (**1**)–(**9**) in *Tabebuia avellanedae* heartwood. (Burnett and Thomson 1967b)

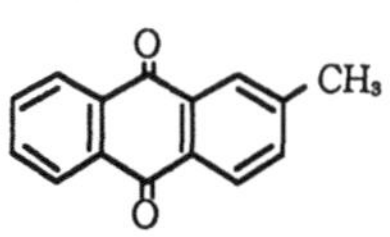

2-Methylanthraquinone (10)

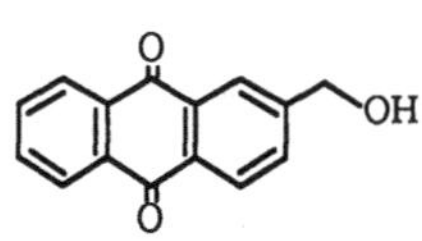

2-Hydroxymethylanthraquinone (11)

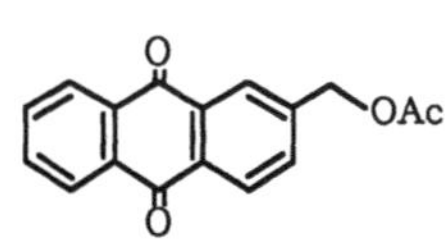

2-Acetoxymethylanthraquinone (12)

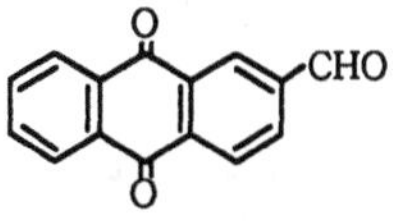

Anthraquinone-2-aldehyde (13)

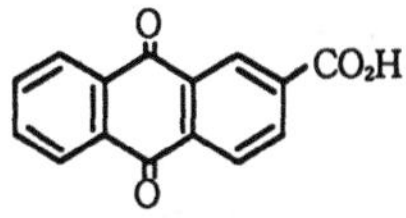

Anthraquinone-2-carboxylic acid (14)

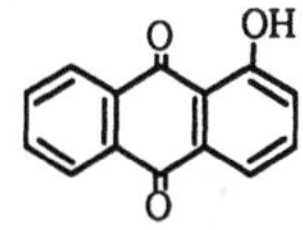

1-Hydroxyanthraquinone (15)

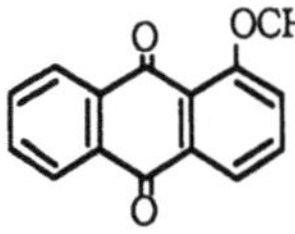

1-Methoxyanthraquinone (16)

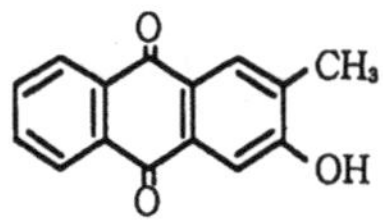

2-Hydroxy-3-methylanthraquinone (17)

**Fig. 3.** Anthraquinones (**10**)–(**17**) in *Tabebuia avellanedae* heartwood. (Burnett and Thomson 1967b) furan-4,9-dione (**21**), and 2,3-dihydro-5-hydroxy-2-(1-methylethenyl) naphtho-[2,3-*b*]furan-4,9-dione (**24**) were isolated along with lapachol (**1**), dehydro-α-lapachone (**7**), and benzo-[*b*]furan-6-carboxaldehyde (**26**) (Wagner et al. 1989; Fig. 4).

From the bark of *Tabebuia avellanedae* ( = *T. impetiginosa*), obtained from Peru in 1983, *T. chrysantha* (Jacq.) Nichols (Tahuari) collected in Peru in March 1987, and *T. rosea* (Bertol.) DC. collected at Cartagena, Colombia, 2-(1-hydroxyethyl) naphtho[2,3-*b*] furan-4,9-dione (**22**) and 2-acetylnaphtho[2,3-*b*]-furan-4,9-dione (**23**) were isolated. (5 or 8)-Hydroxy-2-(1-hydroxyethyl)-naphtho-[2,3-b]furan-4,9-dione was obtained from the bark of *T. rosea*, while from the wood, lapachol (**1**) and dehydro-α-lapachone (**7**) were obtained (Girald et al. 1988). From the methanolic extract of the bark of *T. avellanedae*, three new iridoid glucosides, 6-*O*-(*p*-hydroxybenzoyl)-ajugol, 6-*O*-(3′,4′-dimethoxy-benzoyl)-ajugol and 6-*O*-(*p*-hydroxybenzoyl)-ajugol were isolated (Nakano et al. 1993).

2-Ethylnaphtho[2,3-*b*]furan-4,9-dione was obtained along with 2-iso-propylnaphtho[2,3-*b*]furan-4,9-dione and 2,3-dihydro-2-(1-methylethenyl)-naphtho[2,3-*b*]furan-4, 9-dione (**25**) (dehydro-iso-α-lapachone) from the heart-wood of *Paratecoma peroba* Kuhlm. (Sandermann et al. 1968).

Naphthoquinone (**18**) was first obtained from the wood of *Kigelia pinnata* DC. collected at Heng Chun Tropical Botanic Garden, Taiwan. This compound was designated as "kigelinone", to which structure (**19**) was assigned (Inoue et al. 1981). The same compound was also obtained from the stem bark of *Tabebuia*

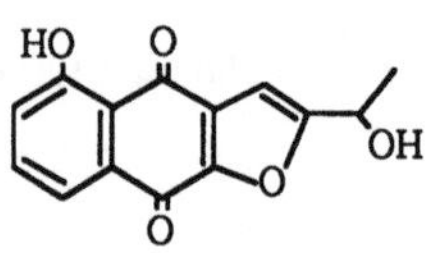


5-Hydroxy-2-(1-hydroxyethyl)-naphtho [2,3-*b*] furan-4,9-dione (18)

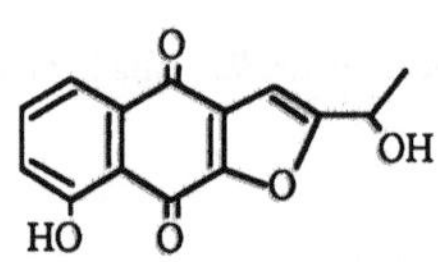


8-Hydroxy-2-(1-hydroxyethyl)-naphtho [2,3-*b*] furan-4,9-dione (19)

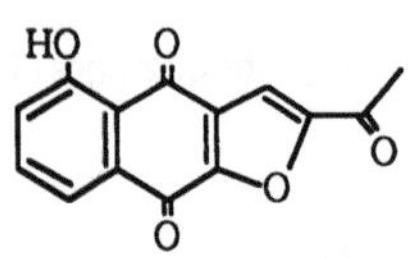


2-Acetyl-5-hydroxynaphtho-[2,3-*b*] furan-4,9-dione (20)

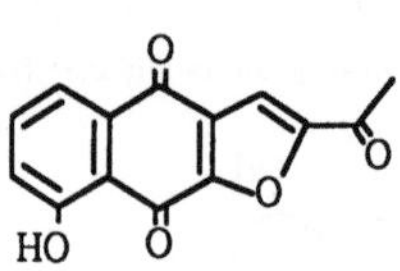


2-Acetyl-8-hydroxynaphtho-[2,3-*b*] furan-4,9-dione (21)

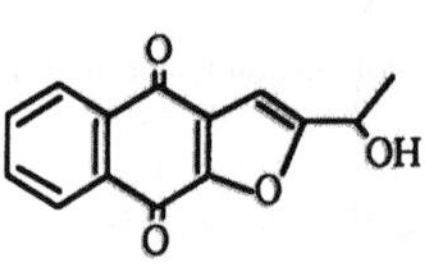


2-(1-hydroxyethyl)-naphtho-[2,3-*b*] furan-4,9-dione (22)

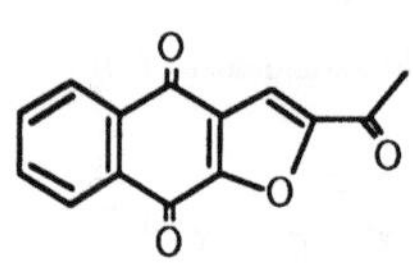


2-Acetylnaphtho [2,3-*b*]-furan-4,9-dione (23)

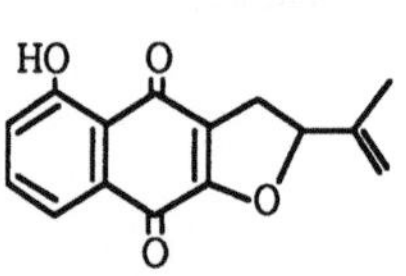


2,3-Dihydro-5-hydroxy-2-(1-methylethenyl)-naphtho-[2,3-*b*] furan-4,9-dione (24)

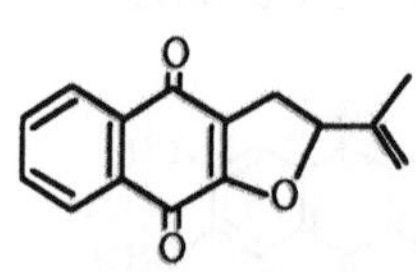


2,3-Dihydro-2-(1-methylethenyl)-naphtho [2,3-*b*] furan-4,9-dione (25)

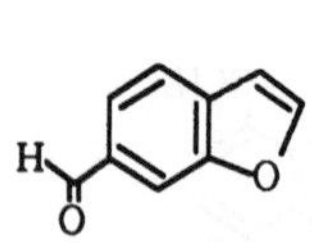


Benzo [*b*] furan-6-carboxaldehyde (26)

**Fig. 4.** Constituents (**18**)–(**26**) of *Tabebuia avellanedae* bark. (Wagner et al. 1989) *cassinoides* (Lam.) DC. An alcoholic extract of the stem bark was found to be cytotoxic in the P-388 assay and the KB cell culture assay (M.M. Rao and Kingston 1982). Almost concurrently, quinones (**18**) and (**20**) were obtained from callus cultures of *Tabebuia caraiba* (Mart.) Bur. ( = *T. argentea* Britton) (Inouye et al. 1982). Recently, the brine shrimp lethality-directed fractionation of the 95% ethanol extract of the bark of *T. barbata* (E. Mey.) Sandwith has led to the isolation of antitumor quinones such as (**19**), which showed potent cytotoxicity to human solid tumors in vitro (A-549, MCF-7, HT-29). The *T. barbata* plant is indigenous to the upper Orinoco and Amazonia rivers (Saizarbitoria et al. 1992). Quite recently, the application of a yeast-based assay, using a genetically engineered yeast strain that lacks the recombination pathway associated with repair of double-stranded breaks and meiotic recombination, known as rad 52, to an extract of a bignoniaceous plant *Crescentia cujete* resulted in the isolation of (**18**), (**22**), (**24**), and (**25**) along with five related naphthoquinones (Heltzel et al. 1993).

### 1.3 Conventional Practices for Propagation, Production of Naphthoquinones, and Their Clinical Applications

Among the Brazilian plants investigated as medicinal resources, *T. avellanedae* is note worthy due to its highly promising therapeutic effects. However, artificial propagation of this plant is very difficult. Wild *T. avellanedae* plants have thus been used for medicinal purpose (Accorsi 1988). Recent findings of the antitumor activity of the extract or the constituent naphtho[2,3-*b*]furan-4,9-dione congener of this plant support their clinical use (M.M. Rao and Kingston 1982; Ueda and Tokuda 1990; Saizarbitoria et al. 1992).

In an effort to find a clinically acceptable antitumor-active compound, lapachol (**1**), a known uncoupler of oxidative phosphorylation (Howland 1963), and its derivatives have been extensively investigated. Lapachol (**1**), obtained from the methanol extract of the powdered wood of *T. avellanedae*, yield 3.18% (Linardi et al. 1975), shows highly significant activity against Walker 256 carcinosarcoma when given intraperitoneally, subcutaneously, intramuscularly, and orally. The best activity and therapeutic index were shown when the drug was given orally twice daily on days 3–6. This treatment regimen resulted in a maximum tolerated dose of 750 mg/kg and a minimum effective dose of 90 mg/kg to yield a therapeutic index of 8.33 and a maximum effectiveness of 0.11 (K.V. Rao et al. 1968; K.V. Rao 1974). Lapachol (1) indicated a significant inhibitory activity against 12-*O*-tetradecanoylphorbol-13-acetate (TPA)-induced Epstein-Barr virus (EBV) activation. Furthermore, this compound inhibited the skin carcinogenesis initiated by 7,12-dimethylbenz(*a*) anthracene (DMBA)-and 12-*O*-tetradecanoylphorbol-13-acetate (TPA)-induced skin carcinogenesis in mice (Ueda et al. 1992). Among almost a 100 synthesized analogs, 2-(3-methyl-2-butenyl)-3-(tetraacetyl-β- D-glucopyranosyl)-1,4-naphthoquinone (2-β-O-glucosyllapachol tetraacetate) extended lapachol activity, becoming effective against mouse lymphocytic leukemia P-388 (Linardi et al. 1975). Recent investigations on the influence of a broad concentration range of naturally occurring naphthoquinones and other cytotoxic or cytostatic agents on human granulocytes and lymphocytes revealed that nearly all substances show the known cytotoxic or immunosuppressive effects at high concentrations (10 $\mu$g ~10 ng/ml), whereas most of the same compounds, except cytostatic vincristine etc., display immunostimulating activities at very low concentrations (10 ng ~10 fg/ml). This observation might be relevant to the antitumoral effects of the naphthoquinone-containing *T. avellanedae* plant (Wagner et al. 1988).

Peroxidation of lapachol (**1**) with 3-chloroperbenzoic acid in the presence of sodium hydrogen carbonate, the epoxide (**27**) was formed, but it rapidly rearranged to (±)-2,3-dihydro-2-(1-hydroxy-1-methylethyl)naphtho[2,3-*b*]-furan-4,9-dione (**28**), yield 49% (Fig. 5). This compound showed enhanced activity against human epidermoid carcinoma of the neopharynx (KB) in comparison with lapachol (**1**) and related compounds (Adams and Lewis 1978).

Lapachol (**1**) occurs in the wood of several species of Bignoniaceae including *T. avellanedae* (Burnett and Thomson 1967a, b), termite-resistant *Tectona grandis* L. fil. (Berbenaceae) (Sandermann and Dietrichs 1957; Sandermann and Simatupang 1966), *Diphysa robinoides* Bent. (Leguminosae) (Sagrero-Nieves 1986), Fijian *Hibiscus tiliaceus* (Malvaceae) (Ali et al. 1980), and in the roots of *Conospermum teretifolium* R. Br. (Proteaceae) (Cannon et al. 1975).

A phase I clinical trial of lapachol (**1**) was carried out at the National Cancer Institute, Baltimore, Maryland, in 1967. The doses ranged from 250 to 3750 mg/day for 5 days. The largest total dose received was 3000 mg/day for 21 days. No toxicity was seen at doses below 1.5 g. Above this level, nausea and vomitting were usual. No hematologic or renal activity was seen at any dosage. At doses above 2 g/day, the prothrombin time was often markedly prolonged and required correction with parenteral vitamin K. This phenomenon was considered to be related to its structural similarity to vitamin K (Block et al. 1974).

2-Acetylnaphtho[2,3-*b*]furan-4,9-dione and (±)-2-(1-hydroxyethyl)-naphtho[2,3-*b*]furan-4,9-dione were synthesized in 36% overall yield starting from furan and phthalic anhydride (Lopes et al. 1984). 7, 8-Dimethoxy-2-acetylnaphtho [2,3-*b*]furan-4,9-dione (**29**) was synthesized in 90% overall yield by a route involving directed ortho metalation. Chromium-pyridine complex oxidation of naturally occurring quinone (**30**) gave compound (**29**) (Fig. 6; de Silva et al. 1979; Zani et al. 1987).

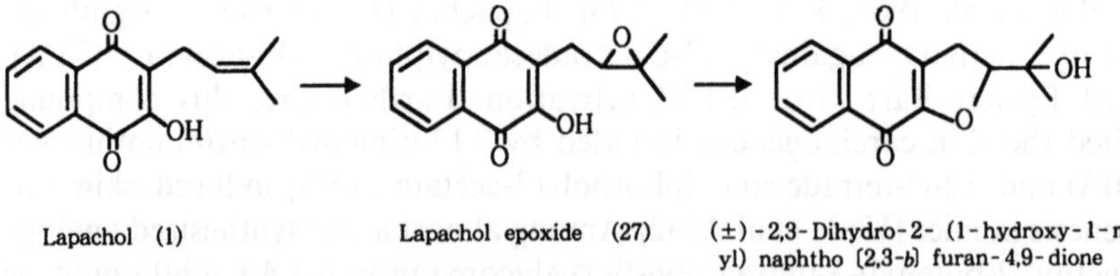

Lapachol (1) Lapachol epoxide (27) (±)-2,3-Dihydro-2-(1-hydroxy-1-methyl) naphtho [2,3-*b*] furan-4,9-dione (28)

**Fig. 5.** Chemical conversion of lapachol into a dihydronaphtho[2,3-*b*]-furan-4,9-dione derivative. (Adams and Lewis 1978)

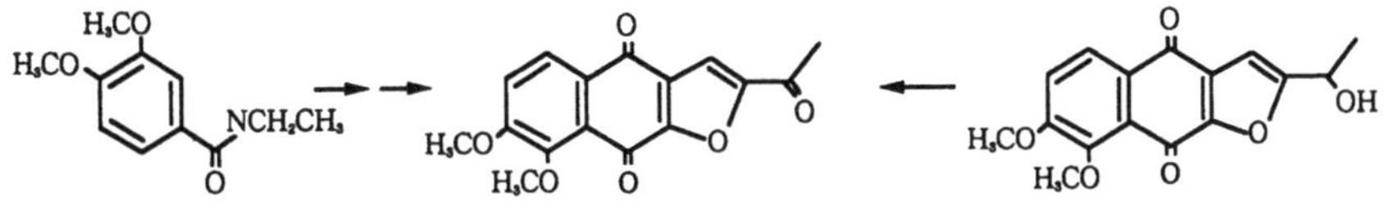

7,8-Dimethoxy-2-acetylnaphtho-[2,3-*b*] furan-4,9-dione (29)

7,8-Dimethoxy-2-(1-hydroxyethyl-naphtho [2,3-*b*] furan-4,9-dione (30)

**Fig. 6.** Synthesis of natural naphtho[2,3-*b*]-furan-4,9-dione. (Zani et al. 1987)

Lapachol (**1**) was commercialized in Brazil as an antitumor drug (Linardi et al. 1975). The inner bark of *T. avellanedae* produced in Brazil is distributed in Asia as a herb tea by TAHEEBO JAPAN Co. Ltd., Osaka, Japan.

# 2 In Vitro Culture Studies

## 2.1 Review of the Tissue Culture/Biotechnology Studies on *Tabebuia avellanedae*

*T. avellanedae* callus tissues (Ueda 1993; Ueda et al. 1994) were induced according to the method applied to *Catalpa ovata* (Ueda et al. 1980). The callus and the cell suspension cultures produce 5-hydroxy-2-(1-hydroxyethyl)naphtho-[2,3-*b*]furan-4,9-dione (**18**) in high yields along with a trace of 2-acetyl-5-hydroxynaphtho-[2,3-*b*]furan-4,9-dione (**20**) (Fig. 7). Both naphthoquinones showed an inhibitory effect in Raji cells on tumor promotion by 12-*O*-tetradecanoylphorbol-13-acetate (TPA) combined with n-butyric acid (Table 1; Ueda and Tokuda 1993; Ueda et al. 1994).

The author established a cell line of *Tabebuia caraiba* (Mart.) Bur. (*T. argentea* Britton) on MS medium (Murashige and Skoog 1962) supplemented with IAA and kinetin either in a solid or liquid state and produced four naphthoquinones, 5-hydroxy-2-(1-hydroxyethyl)naphtho[2,3-*b*]furan-4,9-dione (**18**), 2-acetyl-5-hydroxynaphtho[2,3-*b*]furan-4,9-dione (**20**), 2-acetylnaphtho[2,3-*b*]furan-4,9-dione (**23**), and 5-hydroxy-2-(1-methylethenyl)-naphtho[2,3-*b*]furan-4,9-dione. Pulse administration of precursory 4-(2-carboxyphenyl)-4-oxobutanoic acid (2-succinylbenzoic acid) to the suspension cultures at intervals of 10 days over

5-Hydroxy-2-(1-hydroxyethyl)-naphtho [2,3-*b*] furan-4,9-dione (18)

2-Acetyl-5-hydroxynaphtho-[2,3-*b*] furan-4,9-dione (20)

**Fig. 7.** Naphthoquinones of *Tabebuia avellanedae* cell cultures. (Ueda 1993)

**Table 1.** Inhibitory effects of *Tabebuia avellanedae* naphthoquinones against 12-*O*-tetradecanoyl-phorbol-13-acetate (TPA)-induced Epstein-Barr virus early antigen (EBV-EA) activation

| Compound | Concentrations: molar ratio (test compound/TPA) | | | | | | |
|---|---|---|---|---|---|---|---|
| | 500 | 100 | 10 | 1 | 0.1 | 0.01 | |
| 5-Hydroxy-2-(1-hydroxy-ethyl)-naphtho[2,3-*b*]furan-4,9-dione (**1**) | 0[a](0)[b] | 0(0) | 0(0) | 9.2(20) | 18.9(50) | 65.2(80) | 90.3 |
| 2-Acetyl-5-hydroxynaphtho-[2,3-*b*]-furan-4,9-dione (**2**) | 0(0) | 0(0) | 0(0) | 53.7(70) | 90.4 | 100 | 100 |
| Lapachol | 19.7(80) | 46.2(80) | 68.4 | 100 | 100 | 100 | 100 |
| 1 + Lapachol (molar ratio 1:1) | 0(20) | 0(30) | 36.6(80) | 56.0 | 83.6 | | |

[a] Values are EBV-EA activation (%) ± standard deviation $\sigma$ (± 5%) in the presence of test compound relative to the positive control (100). The activation was caused by TPA (32 pmol).
[b] Values in parentheses represent the viability (%) of Raji cells measured through Trypan Blue staining followed by counting of the surviving cells 48 h after the concomitant treatment of the cells with TPA, n-butyrate, and test substances in a 0.25% phosphate buffer solution (pH 7.2).

a period of 1 month resulted in the production of 2,3-dihydro-2-(1-methyl-ethenyl)naphtho[2,3-*b*]furan-4,9-dione (**25**) in addition to the four quinones described above. *T. caraiba* cell cultures, which are incapable of producing these quinones in the MS or LS medium (Linsmaier and Skoog 1965) supplemented with 2,4-dichlorophenoxyacetic acid, (2,4-D) metabolized exogenous 2-succinyl-benzoic acid to produce quinones (**18**), (**23**), (**25**), and 5-hydroxy-2-(1-methyl-ethenyl)naphtho[2,3-*b*]furan-4,9-dione (Inouye et al. 1982).

On the basis of the structures of the naphthoquinones found in the *Tabebuia avellanedae* and *T. caraiba* cell cultures, the biogenesis of naphthoquinones could be proposed as shown in Fig. 8.

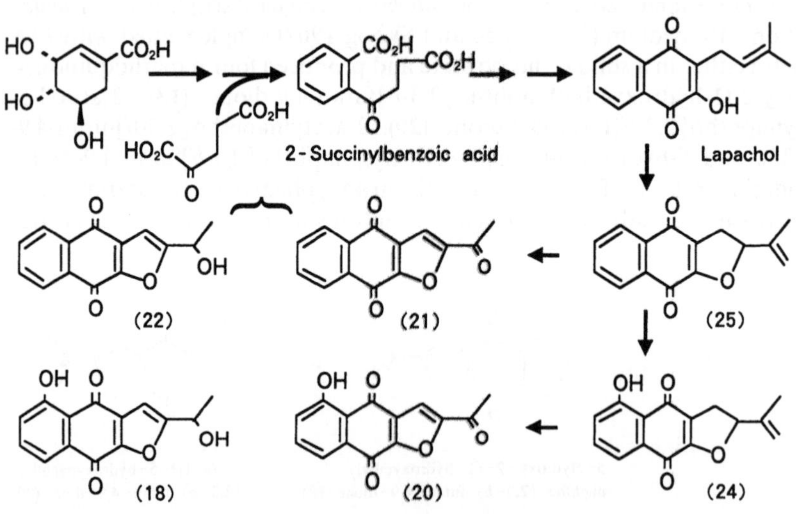


**Fig. 8.** Biogenesis of naphthoquinones in *Tabebuia avellanedae* and *T. caraiba*

## 2.2 Establishment of Tissue Culture/Cell Suspensions

Seeds of *T. avellanedae* were sterilized by soaking them successively in 70% ethanol for 30 s and in sodium hypochlorite solution (1% active chlorine) for 30 min. The seeds, washed five times with sterile water, were placed on 2% agar plates and incubated at 25 °C in the dark. Within 30 days, most of the seeds had germinated. The seedlings of about 5 mm in length were transferred to LS basal agar medium supplemented with $10^{-5}$ mol 2, 4-D and incubated at 25 °C in the dark. Callus tissues induced from the seedlings over 1–3 weeks of incubation were transferred to MS basal agar medium supplemented with $2 \times 10^{-5}$ mol indoleacetic acid (IAA) and $10^{-5}$ mol kinetin, pH 5.8, before autoclaving, at 25 °C in the dark and subcultured every 4–5 weeks. After three to five transfers the agar medium under the callus tissues turned yellow, indicating naphthoquinone production in the callus tissues. The callus tissues were then transferred to a 500-ml flask containing 200 ml of the above described MS medium from which agar was deleted. The liquid culture was stirred at 25 °C in the dark on a reciprocating shaker, 85 strokes/min, 8 cm in amplitude. The cells were subcultured every 4 to 5 weeks.

## 2.3 Extraction and Structure of Naphthoquinones

Lyophilized *Tabebuia avellanedae* callus tissues (10 g) were extracted four times with methanol under reflux for 30 min. The combined methanol extracts were concentrated in vacuo to give a brownish residue which was triturated repeatedly with chloroform. The chloroform extract was washed with water, dried over $MgSO_4$, and concentrated in vacuo to yield a yellowish residue. The residue was applied to thin layer chromatography on silica gel 60 $F_{254}$ plates (0.5 mm in thickness) with toluene–ethylacetate (4:1) as the developing solvent to give two main yellow bands. Extraction of a yellow band (Rf 0.3) with a mixture of chloroform and methanol (9:1) gave 5-hydroxy-2-(1-hydroxyethyl)-naphtho[2,3-*b*]furan-4,9-dione (**18**) (Wagner et al. 1989), yield 5 mg. In the same way, another yellow band (Rf 0.8) gave 2-acetyl-5-hydroxynaphtho[2,3-*b*]furan-4,9-dione (**20**) (Wagner et al. 1989), yield 0.5 mg (Fig. 7). The quinone (**18**) content of this *T. avellanedae* cell line is more than 100 times that of the bark of this plant. 5-Hydroxy-2-(1-hydroxyethyl)-naphtho[2,3-*b*]furan-4,9-dione (**18**), yellow needles (from EtOH), melting point 181 °C; high resolution mass spectrum found molecular ion $[M]^+$ 258.0541 ($C_{14}H_{10}O_5$ requires 258.0525); $[\alpha]_D^{28}$ − 25.0 ° (MeOH); ultraviolet absorption spectrum, absorption maximum, $\lambda$ (MeOH) nm (logε); 204 (4.26), 233 (shoulder, 3.74), 247 (4.05), 300 (3.88), 396 (3.83); (KBr) infrared absorption maximum, ν max $cm^{-1}$ 1670, 1640, 1580, 1460, 1370, 1190, 1170, 970 $^1$H-NMR (220 MHz, $CDCl_3$) δ: 1.65 (3H, doublet coupling constant, J = 6.6Hz), 2.25 (1H, broad singlet disappeared on addition of $D_2O$), 5.05 (1H, multiplet), 6.85 (1H, doublet, J = 0.7), 7.27 (1H, doublets of doublet, J = 8.3, 1.5 Hz), 7.62 (1H, triplet, J = 8.3), 7.82 (1H, doublets of doublet, J = 8.3, 1.5 Hz), 12.18 (1H, singlet, disappeared on addition of $D_2O$).

2-Acetyl-5-hydroxynaphtho[2,3-*b*]furan-4,9-dione (**20**), yellow needles from EtOH, melting point 218–220 °C; UV $\lambda_{max}$. (MeOH) nm (log ε): 200 (4.36), 240 (4.07), 255 (4.30), 275 (4.12), 414 (3.50); ir (KBr) $\nu_{max}$: 2950, 2860, 1700, 1670, 1640, 1580, 1460, 1370, 1250, 1200, 1170, 1110, 970; $^1$H-NMR (220 MHz, $CDCl_3$) δ: 2.67 (3H, singlet), 7.33 (1H, doublets of doublet, J = 8.4, 1.5 Hz), 7.60 (1H, singlet), 7.67 (1H, triplet, J = 7.8 Hz), 7.82 (doublets of doublet J = 7.4, 1.5 Hz), 12.13 (1H, singlet, disappeared on addition of $D_2O$).

## 3 Summary and Conclusions

Callus tissues induced from the sterilized seedlings of *Tabebuia avellanedae* and their cell suspension cultures produced 2-(1-hydroxyethyl)naphtho[2,3-*b*] furan-4,9-dione (**18**) and 2-acetyl-5-hydroxynaphtho [2,3-*b*]furan-4,9-dione (**20**) in high yield. The present success of the induction of callus tissues which produce medicinally promising naphthoquinones at high levels will contribute to the development of effective medicinal substances, including anticancer drugs.

## References


Accorsi WR (1988) Taheebo. Zero Planning, Kobe

Adams JC, Lewis JR (1978) (±)-2,3-Dihydro-2-(1-hydroxy-1-methylethyl) naphtho [2,3-*b*] furan-4,9-quinone; an anti-tumour active peroxidation product of lapachol. J Chem Res (S): 3

Ali S, Singh P, Thomson RH (1980) Naturally occurring quinones. Part 28. Sesquiterpenoid quinones and related compounds from *Hibiscus tiliaceus*. J Chem Soc Perkin I: 257–259

Block JB, Serpick AA, Miller W, Wiernik PH (1974) Early clinical studies with lapachol. Cancer Chemother Rep 2 (4): 27–28

Burnett AR, Thomson RH (1967a) Quinones, Part VIII. Dehydro-α- and -β-lapachone. J Chem Soc (C): 1261–1264

Burnett AR, Thomson RH (1967b) Naturally occurring quinones. Part X. The quinonoid constituents of *Tabebuia avellanedae* (Bignoniaceae). J Chem Soc (C): 2100–2104

Cannon JR, Joshi KR, McDonald IA, Retallack RW, Sierakowski AF, Wong LCH (1975) Structures of nine quinones isolated from two *Conospermum* species. Tetrahedron Lett 2795–2798

da Silva MF, Lisbôa PLB, Lisbôa RCL (1977) Nomes vulgares de plantas amazônicas. Conselho Nacional de Desenvolvimento científico e tecnológico, Inst Nac Pesquisas Amazonia, Manaus, 163 pp

de Silva SO, Watanabe M, Snieckus V (1979) General route to anthraquinone natural products via directed metalation of *N*,*N*-diethylbenzamides. J Org Chem 26: 4802–4808

Domínguez XA, Franco R, Cano G, García F MC, Domínguez S Jr XA, de la Peña M L (1983) Isolation of a new furano-1,4-naphthaquinone, diodantunezone from *Lanthana achyranthifolia*. Planta Med 49: 63

Girald M, Kindack D, Dawson BA, Ethier J-C, Awang DVC (1988) Naphthoquinone constituents of *Tabebuia* spp. J. Nat Pro 51: 1023–1024

Gosálvez M, García-Cañero R, Blanco M, Gurucharri-Lloyd C (1976) Effects and specificity of anticancer agents on the respiration and energy metabolism of tumor cells. Cancer Treat Rep 60: 1–8

Gottlieb OR, Mors WB (1980) Potential utilization of Brazilian wood extractives. J Agric Food Chem 28:196–215
Hartwell JL (1968) Plants used against cancer. A survey. Lloydia 31:71–170
Hashimoto G (1962) Brazil Shokubutu-ki. Teikoku Shoin, Tokyo, 193 pp
Heltzel CE, Gunatilaka AAL, Glass TE, Kingston DGI (1993) Mechanism-based search for anticancer drugs: selective DNA-damaging naphthoquinones from *Crescentia cujete*. Abstr Pap 34th Annu Meet Amer Soc Pharmacognosy, San Diego, CA, 0–8
Howland JL (1963) Uncoupling and inhibition of oxidative phosphorylation by 2-hydroxy-3-alkyl-1,4-naphthoquinones. Biochim Biophys Acta 77:659–662
Inoue K, Inouye H, Chen C-C (1981) A naphthoquinone and a lignan from the wood of *Kigelia pinnata*. Phytochemistry 20:2271–2276
Inouye H, Ueda S, Inoue K, Nayeshiro H, Moritome N (1982) Biosynthesis of prenylnaphthoquinone congeners in cultured cells of *Streptocarpus dunnii* and *Tabebuia argentea*. In: Fujiwara A (ed), Plant Tissue Culture 1982, Maruzen, Tokyo, p 375–376
Joshi KC, Prakash L, Singh P (1973) Quinones and other constituents from *Tabebuia rosea*. Phytochemistry 12:942–943
Joshi KC, Singh P, Sharma MC (1985) Quinones and other constituents of *Markhamia platycalyx* and *Bignonia unguiscati*. J Nat Prod 48:145
Jurd L, Manners GD (1980) Wood extractives as models for the development of new types of pest control agents. J Agric Food Chem 28:183–188
Linardi MCF, Oliveira MM, Sampaio MRP (1975) A lapachol derivative active against mouse lymphocytic leukemia P-388. J Med Chem 18:1159–1161
Linsmaier EM, Skoog F (1965) Organic growth factor requirements of tobacco tissue cultures. Physiol Plant 18:100–127
Lopes CC, Lopes RSC, Pinto AV, Costa PRR (1984) Efficient synthesis of cytotoxic quinones: 2-acetyl-4*H*,9*H*-naphtho[2,3-*b*]furan-4,9-dione and (±)-2-(1-hydroxyethyl)-4*H*,9*H*-naphtho-[2,3-*b*] furan-4,9-dione. J Heterocycl Chem 21:621–622
Mock J, Murphy ST, Ritchie E, Taylor WC (1973) Chemical studies of the Proteaceae VI. Two naphthoquinones from *Stenocarpus salignus*. Aust J Chem 26:1121–1130
Murashige T, Skoog F (1962) A revised medium for rapid growth and bioassays with tobacco tissue cultures. Physiol Plant 15:473–497
Nakano K, Maruyama K, Murakami K, Takaishi Y, Tomimatsu T (1993) Iridoids from *Tabebuia avellanedae*. Phytochemistry 32:371–373
Rao KV (1974) Quinone natural products: streptonigrin (NSC-45383) and lapachol (NSC-11905) structure-activity relationships. Cancer Chemother Rep 2(4) :11–17
Rao KV, McBride TJ, Oleson JJ (1968) Recognition and evaluation of lapachol as an antitumor agent. Cancer Res 28:1952–1954
Rao MM, Kingston DGI (1982) Plant anticancer agents, XII. Isolation and structure elucidation of new cytotoxic quinones from *Tabebuia cassinoides*. J Nat Prod 45:600–604
Sagrero-Nieves L (1986) Isolation of lapachol from *Diphysa robinoides*. J Nat Prod 49:547
Saizarbitoria TC, Anderson JE, McLaughlin JL (1992) Cytotoxic furonaphthaquinones from *Tabebuia barbata* (Bignoniaceae). Abstr Pap 33rd Annu Meet Am Soc Pharmacognosy, Williamsburg, VA, P-22
Sandermann W, Dietrichs HH (1957) Investigations on termiteproof wood species. Holz als Roh- und Werkstoff 15:281–297
Sandermann W, Simatupang MH (1966) On the chemistry and biochemistry of teakwood (*Tectona grandis* L. fil). Holz als Roh- und Werkstoff 24:190–204
Sandermann W, Simatupang MH, Wendeborn W (1968) Neue Chinone aus dem Kernholz von *Paratecoma peroba* Kuhlm. Naturwissenschaften 55:38
Scogin R (1980) Anthocyanins of the Bignoniaceae. Biochem Syst Ecol 8:273–276
Ueda S (1993) Production of furanonaphthoquinones in cell cultures of a South American plant *Tabebuia avellanedae*. Abstr Pap 113th Annu Meet Pharmaceutical Soc Japan, Osaka, No. 2, p181
Ueda S, Tokuda H (1990) Inhibitory effect of *Tabebuia avellanedae* constituents on tumor promotion. Planta Med 56:669–670
Ueda S, Tokuda H (1993) Cancer preventive activity of *Tabebuia avellanedae* extracts and its constituents. Abstr Pap 34th Annu Meet Am Soc Pharmacognosy, San Diego, CA, P-51

Ueda S, Inoue K, Shiobara Y, Kimura I, Inouye H (1980) Quinones and related compounds in higher plants, X. Naphthoquinone derivatives of the callus culture of *Catalpa ovata*. Planta Med 40:168–178

Ueda S, Tokuda H, Nishino H, Iwashima A (1992) Anti-tumor promoting activity of lapachol. Abstr Pap 33rd Annu Meet Am Soc Pharmacognosy, Williamsburg, VA, P-70

Ueda S, Umemura T, Dohguchi K, Matsuzaki T, Tokuda H, Nishino H, Iwashima A (1994) Production of anti-tumor-promoting furanonaphthoquinones in *Tabebuia avellanedae* cell cultures. Phytochemistry 36:323–325

Wagner H, Kreher B, Jurcic K (1988) In vitro stimulation of human granulocytes and lymphocytes by pico- and femtogram quantities of cytostatic agents. Arzneim Forsch 38:273–275

Wagner H, Kreher B, Lotter H, Hamburger MO, Cordell GA (1989) Structure determination of new isomeric naphtho [2,3-*b*] furan-4,9-diones from *Tabebuia avellanedae* by the selective-INEPT technique. Helv Chim Acta 72:659–667

Williams L (1936) Woods of northeastern Peru. In: Dahlgren BE (ed) Field museum of natural history. Botanical Series, vol XV. Field Museum of Natural History, Chicago, pp 22, 444–447, 563

Zani CL, Oliveila AB, Snieckus V (1987) Efficient directed ortho metalation-based route to cytotoxic furanonaphthoquinone natural products. Tetrahedron Lett 28:6561–6564

Zani CL, Oliveira AB, Oliveira GG (1991) Furanonaphthoquinones from *Tabebuia ochracea*. Phytochemistry 30:2379–2381

# XXVIII *Tripterygium wilfordii*: In Vitro Culture and the Production of the Anticancer Compounds Tripdiolide and Triptolide

S. Takayama[1]

## 1 General Account

*Tripterygium* species (family Celastraceae) are distributed in the temperate zone of East Asia, from the east coast of China to Korea, Taiwan, and to the west coast of Japan; four to five species, including *T. wilfordii* and *T. hypoglaucum*, have been recorded (Willis 1973). Most are deciduous, vine-like liana woody shrubs, 2 to 4 m in height, growing in the sunny hilly regions, and covering the vegetation. *Tripterygium* species are important as a Chinese herbal remedy, and are also used as ornamentals, especially in flower arrangements.

The most popular member of the genus, *T. wilfordii* (Fig. 1), grows abundantly in the wild, and has attracted attention because of its various potent physiological activities. The flowering period of this plant is from June to August. It has white panicles with five petals and five anthers, and bears three-winged fruit from August to September.

*Tripterygium* plants produce various physiologically active compounds. The most widely investigated compounds are diterpene triepoxides, tripdiolide, and triptolide, which are potent anticancer substances active against L-1210 and p-388 neoplasms and in vitro cells derived from human carcinoma of the nasopharynx (KB) (Kupchan et al. 1972). The alkaloids, containing a pyridine nucleus, are extracted from the root bark and are active as an insecticide (Feuell 1965). Other important physiological activities include: immunosuppressive activity (Zheng et al. 1989; Li and Weir 1990; Ma et al. 1991; Gu et al. 1992; Kutney et al. 1992; Xiao et al. 1992), antiinflammatory activity (Li 1991; Ma et al. 1991), male contraceptive activity (Kutney et al. 1992; Lan et al. 1992; Tian et al. 1992), anti-HIV activity (Chen et al. 1992a,b), a synergistic insecticidal effect of *Tripterygium* ethanol extract with *Bacillus thuringiensis* (Chiu et al. 1989) or an insect growth-inhibiting effect (Shin 1989), and anti-cancer activity (Kupchan et al. 1972; Kupchan and Schubert 1974; Perdue 1976; Cheng 1981; Zhang et al. 1981; Hori et al. 1987; Wei and Adachi 1991; Zhou 1991).

*Triperygium* plants are easy to propagate by means of seeding, cutting, or layering. Phytohormones, especially auxins, enhance rooting on the proximal end of the stem. Although propagation efficiency is high, especially in the case of

[1] Department of Biological Science and Technology, Tokai University, 317 Nishino, Numazu, Shizuoka 410-03, Japan

Fig. 1. *Tripterygium wilfordii* plant

seeding, the efficiency of biomass production is not high enough to produce a large amount of plant material for extraction of the metabolite(s). In this context, plant tissue culture becomes an important alternative technology for the industrial production of *Tripterygium*.

## 2 In Vitro Culture Studies

### 2.1 Review of the Literature (Table 1)

*Tripterygium wilfordii* is an important producer of various compounds (Table 1), especially the potent anticancer substances, tripdiolide and triptolide. Large-scale extraction of these compounds from plants collected in the wild for toxicological or clinical trials is difficult and, moreover, the contents of active ingredients is quite low (about 0.001%) (Kupchan et al. 1972). Total synthesis of these compounds for the production of samples for toxicological and clinical trials is currently insufficient because extraction processes are too complicated and yields are low (Buckanin et al. 1980; Van Tamelen et al. 1980; Lai et al. 1982).

Production of these compounds by tissue culture is a technique that can be used up to industrial scale. Kutney et al. (1980, 1981) reported the first

**Table 1.** Summary of the work on the production of secondary metabolites in cell cultures of *Tripterygium wilfordii*

| Authors | Cultures | Compounds isolated | Physiological activity |
|---|---|---|---|
| Dujack and Chen (1980) | Cell culture | Tripdiolide, triptolide | Inhibition of DNA synthesis<br>Anticancer |
| Dujack and Chen (1980) | Cell culture | Tripdiolide, triptolide | Inhibition of DNA synthesis<br>Anticancer |
| Kutney et al. (1980) | Cell culture | Tripdiolide, triptolide<br>β-sitosterol and two related compounds.<br>Celastrol | Anticancer |
| Kutney et al. (1981) | Cell culture | Tripdiolide, triptolide<br>Dihydroabietic acid<br>15-Hydroxy-18-norabieta-3,8,11,13-tetraen-3-oicacid methyl ester<br>Celasoltrol, oleanic acid, polpunonic acid | Cytotoxic against KB cells |
| Kutney et al. (1983) | Cell culture | Tripdiolide | Cytotoxic against KB cells |
| Kutney et al. (1992) | Cell culture | 18 Compounds with diterpene and triterpene families | Rheumatoid arthritis<br>skin allergies,<br>male contraception |
| Kutney et al. (1993) | Cell culture<br>elicitation | Oleanane triterpenes | Rheumatoid arthritis |
| Misawa et al. (1983) | Cell culture | Tripdiolide, triptolide | Anticancer |
| Misawa et al. (1985) | Cell culture | Tripdiolide, triptolide | Anticancer |
| Pepin et al. (1991) | Cell culture<br>immobilization | Di- and triterpenes including tripdiolide (approx. 0.04%) | Male contraception, arthritis,<br>skin disorders |
| Takayama et al. (1983) | Cell culture<br>Cell selection | Tripdiolide | Anticancer |
| Takayama (1991) | Cell culture | Tripdiolide, triptolide | Anticancer |

production of tripdiolide from tissue cultures of the *T. wilfordii* plant. The presence of tripdiolide was confirmed by mass spectrometry, 270 MHz NMR, and TLC comparison with authentic samples. The yield of tripdiolide, based on dry cell weight, was 0.003%. Triptolide was also detected on TLC. Several other natural products such as β-sitosterol, dehydroabietic acid, 15-hydroxy-18-norabieta-3,8,11,13-tetraen-3-oic acid methyl ester and two related compounds, and cytotoxic triterpene celastrol were also isolated from tissue cultures of *Tripterygium* (Fig. 2). The presence of the antineoplastic diterpenes tripdiolide and triptolide was also reported by Dujack and Chen (1980) and Misawa et al. (1983, 1985) from tissue culture of *Tripterygium wilfordii*. Dujak and Chen (1980) confirmed the presence of the compounds by TLC and bioassay, and they also

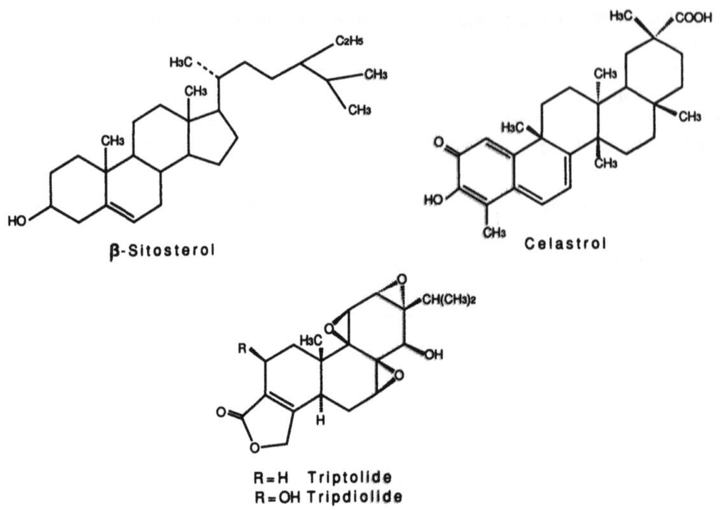


**Fig. 2.** Chemical structures of compounds isolated from cultured tissue of *Tripterygium wilfordii*

**Table 2.** Effect of farnesol on growth and production of tripdiolide by *Tripterygium wilfordii* cells in suspension culture. (Misawa et al. 1985)

| Farnesol added (mg/l) | Cell growth (g/l dry wt.) | Tripdiolide (μ/g dry wt.) |
|---|---|---|
| 0 | 14.3 | 6.3 |
| 30 | 12.7 | 13.0 |
| 100 | 10.1 | 20.0 |

reported that the presence of precursors such as sodium acetate and pyruvic acid in MS-B medium increased biological activity significantly when assayed against tumor cells. Misawa et al. (1983, 1985) detected tripdiolide and triptolide in calli and suspension-cultured cells of *Tripterygium wilfordii*. The highest yield detected was about ten times greater than that observed in the plant. Addition of N-phenyl-N′-(2-chloro-4-pyridyl) urea (a phenyl urea-type cytokinin, abbreviated 4PU) or farnesol (a precursor of diterpene metabolism) stimulated tripdiolide production (Table 2). Although large-scale cell culture of *Tripterygium* has not yet been established, Pepin et al. (1991) developed promising cell culture conditions and also furthered cell immobilization technology.

## 2.2 Propagation by Tissue Culture

Shoot cultures were initiated from shoot segments with axillary buds, shoot tips, or torpedo-shaped embryos taken from seeds. Culture conditions were similar

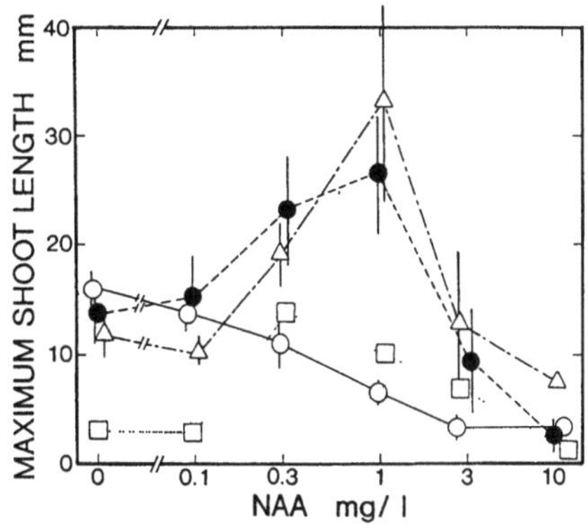


**Fig. 3.** Effect of NAA and kinetin on shoot growth in *Tripterygium wilfordii* embryo cultures. Kinetin concentrations: ○, 0; ●, 0.1; △, 1.0; □, 10 mg/l

for shoot segments, shoot tips, and embryos. Typical results for establishing shoot cultures were obtained with embryo culture. The torpedo-shaped embryos were cultured on MS medium (Murashige and Skoog 1962) supplemented with 30 g/l sucrose and combinations of 0 to 10 mg/l NAA and 0 to 10 mg/l kinetin (Fig. 3). Plantlets with roots were established with 0.3 to 1 mg/l NAA combined with 0.1 to 1 mg/l kinetin. Although higher NAA (3 to 10 mg/l) stimulated root or callus formation, shoot growth was severely repressed. In contrast, higher kinetin stimulated shoot regeneration, but root and callus formation was severely inhibited. The highest kinetin concentration (10 mg/l) stimulated the regeneration of multiple, adventitious shoot formation on the upper surface of expanded cotyledons.

The process for propagation of plantlets in vitro was established by serially transferring the plantlet segments at 3-month intervals. MS medium was used supplemented with 30 g/l sucrose and combinations of 1 mg/l NAA and 0.1 mg/l kinetin. The propagation efficiency of this method was calculated at 650 plantlets a year.

### 2.3 Callus Induction and Growth

*Tripterygium* calli were easily induced from tissue segments of young shoots, leaves, or torpedo-shaped embryos taken from seeds on MS medium supplemented with 1 to 10 mg/l NAA and 1 mg/l kinetin. Induced calli were hard, white to brown and regenerated roots. These calli were transferred and subcultured on MS medium with varying phytohormone combinations for more than 1 year (Table 3). Although a high NAA level (10 mg/l) or addition of 2,4-D (0.5 mg/l) stimulated cell growth, calli became friable, white (or cream-colored) with a low tripdiolide content. In contrast, a low NAA level (1 mg/l) did not stimulate cell growth, cells became hard and had a high tripdiolide content. After serial subculture for more than 1 year on MS medium with 1 mg/l NAA and 1 mg/l kinetin, callus became friable and white, and the tripdiolide content decreased (Takayama et al. 1983).

**Table 3.** Characteristics of *Tripterygium wilfordii* callus induced on MS medium with different phytohormone combinations. (Takayama unpubl.)

| Medium | Callus growth index | Characteristics | Color | Tripdiolide |
|---|---|---|---|---|
| $K_1N_1$ | + | Hard, root-regenerative | Brown-white | + + + |
| $K_1N_{10}$ | + + + | Friable | White | + |
| $K_1D_{0.5}$ | + + | Friable | White | + |
| $K_{0.1}D_3$ | + + + | Friable | White | + |

## 2.4 Production of Antitumor Agents by Cell Culture

The potent antitumor agents, tripdiolide and triptolide, are produced by *Tripterygium* cell cultures. Kutney et al. (1980) obtained tripdiolide and triptolide by cell culture of *Tripterygium wilfordii.* The amount produced was 16 times higher in tripdiolide and 3 times higher in triptolide compared to original plant material (Kutney et al. 1981). The cell strain used in the experiment was first induced from plant material on PRL-4 medium containing 2 mg/l IAA and 100 ml/l coconut milk. Induced calli were further subcultured on PRL medium containing 2 mg/l 2,4-D and 100 ml/l coconut milk, and the high-yielding tripdiolide strain TRP4a was selected. Suspension culture was performed in PRD2Co100 medium (10% inoculum, 27 °C in the dark, rotated at 140 rpm) and subcultured at 3-week intervals (Kutney et al. 1983). Using this strain, the effect of the medium components such as coconut milk, cytokinins, or auxins on productivity was further examined. The greatest tripdiolide production was obtained when 1 mg/l kinetin combined with 2.5 mg/l NAA was added to the medium. In spite of the highly productive characteristics of this medium, cells regenerated adventitious roots and made it difficult to raise large-scale cultures. For this reason, the effects of the medium components were further examined, and a medium supplemented with 0.5 mg/l NAA combined with 0.5 mg/l kinetin was selected. Use of MS medium supplemented with the same phytohormone combination further stimulated the production of tripdiolide (Kutney et al. 1983). Although the productivity of tripdiolide reached 2.6 mg/l in the TRP4a strain when cultured in MS medium supplemented with 0.5 mg/l NAA combined with 0.5 mg/l kinetin, cells became aggregated, and obtained a diameter of ca. 1 cm which hampered the scaling up of the culture. The tripdiolide content decreased markedly in fine cell suspensions selected from TRP4a, so it seems likely that cell morphology is related to the production of tripdiolide.

The author's observations (Misawa et al. 1983, 1985; Takayama et al. 1983; Takayama 1991) were quite similar to these results. Calli were induced from plants collected in Japan and Taiwan, which were subcultured, and callus and cell suspension cultures were established. The established cell strain produced both tripdiolide and triptolide (Table 4). About 500 calli were subcultured individually and extracts of each callus were subjected to tripdiolide assay by KB cells. The high-yielding tripdiolide cell strain was hard and regenerated roots. In contrast, the low-yielding tripdiolide strain produced friable and white

**Table 4.** Composition of tripdiolide and triptolide in ethyl acetate extract of selected and nonselected *Tripterygium wilfordii* callus. (Misawa et al. 1985)

| Cell strain | Characteristics | Tripdiolide | Triptolide |
|---|---|---|---|
| Nonselected callus | Friable, white | 4.8 μg/g dry wt. | 2.1 μg/g dry wt. |
| Selected callus | Hard, root-regenerative | 65.4 | 104.0 |

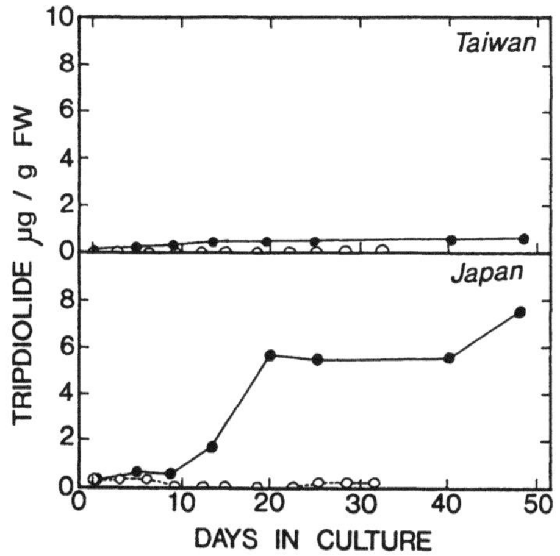


**Fig. 4.** Effect of cell strain and medium composition on tripdiolide production in *Tripterygium wilfordii* cell suspension cultures. ○, MS medium with 0.1 mg/l NAA, 1 mg/l kinetin, and 30 g/l sucrose. ●, 1/4 strength MS medium with 60 g/l sucrose (no phytohormones)

cells without regeneration (Takayama et al. 1983). After serial subcultures, the morphology of the high-yielding tripdiolide cell strain became friable, produced white cells, and the amount of tripdiolide in the cell decreased. In order to maintain productivity of tripdiolide and triptolide, serial selection of hard and root-regenerative callus was required. Cell suspension cultures were also established. After surveying various culture conditions, tripdiolide and triptolide were produced most effectively in a medium having additional 4PU, a high sucrose concentration, a low MS salt strength, and a selected farnesol-resistant cell strain.

Among these factors, MS medium strength, sucrose concentration, and cell strain are the most important for the production of tripdiolide in cell suspension cultures, i.e., 1/4 strength MS medium with 60 g/l sucrose, but without phytohormones, and a selected cell strain of *Tripterygium* originating in Japan (Fig. 4).

## 2.5 Chemical Analysis of Tripdiolide and Triptolide

### *2.5.1 Isolation and Identification* (Fig. 5)

Cultured cells or plant material were dried (air- or freeze-dried), then extracted with 95% ethanol two or three times. The ethanol used for extraction should be neutral or slightly acidic in order to prevent decomposition. Combined extract was condensed with a rotary evaporator, then partitioned into hexane to

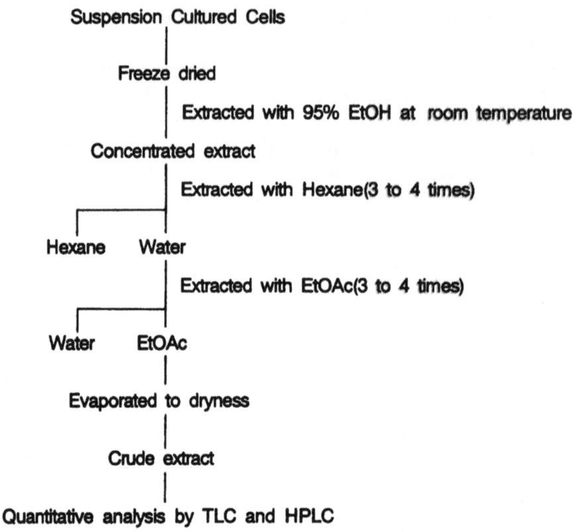


**Fig. 5.** Partition scheme for partial purification of tripdiolide and triptolide from cultured tissue of *Tripterygium wilfordii*

eliminate non-polar lipophilic impurities and to prevent emulsion formation during ethyl acetate-water partitions. The water phase was then partitioned into ethyl acetate from water. The partition ratio of both tripdiolide and triptolide is 50%; partitioning should be repeated three to four times to recover more than 90 to 95% of these components. Ethyl acetate was evaporated to dryness and the extract was subjected to quantitative analysis or further purified by silica gel column chromatography.

Physicochemical data of tripdiolide and triptolide were reported by Kupchan et al. (1972). Tripdiolide: $C_{20}H_{24}O_7$; mp 210–211 °C; $[\alpha]\ ^{25}$D-138°(c 0.139 $CH_2Cl_2$); $UV_{max}$. 217 nm (ε 11 000); ir(KBr) 2.78, 2.88, 5.63, 5.93 μ; mass spectrum (CI) *m/e*377.1621 $(M + 1)^+$ (calcd, 377.1600). Triptolide: $C_{20}H_{24}O_6$; mp 226–227 °C; $[\alpha]\ ^{25}$D-154°(c 0.369, $CH_2Cl_2$; $UV_{max}$. (EtOH) 218 nm (ε 14 000); ir(KBr) 2.89, 5.64, 5.93, 8.05, 8.52 μ; mass spectrum *m/e* 360.1600 $(M^+)$ (calcd, 360.1573). Abbreviations denote: mp: melting point; $[\alpha]^{25}$D: the angle of optical rotatory dispersion at 25 °C; c: concentration in %; ir: infrared absorption; μ: wavelength of infrared absorption; *m/e*: molecular weight; calcd: calculated molecular weight. NMR data for tripdiolide and triptolide have also been provided in the literature.

### *2.5.2 Separation and Quantitative Analysis*

*Thin Layer Chromatography.* Tripdiolide and triptolide were separated by silica gel thin layer chromatography (TLC plate: Merck No. 5729) developed with 30% acetone in hexane, and visualized with the spray reagent of 2% ceric ammonium sulfate in 12% $H_2SO_4$ solution, and heated at 120 °C for 5 min. The

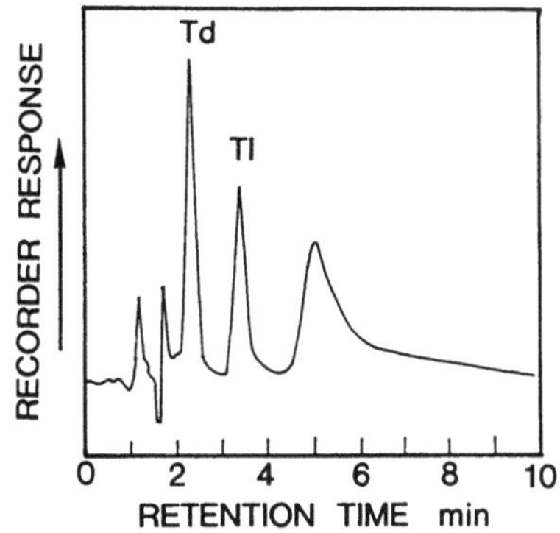


**Fig. 6.** HPLC elution profile of tripdiolide and triptolide. Column: UnisilQC18 (4.6 × 250 mm), GL Science Ltd., Tokyo; solvent: 40% EtOH in water, 2 ml/min; detector: UV 218 nm

sensitivity for both tripdiolide and triptolide was 0.5 μg under visual light, or 0.1 μg under uv illumination (254 or 350 nm). Iodine vapor was also important for visualization and its sensitivity was 0.75 μg for tripdiolide, 1 μg for triptolide. Quantitative analysis of tripdiolide and triptolide was made using a TLC scanner after spraying with 2% ammonium sulfate in 12% $H_2SO_4$ solution and heated at 120 °C for 5 min, and analyzed with an excitation wavelength of 310 nm and an emission wavelength of 500 nm. The sensitivity was 0.03 to 2 μg.

*HPLC.* Tripdiolide and triptolide were determined by HPLC using a UV detector at 218 nm (Fig. 6). A reverse phase column (C18-type column, 4.6 × 150 mm) was used. The solvent system was 40% EtOH in water which was eluted 2 ml/min. The maximum sensitivity was 0.01 μg/injection.

*Bioassay.* Tripdiolide and triptolide are toxic to KB cells (human nasopharynx epidermoid carcinoma) cultured in vitro. KB cells were established about 40 years ago, and were widely used to screen anticancer substances or in fundamental biochemical research. The medium used for KB cell culture contained 9.4 g/l Eagle MEM medium, 17 ml/l of 7.5% sodium bicarbonate solution, 10 ml/l of 200 mM glutamine solution, and 100 ml/l newborn calf serum. The culture medium was inoculated with $4 \times 10^4$/ml KB cells; 3 ml of the KB cells was placed in individual test tubes and injected with 0.01 ml of assay sample dissolved in EtOH. After 3 days of culture, cells were harvested and the cell concentration was determined by the Folin method after alkali decomposition.

The assay range was $3 \times 10^{-3}$ to $3 \times 10^{-1}$ μg/tube for tripdiolide and $3 \times 10^{-4}$ to $1 \times 10^{-1}$ μg/tube for triptolide. Assay results were similar to those of quantitative analysis using HPLC.

## 3 Summary and Conclusion

The tissue culture of *Tripterygium* has attracted the attention of pharmaceutical research, but up to now very few groups have worked with this plant. A protocol

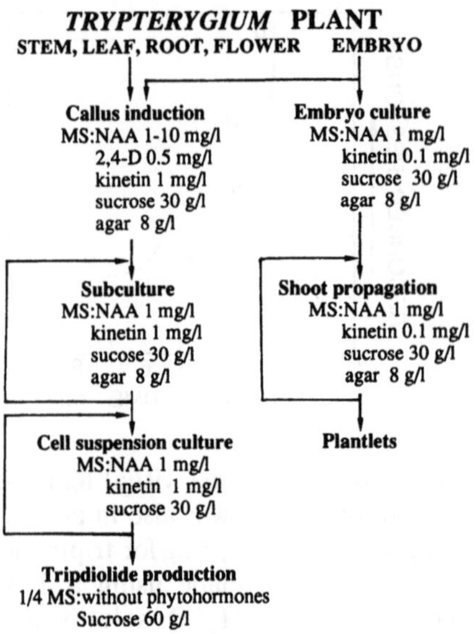


**Fig. 7.** A protocol for tissue culture of *Tripterygium wilfordii*

used for establishing tissue culture of *Trypterygium wilfordii* is summarized in Fig. 7. Callus cultures on agar medium sometimes turned brown to black, and callus growth was either repressed, or in the worst case, the callus died. Even in the case of cell suspension cultures, its establishment takes 2 to 3 years. In order to produce tripdiolide and triptolide effectively, the selection of the culture strain and the use of modified medium (production medium) are the fundamental prerequisites. The addition of 4PU, high sucrose concentrations, low MS salt strength, and the selection of a farnesol-resistant cell strain also stimulate their production. Although the amounts of both tripdiolide and triptolide in the selected cell strain in production medium were ca. 0.01% on a dry weight basis (about ten times higher compared with the amount in the original plant), the production of the compounds on an industrial scale is difficult because of the quite low yield from an industrial viewpoint. In spite of these difficulties, the need for mass production of *Tripterygium* is increasing.

## References

Buckanin RS, Chen SJ, Frieze DM, Sher FT, Berchtold GA (1980) Total synthesis of triptolide and triptonide. J Am Chem Soc 102(3): 1200–1201

Cheng CY (1981) Chinese Celastraceae. Its taxonomy and antineoplastic aspects. Planta Med 39(3): 257–258

Chen K, Shi Q, Fujioka T, Zhang DC, Hu CQ, Jin JQ, Kilkuskie RE, Lee KH (1992a) Anti-aids agents, 4. Tripterifordin a novel anti-HIV principle from *Tripterygium wilfordii* isolation and structural elucidation. J Nat Prod 55(1): 88–92

Chen K, Shi Q, Kashiwada Y, Zhang DC, Hu CQ, Jin JQ, Nozaki H, Kilkuskie RE, Tramontano E (1992b) Anti-aids agents. 6. Salaspermic acid an anti-HIV principle from *Tripterygium wilfordii* and the structure-activity correlation with its related compounds. J Nat Prod 55(3): 340–346

Chiu SF, Huang BQ, Hu MY (1989) Synergistic effect of toosendanin in admixture with *Bacillus thuringiensis* and other insecticides against the cabbage worm *Pieris papae L.* Acta Entomol Sin 32(2): 158–165

Dujack LW, Chen PK (1980) Antineoplastic diterpenes in *Tripterygium wilfordii* tissue culture. The effects of precursors on diterpene production. In Vitro 16: 216

Feuell AJ (1965) Insecticides. Verlag J Cramer, Weinheim, p 43

Gu WZ, Bradnwein SR, Banerjee S (1992) Inhibition of type II collagen induced arthritis in mice by an immunosuppressive extract of *Tripterygium wilfordii* Hook F. J Rheumatol 19(5): 682–68

Hori H, Pang GM, Harimaya K, Iitaka Y, Inayama S (1987) Isolation and structure of regelin and regelinol. New antitumor ursene-type triptepenoids from *Tripterygium wilfordii*. Chem Pharm Bull 35(5): 2125–2128

Kupachan SM, Schubert RM (1974) Selective alkylation a biomimetic reaction of the anti leukemic triptolides. Science 185: 791–793

Kupchan SM, Court WA, Dailey RG, Gilmore CJ, Bryan RF (1972) Triptolide and tripdiolide, novel antileukemic diterpenoid triepoxides from *Tripterygium wilfordii*. J Am Chem Soc 94: 7194–7195

Kutney JP, Beale MH, Salisbury PJ, Sindelar RD, Stuart KL, Worth BR, Townsley PM, Chalmers WT, Donnelly DJ, Nilsson K, Jacoli GG (1980) Tripdiolide from tissue culture of *Tripterygium wilfordii*. Heterocycles 14: 1465–1467

Kutney JP, Hewitt GM, Kurihara T, Salisbury PJ, Sindelar RD, Stuart KL, Townsley PM, Chalmers WT, Jacoli GG (1981) Cytotoxic diterpenes triptolide, tripdiolide and cytotoxic triterpenes form tissue cultures of *Tripterygium wilfordii*. Can J Chem 59: 2677–2683

Kutney JP, Choi LSL, Duffin R, Hewitt G, Kawamura N, Kurihara T, Salisbury P, Sindelar R, Stuart KL (1983) Cultivation of *Tripterygium wilfordii* tissue cultures for the production of the cytotoxic diterpene tripdiolide. Planta Med 48(3) : 158–163

Kutney JP, Hewitt GM, Lee G, Piotrowska K, Roberts M, Rettig SJ (1992) Studies with tissue cultures of the Chinese herbal plant *Tripterygium wilfordii*. Isolation of metabolites of interest in rheumatoid arthritis immunosuppression and male contraceptive activity. Can J Chem 70(5): 1455–1480

Kutney JP, Samija MD, Hewitt GM, Bugante EC, Gu H (1993) Antiinflammatory oleanane triterpenes from *Tripterygium wilfordii* cell suspension cultures by fungal elicitation. Plant Cell Rep 12: 356–359

Lai CK, Buckanin RS, Chen SJ, Zimmerman DF, Sher FT, Berchtold GA (1982) Total synthesis of racemic triptolide and racemic triptonide. J Org Chem 47: 2364–2369

Lan ZJ, Gu ZP, Lu RF, Zhuang LZ (1992) Effects of multiglycosides of *Tripterygium wilfordii* GTW on rat fertility and Leydig and Sertoli cells. Contraception 45(3): 249–261

Li XW, Weir MR (1990) Radix *Tripterygium wilfordii*, a Chinese herbal medicine with potent immunosuppressive properties. Transplantation 50(1): 82–86

Li XY (1991) Immunomodulating Chinese herbal medicines. Mem Inst Oswaldo Cruz Rio J 86(Suppl 2): 159–164

Ma PC, Lu XY, Yang JJ, Zheng QT (1991) 16-Hydroxytriptolide a new active diterpene isolated from *Tripterygium wilfordii*. Acta Pharm Sin 26(10): 459–763

Misawa M, Hayashi M, Takayama S (1983) Production of antineoplastic agents by plant tissue cultures. 1. Induction of callus tissues and detection of the agents in cultured cells. Planta Med 49: 115–119

Misawa M, Hayashi M, Takayama S (1985) Accumulation of antineoplastic agents by plant tissue cultures. In: Neumann KH, Barz W, Reinhard E (eds) Primary and secondary metabolism of plant cell cultures. Springer Berlin Heidelberg, New York, pp 235–246

Murashige T, Skoog F (1962) A revised medium for rapid growth and bioassays with tobacco tissue cultures. Physiol plant 15: 473–497

Pepin MF, Chavarie C, Archambault J (1991) Growth and immobilization of *Tripterygium wilfordii* cultured cells. Biotechnol Bioeng 38: 1285–1291

Perdue RE (1976) Antitumor agents from higher plants. Some practical and economic considerations. Lloydia 39(6): 469

Shin FC (1989) Studies on plants as a source of insect growth regulators for crop protection. J Appl Entomol 107(2): 185–192

Takayama S (1991) Antitumor agents from woody plants. In: Komamine A (ed) Plant cell culture in Japan. CMC, Tokyo, pp 19–26

Takayama S, Misawa M, Hayashi M (1983) Production of anti-cancer agents by tissue culture of *Tripterygium wilfordii*. Selection of tripdiolide and triptolide high yielding cell strain. Proc 1983 Meet Japanese Soc Plant Physiol, 279 pp

Tian J, She Z, Zhou X, Gu H, Zhang D, Jin J (1992) The effect of *Tripterygium wilfordii* on spermatogenic cells and LDH-C4 activity in male rats. Acta Acad Med Shanghai 19(1): 37–40

Van Tamelen EE, Demers JP, Taylor EG, Koller K (1980) Total synthesis of *l*-triptonide and *l*-triptolide. J Am Chem Soc 102: 5424–5425

Wei YS, Adachi I (1991) Inhibitory effect of triptolide on colony formation of breast and stomach cancer cell lines. Acta Pharmacol Sin 12(5): 406–410

Willis JC (1973) A dictionary of flowering plants and ferns, 8th edn. Revised by Airy Shaw HK. University Press, Cambridge

Xiao X, Endo T, Watanabe K, Kumano K, Koshiba K (1992) Immunosuppressive activity of multiglycosides of *Tripterygium wilfordii* GTW on skin allografts in rats. Nephrol Dial Transplant 7(7) : 787

Zhang TM, Chen ZY, Lin C (1981) Antineoplastic action of triptolide and its effect on the immunologic functions in mice. Acta Pharmacol Sin 2(2): 128–131

Zheng YL, Xu Y, Lin JF (1989) Immunosuppressive effects of wilfortrine and euonine. Acta Pharm Sin 24(8): 568–572

Zhou BN (1991) Some progress on the chemistry of natural bioactive terpenoids from Chinese medicinal plants. Mem Inst Oswaldo Cruz Rio J 86 (Suppl 2): 219–226

# Subject Index

# Springer-Verlag and the Environment

We at Springer-Verlag firmly believe that an international science publisher has a special obligation to the environment, and our corporate policies consistently reflect this conviction.

We also expect our business partners – paper mills, printers, packaging manufacturers, etc. – to commit themselves to using environmentally friendly materials and production processes.

The paper in this book is made from low- or no-chlorine pulp and is acid free, in conformance with international standards for paper permanency.

MIX
Papier aus verantwortungsvollen Quellen
Paper from responsible sources
FSC® C105338
FSC
www.fsc.org

If you have any concerns about our products,
you can contact us on
**ProductSafety@springernature.com**

In case Publisher is established outside the EU,
the EU authorized representative is:
**Springer Nature Customer Service Center GmbH**
**Europaplatz 3, 69115 Heidelberg, Germany**

Printed by Libri Plureos GmbH
in Hamburg, Germany